*Current Topics in Membranes, Volume 54*

# Extracellular Nucleotides and Nucleosides: Release, Receptors, and Physiological and Pathophysiological Effects

# Current Topics in Membranes, Volume 54

Series Editors

## Dale J. Benos

*Department of Physiology and Biophysics*
*University of Alabama*
*Birmingham, Alabama*

## Sidney A. Simon

*Department of Neurobiology*
*Duke University Medical Centre*
*Durham, North Carolina*

*Current Topics in Membranes, Volume 54*

# Extracellular Nucleotides and Nucleosides: Release, Receptors, and Physiological and Pathophysiological Effects

*Edited by*

**Erik M. Schwiebert**
*Department of Physiology and Biophysics and Cell Biology*
*Gregory Fleming James Cystic Fibrosis Research Center*
*University of Alabama at Birmingham*
*Birmingham, Alabama*

## ACADEMIC PRESS

An imprint of Elsevier Science

Amsterdam   Boston   Heidelberg   London   New York   Oxford
Paris   San Diego   San Francisco   Singapore   Sydney   Tokyo

Permissionions may be sought directly from Elsevier's Science & Technology Rights Department in Oxford, UK: phone: (+44) 1865 843830, fax: (+44) 1865 853333, e-mail: permissions@elsevier.com.uk. You may also complete your request on-line via the Elsevier Science homepage (http://elsevier.com), by selecting "Customer Support" and then "Obtaining Permissions."

Academic Press
*An Elsevier Science Imprint.*
525 B Street, Suite 1900, San Diego, California 92101-4495, USA
http://www.academicpress.com

Academic Press
84 Theobald's Road, London WC1X 8RR, UK
http://www.academicpress.com

International Standard Book Number: 0-12-153354-9

PRINTED IN THE UNITED STATES OF AMERICA
03  04  05  06  07  08    9  8  7  6  5  4  3  2  1

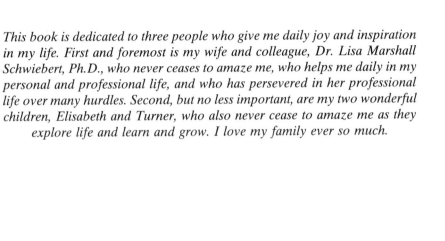

*This book is dedicated to three people who give me daily joy and inspiration in my life. First and foremost is my wife and colleague, Dr. Lisa Marshall Schwiebert, Ph.D., who never ceases to amaze me, who helps me daily in my personal and professional life, and who has persevered in her professional life over many hurdles. Second, but no less important, are my two wonderful children, Elisabeth and Turner, who also never cease to amaze me as they explore life and learn and grow. I love my family ever so much.*

# Contents

Contents

# Contributors

Numbers in parentheses indicate the pages on which the author's contributions begin.

**Edward H. Abraham** (409), Dartmouth Medical School, Dartmouth-Hitchcock Medical Center, Lebanon, New Hampshire 03756

**Matthew A. Bailey** (363), Centre for Nephrology and Department of Physiology, Roayl Free and University College Medical School, University College London, London NW3 2QG, United Kingdom

**Gavin M. Braunstein** (205), Departments of Physiology and Biophysics and Cell Biology, and Gregory Fleming James Cystic Fibrosis Research Center, University of Alabama at Birmingham, Birmingham, Alabama 35294

**Geoffrey Burnstock** (1, 307), Autonomic Neuroscience Institute, Royal Free and University College London Medical School, London NW3 2PF, United Kingdom

**B. R. Cobb** (151), Department of Human Genetics, Department of Pediatrics, University of Alabama at Birmingham, Birmingham, Alabama 35294

**J. P. Clancy** (151), Department of Pediatrics, Gregory Fleming James Cystic Fibrosis Research Center, University of Alabama at Birmingham, Birmingham, Alabama 35294

**Hans H. Detrich** (243), St. Louis University School of Medicine. St. Louis, Missouri 63104

**Terrance M. Egan** (183), Department of Pharmacological and Physiological Science, St. Louis University School of Medicine, St. Louis, Missouri 63104

**Mary L. Ellsworth** (243), St. Louis University School of Medicine. St. Louis, Missouri 63104

**Andrew P. Feranchak** (389), Department of Pediatrics, University of Colorado Health Sciences Center, Denver, Colorado 80262

**J. Gregory Fitz** (389), Department of Medicine, University of Colorado Health Sciences Center, Denver, Colorado 80262

**Thomas Forrester** (269), Department of Pharmacological and Physiological Sciences, St. Louis University School of Medicine, St. Louis, Missouri 63104

**John P. Gleibel** (31), Departments of Surgery and Cellular and Molecular Physiology, Yale University School of Medicine, New Have, Connecticut 06520

**Edward W. Inscho** (447), Department of Physiology, Medical College of Georgi, August, Georgia 30912

**Eduardo R. Lazarowski** (59), Department of Medicine, University of North Carolina School of Medicine, Chapel Hill, North Carolina 27599

**Jens Leipziger** (363), Institute of Physiology, Åarhus University, DK-8000, Denmark

**Keisuke Migita** (183), Department of Pharmacological and Physiological Science, St. Louis University School of Medicine, St. Louis, Missouri 63104

**Eliezer Rapaport** (409), Dartmouth Medical School, Dartmouth-Hitchcock Medical Center, Lebanon, New Hampshire 03756

**Anna Y. Salikhova** (409), Dartmouth Medical School, Dartmouth-Hitchcock Medical Center, Lebanon, New Hampshire 03756

**Erik M. Schwiebert** (31, 97, 205), Departments of Physiology and Biophysics and Cell Biology, and Gregory Fleming James Cystic Fibrosis Research Center, University of Alabama at Birmingham, Birmingham, Alabama 35294

**Randy S. Sprague** (243), St. Louis University School of Medicine. St. Louis, Missouri 63104

**Amanda Taylor Boyce** (97), Department of Cell Biology and Physiology and Biophysics, Gregory Fleming James Cystic Fibrosis Research Center, University of Alabama and Birmingham, Birmingham, Alabama 35294

**Robert J. Unwin** (363), Centre for Nephrology and Department of Physiology, Roayl Free and University College Medical School, University College London, London NW3 2QG, United Kingdom

**Mark M. Voigt** (183), Department of Pharmacological and Physiological Science, St. Louis University School of Medicine, St. Louis, Missouri 63104

**Akos Zsembery** (31), Departments on Physiology and Biophysics and Cell Biology, and Gregory Fleming James Cystic Fibrosis Research Center, University of Alabama at Birmingham, Birmingham, Alabama 35294

# Foreword

## Extracellular Nucleotide and Nucleoside Signaling: First Principles

**Erik M. Schwiebert**
Department of Physiology and Biophysics and of Cell Biology and
Gregory Fleming James Cystic Fibrosis Research Center
University of Alabama at Birmingham
Birmingham, Alabama 35294

It was a joy to participate in the process of assembling this volume of *Current Topics in Membranes*, which focuses on all aspects of extracellular nucleotides and nucleosides: release, receptors, and effects. It was a thrill to have two long-standing experts in the field, Geoff Burnstock from the Royal Free and University College Medical School in London and Tom Forrester from the St. Louis University School of Medicine, contribute outstanding chapters to this book.

I also sought many experts in purinergic signaling in several critical cell, tissue, and organ systems. P2X purinergic receptor channel experts Mark Voigt and Terrance Egan, from the St. Louis University School of Medicine, contributed an extensive chapter on this subfamily of nucleotide receptors. Eduardo Lazarowski, a young investigator at the University of North Carolina School of Medicine at Chapel Hill and an expert in extracellular nucleotide degradation and resynthesis and P2Y G protein-coupled purinergic receptors, wrote an excellent review of P2Y receptors for this book. Two excellent young investigators from the University of Alabama at Birmingham (UAB), Brian Cobb and J. P. Clancy, wrote a first-rate review chapter on adenosine receptors. Another group from the St. Louis University School of Medicine, led by Randy Sprague, authored a chapter on release of ATP from red blood cells (RBCs) and its effects on the RBC and the vasculature. Drew Feranchak and Greg Fitz from the University of Colorado Health Sciences Center, experts in purinergic signaling in liver and its effects on hepatocytes and cholangiocyte cell volume regulation, also contributed a seminal chapter. Two excellent young investigators, Robert

Unwin from the Royal Free and University College Medical School in London and Jens Leipziger from Åarhus University, collaborated on a review of purinergic signaling in the kidney. Our laboratory research group from UAB filled in the gaps with four different book chapters on subjects that included ATP release mechanisms (with John Geibel from University of Alabama at Birmingham), P2X purinergic receptor channels, and extracellular purinergic signaling in epithelial cell models. Finally, Ted Abraham wrote a seminal review on purinergic signaling and its many roles in cancers. All were delighted to participate. It is the hope of the authors and the editor that these chapters will provide a seminal resource for the purinergic field.

In every chapter, a central theme is echoed again and again. It is the concept that, in every paradigm in which extracellular nucleotide and/or nucleoside signaling is examined and discussed, the release, receptors, and effects of active purinergic agonists need to be addressed and studied together in an integrative manner. This is true for any autocrine or paracrine agonist; however, it is particularly true for cAMP, ATP, and adenosine. Geoffrey Burnstock's extensive work, when taken altogether, illustrates how this is critical. In this foreword, I have provided the mother of all cell model figures to illustrate how this integrated signaling system may work. Obviously, there are aspects of this system that may still come into focus or be discovered. However, this model depicts how our laboratory envisions extracellular purinergic signaling in total. From this "big picture," we focus on aspects of the system; however, we try never to lose that big picture.

Figure 1 shows the "first principles" of extracellular nucleotide and nucleoside signaling. In Fig. 1A, all of the key first principles are shown. First, newly synthesized ATP is transported out of the mitochondrion. The steady state cytosolic concentration of ATP is 3–10 m$M$, the fuel for countless metabolic and enzymatic reactions. Second, if a pathway is activated or opened for ATP release, ATP exits the cell down a very favorable chemical concentration gradient. For example, if steady state extracellular ATP is approximately 100 n$M$ under basal conditions and intracellular ATP is 10 m$M$, this gradient is approximately 100,000-fold. It is 10-fold greater yet opposite to the gradient for calcium entry into cells. Third, ATP, released from the same cell or a neighboring cell, is free to bind to specific cell surface receptors, classified as P2 receptors. Alternatively, ADP, a metabolite, may also bind with high affinity to a subset of P2 receptors. UTP and UDP follow similar chemistry; however, the intracellular concentration of UTP is micromolar rather than millimolar. The final metabolite of ATP, adenosine, is also biologically active and has its own subset of P1 receptors. Fourth, ATP is degraded rapidly; thus, it is thought of as a local mediator that acts in an autocrine or paracrine manner with tissues and tissue microenvironments. However, ATP may also be

**A**

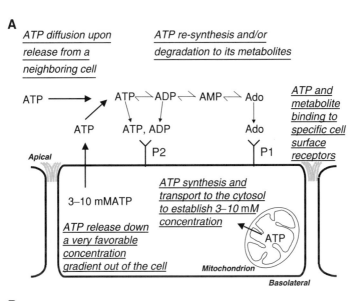

**B**

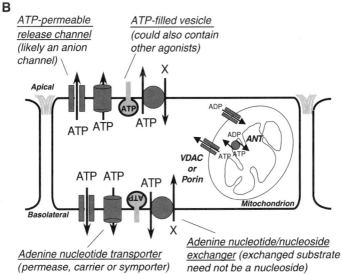

**FIGURE 1**   *(For legend, see p. xx)*

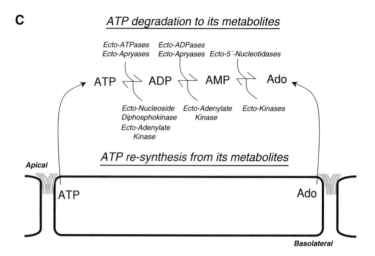

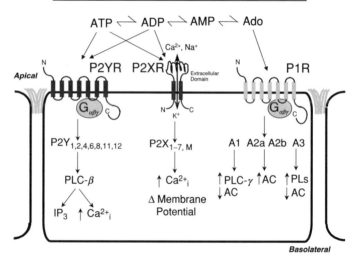

**FIGURE 1** First principles of extracellular nucleotide and nucleoside signaling. (A) First principles. ATP, adenosine 5′ triphosphate, ADP, adenosine diphosphate; AMP, adenosine 5′-monophosphate; Ado, adenosine. (B) Specific mechanisms of ATP release from the cell are shown. ANT, adenine nucleotide transporter; VDAC, voltage-dependent anion channel. (C) Enzymatic degradation and re-synthesis of ATP in the extracellular space by secreted or membrane-bound ecto-enzymes. (D) P2 and P1 receptors and their signaling pathways. $G_{\alpha\beta\gamma}$, heterotrimeric G protein; PLC, phospholipase C; PLs, various phospholipases; AC, adenylyl cyclase; $IP_3$, inositol trisphosphate; $Ca^{2+}_i$, cytosolic calcium.

resynthesized, due to the action of ectokinases. These first principles are revisited again and again in the chapters of this book.

Figure 1B focuses on possible ATP release mechanisms. Adenosine (not pictured) may also be released or taken up by concentrative nucleoside transporters (CNTs) or equilibrative (bidirectional) nucleoside transporters (ENTs). First, newly synthesized ATP is transported across the inner or cisternal membrane of the mitochondrion by an adenine nucleotide transporter (ANT) that exchanges an ATP molecule for its metabolite, ADP. Mitochondrial porin (also known as the voltage-dependent anion channel, VDAC) conducts the ATP across the outer membrane into the cytosol. Second, pictured in the context of a polarized epithelial cell, ATP could be conducted passively down its very favorable chemical concentration gradient by ATP-permeable release channels. These are likely anion channels (plasma membrane forms of VDAC or other chloride channels) that have significant permeability to larger anions, such as gluconate and ATP. Third, concentrative or equilibrative nucleotide transporters may also exist for ATP that are carriers or permeases. The ATP-binding cassette (ABC) transporters such as the cystic fibrosis transmembrane conductance regulator (CFTR), the multidrug resistance proteins (mdr, MRP), and the multiple organic anion transporters (MOATs) may act as nonconductive transporters of ATP, along with a multitude of other substrates. Fourth, exchangers may exist for ATP in the plasma membrane, as they do in the mitochondrial inner membrane, where ADP may be the exchanged substrate. Other substrates such as $Na^+$ or $Cl^-$, which have favorable entry gradients, may be exchanged for ATP. Fifth, ATP-filled vesicles, which may also contain additional agonists or coagonists, may fuse with the plasma membrane releasing ATP. The chapter by Schwiebert, Zsembery, and Geibel (this volume) focuses on these mechanisms and on innovative assays by which to detect released nucleotides and nucleosides.

Once ATP leaves the cell, it is the target of multiple degradative enzymes. The biochemical cascade of degradation is shown in simple terms in Fig. 1C. Ectoenzymes secreted into the extracellular milieu as a secreted protein or membrane bound as an ectoenzyme target ATP. Ectoapyrases cleave the $\gamma$- and $\beta$-phosphates of ATP, yielding ADP and 5'-AMP. Ecto-ATPases and ecto-ADPases also subserve this function. Ecto-5'-nucleotidases can convert 5'-AMP into adenosine, a higher affinity purinergic agonist than ATP for its own class of P1 G protein-coupled receptors. Like any biochemical reaction, these can occur in reverse to resynthesize ATP. Secreted or membrane-bound ectokinases such as adenylate kinase or nucleoside monophospho- and diphosphokinases can phosphorylate nucleosides to remake the triphosphate nucleotide ATP. Eduardo Lazarowski does not write about resynthesis of ATP in extracellular environments; however, he and his mentors, Kendall Harden and Ric Boucher, have published extensively in this field.

Figure 1D shows a schematic breakdown of P2 and P1 receptors within the context of an epithelial cell. Gavin Braunstein and I examine epithelial purinergic signaling in one of the chapters of this book. A question that haunts us is: Why would an epithelial cell express each subtype of purinergic receptor shown in Fig. 1D in the same cell and, sometimes, the same apical membrane domain? Many cells express at least one member of each purinergic receptor subfamily, often multiple subtypes. Several chapters touch on this issue. P2Y and P2X receptors, their subtypes, properties, and cell and tissue distribution are a major focus of the Introduction, by Geoff Burnstock, and is also the focus of four chapters on ATP and adenosine receptors, written from different perspectives.

What is not pictured in Fig. 1 is a cell model summarizing the effects of extracellular nucleotide and nucleoside signaling on cells. There are too many. They have profound effects on cell, tissue, and organ physiology and have roles in pathophysiology, including ischemia/reperfusion, anoxia, pain perception, sensory organs, cystic fibrosis (CF), polycystic kidney disease, and hypertension. The chapters in this book discuss the many effects of extracellular purinergic signaling as they pertain to a particular cell type, tissue, or organ paradigm. It is the sincere hope of the editor that the reader learns a great deal about these topics; the editor certainly did.

**Acknowledgments**
The editor would like to thank the authors for their diligence and hard work and Academic Press for their helpful and frequent communication and expert publication of this volume of *Current Topics in Membranes*.

# Preface

For virtually an entire century, the presence of extracellular nucleotide and nucleoside-dependent signaling has been postulated and proven, proven and postulated. Through the years, the release and signaling in the extracellular milieu of a molecule so precious to the cytoplasm has been met over and over again with hesitance and skepticism. The major purpose of this volume was to make an overwhelming case for the physiological importance of released nucleotides and nucleosides, autocrine and paracrine nucleotide and nucleoside signaling, and nucleotide and nucleoside receptors.

Although you can never hope to cover every issue or topic within a given research field, I believe we have met the charge with some seminal contributions from outstanding investigators. We are privileged to have Drs. Geoff Burnstock and Tom Forrester contribute to this volume. "Purinergic signaling" has been their passion for decades. Dr. Forrester's colleagues at St. Louis University, a hot bed of purinergic research, Drs. Sprague, Egan, Voigt, and others are contributed two additional chapters. This topic is also the central theme of our laboratory. Several of my junior and senior colleagues, as graduate students and postdoctoral fellows, helped me contribute our chapters on ATP release, ATP signaling in epithelia, and a comprehensive review on P2X receptor channels from our perspective. I also included a Foreward for the book regarding the "first principles" of extracellular ATP and adenosine signaling for the uninitiated, hoping that it would help put the chapters that followed in perspective. We are grateful to Drs. Lazarowski, Clancy, and Cobb for chapters on the G protein-coupled receptors utilized by nucleotides and nucleosides, respectively. We also had several chapters on purinergic signaling in kidney (Drs. Unwin, Leipziger, and Inscho), liver (Drs. Feranchak and Fitz), and muscle and vasculature (Drs. Sprague and Forrester). Finally, we had a novel contribution from Dr. Ted Abraham on the role of purinergic signaling in cancer.

Unfortunately, during the compilation of this volume, the field lost a pioneer in adenosine signaling, Dr. Robert Berne from the University of Virginia. I had asked Dr. Berne to provide a chapter on "adenosine signaling in the heart." However, I was unaware of his ill-health at the time of the invitation. This volume is dedicated to him as well as others who were pioneers in this field.

As is always the hope with the extensive work that goes into these volumes, we wish that this volume will stimulate new investigators to probe the importance of extracellular ATP and adenosine signaling in their model systems. Several authors have discussed former a small and focused scientific conference on this subject. Please keep an eye out in the American Physiological Society and/or FASEB literature for a future meeting organized by myself, Dr. Burnstock, and others.

Erik M. Schwiebert

# Previous Volumes in Series

## Current Topics in Membranes and Transport

**Volume 23** Genes and Membranes: Transport Proteins and Receptors*
(1985)
Edited by Edward A. Adelberg and Carolyn W. Slayman

**Volume 24** Membrane Protein Biosynthesis and Turnover (1985)
Edited by Philip A. Knauf and John S. Cook

**Volume 25** Regulation of Calcium Transport across Muscle Membranes
(1985)
Edited by Adil E. Shamoo

**Volume 26** $Na^+-H^+$ Exchange, Intracellular pH, and Cell Function*
(1986)
Edited by Peter S. Aronson and Walter F. Boron

**Volume 27** The Role of Membranes in Cell Growth and Differentiation
(1986)
Edited by Lazaro J. Mandel and Dale J. Benos

**Volume 28** Potassium Transport: Physiology and Pathophysiology*
(1987)
Edited by Gerhard Giebisch

**Volume 29** Membrane Structure and Function (1987)
Edited by Richard D. Klausner, Christoph Kempf, and Jos van Renswoude

**Volume 30** Cell Volume Control: Fundamental and Comparative
Aspects in Animal Cells (1987)
Edited by R. Gilles, Arnost Kleinzeller, and L. Bolis

**Volume 31** Molecular Neurobiology: Endocrine Approaches (1987)
Edited by Jerome F. Strauss, III, and Donald W. Pfaff

**Volume 32** Membrane Fusion in Fertilization, Cellular Transport, and
Viral Infection (1988)
Edited by Nejat Düzgünes and Felix Bronner

*Part of the series from the Yale Department of Cellular and Molecular Physiology*

# CHAPTER 1

# Introduction: ATP and Its Metabolites as Potent Extracellular Agents

## Geoffrey Burnstock

Autonomic Neuroscience Institute, Royal Free and University College London Medical School, London NW3 2PF, United Kingdom

## I. EARLY STUDIES

A seminal paper describing the potent actions of adenine compounds was published by Drury and Szent-Györgyi in 1929. Many years later, ATP was proposed as the transmitter responsible for nonadrenergic, noncholinergic transmission in the gut and bladder and the term "purinergic" was introduced (Burnstock, 1972). Early resistance to this concept appeared to stem from the fact that ATP was recognized first for its important intracellular roles in many biochemical processes and the intuitive feeling was that such a ubiquitous and simple compound was unlikely to be utilized as an extracellular messenger, although powerful extracellular enzymes involved in its breakdown were known to be present.

Implicit in the concept of purinergic neurotransmission was the existence of postjunctional purinergic receptors, and the potent actions of extracellular ATP on many different cell types also implicated membrane receptors. Purinergic receptors were first defined in 1976 (Burnstock, 1976b) and 2 years later a basis for distinguishing two types of purinoceptor, identified as P1 and P2 (for adenosine and ATP/ADP, respectively), was proposed (Burnstock, 1978). At about the same time, two subtypes of the P1 (adenosine) receptor were recognized (Van Calker et al., 1979; Londos et al., 1980), but it was not until 1985 that a proposal suggesting a basis for distinguishing two types of P2 receptor (P2X and P2Y) was made (Burnstock and Kennedy, 1985). A year later, Gordon tentatively named two further P2 purinoceptor subtypes, namely, a P2T receptor selective for ADP on platelets and a P2Z receptor on macrophages (Gordon, 1986). Further subtypes followed, perhaps the most important being the P2U receptor, which could recognize pyrimidines such as UTP as well as ATP (O'Connor et al., 1991). In 1994 Williams made the point at a meeting that a classification of P2 purinoceptors based on a "random walk through the alphabet" was not satisfactory, and Abbracchio and Burnstock (1994), on the basis of studies of transduction mechanisms (Dubyak, 1991) and the cloning of nucleotide receptors (Lustig et al., 1993; Webb et al., 1993; Brake et al., 1994; Valera et al., 1994), proposed that purinoceptors should belong to two major families: a P2X family of ligand-gated ion channel receptors and a P2Y family of G protein-coupled purinoceptors. This nomenclature has been widely adopted and currently seven P2X subtypes and about six P2Y receptor subtypes are recognized, including receptors that are sensitive to pyrimidines as well as purines (see Ralevic and Burnstock, 1998).

It is widely recognized that purinergic signaling is a primitive system (Burnstock, 1996) involved in many nonneuronal as well as neuronal mechanisms (see Abbracchio and Burnstock, 1998), including exocrine and endocrine secretion, immune responses, inflammation, pain, platelet aggregation, and endothelial-mediated vasodilatation (see Gordon, 1986; Olsson and Pearson, 1990; Dubyak and El Moatassim, 1993; Burnstock, 1997, 2001a).

## II. PURINOCEPTOR SUBTYPES

### A. P1 Purinoceptors

Four subtypes of P1 receptors have been cloned, namely, $A_1$, $A_{2A}$, $A_{2B}$, and $A_3$ (see Olah and Stiles, 1995; Ralevic and Burnstock, 1998; Baraldi *et al.*, 2000; Fredholm *et al.*, 2001).

All P1 adenosine receptors couple to G proteins. In common with other G protein-coupled receptors, they have seven putative transmembrane (TM) domains of hydrophobic amino acids, each believed to constitute an $\alpha$ helix of approximately 21 to 28 amino acids. The N terminus of the protein lies on the extracellular side and the C terminus on the cytoplasmic side of the membrane. Typically, the extracellular loop between TM4 and TM5 and the cytoplasmic loop between TM5 and TM6 are extended. The intracellular segment of the receptor interacts with the appropriate G protein, with subsequent activation of the intracellular signal transduction mechanism. It is the residues within the transmembrane regions that are crucial for ligand binding and specificity and, with the exception of the distal (carboxyl) region of the second extracellular loop, the extracellular loops, the C terminus, and the N terminus do not seem to be involved in ligand recognition (Olah and Stiles, 1995). Site-directed mutagenesis of the bovine $A_1$ adenosine receptor suggests that conserved histidine residues in TM6 and TM7 are important in ligand binding.

Specific agonists and antagonists are available for the P1 receptor subtypes (see Table I). For a complete review of adenosine receptors specifically, the reader should refer to the chapter by Cobb and Clancy, entitled Molecular and Cell Biology of Adenosine Receptors (this volume).

### B. P2X Receptors

Members of the existing family of ionotropic $P2X_{1-7}$ receptors show a subunit topology of intracellular N and C termini possessing consensus binding motifs for protein kinases; two transmembrane-spanning regions (TM1 and TM2), the first involved with channel gating and the second lining the ion pore; a large extracellular loop, with 10 conserved cysteine residues forming a series of disulfide bridges; a hydrophobic H5 region close

**TABLE I**

Characteristics of receptors for purines and pyrimidines

| Receptor | Main distribution | Agonists | Antagonists | Transduction mechanisms |
|---|---|---|---|---|
| P1 (adenosine) | | | | |
| $A_1$ | Brain, spinal cord, testis, heart, autonomic nerve terminals | CCPA, CPA | DPCPX, CPX, XAC | $G_i$ (1–3), $\downarrow$cAMP |
| $A_{2A}$ | Brain, heart, lungs, spleen | CGS 21680 | KF17837, SCH58251 | $G_s$, $\uparrow$cAMP |
| $A_{2B}$ | Large intestine, bladder | NECA | Enprofylline | $G_s$, $\uparrow$cAMP |
| $A_3$ | Lung, liver, brain, testis, heart | DB-MECA, DBX MR | MRS1222, L-268,605 | $G_i$ (2, 3), $G_{q/11}$, $\downarrow$cAMP, $\uparrow$IP$_3$ |
| P2X | | | | |
| $P2X_1$ | Smooth muscle, platelets, cerebellum, dorsal horn spinal neurons | $\alpha\beta$meATP = ATP = 2meSATP (rapid desensitization) | TNP-ATP, IP$_5$I, NF023 | Intrinsic cation channel ($Ca^{2+}$ and $Na^+$) |
| $P2X_2$ | Smooth muscle, CNS, retina, chromaffin cells, autonomic and sensory ganglia | ATP $\geq$ ATP$\gamma$S $\geq$ 2mSATP >> $\alpha\beta$meATP (pH + zinc sensitive) | Suramin, PPADS | Intrinsic ion channel (especially $Ca^{2+}$) |
| $P2X_3$ | Sensory neurons, NTS, some sympathetic neurons | 2meSATP $\geq$ ATP $\geq$ $\alpha\beta$meATP (rapid desensitization) | TNP-ATP, suramin, PPADS | Intrinsic cation channel |
| $P2X_4$ | CNS, testis, colon | ATP > $\alpha\beta$meATP | — | Intrinsic ion channel (especially $Ca^{2+}$) |
| $P2X_5$ | Proliferating cells in skin, gut, bladder, thymus, spinal cord | ATP > $\alpha\beta$meATP | Suramin, PPADS | Intrinsic ion channel |
| $P2X_6$ | CNS, motor neurons in spinal cord | —(does not function as homomultimer) | — | Intrinsic ion channel |

| Receptor | Distribution | Agonist potency order | Antagonists | Transduction mechanism |
|---|---|---|---|---|
| P2X$_7$ | Apoptotic cells in immune cells, pancreas, skin, etc. | BzATP > ATP $\geq$ 2meSATP > $\alpha\beta$meATP | KN62, KN04, Coomassie Brilliant Blue | Intrinsic cation channel and a large pore with prolonged activation |
| **P2Y** | | | | |
| P2Y$_1$ | Epithelial and endothelial cells, platelets, immune cells, osteoclasts | 2meSADP > 2meSATP = ADP > ATP | MRS 2279, MRS 2179 | G$_q$/G$_{11}$; PLC$\beta$ activation |
| P2Y$_2$ | Immune cells, epithelial and endothelial cells, kidney tubules, osteoblasts | UTP = ATP | Suramin | G$_q$/G$_{11}$ and possibly G$_i$; PLC$\beta$ activation |
| P2Y$_4$ | Endothelial cells | UTP $\geq$ ATP | Reactive Blue 2, PPADS | G$_q$/G$_{11}$ and possibly G$_i$; PLC$\beta$ activation |
| P2Y$_6$ | Some epithelial cells, placenta, T cells, thymus | UDP > UTP > ATP | Reactive Blue 2, PPADS, suramin | G$_q$/G$_{11}$; PLC$\beta$ activation |
| P2Y$_{11}$ | Spleen, intestine, granulocytes | ARC67085MX > BzATP $\geq$ ATP$\gamma$S > ATP | Suramin, RB2 | G$_q$/G$_{11}$ and G$_s$; PLC$\beta$ activation |
| P2Y$_{12}$ | Platelets, glial cells | ADP | ARC67085MX, ARC69931MX | G$_i$ (2); inhibition of adenylate cyclase |
| P2Y$_{13}$ | Spleen, brain, lymph nodes, bone marrow | ADP = 2me SADP > ATP and 2me SATP | | G$_i$ |
| P2Y$_{14}$ | Placenta, adipose tissue, stomach, intestine, discrete brain regions | UDP glucose = UDP-galactose | | G$_{i/0}$ |

to the pore vestibule, for possible receptor/channel modulation by cations (magnesium, calcium, zinc, copper, and proton ions); and an ATP-binding site, which may involve regions of the extracellular loop adjacent to TM1 and TM2. The $P2X_{1-7}$ receptors show 30–50% sequence identity at the peptide level. The stoichiometry of $P2X_{1-7}$ receptors is now thought to involve three subunits that form a stretched trimer (see Khakh *et al.*, 2001).

It has become apparent that the pharmacology of the recombinant P2X receptor subtypes expressed in oocytes or other cell types is often different from the pharmacology of P2Y-mediated responses in naturally occurring sites. There are several contributing factors to explain these differences. Heteromultimers as well as homomultimers are involved in forming the trimer ion pore. For example, heteromultimers are clearly established for $P2X_{2/3}$ in nodose ganglia (Lewis *et al.*, 1995; Radford *et al.*, 1997), $P2X_{4-6}$ in CNS neurons (Lê *et al.*, 1998), $P2X_{1-5}$ in some blood vessels (Torres *et al.*, 1998; Haines *et al.*, 1999), and $P2X_{2/6}$ in the brainstem (King *et al.*, 2000). $P2X_7$ does not form heteromultimers, and $P2X_6$ will not form a functional homomultimer (Torres *et al.*, 1999; North and Surprenant, 2000). Second, spliced variants of P2X receptor subtypes might play a part. For example, a splice variant of the $P2X_4$ receptor, while it is nonfunctional on its own, can potentiate the actions of ATP through the full-length $P2X_4$ receptors (e.g., Townsend-Nicholson *et al.*, 1999). Third, the presence in tissues of powerful ectoenzymes that rapidly break down purines and pyrimidines is not a factor when examining recombinant receptors (Zimmermann, 1996).

The P2X receptor family shows many pharmacological and operational differences. The kinetics of activation, inactivation, and deactivation also vary considerably among P2X receptors. Calcium permeability is high for some P2X subtypes, a property that may be functionally important. For more specific review of P2X receptor molecular biology, cell biology, physiology, and biophysics, the reader is referred to two subsequent chapters on P2X purinergic receptor channels by Voigt and Egan and by Taylor and Schwiebert (this volume).

### C. P2Y Receptors

Metabotropic $P2Y_{1-14}$ receptors are characterized by a subunit topology of an extracellular N terminus and intracellular C terminus, the latter possessing consensus binding motifs for protein kinases; seven transmembrane-spanning regions, which help to form the ligand-docking pocket; a high level of sequence homology between some transmembrane-spanning regions, particularly TM3, TM6, and TM7; and a structural diversity of intracellular loops and C terminus among P2Y subtypes, so influencing the degree of coupling with $G_{q/11}$, $G_s$, and $G_i$ proteins. Each P2Y receptor binds to a single heterotrimeric G protein (typically $G_{q/11}$), although $P2Y_{11}$ can couple to both $G_{q/11}$ and $G_s$ whereas $P2Y_{12}$ and $P2Y_{13}$ couple to $G_i$ and $P2Y_{14}$ to $G_{i/0}$. P2Y receptors may form homo- and heteromultimeric assemblies under some conditions, and many tissues express multiple P2Y subtypes (see King *et al.*, 2001; King and Burnstock, 2002). P2Y receptors show a low level of sequence homology at the peptide level (19–55% identical) and, consequently, show significant differences in their pharmacological and operational profiles. Some P2Y receptors are activated principally by nucleoside diphosphates ($P2Y_{1,6,12}$), while others are activated mainly by nucleoside triphosphates ($P2Y_{2,4}$). Some P2Y receptors are activated by both purine and pyrimidine nucleotides ($P2Y_{2,4,6}$), and others by purine nucleotides alone ($P2Y_{1,11,12}$). In response to nucleotide activation, recombinant P2Y receptors either activate phospholipase C and release intracellular calcium or affect adenylyl cyclase and alter cAMP levels. There is little evidence to indicate that $P2Y_{5,9,10}$ sequences are nucleotide receptors or affect intracellular signaling cascades. Endogenous P2Y receptors show a greater diversity in intracellular signaling and can activate phospholipases $A_2$, C, and D, major excreted protein (MEP)/ mitogen-activated protein (MAP) kinase, Rho-dependent kinase, and tyrosine kinase, as well as coupling both positively and negatively to adenylyl cyclase.

2-Methylthio-ADP is a potent agonist of mammalian $P2Y_1$ receptors (Hechler *et al.*, 1998), and MRS 2179 is a potent antagonist (Boyer *et al.*, 1998); MRS 2269 and MRS 2286 have been identified as selective antagonists (Brown *et al.*, 2000). At $P2Y_2$ and $P2Y_4$ receptors in the rat, ATP and UTP are equipotent, but the two receptors can be distinguished with antagonists, that is, suramin blocks $P2Y_2$, while Reactive Blue 2 blocks $P2Y_4$ receptors (Bogdanov *et al.*, 1998; King *et al.*, 1998). $P2Y_6$ is UDP selective, while $P2Y_7$ turned out to be a leukotriene receptor (Yokomizo *et al.*, 1997). $P2Y_8$ is a receptor cloned from frog embryos, where all the nucleotides are equipotent (Bogdanov *et al.*, 1997), but no mammalian homolog has been identified to date, apart from a report of $P2Y_8$ mRNA in undifferentiated HL60 cells (Adrian *et al.*, 2000). $P2Y_{11}$ is unusual in that there are two transduction pathways, adenylate cyclase as well as inositol triphosphate, which is the second-messenger system used by the majority of the P2Y receptors. The $P2Y_{12}$ receptor found on platelets was not cloned until more recently (Hollopeter *et al.*, 2001), although it has only 19% homology with the other P2Y receptor subtypes. It seems likely to represent one of a subgroup of P2Y receptors, including $P2Y_{13}$ and $P2Y_{14}$, for which transduction is entirely through adenylate cyclase (Abbracchio *et al.*, 2003; Communi *et al.*, 2001; Zhang *et al.*, 2002). A receptor on C6 glioma cells and possibly a receptor in the midbrain, selective for a diadenosine polyphosphate, also may operate through adenylate cyclase. An interesting question that has arisen by analogy with other G protein-coupled receptors is whether dimers can form between the P2Y subtypes. For a specific review of P2Y receptor cell biology and physiology, the reader should refer to the chapter by Lazarowski (this volume).

Table I summarizes the structure and properties of current receptor subtypes while Table II summarizes the current status of P2 receptor subtype agonists and antagonists.

## III. DISTRIBUTION AND ROLES OF PURINOCEPTOR SUBTYPES

### A. P1 Receptor Subtypes

$A_1$ receptors are widely distributed and mediate diverse biological effects. $A_1$ receptors are particularly ubiquitous within the CNS, with high levels being expressed in the cerebral cortex, hippocampus, cerebellum, thalamus, brainstem, and spinal cord (Rivkees *et al.*, 1995; Dixon *et al.*, 1996). $A_1$ receptor mRNA is widely distributed in peripheral tissues, having been localized in vas deferens, testis, white adipose tissue, stomach, spleen, pituitary, adrenal, heart, aorta, liver, eye, and bladder (Reppert *et al.*, 1991; Dixon *et al.*, 1996). Only low levels of $A_1$ mRNA are present in lung, kidney, and small intestine.

It is well established that adenosine is released from biological tissues during hypoxia and ischemic conditions. One of its effects is to reduce neuronal activity and thereby oxygen consumption; thus it acts as a neuroprotective agent. A significant part of these effects seems to be mediated by the $A_1$ receptor. $A_1$ receptors are located pre- and postsynaptically on cell bodies, and on axons, where they mediate inhibition of neurotransmission by decreasing transmitter release, hyperpolarizing neuronal membranes, reducing excitability and firing rate, and altering axonal transmission. Adenosine can also exert behavioral effects: adenosine actions at $A_1$ receptors have been implicated in sedative, anticonvulsant, anxiolytic, and locomotor depressant effects (Nikodijevic *et al.*, 1991; Stone, 1991; Jain *et al.*, 1995; Malhotra and Gupta, 1997; Kaiser and Quinn, 1999; Williams and Jarvis, 2000). Conversely, xanthine P1 antagonists such as caffeine and theophylline have central stimulatory properties ascribed, at least in part, to inhibition of endogenous adenosine.

$A_1$ receptors mediate cardiac depression through negative chronotropic, dromotropic, and inotropic effects (see Olsson and Pearson, 1990). Slowing of the heart rate occurs via $A_1$ receptors on sinoatrial and atrioventricular nodes, causing bradycardia and heart block, respectively, while the inotropic effects include a decrease in atrial contractility and action potential duration. In the kidney, activation of $A_1$ receptors mediates diverse effects including vasoconstriction (principally of the afferent arteriole), a decrease in glomerular filtration rate, mesangial cell contraction, inhibition of renin secretion, and inhibition of neurotransmitter release (see Ralevic and Burnstock, 1998). Intravenous and intra aortic administration of adenosine in rats decrease water and sodium excretion via $A_1$ receptors cause diuresis and natriuresis. The most important role of $A_1$ receptors with regard to regulation of blood vessel tone appears to be prejunctional modulation of neurotransmitter release (see Burnstock and Ralevic, 1994).

$A_{2A}$ receptors have a wide-ranging, but restricted, distribution that includes immune tissues, platelets, the CNS, vascular smooth muscle, and endothelium. Within the brain, the highest levels of $A_{2A}$ receptors are in the striatum, nucleus accumbens, and olfactory tubercle (Johansson et al., 1997; Rosin et al., 1998). $A_{2A}$ receptors in the CNS and particularly in the peripheral nervous system (PNS) generally facilitate neurotransmitter release. Outside the brain, the most abundant expression of human $A_{2A}$ mRNA is in immune tissues, platelets, eye, and skeletal muscle; heart, lung, bladder, and uterus also show strong expression, with less abundant expression in small intestine, kidney, spleen, stomach, testis, skin, kidney, and liver. In the vasculature, $A_{2A}$ receptors have been described on smooth muscle, where they are associated with vasodilatation. Adenosine also has a mitogenic effect on endothelial cells, which in human endothelial cells is mediated via the $A_{2A}$ receptor and subsequent activation of mitogen-activated protein kinase (MAPK) (Sexl et al., 1997).

$A_{2B}$ receptors are found on practically every cell in most species (Dixon et al., 1996); however, the number of receptors is small and relatively high concentrations of adenosine are generally needed to evoke a response. Northern blot analysis showed relatively high expression of $A_{2B}$ receptors in the cecum, large intestine, and urinary bladder, with lower levels in the brain, spinal cord, lung, vas deferens, and pituitary. Reverse transcriptase-polymerase chain reaction (RT-PCR) revealed the highest expression of $A_{2B}$ receptors in the proximal colon, with lower levels in the eye, lung, uterus, bladder, testis, and skeletal muscle. Functional studies have identified $A_{2B}$ receptors in airway smooth muscle, fibroblasts, glial cells, gastrointestinal tract, and the vasculature.

The $A_3$ receptor is widely distributed, but its physiological role is still largely unknown. $A_3$ mRNA is expressed in testis, lung, kidneys, placenta, heart, brain, spleen, liver, uterus, bladder, jejunum, proximal colon, and eye of rat, sheep, and humans. The highest levels of human $A_3$ mRNA are found in lung and liver, with lower levels in aorta and brain. The $A_3$ receptor on

## TABLE II

Mammalian P2 receptors and assessment of activities of agonists and antagonists (*selective)

| | P2X₁ | P2X₂ | P2X₃ | P2X₄ | P2X₅ | P2X₆ | P2X₇ | P2X₂/₃ | P2X₁/₅ | P2X₄/₆ | P2Y₁ | P2Y₂ | P2Y₄ | P2Y₆ | P2Y₁₁ | P2Y₁₂ | P2Y₁₃ | P2Y₁₄ |
|---|---|---|---|---|---|---|---|---|---|---|---|---|---|---|---|---|---|---|
| *Agonists* | | | | | | | | | | | | | | | | | | |
| ATP | ✓✓✓ | ✓✓ | ✓✓✓ | ✓✓ | ✓ | ✓ | ✓ | ✓✓✓ | ✓✓✓ | ✓✓ | ✓ | ✓✓✓ | ✓✓ | – | ✓ | | | |
| ADP | ✓ | – | ✓ | – | – | – | | | | | ✓✓ | ✓ | ✓ | – | ✓ | ✓✓ | ✓✓ | |
| 2-MeSATP | ✓✓✓ | ✓✓ | ✓✓✓ | ✓✓ | ✓ | – | ✓ | | | | ✓✓ | ✓ | | ✓ | | ✓✓ | ✓ | |
| PAPET-ATP | ✓✓ | ✓ | ✓✓✓ | ✓ | – | ✓ | | | | | ✓✓✓* | | | | | | | |
| 2-MeSADP | ✓✓ | | ✓✓ | ✓ | | | | | | | ✓✓✓* | | | ✓✓ | | ✓✓✓ | ✓✓ | |
| HT-AMP | ✓✓ | – | ✓✓✓ | ✓✓ | | | | | | | ✓✓ | | | | | | | |
| α,β-meATP | ✓✓✓* | ✓✓✓* | ✓✓✓* | ✓ | ✓ | – | – | ✓✓✓ | ✓✓ | ✓ | ✓ | ✓ | – | | | | | |
| β,γ-meATP | ✓✓ | – | ✓✓ | ✓✓ | ✓ | – | – | | | | ✓✓ | ✓ | – | | | | | |
| ATPγS | ✓✓ | ✓✓ | ✓ | ✓✓ | ✓ | | | | | | ✓✓ | ✓ | ✓ | ✓ | ✓✓✓ | ✓ | | |
| ATPβS | ✓✓ | ✓✓ | ✓✓✓ | ✓✓ | ✓✓ | | | | | | ✓✓ | ✓ | ✓ | ✓✓✓ | ✓✓✓ | ✓✓ | | |
| Bz-ATP | ✓✓✓ | ✓✓ | ✓ | ✓✓✓ | ✓✓ | ✓✓ | | | | | | | | | | | | |
| UDP-glucose | | | | | | | | | | | | | ✓ [antag] | | | | | ✓✓✓* |
| UDPβS | | | | | | | | | | | | | | ✓✓✓* | | | | |
| 2-dATP | | | | | | | | | | | | | | | ✓✓ | | | |
| Ap₄A | ✓✓✓ | ✓ | ✓✓✓ | ✓ | ✓ | | | | | | | ✓✓ | ✓✓ | ✓ | | | | |
| UTP | – | – | – | – | – | | | | | | | ✓✓✓ | ✓✓✓ | | | | | |
| UTPγS | | | | | | | | | | | | ✓✓✓* | ✓ | | | | | |
| UDP | ✓ | ✓ | – | – | – | | | | | | | ✓ | ✓✓ | ✓✓✓ | | | | |
| CTP | ✓ | ✓ | ✓ | ✓ | ✓ | | | | | | | | ✓✓ | | | | | |
| *Antagonists* | | | | | | | | | | | | | | | | | | |
| PPADS | ✓✓ | ✓✓ | ✓✓ | – | ✓✓ | ✓ | ✓ | | | | ✓✓ | | ✓ | ✓✓ | | | | |
| isoPPADS | ✓✓ | ✓✓ | ✓✓ | ✓✓ | ✓ | | | | | | | | | | | | | |
| PPNDS | ✓✓* | | | | | | | | | | | | | | | | | |
| Suramin | ✓✓ | ✓ | ✓ | – | ✓✓ | – | ✓✓ | | | | ✓ | ✓ | – | – | ✓ | ✓ | ✓ | |
| NF023 | ✓✓* | ✓ | ✓ | | | | | | | | | | | | ✓✓ | ✓ | | |
| Reactive blue 2 | ✓ | ✓✓ | ✓ | ✓✓ | ✓ | – | – | | | | ✓ | ✓ | ✓ | ✓✓ | ✓ | ✓✓ | | ✓✓ |

| | | | | | | | | | |
|---|---|---|---|---|---|---|---|---|---|
| MRS 2179 | ✓ | — | ✓ | — | | | | ✓✓✓ | — |
| MRS 2279 | — | — | | | | | | ✓✓✓* | |
| TNP-ATP | ✓✓✓* | ✓ | ✓✓✓* | ✓ | ✓ | | | | |
| KN-62 | | | | | | ✓ | | | |
| AR-C67085 MX | | | | | | ✓✓✓(h) | | | |
| 2-MeSAMP | | | | | | | | ✓✓*[ag] | ✓✓✓* ✓✓ |
| Brilliant Blue G | ✓✓✓* | ✓ | ✓ | — | ✓ | — | ✓✓✓(r) | | |
| Ip₅I | ✓✓✓* | — | ✓ | — | | | | | |
| MRS 2257 | ✓✓✓* | ✓✓* | — | | | | | | |
| NF279 | ✓✓✓* | ✓ | — | ✓ | | | | | |

Number of ticks (✓) indicates relative potency with respect to agonist/antagonist concentration. Agonists: ✓✓✓ < 1 μM, ✓✓ 1–10 μM, ✓ > 10 μM, –ν virtually inactive. Antagonists: ✓✓✓ < 10nM, ✓✓ 10nM–300nM, ✓ > 300nM, – virtually inactive. *, selective agonist or antagonist. *h* human, *r* rat. ATP, Adenine-5′-triphosphate; ADP, adenosine-5′-diphosphate; 2-MeSATP, 2-methylthioadenosine 5′-triphosphate; PAPET, 2-[2-(4-aminophenyl)ethyl-thio]adenosine 5′-triphosphate; 2-MeSADP, 2-methylthio ADP; HT-AMP, 2-(hexylthio) adenosine 5′-monophosphate; α,β-meATP, α,β methylene ATP; β,γ-meATP, β,γ methylene ATP; BzATP, benzoyl benzoyl ATP; 2-dATP, deoxyATP; AP₄A, P¹, P⁴-Di-(adenosine-5′)tetraphosphate ammonium; UTP, uridine triphosphate, UDP, uridine diphosphate; CTP, cytidine triphosphate; PPADS, pyridoxal-phosphate-6-azophenyl-2′,4′-disulphonic acid; iso-PPADS, pyridoxal-phosphate-6-azophenyl-2′,5′-disulphonic acid; PPNDS, pyridox-al-5′-phosphate-6-(2′-napthylazo-6′-nitro-4′,8′ disulfonate; NF023, 8, 8′-[carbonylbis (imino-3,1-phenylenecarbonyl-imino]] bis-(1,3,5-napthalene trisulphonate; MRS 2179, N⁶-methyl-2′deoxyadenosine 3′,5′-bisphosphate; MRS 2269, N⁶-methyl-1,5-anhydro-2-(adenin-9-yl)-2,3-dideoxy-D-arabino-hexitol-4,6-bis (diammonium phosphate; TNP-ATP, trinitrophenol-ATP, KN-62, 1-[N, O-bis (5-isoquinolinesulfonyl)-N-methyl-L-tyrosyl]-4-phenylpiperazine; AR-C67085 MX, 2-propythio-D-β-γ-dichloromethylene ATP; IP₅I; diinosine pentaphosphate, MRS 2257, pyridoxal-5′-phosphonate-6-azophenyl-3′,5′-bismethyl phosphonate; NF279, (8,8′-(carbonylbis(i-mino-4,1-phenylene carbonylimino-4,1-phenylene carbonyl-imino))bis (1,3,5-naphthalenetrisulphonic acid).

11

mast cells facilitates the release of allergic mediators including histamine, suggesting a role in inflammation. $A_3$ receptors may be involved in the cardioprotective effect of adenosine in ischemia and preconditioning during ischemia reperfusion injury.

$A_1$, $A_{2A}$, $A_{2B}$, and $A_3$ adenosine receptors have distinct but frequently overlapping tissue distributions. The fact that more than one adenosine/P1 receptor subtype may be expressed by the same cell raises questions about the functional significance of this colocalization.

### B. P2X Receptors

Early studies of the distribution of P2X receptor subtypes based on Northern blot and *in situ* hybridization studies (Collo *et al.*, 1996) have been extended substantially, after antibodies to these receptors became available, by immunohistochemical localization at both light microscopic (Vulchanova *et al.*, 1996; Bradbury *et al.*, 1998; Chan *et al.*, 1998; Xiang *et al.*, 1998a,b, 1999; Bo *et al.*, 1999; Gröschel-Stewart *et al.*, 1999a,b; Bardini *et al.*, 2000; Brouns *et al.*, 2000; Lee *et al.*, 2000a,b) and electron microscopic levels (Llewellyn-Smith and Burnstock, 1998; Loesch and Burnstock, 1998, 2000; Loesch *et al.*, 1999). For example, while it was originally thought that smooth muscle contained only $P2X_1$ receptors, there is now evidence of the presence of $P2X_2$, $P2X_4$, and probably $P2X_5$ receptors as both homo-multimers and heteromultimers (Nori *et al.*, 1998; Hansen *et al.*, 1999; Lewis and Evans, 2000).

$P2X_1$ receptors, which, in earlier studies, were not considered to be present in the brain, now have been found at postjunctional sites in synapses in the cerebellum (Loesch and Burnstock, 1998). The $P2X_1$ receptor is characterized by rapid desensitization and potent actions of $\alpha\beta$-methylene ATP ($\alpha\beta$meATP), and there are now potent selective antagonists for this receptor, such as trinitrophenyl ATP (Lewis *et al.*, 1998) and diinosine pentaphosphate (King *et al.*, 1999).

The $P2X_2$ receptor is widespread in the central nervous system and has been found at both pre- and postsynaptic sites in the hypothalamus (Loesch *et al.*, 1999). A feature of the $P2X_2$ receptor is its lack of fast desensitization and its high sensitivity to acidity and $Zn^{2+}$ (King *et al.*, 1996; Wildman *et al.*, 1999). $P2X_3$ receptors are interesting in that they are predominantly localized on sensory nerves, particularly the small nociceptive neurons in the dorsal root, trigeminal, and nodose ganglia (see Burnstock, 2000). The central projections are located in the inner lamina II of the dorsal horn of the spinal cord, and peripheral extensions have been noted in the skin, tongue (Rong *et al.*, 2000), and more recently in tooth pulp (Alavi *et al.*,

2001) and bladder (Vlaskovska *et al.*, 2001). There is evidence of $P2X_2$ and $P2X_3$ labeling of endothelial cells of microvessels in brain, thymus, thyroid, and gut and also in epithelial cells in the thyroid (Gröschel-Stewart *et al.*, 1999b; Loesch and Burnstock, 2000; Glass *et al.*, 2000; Glass and Burnstock, 2001). $P2X_4$ and $P2X_6$ receptors are prominent in the central nervous system and are unique among P2X subtypes in that responses to purines mediated by these receptors are not blocked alone, even potentiated by agents commonly used as P2 antagonists (see Burnstock, 2001a). $P2X_5$ receptors have been shown to be associated with proliferating and differentiating epithelial cells in the skin and hair follicles (Gröschel-Stewart *et al.*, 1999a) and in the bladder, ureter, and vagina (Lee *et al.*, 2000a,b; Bardini *et al.*, 2000). $P2X_5$ receptors are also prominent in a number of cell types in embryonic development (Meyer *et al.*, 1999a; Ryten *et al.*, 2001).

$P2X_7$ receptors are unique in that, as well as a cation pore, a large, 4-nm pore can be formed, which appears to be linked with apoptosis, perhaps associated with the elongated C terminus of this receptor (Surprenant *et al.*, 1996). The $P2X_7$ receptor is often internalized in cells, although under pathological conditions, such as ischemia and cancer, it becomes externalized, leading to apoptosis. Fluorescent green protein coupled to $P2X_7$ receptors will provide a valuable technique for observing the movement of receptors in living cells (see Dutton *et al.*, 2000).

The proposal in 1976 that more than one transmitter can be released from nerve terminals (the cotransmitter hypothesis; Burnstock, 1976a,b) is now widely accepted, and the focus has been on defining the "chemical coding" of various nerve types, that is, to describe the combination of neurotransmitters in these nerves and their projections to various sites. ATP appears to be a cotransmitter in many nerve types, probably reflecting the primitive nature of purinergic signaling (Burnstock, 1996, 1999). ATP is a cotransmitter with noradrenaline and neuropeptide Y in sympathetic nerves, with acetylcholine and vasoactive intestinal peptide in some parasympathetic nerves, with nitric oxide and vasoactive intestinal peptide in enteric nonadrenergic, noncholinergic (NANC) inhibitory nerves, and with calcitonin gene-related peptide and substance P in sensory motor nerves. There is also evidence of ATP with $\gamma$-aminobutyric acid in retinal nerves and of ATP with glutamate or with dopamine in nerves in the brain. In sympathetic nerves supplying visceral and vascular smooth muscle and in parasympathetic nerves supplying the urinary bladder, ATP provokes a fast, short-lasting twitch response via P2X receptors, while the slower component is mediated by G protein-coupled adrenoceptors and muscarinic receptors, respectively (see Burnstock and Ralevic, 1996; Burnstock 2001c). In the gut, ATP released from NANC inhibitory nerves produces the fastest response, nitric oxide gives a less rapid response, and vasoactive intestinal peptide

**14**                                                            Geoffrey Burnstock

produces slow tonic relaxations. In all cases of cotransmission, there are considerable differences in the proportion of the cotransmitters in nerves supplying different regions of the gut or vasculature and between species (see Burnstock, 2001d).

The first clear evidence of nerve–nerve purinergic synaptic transmission was published in 1992. Excitatory postsynaptic potentials in the celiac ganglion were reversibly antagonized by suramin (Evans *et al.*, 1992; Silinsky *et al.*, 1992) and similar experiments were carried out in the medial habenula in the brain, showing reversible block of excitatory postsynaptic potentials by suramin (Edwards *et al.*, 1992). Since then, other examples of purinergic sympathetic transmission have been identified and there have been many articles describing the distribution of various P2 receptor subtypes in the brain and spinal cord and electrophysiological studies of the effects of purines in brain slices, isolated nerves, and glial cells (see Chapter 10 in this volume). Synaptic transmission also has been found in the myenteric plexus and in various sensory and sympathetic and pelvic ganglia (see Dunn *et al.*, 2001).

In addition to the examples of short-term signaling, there are now many examples of purinergic signaling concerned with long-term (trophic) events, such as development and regeneration, proliferation, and cell death (Abbracchio *et al.*, 1996; Neary *et al.*, 1996; Abbracchio and Burnstock, 1998). For example, $\alpha\beta$meATP produces proliferation of glial cells, whereas adenosine inhibits proliferation. $P2X_5$ and $P2X_6$ receptors have been implicated in the development of chick skeletal muscle (Meyer *et al.*, 1999a). In studies of purinoceptor expression in mouse myotubes, we have shown progressive expression of $P2X_5$ [from embryonic day 14 (E14) to E18], $P2X_6$ (from E16 to E18), and $P2X_2$ (from E18 to postnatal day 7; Ryten *et al.*, 2001).

### C. P2Y Receptors

There are now many examples of P2Y purinoceptor-mediated responses in nonneuronal and nonmuscular cell types, as well as in the nervous system. Such examples include endothelial cells, which express $P2Y_1$, $P2Y_2$, and probably $P2Y_4$ receptors that, when occupied, release nitric oxide leading to vasodilatation. P2Y receptors in pancreatic $\beta$ cells are involved in insulin secretion (Loubatières-Mariani and Chapal, 1988). $P2Y_2$ receptors have been identified in hepatocytes (Schöfl *et al.*, 1999); $P2Y_T$, $P2X_1$, and $P2Y_1$ receptors in platelets (Kunapuli and Daniel, 1998); $P2Y_2$ receptors on myelinating Schwann cells; and $P2Y_1$ receptors on nonmyelinating Schwann cells (Mayer *et al.*, 1998). P2Y receptors also are involved in signaling to endocrine cells, leading to hormone secretion (Chen *et al.*, 1995; Lee *et al.*, 1996; Tomic *et al.*, 1996; Sperlágh *et al.*, 1999; Glass *et al.*, 2000; Vainio and Törnquist, 2000). There is a widespread presence of P2Y receptor subtypes in different regions of kidney glomeruli, tubules, and collecting ducts (Eltze and Ullrich, 1996; Huber-Lang *et al.*, 1997; Cha *et al.*, 1998; van der Weyden *et al.*, 2000; Bailey *et al.*, 2000; Cuffe *et al.*, 2000; Dockrell *et al.*, 2001). P2Y receptors are strongly represented in the brain, both at presynaptic sites and on glial cells (see Chapter 10 in this volume).

There are also examples of P2Y purinergic signaling concerned with long-term events such as development and regeneration, and proliferation. A $P2Y_8$ receptor, cloned from frog embryo, appears to be involved in the development of the neural plate (Bogdanov *et al.*, 1997). $P2Y_1$ receptors seem to have a role in cartilage development in limb buds and in the development of the mesonephros (Meyer *et al.*, 1999b).

## IV. PATHOPHYSIOLOGICAL ROLES OF PURINOCEPTORS AND THERAPEUTIC POTENTIAL

An increasing number of pathophysiological roles for purinoceptors are emerging, some of which have therapeutic potential (see Abbracchio and Burnstock, 1998; Fischer, 1999; Agteresch *et al.*, 1999; Williams and Jarvis, 2000; Burnstock and Williams, 2000; Communi *et al.*, 2000).

### A. Bladder Incontinence

Urinary bladder function is regulated by parasympathetic nerve stimulation, resulting in bladder contraction via P2X receptors on the smooth muscle (Burnstock *et al.*, 1978). Detrusor malfunction results in urinary incontinence. Studies of $P2X_3$ knockout mice have shown bladder hyporeflexia as well as reduced pain (Cockayne *et al.*, 2000; Vlaskovska *et al.*, 2001). This and other results have been reviewed in relation to potential treatment of interstitial cystitis and of obstructive and neurogenic bladder (see Burnstock, 2001c).

### B. Contraception and Fertility

Several findings suggest a potential role for purines in contraception and fertility. For example, in male rat genitalia, antibodies to $P2X_1$ and $P2X_2$ subunits show immunoreactivity on the membranes of the smooth muscle of the vas deferens (Lee *et al.*, 2000b). In male P2X$_1$ receptor knockout mice, fertility is reduced by 90% without affecting copulatory performance. This is due to a decreased sperm count in the ejaculate caused by a 60% reduction in the sensitivity of the vas deferens to sympathetic nerve stimulation (Mulryan *et al.*, 2000). P2X$_1$ and P2X$_2$ receptors have also been implicated in erectile function, especially in diabetes (Gür and Öztürk, 2000). A study of the rat testis has shown involvement of several P2X receptor subtypes in spermatogenesis (Glass *et al.*, 2001).

## C. Skin Diseases

In stratified epithelium of skin, P2X$_5$ receptors are associated with proliferating and differentiating cells, while P2X$_7$ receptors label apoptotic cells (Gröschel-Stewart *et al.*, 1999a). P2X$_5$ and P2X$_7$ receptor agonists and antagonists may have potential in the treatment of psoriasis, scleroderma, and basal cell carcinoma (as well as restenosis following angioplasty).

## D. Diabetes

ATP stimulates pancreatic insulin release via a glucose-dependent, P2Y receptor-mediated mechanism (Loubatières-Mariani *et al.*, 1997) and also modulates insulin secretion by interactions with ATP-sensitive potassium channels in islet $\beta$ cells.

## E. Thrombosis

ADP produces platelet aggregation and three purinoceptor subtypes are present on platelets: P2Y$_1$, P2X$_1$, and the more recently cloned P2Y$_{12}$ receptors (Hollopeter *et al.*, 2001). An orally active antagonist is being developed as an antithrombotic agent (Humphries *et al.*, 1995).

## F. Gut Motility Disorders

ATP was recognized early as a neurotransmitter in NANC inhibitory nerves in the gut (Burnstock, 1972). Purinergic synaptic transmission is now also established in both myenteric and submucosa plexuses and abnormalities in motility are being explored as potential targets for purinergic agents (see Burnstock, 2001d).

## G. Cardiopulmonary Diseases

For cardiopulmonary function, ATP is a mediator of vagal reflexes in the heart and lung (see Burnstock, 2000). In anesthetized rats, P2X receptors have been implicated in evoking a Bezold–Jarisch response (hyperventilation, bradycardia, hypotension, apnea). Adenosine is being used to treat supraventricular tachycardia and adenosine receptor antagonists are being used for the treatment of bradyarrhythmias. $A_1$ receptors have been widely reported to mediate the protective effects of adenosine in preconditioning and during ischemia or reperfusion injury in the heart (see Liang and Jacobson, 1999; Ganote and Armstrong, 2000).

ATP and UTP, acting via $P2Y_2$ receptors, stimulate chloride secretion in airway epithelium and mucin glycoprotein release from epithelial goblet cells (Stutts and Boucher, 1999), enhancing mucociliary clearance and reflecting a potential treatment for cystic fibrosis and chronic bronchitis (Yerxa, 2000). ATP may also have a direct role in asthma via its actions on bronchial innervation. Selective inhibition of the synthesis of $A_1$ receptors with antisense oligonucleotides confirmed that these receptors are involved in an animal model of asthma (Nyce and Metzger, 1997). There was a marked reduction in the number of $A_1$ receptors in the lung and attenuation of airway constriction to adenosine, histamine, and dust mite allergen (Nyce and Metzger, 1997). Although the site of action remains to be determined, selective antagonism of $A_1$ receptors offers a possible new approach in asthma therapy.

## H. Cancer

The anticancer activity of adenine nucleotides was first described by Rapaport (1983). Intraperitoneal injection into tumor-bearing mice resulted in significant anticancer activity against several fast-growing aggressive carcinomas (Fang et al., 1992; Spungin and Friedberg, 1993; Rapaport, 1997; Agteresch et al., 2000).

## I. Diseases of the Ear

In the auditory system, ATP, acting via P2Y receptors, depresses sound-evoked gross compound action potentials in the auditory nerve and the distortion product otoacoustic emission, the latter being a measure of the active process of the outer hair cells (see Housley, 2000). Both P2X and P2Y receptors have been identified in the vestibular system. P2X splice variants were found on the endolymphatic surface of the cochlear endothelium, an area associated with sound transduction. It has been suggested that ATP may regulate fluid homeostasis, cochlear blood flow, hearing sensitivity, and development, and thus may be useful in the treatment of Ménière's disease, tinnitus, and sensorineural deafness.

### J. Diseases of the Eye

In the eye, ATP, acting via both P2X and P2Y receptors, modulates retinal neurotransmission affecting retinal blood flow and intraocular pressure. The ATP analog $\beta,\gamma$-methylene ATP ($\beta\gamma$meATP) has greater efficacy in reducing intraocular pressure (40%) than do muscarinic agonists such as pilocarpine (25%) or $\beta$-adrenoceptor blockers (30%) (Pintor, 1999). In the ocular mucosa, $P2Y_2$ receptor activation increases salt, water, and mucous secretion and thus represents a potential treatment for dry eye disease (Yerxa, 2000). In the retinal pigmented layer, $P2Y_2$ receptor activation promotes fluid absorption and may be involved in retinal detachment (Pintor and Peral, 2001).

### K. Behavioral Disorders

Microinjection of ATP analogs into the brain prepiriform cortex induced generalized motor seizures similar to those seen with N-methyl-D-aspartate and bicuculline (Knutsen and Murray, 1997). $P2X_2$, $P2X_4$, and $P2X_6$ receptors are expressed in the prepiriform cortex, suggesting that a P2X receptor antagonist may have potential as an antiepileptic (Collo et al., 1997). P1 ($A_{2A}$) receptor antagonists are being explored for the treatment of Parkinson's disease (Morelli and Pinna, 2001). ATP, in combination with growth factors, can act to stimulate astrocyte proliferation, contributing to the process of reactive astrogliosis, a hypertrophic/hyperplastic response that is associated with brain trauma, stroke/ischemia, seizure, and neurodegenerative disorders (Neary, 1996).

### L. Bone Disorders

Purinergic receptors have a strong presence in bone cells (Morrison et al., 1998; Dixon and Sims, 2000; Hoebertz et al., 2000). P2X and P2Y receptors are present on osteoclasts, while P2Y receptors are present only on osteoblasts. ATP, but not adenosine, stimulates the formation of osteoclasts and their resorptive actions *in vitro* (Morrison *et al.*, 1998) and can inhibit osteoblast-dependent bone formation. The bisphosphonate clodronate, which is used in the treatment of Paget's disease and tumor-induced osteolysis, may act via osteoclast P2 receptors (Dixon and Sims, 2000). Modulation of P2 receptor function may have potential in the treatment of osteoporosis, rheumatoid arthritis, periodontitis, and osteopenia. One study has shown that low (nanomolar) concentrations of ADP acting through $P2Y_1$ receptors turn on osteoclast activity (Hoebertz *et al.*, 2001).

## M. Pain

ATP elicits pain responses via $P2X_3$ or $P2X_{2/3}$ receptors and may contribute to the pain associated with causalgia, reflex sympathetic dystrophy, angina, and migraine, as well as visceral pain and cancer pain (see Burnstock, 2000, 2001b). The nucleotide is also a key mediator of neurogenic inflammation via its actions on neutrophils, macrophages, and monocytes, activation of which results in cytokine production and release (Dubyak and El Moatassim, 1993). For visceral pain, a purinergic mechanosensory transduction mechanism has been proposed (Burnstock, 2001b) whereby distension of tubes, such as the ureter and gut, salivary and bile ducts, and sacs including the urinary and gall bladder causes ATP release from the lining epithelial cells to act on $P2X_3$ receptors located on the subepithelial sensory nerve plexus to relay nociceptive signals to the CNS.

## References

Abbracchio, M. P., and Burnstock, G. (1994). Purinoceptors: Are there families of $P_{2X}$ and $P_{2Y}$ purinoceptors? *Pharmacol. Ther.* **64,** 445–475.

Abbracchio, M. P., and Burnstock, G. (1998). Purinergic signalling: Pathophysiological roles. *Jpn. J. Pharmacol.* **78,** 113–145.

Abbracchio, M. P., Ceruti, S., Bolego, C., Puglisi, L., Burnstock, G., and Cattabeni, F. (1996). Trophic roles of P2-purinoceptors in central nervous system astroglial cells. *In* "P2 Purinoceptors: Localization, Function and Transduction Mechanisms. Ciba Foundation Symposium 198," pp. 142–148. John Wiley & Sons, Chichester.

Abbracchio, M. P., Boeynaems, J.-M., Barnard, E. A., Boyer, J. L., Kennedy, C., Miras-Portugal, M. T., King, B. F., Gachet, C., Jacobsen, K. A., Weisman, G. A., and Burnstock, G. (2003). Characterization of the UDP-glucose receptor (re-named here the $P2Y_{14}$ receptor) adds diversity to the P2Y receptor family. *Trends. Pharmacol. Sci.* **24,** 52–55.

Adrian, K., Bernhard, M. K., Breitinger, H.-G., and Ogilvie, A. (2000). Expression of purinergic receptors (ionotropic P2X1-7 and metabotropic P2Y1-11) during myeloid differentiation of HL60 cells. *Biochim. Biophys. Acta.* **1492,** 127–138.

Agteresch, H. J., Dagnelie, P. C., van den Berg, J. W. O., and Wilson, J. H. (1999). Adenosine triphosphate: Established and potential clinical applications. *Drugs* **58,** 211–232.

Agteresch, H. J., Dagnelie, P. C., van der Gast, A., Stijnen, T., and Wilson, J. H. (2000). Randomized clinical trial of adenosine 5′-triphosphate in patients with advanced non-small-cell lung cancer. *J. Natl. Cancer Inst.* **92**, 321–328.

Alavi, A. M., Dubyak, G. R., and Burnstock, G. (2001). Immunohistochemical evidence for ATP receptors in human dental pulp. *J. Dent. Res.* **80**, 476–483.

Bailey, M. A., Imbert-Teboul, M., Turner, C., Marsy, S., Srai, K., Burnstock, G., and Unwin, R. J. (2000). Axial distribution and characterization of basolateral P2Y receptors along the rat renal tubule. *Kidney Int.* **58**, 1893–1901.

Baraldi, P. G., Cacciari, B., Romagnoli, R., Merighi, S., Varani, K., Borea, P. A., and Spalluto, G. (2000). A3 adenosine receptor ligands: History and perspectives. *Med. Res. Rev.* **20**, 103–128.

Bardini, M., Lee, H. Y., and Burnstock, G. (2000). Distribution of P2X receptor subtypes in the rat female reproductive tract at late pro-oestrus/early oestrus. *Cell Tissue Res.* **299**, 105–113.

Bo, X., Alavi, A., Xiang, Z., Oglesby, I., Ford, A., and Burnstock, G. (1999). Localization of ATP-gated $P2X_2$ and $P2X_3$ receptor immunoreactive nerves in rat taste buds. *Neuroreport* **10**, 1107–1111.

Bogdanov, Y. D., Dale, L., King, B. F., Whittock, N., and Burnstock, G. (1997). Early expression of a novel nucleotide receptor in the neural plate of *Xenopus* embryos. *J. Biol. Chem.* **272**, 12583–12590.

Bogdanov, Y. D., Wildman, S. S., Clements, M. P., King, B. F., and Burnstock, G. (1998). Molecular cloning and characterization of rat $P2Y_4$ nucleotide receptor. Special Report. *Br. J. Pharmacol.* **124**, 428–430.

Boyer, J. L., Mohanram, A., Camaioni, E., Jacobson, K. A., and Harden, T. K. (1998). Competitive and selective antagonism of $P2Y_1$ receptors by $N^6$-methyl 2′-deoxyadenosine 3′,5′-bisphosphate. *Br. J. Pharmacol.* **124**, 1–3.

Bradbury, E. J., Burnstock, G., and McMahon, S. B. (1998). The expression of $P2X_3$ purinoceptors in sensory neurons: Effects of axotomy and glial-derived neurotrophic factor. *Mol. Cell. Neurosci.* **12**, 256–268.

Brake, A. J., Wagenbach, M. J., and Julius, D. (1994). New structural motif for ligand-gated ion channels defined by an ionotropic ATP receptor. *Nature* **371**, 519–523.

Brouns, I., Adriaensen, D. D., Burnstock, G., and Timmermans, J.-P. (2000). Intraepithelial vagal sensory nerve terminals in rat pulmonary neuroepithelial bodies express $P2X_3$ receptors. *Am. J. Respir. Cell Mol. Biol.* **23**, 52–61.

Brown, S. G., King, B. F., Kim, Y.-C., Jang, S. Y., Burnstock, G., and Jacobson, K. A. (2000). Activity of novel adenine nucleotide derivatives as agonists and antagonists at recombinant rat P2X receptors. *Drug Dev. Res.* **49**, 253–259.

Burnstock, G. (1972). Purinergic nerves. *Pharmacol. Rev.* **24**, 509–581.

Burnstock, G. (1976a). Purinergic receptors. *J. Theor. Biol.* **62**, 491–503.

Burnstock, G. (1976b). Do some nerve cells release more than one transmitter? *Neuroscience* **1**, 239–248.

Burnstock, G. (1978). A basis for distinguishing two types of purinergic receptor. *In* "Cell Membrane Receptors for Drugs and Hormones: A Multidisciplinary Approach" (R. W. Straub and L. Bolis, eds.), pp. 107–118. Raven Press, New York.

Burnstock, G. (1996). Purinoceptors: Ontogeny and phylogeny. *Drug Dev. Res.* **39**, 204–242.

Burnstock, G. (1997). The past, present and future of purine nucleotides as signalling molecules. *Neuropharmacology* **36**, 1127–1139.

Burnstock, G. (1999). Purinergic cotransmission. *Brain Res. Bull.* **50**, 355–357.

Burnstock, G. (2000). P2X receptors in sensory neurones. *Br. J. Anaesth.* **84**, 476–488.

Burnstock, G. (2001a). Expanding field of purinergic signaling. *Drug Dev. Res.* **52**, 1–10.

Burnstock, G. (2001b). Purine-mediated signalling in pain and visceral perception. *Trends Pharmacol. Sci.* **22**, 182–188.

Burnstock, G. (2001c). Purinergic signalling in lower urinary tract. In "Handbook of Experimental Pharmacology," Vol. 151/I: "Purinergic and Pyrimidinergic Signalling I: Molecular, Nervous and Urinogenitary System Function" (M. P. Abbracchio and M. Williams, eds.), pp. 423–515. Springer-Verlag, Berlin.

Burnstock, G. (2001d). Purinergic signalling in gut. In "Handbook of Experimental Pharmacology," Vol. 151/II: "Purinergic and Pyrimidinergic Signalling II: Cardiovascular, Respiratory, Immune, Metabolic and Gastrointestinal Tract Function" (M. P. Abbracchio and M. Williams, eds.), pp. 141–238. Springer-Verlag, Berlin.

Burnstock, G., and Kennedy, C. (1985). Is there a basis for distinguishing two types of $P_2$-purinoceptor? *Gen. Pharmacol.* **16**, 433–440.

Burnstock, G., and Ralevic, V. (1994). New insights into the local regulation of blood flow by perivascular nerves and endothelium. *Br. J. Plast. Surg.* **47**, 527–543.

Burnstock, G., and Ralevic, V. (1996). Cotransmission. In "The Pharmacology of Vascular Smooth Muscle" (C. J. Garland and J. A. Angus, eds.), pp. 210–232. Oxford University Press, Oxford.

Burnstock, G., and Williams, M. (2000). P2 purinergic receptors: Modulation of cell function and therapeutic potential. *J. Pharmacol. Exp. Ther.* **295**, 862–869.

Burnstock, G., Cocks, T., Crowe, R., and Kasakov, L. (1978). Purinergic innervation of the guinea-pig urinary bladder. *Br. J. Pharmacol.* **63**, 125–138.

Cha, S. H., Sekine, T., and Endou, H. (1998). P2 purinoceptor localization along rat nephron and evidence suggesting existence of subtypes P2Y1 and P2Y2. *Am. J. Physiol.* **274**, F1006–F1014.

Chan, C. M., Unwin, R. J., Bardini, M., Oglesby, I. B., Ford, A. P. D. W., Townsend-Nicholson, A., and Burnstock, G. (1998). Localization of the $P2X_1$ purinoceptors by autoradiography and immunohistochemistry in rat kidneys. *Am. J. Physiol.* **274**, F799–F804.

Chen, Z.-P., Kratzmeier, M., Levy, A., McArdle, C. A., Poch, A., Day, A., Mukhopadhyay, A. K., and Lightman, S. L. (1995). Evidence for a role of pituitary ATP receptors in the regulation of pituitary function. *Proc. Natl. Acad. Sci. USA* **92**, 5219–5223.

Cockayne, D. A., Hamilton, S. G., Zhu, Q.-M., Dunn, P. M., Zhong, Y., Novakovic, S., Malmberg, A. B., Cain, G., Berson, A., Kassotakis, L., Hedley, L., Lachnit, W. G., Burnstock, G., McMahon, S. B., and Ford, A. P. D. W. (2000). Urinary bladder hyporeflexia and reduced pain-related behaviour in P2X3-deficient mice. *Nature* **407**, 1011–1015.

Collo, G., North, R. A., Kawashima, E., Merlo-Pich, E., Neidhart, S., Surprenant, A., and Buell, G. (1996). Cloning of P2X5, and P2X6 receptors and the distribution and properties of an extended family of ATP-gated ion channels. *J. Neurosci.* **16**, 2495–2507.

Collo, G., Neidhart, S., Kawashima, E., Kosco-Vilbois, M., North, R. A., and Buell, G. (1997). Tissue distribution of the P2X7 receptor. *Neuropharmacology* **36**, 1277–1283.

Communi, D., Janssens, R., Suarez-Huerta, N., Robaye, B., and Boeynaems, J.-M. (2000). Advances in signalling by extracellular nucleotides. The role and transduction mechanisms of P2Y receptors. *Cell Signal* **12**, 351–360.

Communi, D., Gonzalez, N. S., Detheux, M., Brezillon, S., Lannoy, V., Parmentier, M., and Boeynaems, J. M. (2001). Identification of a novel human ADP receptor coupled to G(i). *J. Biol. Chem.* **276**, 41479–41485.

Cuffe, J. E., Bielfeld-Ackermann, A., Thomas, J., Leipziger, J., and Korbmacher, C. (2000). ATP stimulates $Cl^-$ secretion and reduces amiloride-sensitive $Na^+$ absorption in M-1 mouse cortical collecting duct cells. *J. Physiol.* **524**, 77–90.

Dixon, A. K., Gubitz, A. K., Sirinathsinghji, D. J. S., Richardson, P. J., and Freeman, T. C. (1996). Tissue distribution of adenosine receptor mRNAs in the rat. *Br. J. Pharmacol.* **118,** 1461–1468.

Dixon, S. J., and Sims, S. M. (2000). P2 purinergic receptors on osteoblasts and osteoclasts: Potential targets for drug development. *Drug Dev. Res.* **49,** 187–200.

Dockrell, M. E., Noor, M. I., James, A. F., and Hendry, B. M. (2001). Heterogeneous calcium responses to extracellular ATP in cultured rat renal tubule cells. *Clin. Chim. Acta.* **303,** 133–138.

Drury, A. N., and Szent-Györgyi, A. (1929). The physiological activity of adenine compounds with special reference to their action upon the mammalian heart. *J. Physiol.* **68,** 213–237.

Dubyak, G. R. (1991). Signal transduction by $P_2$-purinergic receptors for extracellular ATP. *Am. J. Respir. Cell Mol. Biol.* **4,** 295–300.

Dubyak, G. R., and El Moatassim, C. (1993). Signal transduction via P2-purinergic receptors for extracellular ATP and other nucleotides. *Am. J. Physiol.* **265,** C577–C606.

Dunn, P. M., Zhong, Y., and Burnstock, G. (2001). P2X receptors in peripheral neurones. *Prog. Neurobiol.* **65,** 107–134.

Dutton, J. L., Poronnik, P., Li, G. H., Holding, C. A., Worthington, R. A., Vandenberg, R. J., Cook, D. I., Barden, J. A., and Bennett, M. R. (2000). $P2X_1$ receptor membrane redistribution and down-regulation visualized by using receptor-coupled green fluorescent protein chimeras. *Neuropharmacology* **39,** 2054–2066.

Edwards, F. A., Gibb, A. J., and Colquhoun, D. (1992). ATP receptor-mediated synaptic currents in the central nervous system. *Nature* **359,** 144–147.

Eltze, M., and Ullrich, B. (1996). Characterization of vascular P2 purinoceptors in the rat isolated perfused kidney. *Eur. J. Pharmacol.* **306,** 139–152.

Evans, R. J., Derkach, V., and Surprenant, A. (1992). ATP mediates fast synaptic transmission in mammalian neurons. *Nature* **357,** 503–505.

Fang, W. G., Pirnia, F., Bang, Y. J., Myers, C. E., and Trepel, J. B. (1992). $P_2$-purinergic receptor agonists inhibit the growth of androgen-independent prostate carcinoma cells. *J. Clin. Invest.* **89,** 191–196.

Fischer, B. (1999). Therapeutic applications of ATP-(P2)-receptors agonists and antagonists. *Exp. Opin. Ther. Patents* **9,** 385–399.

Fredholm, B. B., Arslan, G., Halldner, L., Kull, B., Schulte, G., Ådén, U., and Svenningsson, P. (2001). Adenosine receptor signaling *in vitro* and *in vivo. Drug Dev. Res.* **52,** 274–282.

Ganote, C. E., and Armstrong, S. C. (2000). Adenosine and preconditioning in the rat heart. *Cardiovasc. Res.* **45,** 134–140.

Glass, R., and Burnstock, G. (2001). Immunohistochemical identification of cells expressing ATP-gated cation channels (P2X receptors) in the adult rat thyroid. *J. Anat.* **198,** 569–579.

Glass, R., Townsend-Nicholson, A., and Burnstock, G. (2000). P2 receptors in the thymus: Expression of P2X and P2Y receptors in adult rats, an immunohistochemical and *in situ* hybridisation study. *Cell Tissue Res.* **300,** 295–306.

Glass, R., Bardini, M., Robson, T., and Burnstock, G. (2001). The expression of nucleotide P2X receptor subtypes during spermatogenesis in the adult rat testis. *Cells Tissues Organs* **169,** 377–387.

Gordon, J. L. (1986). Extracellular ATP: Effects, sources and fate. *Biochem. J.* **233,** 309–319.

Gröschel-Stewart, U., Bardini, M., Robson, T., and Burnstock, G. (1999). Localisation of $P2X_5$ and $P2X_7$ receptors by immunohistochemistry in rat stratified squamous epithelia. *Cell Tissue Res.* **296,** 599–605.

Gröschel-Stewart, U., Bardini, M., Robson, T., and Burnstock, G. (1999). P2X receptors in the rat duodenal villus. *Cell Tissue Res.* **297,** 111–117.

Gür, S., and Öztürk, B. (2000). Altered relaxant responses to adenosine and adenosine 5'-triphosphate in the corpus cavernosum from men and rats with diabetes. *Pharmacology* **60**, 105–112.

Haines, W. R., Torres, G. E., Voigt, M. M., and Egan, T. M. (1999). Properties of the novel ATP-gated ionotropic receptor composed of the P2X$_1$ and P2X$_5$ isoforms. *Mol. Pharmacol.* **56**, 720–727.

Hansen, M. A., Dutton, J. L., Balcar, V. J., Barden, J. A., and Bennett, M. R. (1999). P$_{2X}$ (purinergic) receptor distributions in rat blood vessels. *J. Auton. Nerv. Syst.* **75**, 147–155.

Hechler, B., Vigne, P., Leon, C., Breittmayer, J. P., Gachet, C., and Frelin, C. (1998). ATP derivatives are antagonists of the P2Y$_1$ receptor: Similarities to the platelet ADP receptor. *Mol. Pharmacol.* **53**, 727–733.

Hoebertz, A., Townsend-Nicholson, A., Glass, R., Burnstock, G., and Arnett, T. R. (2000). Expression of P2 receptors in bone and cultured bone cells. *Bone* **27**, 503–510.

Hoebertz, A., Meghji, S., Burnstock, G., and Arnett, T. R. (2001). Extracellular ADP is a powerful osteolytic agent: Evidence for signaling through the P2Y$_1$ receptor on bone cells. *FASEB J.* **15**, 1139–1148.

Hollopeter, G., Jantzen, H.-M., Vincent, D., Li, G., England, L., Ramakrishnan, V., Yang, R.-B., Nurden, P., Nurden, A., Julius, D., and Conley, P. B. (2001). Identification of the platelet ADP receptor targeted by antithrombotic drugs. *Nature* **409**, 202–207.

Housley, G. D. (2000). Physiological effects of extracellular nucleotides in the inner ear. *Clin. Exp. Pharmacol. Physiol.* **27**, 575–580.

Huber-Lang, M., Fischer, K. G., Gloy, J., Schollmeyer, P., Kramer-Guth, A., Greger, R., and Pavenstadt, H. (1997). UTP and ATP induce different membrane voltage responses in rat mesangial cells. *Am. J. Physiol.* **272**, F704–F711.

Humphries, R. G., Robertson, M. J., and Leff, P. (1995). A novel series of P$_{2T}$ purinoceptor antagonists: Definition of the role of ADP in arterial thrombosis. *Trends Pharmacol. Sci.* **16**, 179–181.

Jain, N., Kemp, N., Adeyemo, O., Buchanan, P., and Stone, T. W. (1995). Anxiolytic activity of adenosine receptor activation in mice. *Br. J. Pharmacol.* **116**, 2127–2133.

Johansson, B., Georgiev, V., and Fredholm, B. B. (1997). Distribution and postnatal ontogeny of adenosine A$_{2A}$ receptors in rat brain: Comparison with dopamine receptors. *Neuroscience* **80**, 1187–1207.

Kaiser, S. M., and Quinn, R. J. (1999). Adenosine receptors as potential therapeutic targets. *Drug Discov. Today* **4**, 542–551.

Khakh, B. S., Burnstock, G., Kennedy, C., King, B. F., North, R. A., Séguéla, P., Voigt, M., and Humphrey, P. P. A. (2001). International Union of Pharmacology. XXIV. Current status of the nomenclature and properties of P2X receptors and their subunits. *Pharmacol. Rev.* **53**, 107–118.

King, B. F., and Burnstock, G. (2002). Purinergic receptors. *In* M. Pangalos and C. Davies (eds) pp. 422–438. Oxford University Press, Oxford.

King, B. F., Ziganshina, L. E., Pintor, J., and Burnstock, G. (1996). Full sensitivity of P2X$_2$ purinoceptor to ATP revealed by changing extracellular pH. *Br. J. Pharmacol.* **117**, 1371–1373.

King, B. F., Townsend-Nicholson, A., and Burnstock, G. (1998). Metabotropic receptors for ATP and UTP: Exploring the correspondence between native and recombinant nucleotide receptors. *Trends Pharmacol. Sci.* **19**, 506–514.

King, B. F., Liu, M., Pintor, J., Gualix, J., Miras-Portugal, M. T., and Burnstock, G. (1999). Diinosine pentaphosphate (IP$_5$ I) is a potent antagonist at recombinant rat P2X$_1$ receptors. *Br. J. Pharmacol.* **128**, 981–988.

King, B. F., Townsend-Nicholson, A., Wildman, S. S., Thomas, T., Spyer, K. M., and Burnstock, G. (2000). Coexpression of rat P2X$_2$ and P2X$_6$ subunits in *Xenopus* oocytes. *J. Neurosci.* **20,** 4871–4877.

King, B. F., Burnstock, G., Boyer, J. L., Boeynaems, J.-M., Weisman, G. A., Kennedy, C., Jacobson, K. A., Humphries, R. G., Abbracchio, M. P., and Miras-Portugal, M. T. (2001). The P2Y receptors. *In* "The IUPHAR Compendium of Receptor Characterization and Classification" (D. Girdlestone, ed.), IUPHAR Media, London.

Knutsen, L. J. S., and Murray, T. F. (1997). Adenosine and ATP in epilepsy. *In* "Purinergic Approaches in Experimental Therapeutics" (K. A. Jacobson and M. F. Jarvis, eds.), pp. 423–447. Wiley-Liss, New York.

Kunapuli, S. P., and Daniel, J. L. (1998). P2 receptor subtypes in the cardiovascular system. *Biochem. J.* **336,** 513–523.

Lê, K. T., Babinski, K., and Séguéla, P. (1998). Central P2X$_4$ and P2X$_6$ channel subunits coassemble into a novel heteromeric ATP receptor. *J. Neurosci.* **18,** 7152–7159.

Lee, P. S., Squires, P. E., Buchan, A. M., Yuen, B. H., and Leung, P. C. (1996). P2-purinoreceptor evoked changes in intracellular calcium oscillations in single isolated human granulosa-lutein cells. *Endocrinology* **137,** 3756–3761.

Lee, H.-Y., Bardini, M., and Burnstock, G. (2000). Distribution of P2X receptors in the urinary bladder and ureter of the rat. *J. Urol.* **163,** 2002–2007.

Lee, H. Y., Bardini, M., and Burnstock, G. (2000). P2X receptor immunoreactivity in the male genital organ of the rat. *Cell Tissue Res.* **300,** 321–330.

Lewis, C., Neidhart, S., Holy, C., North, R. A., Buell, G., and Surprenant, A. (1995). Coexpression of P2X$_2$ and P2X$_3$ receptor subunits can account for ATP-gated currents in sensory neurons. *Nature* **377,** 432–435.

Lewis, C. J., and Evans, R. J. (2000). Comparison of P2X receptors in rat mesenteric, basilar and septal (coronary) arteries. *J. Auton. Nerv. Syst.* **81,** 69–74.

Lewis, C. J., Surprenant, A., and Evans, R. J. (1998). 2′,3′-O-(2,4,6-Trinitrophenyl) adenosine 5′-triphosphate (TNP-ATP)—a nanomolar affinity antagonist at rat mesenteric artery P2X receptor ion channels. *Br. J. Pharmacol.* **124,** 1463–1466.

Liang, B. T., and Jacobson, K. A. (1999). Adenosine and ischemic preconditioning. *Curr. Pharm. Des.* **5,** 1029–1041.

Llewellyn-Smith, I. J., and Burnstock, G. (1998). Ultrastructural localization of P2X$_3$ receptors in rat sensory neurons. *Neuroreport* **9,** 2245–2250.

Loesch, A., and Burnstock, G. (1998). Electron-immunocytochemical localization of the P2X$_1$ receptors in the rat cerebellum. *Cell Tissue Res.* **294,** 253–260.

Loesch, A., and Burnstock, G. (2000). Ultrastructural localisation of ATP-gated P2X$_2$ receptor immunoreactivity in vascular endothelial cells in rat brain. *Endothelium* **7,** 93–98.

Loesch, A., Miah, S., and Burnstock, G. (1999). Ultrastructural localisation of ATP-gated P2X$_2$ receptor immunoreactivity in the rat hypothalamo–neurohypophysial system. *J. Neurocytol.* **28,** 495–504.

Londos, C., Cooper, D. M., and Wolff, J. (1980). Subclasses of external adenosine receptors. *Proc. Natl. Acad. Sci. USA* **77,** 2551–2554.

Loubatières-Mariani, M. M., and Chapal, J. (1988). Purinergic receptors involved in the stimulation of insulin and glucagon secretion. *Diabetes Metab.* **14,** 119–126.

Loubatières-Mariani, M. M., Hillaire-Buys, D., Chapal, J., Bertrand, G., and Petit, P. (1997). P2 purinoceptor agonists: New insulin secretagogues potentially useful in the treatment of non-insulin-dependent diabetes mellitus. *In* "Purinergic Approaches in Experimental Therapeutics" (K. A. Jacobson and M. F. Jarvis, eds.), pp. 253–260. Wiley-Liss, New York.

Lustig, K. D., Shiau, A. K., Brake, A. J., and Julius, D. (1993). Expression cloning of an ATP receptor from mouse neuroblastoma cells. *Proc. Natl. Acad. Sci. USA* **90,** 5113–5117.

Malhotra, J., and Gupta, Y. K. (1997). Effect of adenosine receptor modulation on pentylenetetrazole-induced seizures in rats. *Br. J. Pharmacol.* **120**, 282–288.

Mayer, C., Quasthoff, S., and Grafe, P. (1998). Differences in the sensitivity to purinergic stimulation of myelinating and non-myelinating Schwann cells in peripheral human and rat nerve. *Glia* **23**, 374–382.

Meyer, M. P., Gröschel-Stewart, U., Robson, T., and Burnstock, G. (1999a). Expression of two ATP-gated ion channels, P2X$_5$, and P2X$_6$, in developing chick skeletal muscle. *Dev. Dyn.* **216**, 442–449.

Meyer, M. P., Clarke, J. D. W., Patel, K., Townsend-Nicholson, A., and Burnstock, G. (1999b). Selective expression of purinoceptor cP2Y$_1$ suggests a role for nucleotide signalling in development of the chick embryo. *Dev. Dyn.* **214**, 152–158.

Morelli, M., and Pinna, A. (2001). Modulation by adenosine A$_{2A}$ receptors of dopamine-mediated motor behaviour as a basis for antiparkinson's disease drugs. *Drug Dev. Res.* **52**, 387–393.

Morrison, M. S., Turin, L., King, B. F., Burnstock, G., and Arnett, T. R. (1998). ATP is a potent stimulator of the activation and formation of rodent osteoclasts. *J. Physiol.* **511**, 495–500.

Mulryan, K., Gitterman, D. P., Lewis, C. J., Vial, C., Leckie, B. J., Cobb, A. L., Brown, J. E., Conley, E. C., Buell, G., Pritchard, C. A., and Evans, R. J. (2000). Reduced vas deferens contraction and male infertility in mice lacking P2X$_1$ receptors. *Nature* **403**, 86–89.

Neary, J. T. (1996). Trophic actions of extracellular ATP on astrocytes, synergistic interactions with fibroblast growth factors and underlying signal transduction mechanisms. *In* "P2 Purinoceptors: Localization, Function and Transduction Mechanisms" (D. J. Chadwick and J. A. Goode, eds.), pp. 130–141. John Wiley & Sons, Chichester.

Neary, J. T., Rathbone, M. P., Cattabeni, F., Abbracchio, M. P., and Burnstock, G. (1996). Trophic actions of extracellular nucleotides and nucleosides on glial and neuronal cells. *Trends Neurosci.* **19**, 13–18.

Nikodijevic, O., Sarges, R., Daly, J. W., and Jacobson, K. A. (1991). Behavioral effects of A1- and A2-selective adenosine agonists and antagonists: Evidence for synergism and antagonism. *J. Pharmacol. Exp. Ther.* **259**, 286–294.

Nori, S., Fumagalli, L., Bo, X., Bogdanov, Y., and Burnstock, G. (1998). Coexpression of mRNAs for P2X$_1$, P2X$_2$ and P2X$_4$ receptors in rat vascular smooth muscle: An in situ hybridization and RT-PCR study. *J. Vasc. Res.* **35**, 179–185.

North, R. A., and Surprenant, A. (2000). Pharmacology of cloned P2X receptors. *Annu. Rev. Pharmacol. Toxicol.* **40**, 563–580.

Nyce, J. W., and Metzger, W. J. (1997). DNA antisense therapy for asthma in an animal model. *Nature* **385**, 721–725.

O'Connor, S. E., Dainty, I. A., and Leff, P. (1991). Further subclassification of ATP receptors based on agonist studies. *Trends Pharmacol. Sci.* **12**, 137–141.

Olah, M. E., and Stiles, G. L. (1995). Adenosine receptor subtypes: Characterization and therapeutic regulation. *Annu. Rev. Pharmacol. Toxicol.* **35**, 581–606.

Olsson, R. A., and Pearson, J. D. (1990). Cardiovascular purinoceptors. *Physiol. Rev.* **70**, 761–845.

Pintor, J. (1999). Purinergic signaling in the eye. *In* "The Autonomic Nervous System," Vol. 13: "Nervous Control of the Eye" (G. Burnstock, ed.), pp. 171–210. Harwood Academic, Amsterdam.

Pintor, J., and Peral, A. (2001). Therapeutic potential of nucleotides in the eye. *Drug Dev. Res.* **52**, 190–195.

Radford, K. M., Virginio, C., Surprenant, A., North, R. A., and Kawashima, E. (1997). Baculovirus expression provides direct evidence for heteromeric assembly of P2X$_2$ and P2X$_3$ receptors. *J. Neurosci.* **17**, 6529–6533.

Ralevic, V., and Burnstock, G. (1998). Receptors for purines and pyrimidines. *Pharmacol. Rev.* **50**, 413–492.

Rapaport, E. (1983). Treatment of human tumor cells with ADP or ATP yields arrest of growth in the S phase of the cell cycle. *J. Cell Physiol.* **114**, 279–283.

Rapaport, E. (1997). ATP in the treatment of cancer. In "Purinergic Approaches in Experimental Therapeutics" (K. A. Jacobson and M. F. Jarvis, eds.), pp. 545–553. Wiley-Liss, New York.

Reppert, S. M., Weaver, D. R., Stehle, J. H., and Rivkees, S. A. (1991). Molecular cloning and characterization of a rat A1-adenosine receptor that is widely expressed in brain and spinal cord. *Mol. Endocrinol.* **5**, 1037–1048.

Rivkees, S. A., Price, S. L., and Zhou, F. C. (1995). Immunohistochemical detection of A1 adenosine receptors in rat brain with emphasis on localization in the hippocampal formation, cerebral cortex, cerebellum, and basal ganglia. *Brain Res.* **677**, 193–203.

Rong, W., Burnstock, G., and Spyer, K. M. (2000). P2X purinoceptor-mediated excitation of trigeminal lingual nerve terminals in an *in vitro* intra-arterially perfused rat tongue preparation. *J. Physiol.* **524**, 891–902.

Rosin, D. L., Robeva, A., Woodard, R. L., Guyenet, P. G., and Linden, J. (1998). Immunohistochemical localization of adenosine $A_{2A}$ receptors in the rat central nervous system. *J. Comp. Neurol.* **401**, 163–186.

Ryten, M., Hoebertz, A., and Burnstock, G. (2001). Sequential expression of three receptor subtypes for extracellular ATP in developing rat skeletal muscle. *Dev. Dyn.* **221**, 331–341.

Schöfl, C., Ponczek, M., Mader, T., Waring, M., Benecke, H., von zur Mühlen, A., Mix, H., Cornberg, M., Böker, K. H. W., Manns, M. P., and Wagner, S. (1999). Regulation of cytosolic free calcium concentration by extracellular nucleotides in human hepatocytes. *Am. J. Physiol.* **276**, G164–G172.

Sexl, V., Mancusi, G., Holler, C., Gloria-Maercker, E., Schutz, W., and Freissmuth, M. (1997). Stimulation of the mitogen-activated protein kinase via the A2A-adenosine receptor in primary human endothelial cells. *J. Biol. Chem.* **272**, 5792–5799.

Silinsky, E. M., Gerzanich, V., and Vanner, S. M. (1992). ATP mediates excitatory synaptic transmission in mammalian neurones. *Br. J. Pharmacol.* **106**, 762–763.

Sperlágh, B., Mergl, Z., Jurányi, Z., Vizi, E. S., and Makara, G. B. (1999). Local regulation of vasopressin and oxytocin secretion by extracellular ATP in the isolated posterior lobe of the rat hypophysis. *J. Endocrinol.* **160**, 343–350.

Spungin, B., and Friedberg, I. (1993). Growth inhibition of breast cancer cells induced by exogenous ATP. *J. Cell. Physiol.* **157**, 502–508.

Stone, T. W. (1991). *"Adenosine in the Nervous System."* Academic Press, London.

Stutts, M. J., and Boucher, R. C. (1999). Cystic fibrosis gene and functions of CFTR implications of dysfunctional ion transport for pulmonary pathogenesis. In "Cystic Fibrosis in Adults" (J. R. Yankaskas and M. R. Knowles, eds.), pp. 3–25. Lippincott-Raven, Philadelphia, PA.

Surprenant, A., Rassendren, F., Kawashima, E., North, R. A., and Buell, G. (1996). The cytolytic $P_{2Z}$ receptor for extracellular ATP identified as a $P_{2X}$ receptor (P2X7). *Science* **272**, 735–738.

Tomic, M., Jobin, R. M., Vergara, L. A., and Stojilkovic, S. S. (1996). Expression of purinergic receptor channels and their role in calcium signaling and hormone release in pituitary gonadotrophs: Integration of P2 channels in plasma membrane- and endoplasmic reticulum-derived calcium oscillations. *J. Biol. Chem.* **271**, 21200–21208.

Torres, G. E., Haines, W. R., Egan, T. M., and Voigt, M. M. (1998). Co-expression of P2X1 and P2X5, receptor subunits reveals a novel ATP-gated ion channel. *Mol. Pharmacol.* **54**, 989–993.

Torres, G. E., Egan, T. M., and Voigt, M. M. (1999). Hetero-oligomeric assembly of P2X receptor subunits: Specificities exist with regard to possible partners. *J. Biol. Chem.* **274**, 6653–6659.

Townsend-Nicholson, A., King, B. F., Wildman, S. S., and Burnstock, G. (1999). Molecular cloning, functional characterization and possible cooperativity between the murine P2X$_4$ and P2X$_{4a}$ receptors. *Brain Res. Mol. Brain Res.* **64**, 246–254.

Vainio, M., and Törnquist, K. (2000). The role of adenosine A$_1$ receptors in the ATP-evoked Ca$^{2+}$ response in rat thyroid FRTL-5 cells. *Eur. J. Pharmacol.* **390**, 43–50.

Valera, S., Hussy, N., Evans, R. J., Adani, N., North, R. A., Surprenant, A., and Buell, G. (1994). A new class of ligand-gated ion channel defined by P$_{2X}$ receptor for extra-cellular ATP. *Nature* **371**, 516–519.

Van Calker, D., Müller, M., and Hamprecht, B. (1979). Adenosine regulates via two different types of receptors, the accumulation of cyclic AMP in cultured brain cells. *J. Neurochem.* **33**, 999–1005.

van der Weyden, L., Adams, D. J., and Morris, B. J. (2000). Capacity for purinergic control of renin promoter via P2Y$_{11}$ receptor and cAMP pathways. *Hypertension* **36**, 1093–1098.

Vlaskovska, M., Kasakov, L., Rong, W., Bodin, P., Bardini, M., Cockayne, D. A., Ford, A. P. D. W., and Burnstock, G. (2001). P2X$_3$ knockout mice reveal a major sensory role for urothelially released ATP. *J. Neurosci.* **21**, 5670–5677.

Vulchanova, L., Arvidsson, U., Riedl, M., Wang, J., Buell, G., Surprenant, A., North, R. A., and Elde, R. (1996). Differential distribution of two ATP-gated channels (P2X receptors) determined by imunocytochemistry. *Proc. Natl. Acad. Sci. USA* **93**, 8063–8067.

Webb, T. E., Simon, J., Krishek, B. J., Bateson, A. N., Smart, T. G., King, B. F., Burnstock, G., and Barnard, E. A. (1993). Cloning and functional expression of a brain G-protein-coupled ATP receptor. *FEBS Lett.* **324**, 219–225.

Wildman, S. S., King, B. F., and Burnstock, G. (1999). Modulatory activity of extracellular H$^+$ and Zn$^{2+}$ on ATP-responses at rP2X$_1$ and rP2X$_3$ receptors. *Br. J. Pharmacol.* **128**, 486–492.

Williams, M., and Jarvis, M. F. (2000). Purinergic and pyrimidinergic receptors as potential drug targets. *Biochem. Pharmacol.* **59**, 1173–1185.

Xiang, Z., Bo, X., and Burnstock, G. (1998). Localization of ATP- gated P2X receptor immunoreactivity in rat sensory and sympathetic ganglia. *Neurosci. Lett.* **256**, 105–108.

Xiang, Z., Bo, X., Oglesby, I. B., Ford, A. P. D. W., and Burnstock, G. (1998). Localization of ATP-gated P2X$_2$ receptor immunoreactivity in the rat hypothalamus. *Brain Res.* **813**, 390–397.

Xiang, Z., Bo, X., and Burnstock, G. (1999). P2X receptor immunoreactivity in the rat cochlea, vestibular ganglion and cochlear nucleus. *Hear. Res.* **128**, 190–196.

Yerxa, B. R. (2000). Therapeutic use of nucleotides in respiratory and ophthalmic diseases. *Drug Dev. Res.* **50**, 22.

Yokomizo, T., Izumi, T., Chang, K., Takuwa, Y., and Shimizu, T. (1997). A G-protein-coupled receptor for leukotriene B$_4$ that mediates chemotaxis. *Nature* **387**, 620–624.

Zhang, F. L., Luo, L., Gustafson, E., Palmer, K., Qiao, X., Fan, X., Yang, S., Laz, T. M., Bayne, M., and Monsma, F., Jr. (2002). P2Y$_{13}$: identification and characterization of a novel Galphai-coupled ADP receptor from human and mouse. *J. Pharmacol. Exp. Ther.* **301**, 705–713.

Zimmermann, H. (1996). Extracellular purine metabolism. *Drug Dev. Res.* **39**, 337–352.

# PART I

Molecular Biology, Cell Biology, and
Biochemistry

# CHAPTER 2

# Cellular Mechanisms and Physiology of Nucleotide and Nucleoside Release from Cells: Current Knowledge, Novel Assays to Detect Purinergic Agonists, and Future Directions

**Erik M. Schwiebert,**[*]**Akos Zsembery,**[*] **and John P. Geibel**[†]

[*]Departments of Physiology and Biophysics and Cell Biology and Gregory Fleming James Cystic Fibrosis Research Center, University of Alabama at Birmingham, Birmingham, Alabama 35294

[†]Departments of Surgery and Cellular and Molecular Physiology, Yale University School of Medicine, New Haven, Connecticut 06520

## I. INTRODUCTION

Extracellular nucleotide and nucleoside signaling is initiated only with the release of ATP and adenosine from physiological sources. Those physiological sources are cells that synthesize ATP as their chief intracellular fuel. In this volume, each specific chapter on extracellular ATP signaling in cells and tissues discusses the sources of released ATP; however, it can be confidently argued that all cells have the potential to release ATP. Table I shows a current list of cell types implicated or studied for their ATP release capacity. Although it is true that bacterial adherence, anoxia, ischemia, apoptosis, and necrosis can lead to partial or complete cell lysis and subsequent ATP release, there are multiple physiological stimuli that activate physiological mechanisms that potentiate ATP release from cells. The aforementioned lytic stimuli are pathophysiological triggers. This chapter and other chapters herein focus on physiological ATP release mechanisms and their effects on cells and tissues.

As in all physiological systems, there are first principles. These are illustrated in Fig. 1. There is an enormous chemical concentration gradient for ATP release from cells. For example, if the steady state extracellular concentration is approximately 50 n$M$ under basal conditions, then a 5 m$M$ intracellular concentration provides a 100,000-fold gradient for ATP efflux from cells. Because no assay has measured a steady state extracellular concentration above 10 $\mu M$ under stimulated conditions (likely due to degradative enzymes competitively clearing a portion of the released ATP), a mammalian cell releases only 0.1% of its total intracellular ATP pool. Thus, again, extracellular nucleotide signaling does not compromise cellular metabolism. The 50% effective concentration (EC$_{50}$) for stimulation of all nucleotide receptors is approximately 1 $\mu M$, with a notable exception being the P2X$_7$/P2Z receptor, which requires high micromolar to millimolar ATP. Adenosine receptors have EC$_{50}$ values that are equal to or less than 1 $\mu M$. Thus, the ATP and adenosine that are released from cells and the concentrations that have been measured in detection assays are sufficient to stimulate any and all purinergic receptors expressed by that cell, with the metabolism of that cell proceeding unaltered.

This is a physiological autocrine and paracrine signaling system. This fact is echoed again and again in the following chapters in this volume. Although not an endocrine signaling system in the general circulation, ATP and

**TABLE I**

Cell Types Implicated in ATP Release[a]

| Cell type | Ref. |
|---|---|
| Hematopoietic cells | |
| Red blood cells | Bergfeld and Forrester (1992) |
| | Sprague *et al.* (1998, 2001a,b) |
| | Edward *et al.* (2001) |
| | Abraham *et al.* (2001) |
| | Hoffman (1997) |
| | Light *et al.* (1999) |
| | H. R. de Jonge (unpublished observations) |
| Macrophages and monocytes | Beigi *et al.* (1999) |
| Platelets | Beigi *et al.* (1999) |
| Epithelial cells | |
| Pulmonary | Schwiebert *et al.* (1995) |
| | Taylor *et al.* (1998) |
| | Braunstein *et al.* (2001) |
| | Schneider *et al.* (1999) |
| | Watt *et al.* (1998) |
| | Donaldson *et al.* (2000) |
| | Homolya *et al.* (2000) |
| | Walsh *et al.* (2000) |
| | Huang *et al.* (2001) |
| | Musante *et al.* (1999) |
| Intestinal | Hazama *et al.* (1999) |
| | Sabirov *et al.* (2001) |
| Pancreatic | Sorensen and Novak (2001) |
| Hepatic | Wang *et al.* (1996) |
| | Roman *et al.* (1999) |
| | Frame and de Feijter (1997) |
| Renal | Wilson *et al.* (1999) |
| | Schwiebert *et al.* (2002b) |
| | Ostrom *et al.* (2000) |
| Macula densa of kidney | Bell *et al.* (2000) |
| Urothelium | Knight *et al.* (2002) |
| Ocular epithelium | Mitchell (2001) |
| Retinal pigment epithelium | Mitchell *et al.* (1998) |
| Endothelial cells | Schwiebert *et al.* (2002a) |
| | Shinozuka *et al.* (2001) |
| | Yamamoto *et al.* (2000) |
| Cardiac myocytes | Lader *et al.* (2000) |
| Neurons, astrocytes, glia | G. M. Braunstein, E. M. Schwiebert, |
| (mixed primary culture) | and L.L. McMahon |
| | (unpublished observations) |

(*Continued*)

**TABLE I** (*Continued*)

| Cell type | Ref. |
|---|---|
| Pancreatic beta cells | Hazama *et al.* (1998) |
| Prostate cancer cells | Sauer *et al.* (2000) |
| *Xenopus laevis* oocytes | Jiang *et al.* (1998) |
| | Maroto and Hamill (2001) |
| Heterologous cells | |
| C127 cells | Hazama *et al.* (2000) |
| | Prat *et al.* (1996) |
| COS-7 cells | Braunstein *et al.* (2001) |
| NIH 3T3 fibroblasts | Roman *et al.* (2001) |
| | Braunstein *et al.* (2001) |

[a]Illustrated is the range of different cell types implicated in ATP release. The list is not meant to be comprehensive; however, it shows that ATP release may be a physiological process that is ubiquitous.

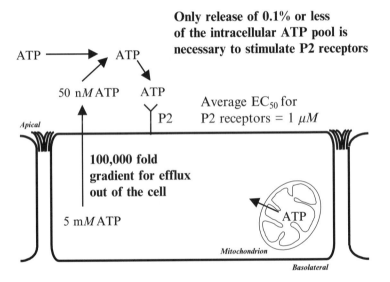

FIGURE 1  First principles of ATP release: Some simple principles of ATP release cell biology that are often overlooked; yet, they are critical to understanding ATP release as a biological process.

adenosine are local blood-borne mediators that affect the microcirculation within vascular beds (see the chapters Purinergic Signaling in Hematopoietic Cells and the Vasculature, and Purinergic Signaling Driven by the Red Blood Cells, elsewhere in this volume). Microenvironments within tissues and surrounding cells are likely areas where potent autocrine and paracrine purinergic signaling occurs.

## II. POSSIBLE CELLULAR MECHANISMS OF ATP RELEASE FROM CELLS

ATP is an anion that carries multiple negative charges up to and including the species $ATP^{4-}$. Thus, ATP transport mechanisms can be modeled after other anions that are more traditionally thought of as being transported across biological membranes. These anions include $Cl^-$, $HCO_3^-$, and glutathione, to name a few. Hints derive from the inner and outer membranes of the mitochondrion, which has adenine nucleotide transporters (ANTs) to move newly synthesized ATP near the outer membrane and the voltage-dependent anion channel (VDAC) or porin to conduct the ATP out into the cytoplasm. Hints also derive from neuroendocrine cells. ATP is also an agonist with a charge, like any other peptide hormone or neurotransmitter that has a net charge because of its chemical or amino acid content. As such, ATP is packaged in vesicles alone or together with other agonists. ATP is released with histamine from secretory granules of the mast cell, with epinephrine in chromaffin granules of the adrenal medulla, and with many different neurotransmitters from presynaptic vesicles at synapses. Not only can ATP be transported as an anion but it can also be released via exocytosis. Below are descriptions of the possible cellular mechanisms of ATP release from mammalian cells and pharmacological and cell biological maneuvers to discriminate between them.

### A. ATP-Permeable Release Channel: ATP-Permeable Anion Channel

The existence of ion channels that are permeated by ATP anions was underscored by many laboratories investigating whether the cystic fibrosis transmembrane conductance regulator (CFTR) conducted $ATP^-$ itself in addition to $Cl^-$ or regulated a separate ion channel that was permeable to $ATP^-$. One group suggested that CFTR and the multidrug resistance transporter (mdr or P-glycoprotein) conducted $ATP^-$ (Abraham *et al.*, 1993; Reisin, *et al.*, 1994). Guggino and co-workers showed that CFTR potentiated autocrine ATP release to upregulate a separate anion channel, the outwardly rectifying $Cl^-$ channel (ORCC) (Schwiebert, *et al.*, 1995). The simplest interpretation was that CFTR conducted $ATP^-$; however, this group was careful to point out that CFTR, as a conductance regulator, might also regulate an ATP release channel, transporter, or other release pathway (Schwiebert *et al.*, 1995). Subsequently, several other laboratories failed to show conduction of ATP⁻ through CFTR (Reddy, *et al.*, 1996; Grygorczyk and Hanrahan, 1997; Watt *et al.*, 1998; Li *et al.*, 1996); each laboratory performed a limited number of experiments in their system of choice and stopped early on in their study when the negative data mounted. Studies by Schwiebert and co-workers (Braunstein *et al.*, 2001) and by Foskett and colleagues (Sugita *et al.*, 1998) have shown that CFTR is likely not an ATP channel itself; however, it is closely associated with a separate ATP-permeable channel that it regulates positively. An invited minireview addressed this area of research in detail (Schwiebert, 1999).

A handful of laboratories have performed additional research to begin to identify ATP release pathways in cells, to understand why CFTR would potentiate ATP release from cells and in response to the best stimuli, and to identify more specifically any and all ATP-permeable release channels. An initial assumption is that ATP-permeable release channels are anion channels. ATP release assays in hepatocytes and cholangiocytes by Fitz and co-workers, in airway and renal epithelial cells by Schwiebert and co-workers, and in heterologous cells by both of these laboratories have identified the lanthanides, gadolinium and lanthanum, as blockers of ATP release under basal and stimulated conditions (Taylor *et al.*, 1998; Wilson *et al.*, 1999; Roman *et al.*, 1999a; Braunstein *et al.*, 2001). The lanthanides are known blockers of mechanosensitive ion channels that are permeable to cations as well as anions. Schwiebert and co-workers has also shown that the disulfonic stilbene, DIDS (disothio cyanodisulfonic stilbene), is a blocker of ATP release that is even more potent than gadolinium (Braunstein *et al.*, 2001). Taken together, blockade of ATP release in biological cells by these agents suggests that ATP release channels are important conduits for ATP release.

One assumption is that some specific anion channels may also be permeable to ATP⁻. Who are the most likely candidates for ATP-permeable anion channels? Thinnes and colleagues, among others, have documented that plasma membrane forms of porins or voltage-dependent anion channels (PL-VDACs) exist in a variety of cell types (Reymann *et al.*, 1995; Thinnes *et al.*, 2000). In one study, they show colocalization of CFTR with a PL-VDAC (Reymann *et al.*, 1995). It is known that "maxi" Cl⁻ channels (or large conductance anion channels) with voltage-dependent inactivation characteristics similar to VDAC have been found by numerous investigators in numerous cell types. In fact, Fitz and co-workers and Schwiebert and colleagues have studied such maxi anion channels in the past in Chinese hamster ovary (CHO) cells (Mangel *et al.*, 1993; McGill *et al.*, 1993) and renal cortical collecting duct (CCD) cells (Light *et al.*, 1990; Schwiebert *et al.*, 1994), respectively. Because VDAC is a major conduit for the release of newly synthesized ATP in the outer membrane of the mitochondrion

(Benz *et al.*, 1988), an ATP-permeable VDAC in the plasma membrane may be a major conductive pathway for ATP exit into the extracellular milieu. Schwiebert and Stanton showed that the maxi anion channel in RCCT-28A cells, an A-type intercalated cell line from rabbit CCD, had significant permeability to gluconate (Cl:gluconate, 8:1) and bicarbonate (Cl:$HCO_3$, 2:1) (Light *et al.*, 1990; Schwiebert *et al.*, 1994). Okada and coworkers have shown that an ATP-permeable maxi anion channel is expressed in mouse mammary carcinoma (C127) cells (Sabirov *et al.*, 2001). Preliminary work from Bell and co-workers has implicated a maxi anion channel in ATP release from the macula densa induced by changes in NaCl concentration in the cortical thick ascending limb (Bell *et al.*, 2000). It is postulated that this ATP-permeable maxi anion channel is responsible for release from the macula densa of ATP that diffuses to and binds to P2 receptors in the glomerulus as an important signal for tubuloglomerular feedback, a form of single-nephron "autoregulation." A similar maxi anion channel (approximately 400 pS) has been found in cell-attached and excised patch-clamp recordings from non-CF and CF airway epithelial primary cultures and cell lines and is permeable to a multitude of different anions including ATP (E. M. Schwiebert, unpublished observations). Taken together, it is clear that the VDAC-like maxi anion channels are a leading candidate for ATP release channels in the plasma membrane of many cell types and tissues.

It is clear that ATP-permeable anion channels may not be limited to maxi anion channels. At least two other candidates have been implicated indirectly in published studies. These include, but are not limited to, the outwardly rectifying $Cl^-$ channel (or ORCC) and the volume-sensitive organic anion channel (or VSOAC). In a study by Braunstein *et al.*, planar lipid bilayer experiments were performed in which bovine tracheal epithelial vesicle protein was fused with the bilayers containing or immunodepleted of CFTR protein (Braunstein *et al.*, 2001). Under each condition, an ATP conductance of approximately 20 pS was recorded that was potentiated by hydrostatic pressure applied to the bilayer and was inhibited by DIDS, gadolinium, and diphenylamine-2-carboxylate (DPC). Under the condition of CFTR protein immunodepletion, the ORCC was also present, as determined by removing ATP-containing solutions and replacing ATP with Cl-containing solutions. The simplest interpretation was that the 40- to 80-pS ORCC $Cl^-$ channel had a partial permeability to ATP. Because ORCCs may be conferred by CLC-3 and/or CLC-5 channels (or a heteromeric mixture of these two CLC channels along with additional CLC subtypes), CLC channels may also be candidates for ATP release channels. Another, equally valid, interpretation was that the 20-pS ATP channel was conferred by a separate protein. Another candidate for this

ATP channel is the VSOAC channel first described and named as such by Strange and co-workers. In an initial characterization, Jackson and Strange showed that intracellular millimolar ATP concentrations supported VSOAC currents, while supraphysiologic external concentrations of millimolar ATP actually blocked macroscopic VSOAC currents carried by $Cl^-$ or by organic osmolytes such as betaine or taurine (Jackson and Strange, 1995). Because this external block was concentration and voltage dependent, ATP can be deduced to be a possible pore blocker of VSOAC and, thus, a putative permeable species along with large anions such as taurine and betaine.

Therefore, there are candidates for ATP release channels. It is important to note that these release channels may differ in their expression as well as their polarity of expression (if studied, in epithelial or endothelial cells or in neurons). The next few years of work by several laboratories should uncover much new information in the arena of ATP-release channels. Arguably, this field was brought to light via the controversy concerning CFTR, mdr, and ATP permeability through these ATP-binding cassette (ABC) transporters. Without that series of articles for and against this phenomenon, it is possible that this field may not have emerged this rapidly.

## B. Adenine Nucleotide Transporter: Permease, Carrier, Symporter

Equally viable is the concept that adenine nucleotide transporters (ATP transporters or ANTs) may also confer ATP permeability or release across biological membranes. Even though evidence is emerging that ABC transporters do not conduct ATP at rates consistent with an ionic channel, these transporters may transport ATP across membranes at rates that may lie in a range more consistent with a transporter. The fact that glibenclamide, tamoxifen, cyclosporin A, and verapamil inhibit ATP release in several systems underscores this possibility (Roman *et al.*, 2001). Indeed, Linsdell and Hanrahan have shown that CFTRs transport glutathione at rates that border on channel conduction rates ($10^7$ ions/s and above) versus transporter rates ($10^6$ molecules/s or below) (Linsdell and Hanrahan, 1998). In fact, symport of glutathione and ATP or of ATP with another substrate could occur. KCl symporters are critical in red blood cell and hematopoietic cell volume regulation (Fujise *et al.*, 2001); a KATP symporter is not out of the realm of possibility. Moreover, the promiscuity of ABC transporters [mdr, multidrug resistance-related protein (MRP)] and the substrates that they move across membranes suggests that ATP could be carried along for the ride with these countless substrates (Roman *et al.*, 2001). Multiple organic anion transporters (the MOATs) are also possible candidates

(Roman *et al.*, 2001; Zeng *et al.*, 2000). Investigators should keep an open mind to this possible mechanism of ATP release. In relation to vesicular ATP release (see below), this ATP needs to be loaded into the vesicle before fusion with the plasma membrane, much like what occurs with neurotransmitters or neuroendocrine agonists.

## C. Adenine Nucleotide Transporter: ATP/ADP Exchanger

As with VDAC or porin channels in outer mitochondrial membrane, ANTs, adenine nucleotide transporters that transport newly synthesized ATP across the inner mitochondrial membrane to the exchange of ADP, also play a critical role in intracellular ATP transport (Fig. 2) (Brustovetsky *et al.*, 1996). Once thought to be restricted to the mitochondrion, emerging evidence suggests that ANTs may also be expressed in endoplasmic reticulum, the nuclear envelope membrane, and in brain synaptic vesicle (Gualix *et al.*, 1999; Guillen and Hirschberg, 1995; Bankston and Guidotti, 1996). The latter result suggests that ANTs may reside, at least for a brief period of time, in the plasma membrane of neurons once synaptic vesicles fuse to release their contents. A reason for addressing possible expression in synaptic vesicles is the known fact that ATP is present in high (millimolar) concentrations along with neurotransmitters or neuroendocrine agonists such as histamine and epinephrine. Such exchange of ATP released into the extracellular milieu with a metabolite of ATP transported back into the cell for resynthesis to ATP is also a viable hypothesis. Atractyloside, a specific inhibitor of the ANTs, can be used as an extracellular inhibitor of ATP release to probe the role of ANTs in autocrine ATP release and signaling.

## D. ATP-Filled Vesicle

It is a known fact, dating back to the work of Burnstock and Gordon and their seminal reviews on the topic of extracellular ATP or "purinergic" signaling (Gordon, 1986; Burnstock *et al.*, 1970), that a high content of ATP is found in synaptic vesicles, histamine granules, and chromaffin granules. ATP and its metabolites are known cotransmitters that modulate the effect of excitatory or inhibitory neurotransmitters. ADP and ATP are released by platelets via exocytosis, creating a "self-aggregation" signal at the clotting zone. Release of ATP-filled vesicles is the principal transmitter signal for purinergic nerves in the gut, the nociceptive nodose neurons in

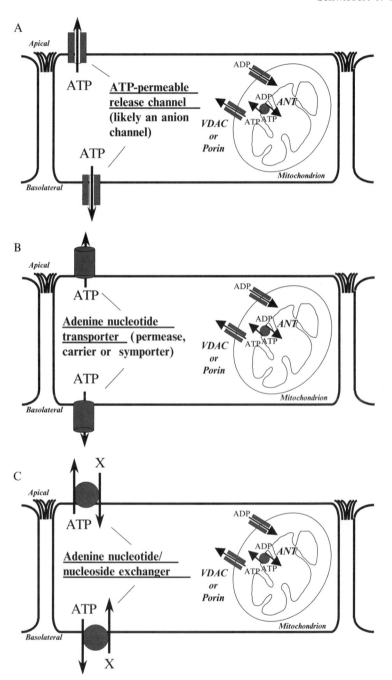

**FIGURE 2**   (*For legend, see opposite page.*)

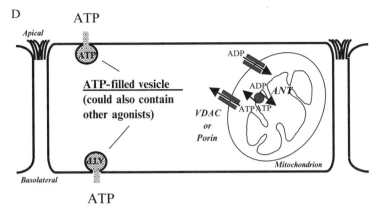

**FIGURE 2** Putative ATP release mechanisms. (A) ATP-permeable release channel. (B) Adenine nucleotide transporter: Possible inhibitors are shown for this putative ATP release pathway include dipyridamole and NBTI (nitrobenzoylthioinosine) are listed as well since they inhibit ATP release from red blood cells; however, they are more well known as inhibitors of adenine nucleoside transporters. (C) Adenine nucleotide/nucleoside exchangers: As with the transporters described above, atractyloside may be the only drug specific for this type of transporter. (D) ATP-filled vesicle: This is the most elusive ATP release mechanism. Cooling of cells or monolayers to 4°C may be the only recourse to block all vesicle traffic and fusion, even preformed and prefilled vesicles poised to fuse with the plasma membrane. Pharmacological agents such as brefeldin A block only the early parts of the secretory pathway, and microtubule-disrupting reagents may only block secretory vesicles on their way to the plasma membrane from the Golgi. Monensin, an inhibitor of H⁺-ATPase, may be effective in some systems; however, specific use of botulinum toxin and tetanus toxin, which block the molecular mechanisms of vesicle docking and fusion, is best (however, they need to be loaded in some way into non-excitable cells that lack the receptors for these toxins).

the dorsal root ganglia of the spinal cord, and in specialized nerves in the brain. Thus, exocytosis of ATP-filled vesicles, preformed and poised to fuse beneath the plasma membrane, should be considered a major release mechanism. One critical question regarding the presence of ATP in vesicles is as follows: How does the ATP become loaded and concentrated in these vesicles to high millimolar concentrations? We hypothesize that ATP conductive channels and transporters may function to transport ATP into these vesicles and concentrate the ATP within. As an anion, ATP transport into the vesicle could be facilitated by $H^+$-ATPases, which would transport and concentrates $H^+$ in the vesicle lumen. ATP would follow to preserve macroscopic electroneutrality across the vesicle membrane. Taken together, it is important to note that, once a vesicle fuses and releases its quanta of ATP, that ATP channels and/or transporters would then be inserted and present in the plasma membrane to drive ATP release further.

## III. INDIRECT PHARMACOLOGICAL SCREENING FOR EXTRACELLULAR NUCLEOTIDE AND NUCLEOSIDE RELEASE AND SIGNALING

### A. Nucleotide and Nucleoside Scavengers

A potent and effective way of blocking autocrine and paracrine signaling by nucleotides and nucleosides is by inclusion of scavengers for purinergic agonists (Fig. 3). Hexokinase (with inclusion of 5 m$M$ glucose) will scavenge ATP in preparations, yielding ADP and donating the terminal or γ-phosphate of ATP to glucose, forming glucose 6-phosphate. This maneuver eliminates ATP from solutions and prevents any endogenous ATP released in a cell or tissue preparations from interacting with P2 receptors that prefer ATP. However, this enzymatic reaction yields ADP. ADP is a better agonist at some P2Y receptors and a weak agonist for some P2X receptors. As such, in lieu of hexokinase, apyrase, an ATPase/ADPase, would eliminate ATP and ADP from solutions rapidly. This would effectively prevent any P2 receptors from being occupied by endogenous ATP or ADP. However, apyrase generates a precursor of adenosine, 5′-AMP. If a cell or tissue preparation expresses ecto-5′-nucleotidase, then adenosine could be generated extracellularly and activate

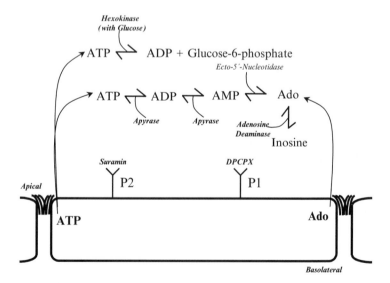

**FIGURE 3** Chemistry of blockade of endogenous purinergic signaling. Multiple chemical and pharmacological maneuvers can be employed separately or together to block endogenous purinergic signaling in a biological system.

P1 adenosine receptors. A final maneuver of inclusion of adenosine deaminase, which converts active adenosine into inert inosine, would eliminate endogenous adenosine and any possibility of activating adenosine receptors. A cocktail of all three scavengers at approximately 0.1 unit/ml would likely eliminate all purinergic agonists from an experimental preparation.

### B. Global P2 and P1 Receptor Antagonists

An alternative maneuver to prevent autocrine or paracrine activation of purinergic receptors is to use nonselective or global antagonists that block most if not all P2 or P1 receptors. To verify the effect of the scavengers described above, this approach should also be performed to solidify evidence of autocrine or paracrine purinergic signaling affecting a biological end point. The best approach for broad blockade of P2 receptors is suramin. It is not perfect, in that a subset of P2X receptor channels is blocked poorly by this drug. In this case, pyridoxal phosphate-6-azophenyl 2,4-disulfonic acid (PPADS) can be used together with suramin in a cocktail. One caveat is that PPADS and TNP-ATP (trinitrophenyl-ATP), another P2X-selective agonist, change the color of the Ringer's solution and cannot be used in fluorescence experiments (see Chapters 3 and 4 in this volume for more information about pharmacology).

A similar approach can be taken for adenosine receptors. CPX (also known as DPCPX or 8-cyclopentyl-1,3-dipropylxanthine) is an antagonist of P1 receptors. The 50% inhibitory concentration ($IC_{50}$) for A1 receptors is less than 1 n$M$, while the $IC_{50}$ for A2 receptors is approximately 500 n$M$. As such, in most tissues that do not also express a functional A3 receptor, DPCPX at 1 $\mu M$ would antagonize A1 and A2 receptors that are normally expressed by most cells (see the chapter Molecular and Cell Biology of Adenosine Receptors, in this volume, for more information about receptor classification). This maneuver would effectively block autocrine or paracrine adenosine signaling endogenous to a cell or tissue preparation. This approach should also be taken to verify the role of endogenous adenosine signaling in a biological paradigm.

### IV. NOVEL ASSAYS TO DETECT ATP RELEASE FROM CELLS

Ultimately, direct measurement of extracellular ATP in solution is the preferred course of action to study whether a cell or tissue is releasing ATP in a physiological or biological manner. There has been a revolution in the

A

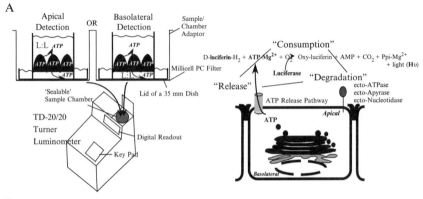

B     "Consume the ATP as It Is Released"          "Consume the ATP before It Is Degraded"

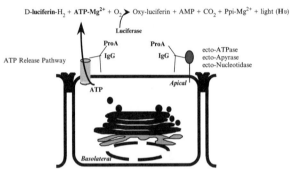

"Either Way, Consume the ATP in the Cell Surface Microenvironment"

C

"Image the ATP as It Is Released
In the AcinarSecretions via Confocal
Microscopy of the Bioluminescence"

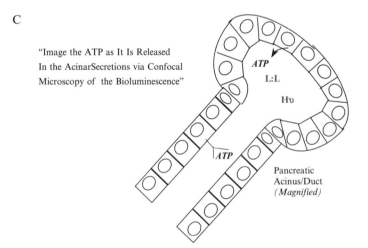

**FIGURE 4**   *(For legend, see opposite page.)*

establishment of assays that can detect extracellular ATP signaling in cells and tissues (Fig. 4). This has involved either adaptation of the luciferase: luciferin bioluminescence detection assay from the test tube to "sense" ATP release in biological preparations, the linkage of luciferase to cells via protein A and extracellular epitope antibodies to sense ATP release in the cell surface microenvironment, the development of the PC-12 pheochromo-cytoma cell as a "biosensor" that can measure ATP, and the invention of an atomic force microscopy probe coated with myosin to sense ATP.

### A. Adaptation of the Luciferase:Luciferin Bioluminescence Detection Assay to Study ATP Release from Cells and Cell Monolayers in Real Time

Firefly and *Renilla* luciferase and luciferin have been used for decades to measure ATP in preparations. Laboratories have taken aliquots of extracellular solution bathing cells and tissues, mixed these samples with luciferase and luciferin, and measured bioluminescence in a multistep process. This has yielded much interesting and novel information. However, a handful of laboratories have included the luciferase:luciferin detection reagent directly in the medium or solution bathing the cells, cell monolayer, or tissue and measured ATP release in a luminometer or by imaging of the bioluminescence in real time. Schwiebert and co-workers developed such an

---

FIGURE 4   Luciferase-based ATP release detection assays. (A) Luciferase-based detection of released ATP from epithelial and endothelial monolayers in real time. This schematic shows that the epithelial monolayer is bathed in a serum-free medium devoid of ATP, placed on the lid of a 35-mm dish on a sample chamber adaptor platform, and lowered into the luminescence chamber within the luminometer on that platform. ATP release (measured as bioluminescence) is measured in real time in consecutive 15–s intervals. The assay is "triangular" (see *right*) in that it depends on the rate of release and the relative consumption by the luciferase "sensor/detector" and ecto-ATPases that degrade the ATP before it can be sensed. (B) Luciferase-based detection of released ATP in real time, in which the luciferase "sensor" is bound or tethered to the cell surface via an antibody recognizing an extracellular epitope on a transmembrane protein and protein A linked to luciferase. Released ATP is then measured in the microenvironment just above the cell surface and, possibly, before it is degraded by ecto-ATPases. An added bonus would be tethering the conjugate bearing luciferase to the transmembrane protein that transported the ATP out of the cell (it would be detected as soon as it left the cell). (C) Luciferase-based detection of ATP release by infusion of a luciferase-containing medium into the acinus of a freshly dissected part of the pancreas to detect bioluminescence in real time by confocal imaging. The advantage is that this is an "*in vivo*-like" preparation in which mechanical, hypotonic, and carbachol-induced ATP release can be measured in real time as the acinus performs its normal function, secretion. L:L, Luciferase: luciferin.

assay for detecting or "sensing" ATP release in real time from nonpolarized cells in 35-mm culture dishes and, more importantly, for polarized epithelial and endothelial cell monolayers grown on 12-mm-diameter filter supports (Taylor *et al.*, 1998; Wilson *et al.*, 1999; Braunstein *et al.*, 2001; Schwiebert *et al.*, 2002a,b). The adaptation of the assay was described in a 1998 Special Communication in the *American Journal of Physiology: Cell Physiology* (Taylor *et al.*, 1998). The same real-time assay has been used extensively by Fitz and co-workers to study ATP release from hepatocytes and cholangiocytes (Roman *et al.*, 1999a,b, 2001; Feranchak *et al.*, 2000). Both laboratories have demonstrated that cells release ATP constitutively under basal or unstimulated conditions. Both laboratories have also shown that elevation of cyclic AMP, increases in cytosolic calcium, and hypotonic challenge trigger ATP release, while hypertonicity inhibits ATP release. Another obvious application of this technique was to show real-time differences in ATP release between non-CF and CF epithelial cell monolayers (Taylor *et al.*, 1998; Braunstein *et al.*, 2001) as well as between normal and polycystic kidney disease (PKD) epithelial cell monolayers (Wilson *et al.*, 1999; Schwiebert *et al.*, 2002b). Via this assay, it was shown that ATP release and signaling are lost in the apical microenvironment of CF monolayers versus controls, while ATP release and signaling in PKD monolayers were as great or greater in the apical microenvironment and greater in the basolateral environment. Another important application of the assay has been the dissection of ATP release mechanisms (described above). As long as pharmacological and cell biological maneuvers do not affect luciferase:luciferin detection activity, inhibitors of channel-, transporter-, and exocytic-driven ATP release can be used in real time to assess their relative roles in ATP release from cells, cell monolayers, and, ultimately, tissues.

## B. Adaptation of the Luciferase:Luciferin Bioluminescence Detection Assay to Perform Confocal Imaging of ATP Release from Cells and Tissues in Real Time

A groundbreaking study has applied luciferase:luciferin bioluminescence detection technology to a tissue preparation. Sorensen and Novak used confocal imaging of the bioluminescence generated from ATP being catalyzed by luciferase (Sorensen and Novak, 2001). They studied individually dissected pancreatic acini to measure ATP release and signaling in this "*in vivo*-like" preparation. They showed elegantly that mechanical stimuli, hypotonic challenge, and carbachol, an agonist that increases cytosolic calcium in this tissue, all increased ATP release. They were also able to show intracellular ATP store depletion through caged

ATP as well as quinacrine, which labelled a similar ATP pool. It was estimated that ATP release on cholinergic stimulation generated an ATP concentration of 9 $\mu M$ in the acinus. It was postulated that this acinus-derived ATP not only could modulate acinar cell function in an autocrine manner but could also diffuse to the pancreatic ducts in acinar secretions to stimulate P2 receptors. This laboratory as well as our own are trying to take this technology to more *in vivo*-like tissue preparations to relate the *in vitro* findings in cultured cells to freshly dissected tissues. One important issue is cell damage during the dissection. Soresen and Novak showed elegantly via multiple lines of evidence that this possibility was eliminated carefully in their work.

## C. Linkage of a Luciferase:Protein A Conjugate to an Antibody-Tagged Cell Surface Protein for the Study of ATP Release at the Cell Surface in Real Time

A limitation of the above-described methods of extracellular ATP "sensing" or detection is the fact that the luciferase:luciferin detection reagent is solubilized in the solution or medium; however, because of unstirred layers both in the cell cultures and cell monolayers *in vitro* and the freshly dissected tissues, the assay may miss ATP released into the microenvironment immediately above the cell surface. Dubyak and co-workers were aware of this issue and of the fact that the ATP might be degraded by ecto-ATPases (ecto-apyrases) on the cell surface before it could be detected in bioluminescence assays (Beigi *et al.*, 1999). As such, they used molecular biology to develop a conjugate that linked luciferase to protein A. This conjugate, through protein A chemistry, could be linked to any antibody that could be bound to an extracellular epitope of any cell surface antigen, usually a transmembrane protein. The Dubyak laboratory was interested in extracellular ATP signaling and how it modulated the aggregation mechanisms and the immune response in platelets, monocytes, and macrophages. They took advantage of the fact that many extracellular epitope antibodies have been created for cell surface antigens on hematopoietic cells. They showed that they could detect micromolar amounts of ATP being released under basal conditions and that inhibition of ecto-ATPases could prolong the bioluminescence signal. The long-term goals of the Dubyak laboratory are to understand how engagement of P2Y and/or P2X receptors by this autocrine or paracrine ATP signal affects the function of hematopoietic cells critical for the immune response. Theoretically, this luciferase:protein A conjugate could be targeted to any cell surface antigen that has a well-characterized and high-affinity antibody that recognizes an extracellular epitope. Targets may include

CFTR or other ABC transporters implicated in the facilitation of ATP release, CD39, and other ecto-ATPases where the competition between degradation and detection could be studied, or P2 receptors themselves to sense what the local ATP concentration is at the receptor when it is finally engaged and activated. Development of this reagent was an important advance in this area of research.

## D. The P2X Receptor-Expressing PC-12 Cell as a Biosensor for the Study of ATP Release from Cells and Tissues in Real Time

Okada and co-workers used much ingenuity in developing a technique by which PC-12, a pheochromocytoma cell line used predominantly to study nerve growth factor (NGF) signaling, was exploited to be used as a "biosensor" because it expressed the $P2X_2$ receptor and, possibly, other P2 receptors as well (Hazama *et al.*, 1998). The initial paradigm developed involved establishing a whole cell patch-clamp recording designed to record P2X receptor channel currents. The PC-12 cells were then detached from the substrate (PC-12 cells do not attach well to tissue culture plastic as it is), and a single PC-12 cell, still in whole cell configuration, was moved into close proximity to a cultured cell of interest. If P2X receptor currents opened and were recorded, then this would suggest that the cultured cell was releasing ATP. Okada and colleagues did this originally with pancreatic beta cells to show that micromolar ATP was released along with insulin in response to glucose (Hazama *et al.*, 1998). Okada and co-workers have also applied this assay to non-CF versus CF cells as well as intestinal 407 cells (Hazama *et al.*, 2000). They have shown that ATP release is impaired in CF cells or in cells lacking CFTR when compared with controls.

Bell and co-workers, collaborating closely with Okada and colleagues, have extended this assay to study macula densa signaling in an *in vivo*-like preparation (Bell *et al.*, 2000) (Fig. 5). Peti-Peterdi in Bell's group dissected a nephron preparation in which the cortical thick ascending limb (cTAL) with a small patch of macula densa cells is attached to the glomerulus of the same nephron. Using the PC-12 biosensor cell either in whole cell patch-clamp mode or loaded with Fura-2 to measure calcium influx through P2X receptor channels, this group has shown that ATP is released from the serosal or basolateral aspect of the macula densa cells. This ATP release is triggered by changes in lumenal NaCl concentration in the cTAL. Bell's group argues that the macula densa cells "sense" lumenal NaCl concentration, possibly by subsequent changes in macula densa cell volume, triggering ATP release. Patch-clamp evidence suggests the presence of a maxi anion channel, with significant permeability to ATP, as the principal ATP release mechanism in

**Examining the ATP Release Mechanism in the Basolateral Membrane of Macula Densa**

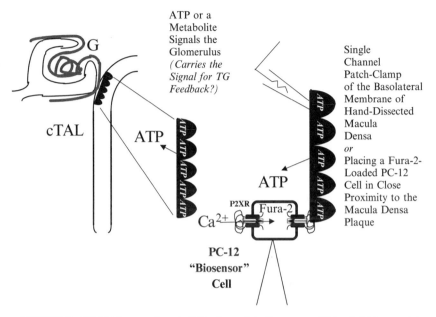

**FIGURE 5**  PC-12 biosensor-based ATP release detection assays. Either held by a patch-clamp electrode in a whole cell patch-clamp configuration linked to a headstage and patch-clamp amplifier or held by a manipulating pipette after being loaded with the calcium-sensitive dye, Fura-2/AM, this PC-12 cell is now a biosensor for ATP, because it expresses a $P2X_2$ receptor (and, possibly, other P2XR subtypes) that conducts cation current and/or allows influx of calcium from extracellular stores on binding of its extracellular ligand, ATP. Any ATP binding and stimulating of this P2X receptor on the PC-12 cell would, therefore, have been released by the cells of interest (in this example, the plaque of macula densa cells normally found in the cortical thick ascending limb in close juxtaposition with the glomerulus of the same nephror). G, Glomerulus; cTAL, cortical thick ascending limb; TG, tubuloglomerular feedback.

macula densa. As with the atomic force microscopy method below, the PC-12 cell biosensor method provides a probe that could be used in many other freshly dissected tissue culture preparations.

### E. A Myosin-Coated Atomic Force Microscopy Probe to Detect ATP Release from Cells and Tissues in Real Time

We have been able to develop a functional biosensor for real-time ATP measurements that allows for the direct detection of ATP on the surface of living cells (Schneider *et al.*, 1999). This sensor incorporates ultrahigh surface topography measurements, using the principles of atomic force microscopy (AFM) and, at the same time, allows us to scan for changes in ATP concentrations only angstroms from the surface of cells. With the implementation of this new technology, we now have a functional "bioprobe" that allows us to obtain accurate and reproducible recordings from living cells of both the topography and surface ATP concentrations in real time. By developing the concept of a biosensor attached to the surface of an AFM cantilever, it is now possible to generate specific biosensors that can probe the surface environment of living cells under resting and experimental conditions and record changes in both the topography and the microenvironment surrounding the cell membrane.

The AFM gives ultrahigh-resolution images of surface topography with potential resolutions down to the atomic scale. The rate-limiting step in developing topography profiles of cells and other biological structures is the surface that is being scanned, and the diameter of the scanning tip. Because the probes use the complex interactions of van der Waal forces from the structure being examined to both attract (when the tip is far from the surface) and repel (when the tip is very close to the surface) the tip, it is possible to gain high-resolution images of the surfaces of living cells. However, because cells are made of fluid-filled compartments, the interactions of the tip with the surface can lead to some distortions of both the membrane and the cytosol at ultrahigh resolution (attempts to scan at the atomic scale). In an attempt to counter these biological distortions as much as possible, minimal force (in nanonewtons) is applied to the tip to maintain a contact profile with the surface. We chose to take advantage of this technology to construct a functional biosensor that combined the principals of the van der Waal forces used in causing tip deflections of the scanning probe of the AFM with the biochemical interactions of a myosin molecule with ATP.

## 1. Construction of the Sensor

Individual cantilevers were selected and pretreated by immersion of the AFM cantilever ($Si_3Ni_4$) silicon tip into a solution containing the S1 chain ($1 \ \mu M$ for 24 h) of the myosin molecule. We chose this tip immersion technique so that the biologically reactive area was localized to the tip of the cantilever that would be coming into closest proximity to the surface of the cell and would have the best chance to respond to localized areas of ATP along the surface of the cell.

To determine that the immersion created a functional biosensor unit, all tips were first tested in a perfusion chamber while scanning with the AFM. After establishing contact with the chamber bottom and developing a topographical scan, we raised the tip off the surface and continued to scan, using a correction method that allowed us to obtain scans in both a contact and height mode. As the tip was scanning, solutions were exchanged in the chamber so that both ATP-containing and non-ATP-containing solutions entered the chamber. We conducted the studies at constant temperature to remove potential interference with the tip. The results of these studies demonstrated that, on introduction of ATP into the chamber, ATP had no effect on noncoated tips. Only tips with a myosin coating responded to ATP by giving a positive deflection away from the surface (see Fig. 6). This test allowed us to determine that we had a "functionalized" tip prior to attempting to scan the surface of living cells. Also illustrated in Fig. 6, the cells, which were grown on glass coverslips, were introduced into the chamber. A functionalized tip was added to the scanner, and we proceeded to collect images of the cells prior to assaying for ATP. Because the tip could be forced to maintain contact with the cell surface, it was possible to obtain images of the cells and determine that we had an intact tip.

## 2. ATP Measurements in Cells Containing Functional CFTR Protein

We have been able to demonstrate that when S9 lung cells, CF airway cells stably complemented with wild-type CFTR, are grown to confluency on glass coverslips and are scanned with our biosensor, we were able to detect localized changes in ATP concentration (Schneider *et al.*, 1999). These changes in ATP appeared as lines across the surface of the cells that were actively secreting ATP. This aspect was intriguing to us, as it might have been expected that the cell would have single "hot spots" where the ATP was being released. By calculating the small volume that the tip actually occupies, we could determine that we would need about four molecules of ATP to be present to activate the tip (Schneider, *et al.*, 1999). We were able to determine that, as the tip is not static but rather is constantly moving, when an area of ATP was detected there was a rise and then a fall in the profile curve. This rise and fall resulted in a single scan line of ATP being detected. When we reduced the area of interest to focus only on a small area ($2 \times 2 \ \mu$m) within the initial line of activity, we found that we had continuous stimulation of the tip.

As it is now possible to attach biologically active molecules to the tip of the AFP, it may now be possible to develop a series of biosensors that can detect other reactive effluents from cells. We have begun the development of a tip that would allow us to incorporate luciferase on the surface of the tip and, by the relative changes in fluorescence of the tip as well as the deflection due to the heat of the luciferace reaction following exposure to ATP, directly calculate the level of ATP being released per unit time. This new system will require us to simultaneously scan with the AFM while collecting

A

## Myosin-coated Atomic Force Microscopy Probe "Biosensor"
## BioScope

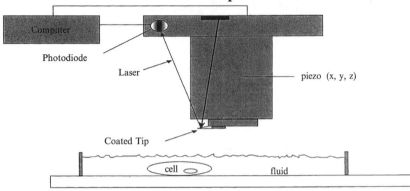

B

### Myosin-coated Atomic Force Microscopy Probe "Biosensor"

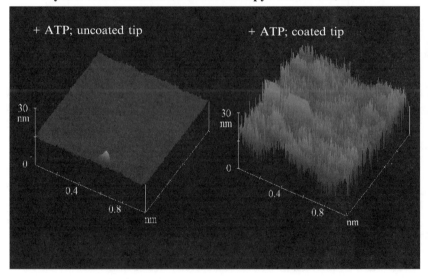

**FIGURE 6**    *(For legend, see opposite page.)*

high-resolution video images that will give us the changes in the fluorescent signal that can be calibrated to give concentrations of ATP. Another direction would be to create antibody-doped tips that would be generated against the extracellular domain of CFTR. With this technique, we would

C

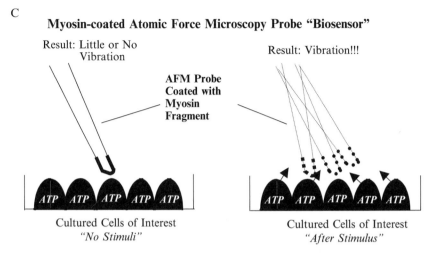

**Myosin-coated Atomic Force Microscopy Probe "Biosensor"**

**FIGURE 6**  Myosin-coated atomic force microscopy probe biosensor detection method. (A) Schematic of the Geibel and co-workers atomic force microscopy-based "bioprobe" system. (B) Three-dimensional plots of sample data generated from an S1 myosin-coated probe versus a noncoated probe. (C) Schematic of paradigm used to study ATP release from CF airway epithelial cells versus the same CF cells complemented with wild-type CFTR.

potentially be able to look at areas of attraction as the antibody binds to the protein and these negative deflections would give us some idea as to the density of the protein on the surface of the cell. A final tip that we would like to create would link a $Cl^-$-sensitive probe to the tip so that we could determine the sites of $Cl^-$ release in real time and generate a profile of channel density per area. As we presently have a great deal of data with patch-clamp recordings, direct correlations of the biosensor could be made. Should this technique work, we could then examine perturbations to the extracellular and intracellular environment that might change channel density and could correlate this over time with the biosensor.

## V. FUTURE DIRECTIONS

It is not difficult to anticipate where the field of ATP release is going. Soon, multiple pathways of ATP release that involve conductive anion channel-mediated transport, nonconductive facilitated transport by permease or exchanger, and ATP-filled vesicles will emerge in several different biological systems. In the other chapters of this volume, the reader will see that some of these mechanisms are already being worked out in many biological systems. A hint of another future direction comes from the work of Novak and co-workers and Bell and co-workers. Application of the many different ATP detection reagents and assays to *in vivo*-like preparations of freshly dissected tissue, especially transgenic animals or models, is necessary. Having said that, the real-time assays in *in vitro* systems will continue to provide novel information that can be adapted or applied to tissue preparations. Finally, the revelation that endogenous and constitutive ATP release from cells or tissues occurs under basal conditions (Roman *et al.*, 1999b; Schwiebert, *et al.*, 2002a), and that signaling affects the set point for many different signal transduction pathways, underscores the need to address autocrine transduction purinergic signaling in all biological paradigms. We have learned that how epithelial cells are handled when they are lysed to prepare membrane proteins for Western blot analysis of epithelial P2Y and P2X receptors can affect the overall expression of these proteins in the plasma membrane. Extensive washing of the preparation can release ATP, which then binds and activates the receptors and may cause their internalization and loss from the membrane protein pool. If ATP receptor-triggered signal transduction is being studied, extensive washing before lysis of cells to study intracellular signal transduction molecules could already activate those receptors and down-stream effectors during the washes, preventing clean analysis of the signal transduction pathways.

The intent of this chapter was to convince the reader that ATP release is a biological process of physiologic importance, not merely a by-product of cell damage. The study of ATP receptors has leaped ahead of the study of ATP release, without any knowledge of where the ATP that would activate the many different P2 receptor subtypes was coming from. However, the study of purinergic signaling and its detection is advancing rapidly. The most meaningful future work will be the integrated analysis of ATP release processes, the relevant ATP receptors in that same cell or tissue, the possible involvement of ATP metabolites (ADP and adenosine) in a cell or tissue microenvironment, and the effect of ATP, via its receptors, on cell function. In many respects, this integrated analysis paradigm is followed by many of the authors of the subsequent chapters in this volume.

### Acknowledgments
Work on ATP release, its mechanisms, a real-time luciferase-based assay for detection, and the role of ATP signaling in CF and PKD by E.M.S. has been funded by the CF Foundation, the American Heart Association (Southern Research Consortium), and the Polycystic Kidney Disease Research Foundation. Presently, E.M.S. is funded to study ATP release mechanisms in epithelial cells by an NIH grant (R01 DK54367).

## References

Abraham, E. H., Prat, A. G., Gerweck, L., Seneveratne, T., Arceci, R. J., Kramer, R., Guidotti, G., and Cantiello, H. F. (1993). The multidrug resistance (mdr1) gene product functions as an ATP channel. *Proc. Natl. Acad. Sci. USA* **90**, 312–316.

Abraham, E. H., Sterling, K. M., Kim, R. J., Salikhova, A. Y., Huffman, H. B., Crockett, M. A., Johnston, N., Parker, H. W., Boyle, W. E., Jr., Hartov, A., Demidenko, E., Efird, J., Kahn, J., Grubman, S. A., Jefferson, D. M., Robson, S. C., Thakar, J. H., Lorico, A., Rappa, G., Sartorelli, A. C., and Okunieff, P. (2001). Erythrocyte membrane ATP binding cassette (ABC) proteins: MRP1 and CFTR as well as CD39 (ecto-apyrase) involved in RBC ATP transport and elevated blood plasma ATP of cystic fibrosis. *Blood Cells Mol. Dis.* **27**, 165–180.

Bankston, L. A., and Guidotti, G. (1996). Characterization of ATP transport into chromaffin granule ghosts: Synergy of ATP and serotonin accumulation in chromaffin granule ghosts. *J. Biol. Chem.* **271**, 17132–17138.

Beigi, R., Kobatake, E., Aizawa, M., and Dubyak, G. R. (1999). Detection of local ATP release from activated platelets using cell surface-attached firefly luciferase. *Am. J. Physiol. Cell Physiol.* **276**, C267–C278.

Bell, P. D., Lapointe, J.-Y., Subirov, R., Hayashi, S., Peti-Peterdi, J., Manabe, K., Kovacs, G., and Okada, Y. (2003). Macula densa cells signaling involves ATP release through a maxi anion channel. *Proc. Natl. Acad. Sci.* (In press).

Benz, R., Wojtczak, L., Bosch, W., and Brdiczka, D. (1988). Inhibition of adenine nucleotide transport through the mitochondrial porin by a synthetic polyanion. *FEBS Lett.* **231**, 75–80.

Bergfeld, G. R., and Forrester, T. (1992). Release of ATP from human erythrocytes in response to a brief period of hypoxia and hypercapnia. *Cardiovasc. Res.* **26**, 40–47.

Braunstein, G. M., Roman, R. M., Clancy, J. P., Kudlow, B. A., Taylor, A. L., Shylonsky, V. G., Jovov, B., Peter, K., Jilling, T., Ismailov, I. I., Benos, D. J., Schwiebert, L. M., Fitz, J. G., and Schwiebert, E. M. (2001). Cystic fibrosis transmembrane conductance regulator facilitates ATP release by stimulating a separate ATP release channel for autocrine control of cell volume regulation. *J. Biol. Chem.* **276**, 6621–6630.

Brustovetsky, N., Becker, A., Klingenberg, M., and Bamberg, E. (1996). Electrical currents associated with nucleotide transport by the reconstituted mitochondrial ADP/ATP carrier. *Proc. Natl. Acad. Sci. USA* **93**, 664–668.

Burnstock, G., Campbell, G., Satchell, D., and Smythe, A. (1970). Evidence that adenosine triphosphate or a related nucleotide is the transmitter substance released by non-adrenergic inhibitory nerves in the gut. *Br. J. Pharmacol.* **40**, 668–688.

Donaldson, S. H., Lazarowski, E. R., Picher, M., Knowles, M. R., Stutts, M. J., and Boucher, R. C. (2000). Basal nucleotide levels, release, and metabolism in normal and cystic fibrosis airways. *Mol. Med.* **6**, 969–982.

Edward, J., Sprung, R., Sprague, R., and Spence, D. (2001). Chemiluminescence detection of ATP release from red blood cells upon passage through microbore tubing. *Analyst.* **126**, 1257–1260.

Feranchak, A. P., Fitz, J. G., and Roman, R. M. (2000). Volume-sensitive purinergic signaling in human hepatocytes. *J. Hepatol.* **33**, 174–182.

Frame, M. K., and de Feijter, A. W. (1997). Propagation of mechanically induced intercellular calcium waves via gap junctions and ATP receptors in rat liver epithelial cells. *Exp. Cell Res.* **230**, 197–207.

Fujise, H., Higa, K., Kanemaru, T., Fukuda, M., Adragna, N. C., and Lauf, P. K. (2001). Glutathione (GSH) depletion, KCl cotransport, and regulatory volume decrease in high K/high GSH dog red blood cells. *Am. J. Physiol. Cell Physiol.* **281**, C2003–C2009.

Gordon, J. L. (1986). Extracellular ATP: Effects, sources, and fates. *Biochem. J.* **233**, 309–319.

Grygorczyk, R., and Hanrahan, J. W. (1997). CFTR-independent ATP release from epithelial cells triggered by mechanical stimuli. *Am. J. Physiol. Cell Physiol.* **272,** C1058–C1066.

Gualix, J., Pintor, J., and Miras-Portugal, M. T. (1999). Characterization of nucleotide transport into rat brain synaptic vesicles. *J. Neurochem.* **73,** 1098–1104.

Guillen, E., and Hirschberg, C. B. (1995). Transport of adenosine triphosphate into endoplasmic reticulum proteoliposomes. *Biochemistry* **34,** 5472–5476.

Hazama, A., Hayashi, S., and Okada, Y. (1998). Cell surface measurement of ATP release from single pancreatic $\beta$ cells using a novel biosensor technique. *Pflugers Arch.* **437,** 31–35.

Hazama, A., Shimizu, T., Ando-Akatsuka, Y., Hayashi, S., Tanaka, S., Maeno, E., and Okada, Y. (1999). Swelling-induced, CFTR-independent ATP release from a human epithelial cell line: Lack of correlation with volume-sensitive chloride channels. *J. Gen. Physiol.* **114,** 525–533.

Hazama, A., Fan, H. T., Abdullaev, I., Maeno, E., Tanaka, S., Ando-Akatsuka, Y., and Okada, Y. (2000). Swelling-activated, cystic fibrosis transmembrane conductance regulator-augmented ATP release and chloride conductances in murine C127 cells. *J. Physiol.* **523,** 1–11.

Hoffman, J. F. (1997). ATP compartmentalization in human erythrocytes. *Curr. Opin. Hematol.* **4,** 112–115.

Homolya, L., Steinberg, T. H., and Boucher, R. C. (2000). Cell to cell communication in response to mechanical stress via bilateral release of ATP and UTP in polarized epithelia. *J. Cell Biol.* **150,** 1349–1360.

Huang, P., Lazarowski, E. R., Tarran, R., Milgram, S. L., Boucher, R. C., and Stutts, M. J. (2001). Compartmentalized autocrine signaling to cystic fibrosis transmembrane conductance regulator at the apical membrane of airway epithelial cells. *Proc. Natl. Acad. Sci. USA* **98,** 14120–14125.

Jackson, P. S., and Strange, K. (1995). Characterization of the voltage-dependent properties of a volume-sensitive anion conductance. *J. Gen. Physiol.* **105,** 661–667.

Jiang, Q., Mak, D., Devidas, S., Schwiebert, E. M., Bragin, A., Zhang, Y., Skach, W. R., Guggino, W. B., Foskett, J. K., and Engelhardt, J. F. (1998). CFTR associated ATP release is controlled by a "chloride sensor" within the CFTR pore. *J. Cell Biol.* **143,** 645–657.

Knight, G. E., Bodin, P., De Gront, W. C., and Burnstock, G. (2002). ATP is released from guinea pig ureter epithelium on distension. *Am. J. Physiol. Renal Physiol.* **282,** F281–F288.

Lader, A. S., Xiao, Y. F., O'Riordan, C. R., Prat, A. G., Jackson, G. R., Jr., and Cantiello, H. F. (2000). Cyclic AMP activates an ATP-permeable pathway in neonatal rat cardiac myocytes. *Am. J. Physiol. Cell Physiol.* **279,** C173–C187.

Li, C., Ramjeesingh, M., and Bear, C. E. (1996). Purified cystic fibrosis transmembrane conductance regulator (CFTR) does not function as an ATP channel. *J. Biol. Chem.* **271,** 11623–11626.

Light, D. B., Schwiebert, E. M., Fejes-Toth, G., Naray-Fejes-Toth, A., Karlson, K. H., McCann, F. V., and Stanton, B. A. (1990). Chloride channels in the apical membrane of cortical collecting duct cells. *Am. J. Physiol. Renal Physiol.* **258,** F273–F280.

Light, D. B., Capes, T. L., Gronau, R. T., and Adler, M. R. (1999). Extracellular ATP stimulates volume decrease in *Necturus* red blood cells. *Am. J. Physiol. Cell Physiol.* **277,** C480–C491.

Linsdell, P., and Hanrahan, J. W. (1998). Adenosine triphosphate-dependent asymmetry of anion permeation in the cystic fibrosis transmembrane conductance regulator chloride channel. *J. Gen. Physiol.* **111,** 601–614.

Mangel, A. W., Raymond, J. R., and Fitz, J. G. (1993). Regulation of high conductance anion channels by G proteins and 5-HT1A receptors in CHO cells. *Am. J. Physiol. Renal Physiol.* **264,** F490–F495.

Maroto, R., and Hamill, O. P. (2001). Brefeldin A block of integrin-dependent mech-anosensitive ATP release from *Xenopus* oocytes reveals a novel mechanism of mechanotransduction. *J. Biol. Chem.* **276**, 23867–23872.

McGill, J. M., Gettys, T. W., Basavappa, S., and Fitz, J. G. (1993). GTP-binding proteins regulate high conductance anion channels in rat bile duct epithelial cells. *J. Membr. Biol.* **133**, 253–261.

Mitchell, C. H. (2001). Release of ATP by a human pigment epithelial cell line: Potential for autocrine stimulation through subretinal space. *J. Physiol.* **534**, 193–202.

Mitchell, C. H., Carre, D. A., McGlinn, A. M., Stone, R. A., and Civan, M. M. (1998). A release mechanism for stored ATP in ocular ciliary epithelial cells. *Proc. Natl. Acad. Sci. USA* **95**, 7174–7178.

Musante, L., Zegarra-Moran, O., Montaldo, P. G., Ponzoni, M., and Galietta, L. J. (1999). Autocrine regulation of volume-sensitive anion channels in airway epithelial cells by adenosine. *J. Biol. Chem.* **274**, 11701–11707.

Ostrom, R. S., Gregorian, C., and Insel, P. A. (2000). Cellular release of and response to ATP as key determinants of the set-point of signal transduction pathways. *J. Biol. Chem.* **275**, 11735–11739.

Prat, A. G., Reisin, I. L., Ausiello, D. A., and Cantiello, H. F. (1996). Cellular ATP release by the cystic fibrosis transmembrane conductance regulator. *Am. J. Physiol. Cell Physiol.* **270**, C538–C545.

Reddy, M. M., Quinton, P. M., Haws, C., Wine, J. J., Grygorczyk, R., Tabcharani, J. A., Hanrahan, J. W., Gunderson, K. L., and Kopito, R. R. (1996). Failure of the cystic fibrosis transmembrane conductance regulator to conduct ATP. *Science* **271**, 1876–1879.

Reisin, I. L., Prat, A. G., Abraham, E. H., Amara, J. F., Gregory, R. J., Ausiello, D. A., and Cantiello, H. F. (1994). The cystic fibrosis transmembrane conductance regulator is a dual ATP and chloride channel. *J. Biol. Chem.* **269**, 20584–20591.

Reymann, S., Florke, H., Heiden, M., Jakob, C., Stadtmuller, U., Steinacker, P., Lalk, V. E., Pardowitz, I., and Thinnes, F. P. (1995). Further evidence for multitopological localization of mammalian porin (VDAC) in the plasmalemma forming part of a chloride channel complex affected in cystic fibrosis and encephalomyopathy. *Biochem. Mol. Med.* **54**, 75–87.

Roman, R. M., Feranchak, A. P., Davison, A. K., Schwiebert, E. M., and Fitz, J. G. (1999a). Evidence for $Gd^{3+}$ as an inhibitor of membrane ATP permeability and purinergic signaling. *Am. J. Physiol. Gastrointest. Liver Physiol.* **277**, G1222–G1230.

Roman, R. M., Feranchak, A. P., Salter, K. D., Wang, Y., and Fitz, J. G. (1999b). Endogenous ATP regulates chloride secretion in cultured human and rat biliary epithelial cells. *Am. J. Physiol. Gastrointest. Liver Physiol.* **276**, G1391–G1400.

Roman, R. M., Lomri, N., Braunstein, G., Feranchak, A. P., Simeoni, L. A., Davison, A. K., Mechetner, E., Schwiebert, E. M., and Fitz, J. G. (2001). Multidrug resistance P glycoproteins regulate cellular ATP permeability. *J. Membr. Biol.* **183**, 165–173.

Sabirov, R. Z., Dutta, A. K., and Okada, Y. (2001). Volume-dependent large-conductance anion channel as a pathway for swelling-induced ATP release. *J. Gen. Physiol.* **118**, 251–266.

Sauer, H., Hescheler, J., and Wartenberg, M. (2000). Mechanical strain-induced calcium waves in human prostate cancer cells are propagated via ATP release and purinergic receptor activation. *Am. J. Physiol. Cell Physiol.* **279**, C295–C307.

Schneider, S. W., Egan, M. E., Vena, B. P., Guggino, W. P., Oberleithner, H., and Geibel, J. P. (1999). Continuous detection of extracellular ATP on living cells by using atomic force microscopy. *Proc. Natl. Acad. Sci. USA* **96**, 12180–12185.

Schwiebert, E. M. (1999). ABC transporter-facilitated ATP transport. *Am. J. Physiol. Cell Physiol.* **276**, C1–C8.

Schwiebert, E. M., Mills, J. W., and Stanton, B. A. (1994). Actin-based cytoskeleton regulates a chloride channel and cell volume in a renal cortical collecting duct cell line. *J. Biol. Chem.* **269**, 7081–7089.

Schwiebert, E. M., Egan, M. E., Hwang, T. H., Fulmer, S. B., Allen, S. S., Cutting, G. R., and Guggino, W. B. (1995). CFTR regulates outwardly rectifying chloride channels through an autocrine mechanism involving ATP. *Cell* **81,** 1063–1073.

Schwiebert, L. M., Rice, W. C., Kudlow, B. A., Taylor, A. L., and Schwiebert, E. M. (2002a). Extracellular ATP signaling and P2X nucleotide receptors in monolayers of primary human vascular endothelial cells. *Am. J. Physiol. Cell Physiol.* **282,** C282–C301.

Schwiebert, E. M., Wallace, D. P., Braunstein, G. M., King, S. R., Peti-Peterdi, J., Hanaoka, K., Guggino, W. B., Guay-Woodford, L. M., Bell, P. D., Sullivan, L. P., Grantham, J. J., and Taylor, A. L. (2002b). Autocrine extracellular purinergic signaling in epithelial cells derived from polycystic kidneys. *Am. J. Physiol. Renal Physiol.* **282,** F763–F775.

Shinozuka, K., Tanaka, N., Kawasaki, K., Mizuno, H., Kubota, Y., Nakamura, K., Hashimoto, M., and Kunitomo, M. (2001). Participation of ATP in cell volume regulation in the endothelium after hypotonic stress. *Clin. Exp. Pharmacol. Physiol.* **28,** 799–803.

Sorensen, C. E., and Novak, I. (2001). Visualization of ATP release in pancreatic acini in response to cholinergic stimulus: Use of fluorescent probes and confocal microscopy. *J. Biol. Chem.* **276,** 32925–32935.

Sprague, R. S., Ellsworth, M. L., Stephenson, A. H., Kleinhenz, M. E., and Lonigro, A. J. (1998). Deformation-induced ATP release from red blood cells requires CFTR activity. *Am. J. Physiol. Heart Circ. Physiol.* **275,** H1726–H1732.

Sprague, R. S., Ellsworth, M. L., Stephenson, A. H., and Lonigro, A. J. (2001a). Participation of cAMP in a signal transduction pathway relating erythrocyte deformation to ATP release. *Am. J. Physiol. Cell Physiol.* **281,** C1158–C1164.

Sprague, R. S., Stephenson, A. H., Ellsworth, M. L., Keller, C., and Lonigro, A. J. (2001b). Impaired release of ATP from red blood cells of humans with primary pulmonary hypertension. *Exp. Biol. Med. (Maywood)* **226,** 434–439.

Sugita, M., Yue, Y., and Foskett, J. K. (1998). CFTR chloride channels and CFTR-associated ATP channels: Distinct pores regulated by common gates. *EMBO J.* **17,** 898–908.

Taylor, A. L., Kudlow, B. A., Marrs, K. L., Gruenert, D. C., Guggino, W. B., and Schwiebert, E. M. (1998). Bioluminescence detection of ATP release mechanisms in epithelia. *Am. J. Physiol. Cell Physiol.* **275,** C1391–C1406.

Thinnes, F. P., Hellman, K. P., Hellmann, T., Merker, R., Brockhaus-Pruchniewicz, U., Schwarzer, C., Walter, G., Gotz, H., and Hilschmann, N. (2000). Studies of human porin XXII: Cell membrane integrated human porin channels are involved in regulatory volume decrease (RVD) of HeLa cells. *Mol. Genet. Metab.* **69,** 331–337.

Walsh, D. E., Harvey, B. J., and Urbach, V. (2000). CFTR regulation of intracellular calcium in normal and cystic fibrosis human airway epithelia. *J. Membr. Biol.* **177,** 209–219.

Wang, Y., Roman, R., Lidofsky, S. D., and Fitz, J. G. (1996). Autocrine signaling through ATP release represents a novel mechanism for cell volume regulation. *Proc. Natl. Acad. Sci. USA* **93,** 12020–12025.

Watt, W. C., Lazarowski, E. R., and Boucher, R. C. (1998). Cystic fibrosis transmembrane regulator-independent release of ATP: Its implications for the regulation of P2Y2 receptors in airway epithelia. *J. Biol. Chem.* **273,** 14053–14058.

Wilson, P. D., Hovater, J. S., Casey, C. C., Fortenberry, J. A., and Schwiebert, E. M. (1999). ATP release mechanisms in primary cultures of epithelia derived from the cysts of polycystic kidneys. *J. Am. Soc. Nephrol.* **10,** 218–229.

Yamamoto, K., Korenaga, R., Kamiya, A., and Ando, J. (2000). Fluid shear stress activates calcium influx into human endothelial cells via P2X4 purinoceptors. *Circ. Res.* **87,** 385–391.

Zeng, H., Liu, G., Rea, P. A., and Kruh, G. D. (2000). Transport of amphipathic anions by human multidrug resistance protein 3. *Cancer Res.* **60,** 4779–4784.

# CHAPTER 3

# Molecular and Biological Properties of P2Y Receptors

**Eduardo R. Lazarowski**

Department of Medicine, University of North Carolina School of Medicine, Chapel Hill, North Carolina 27599

## I. INTRODUCTION

Some time has elapsed since the proposal by Burnstock of the existence of purinergic terminals and purinergic receptors was expanded to reflect the notion that a subset of $P_{2Y}$- and $P_{2X}$-purinergic receptors could account for the physiological actions of extracellular adenine nucleotides (Burnstock, 1972, 1978; Burnstock and Kennedy, 1985). The $P_{2Y}$-purinergic moniker was adopted for the receptor that mediates vasodilatation and relaxation of visceral smooth muscles in response to the base-modified adenine nucleotide 2-methylthio-ATP (2MeSATP), as opposed to the $P_{2X}$-purinergic receptor that mediates vasoconstriction and smooth muscle contraction in response to phosphate-modified adenine nucleotides such as $\alpha,\beta$-methylene ATP ($\alpha,\beta$-MetATP).[1] During subsequent years it became evident that (1) $P_{2Y}$-purinergic receptor-mediated actions were secondary to the activation of guanine nucleotide-binding proteins (G proteins), and that (2) additional nucleotide-activated G protein-coupled receptors exist that, as a family now denoted with the P2Y acronym, transduce the actions of extracellular adenine and uridine nucleotides, dinucleotide polyphosphates, and nucleotide sugars into biological responses in nearly all tissues. P2Y receptor-mediated responses include neural signaling transmission and reactive gliosis, modulation of cardiomyocyte glucose transport, vascular tone and vascular smooth muscle proliferation, platelet aggregation, immunocyte differentiation, granulocyte activation, and epithelial electrolyte transport. In the present chapter, the most distinguishing properties of P2Y receptors are reviewed, focusing attention on agonist specificities, second-messenger production, and downstream signaling revealed by studies with recombinant proteins. The relationship between the biological actions of extracellular nucleotides on native tissues and cloned P2Y receptors is discussed in cases in which compelling evidence has been obtained. Excellent reviews covering different aspects of P2 receptors have been published and they have been a helpful source of information for the present work (Harden *et al.*, 1995; King *et al.*, 1998; Ralevic and Burnstock, 1998; Communi *et al.*, 2000a).

## II. OVERVIEW OF THE PHARMACOLOGICAL PROPERTIES OF P2Y ZRECEPTORS

### A. Nomenclature and Structural Features

The incorporation of molecular biology approaches has been a key factor in the identification not only of G protein-coupled nucleotide receptors previously revealed by functional criteria, but also receptors whose existence was not predicted by pharmacological studies. However, the impressive growth to more than a dozen gene products tentatively assigned to the P2Y receptor family was not always accompanied by meticulous evaluation through functional analysis, and nomenclatural confusion has resulted. In this regard, the International Union of Pharmacology Nomenclature

---

[1]For specific reviews on P2X purinergic receptors, the reader is referred to the two chapters by Egun and Voigt and by Taylor and Schwiebert in this volume.

Committee has recommended that the notation P2Y be used to denote cloned functionally characterized G protein-coupled mammalian nucleotide receptors, and that lower case notation (p2y) be used both for nonmammalian nucleotide receptors and for cloned orphan receptors with predicted primary structures similar to the P2Y receptors but that have not been demonstrated to be functional nucleotide-activated receptors (Vanhoutte et al., 1996; King et al., 2000).

The G protein-coupled P2Y receptor family is composed of seven mammalian species whose primary amino acid sequences and pharmacological profiles for agonist-mediated second-messenger production have been unambiguously defined. They are the $P2Y_1$, $P2Y_2$, $P2Y_4$, $P2Y_6$, $P2Y_{11}$, $P2Y_{12}$, and an additional species that is activated by UDP-glucose, and referred to hereafter as $P2Y_{UDPG}$ receptor (Table I). Sequence analyses revealed that P2Y receptor transcripts encode peptides of 308–377 amino acids with predicted molecular masses of 41–53 kDa after glycosylation. The coding sequences of the $P2Y_1$, $P2Y_2$, $P2Y_4$, $P2Y_6$, $P2Y_{11}$, $P2Y_{12}$, and $P2Y_{UDPG}$ receptors exhibit an overall 20 to 50% homology at the amino acid level within receptor subtypes. The highest degree of homology is observed between $P2Y_2$ and $P2Y_4$ receptors (54%), and between $P2Y_{12}$ and $P2Y_{UDPG}$ receptors (44%). More than 70% homology is observed within

**TABLE I**

Nucleotide Selectivity and Second-Messenger Production of Seven Cloned Human P2Y Receptors

| Receptor | Agonists | Antagonists | Signaling |
|---|---|---|---|
| $P2Y_1$ | 2MeSADP > 2MeSATP > ADP > ATP | MRS2179 > A3P5PS > PPADS | $G_q \to PLC\text{-}\beta \to DAG + InsP_3/Ca^{2+}$ |
| $P2Y_2$ | UTP = ATP > UTP$\gamma$S, Up$_4$U > ATP$\gamma$S > Ap$_4$A | Suramin | $G_q \to PLC\text{-}\beta \to DAG + InsP_3/Ca^{2+}$ |
| $P2Y_4$ | UTP > UTP$\gamma$S, Up$_4$U | Reactive Blue 2 | $G_q \to PLC\text{-}\beta \to DAG + InsP_3/Ca^{2+}$ |
| $P2Y_6$ | UDP $\gg$ ADP | Reactive Blue 2, PPADS, suramin | $G_q \to PLC\text{-}\beta \to DAG + InsP_3/Ca^{2+}$ |
| $P2Y_{11}$ | ARC67085 > ATP$\gamma$S > BzATP > dATP > ATP $\gg$ 2MeSATP | Suramin | $G_q \to PLC\text{-}\beta \to DAG + InsP_3/Ca^{2+}$ |
| $P2Y_{12}$ | 2MeSADP>ADP | 2MeSAMP, C1330-7, ARC67085, clopidogrel | $G_s \to AC \to \uparrow cAMP$ $G_i \to AC \to \downarrow cAMP$ |
| $P2Y_{UDPG}$ | UDP-glucose > UDP-galactose | | $G_i \to AC \to \downarrow cAMP$ |

species orthologs. The P2Y receptors have a predicted structure of seven hydrophobic α-helical arrays of 20–26 amino acids each that span the plasma membrane between the extracellular amino terminus and the cytosolic carboxyl-terminal domain. Three extracellular and three intracellular 10- to 50-amino acid-long loops interconnect the putative seven transmembrane domains. Consensus sites for extracellular N-linked glycosylation, and for phosphorylation by protein kinases in the third intracellular loop and in the carboxyl terminus, have been identified in most P2Y receptors (Fig. 1).

G protein-coupled receptors closely related to the mammalian P2Y receptors have been identified in avian, amphibian, and lower organisms. A receptor that is 60% identical to the P2Y$_1$ receptors was isolated from *Raja erinacea* (skate) and shown to hybridize with genomic DNA from other lower vertebrates and nonvertebrates, but not with avian or mammalian DNA, suggesting that the skate P2Y receptor is an evolutionary ancestor of

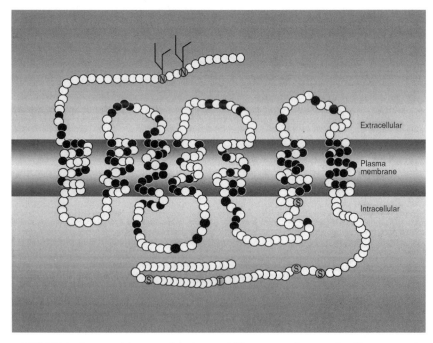

**FIGURE 1**   Structural features of the human P2Y$_2$ receptor. Conserved residues among the human P2Y$_2$, P2Y$_4$, and P2Y$_6$ receptors are represented by dark circles. Extracellular N-linked glycosylation sites (circled N's) and intracellular consensus sites for phosphorylation by protein kinases (circled S's and T's) are indicated.

P2Y$_1$ receptors (Dranoff *et al.*, 2000). The p2y$_3$ receptor, which was cloned out of an embryonic chick brain cDNA library (Webb *et al.*, 1993), is relatively highly homologous (>60%), and has pharmacological properties similar to those of the P2Y$_6$ receptor, likely representing the avian ortholog of the P2Y$_6$ receptor. The p2y$_8$ receptor identified in *Xenopus laevis* (Bogdanov *et al.*, 1997) shares ~50–60% homology along the transmembrane domains with the P2Y$_2$ and P2Y$_4$ receptors. A turkey G protein-coupled nucleotide receptor (p2y$_{turkey}$) has ~50–60% homology with the mammal P2Y$_2$ and P2Y$_4$ receptors and with the *Xenopus* p2y$_8$ receptor through overlapping sequences (Boyer *et al.*, 1997). The p2y$_8$ and p2y$_{turkey}$ receptors, which are nonspecific nucleoside triphosphate-recognizing receptors, may represent early evolutionary ancestors of the P2Y$_2$ and P2Y$_4$ receptors.

A G protein-coupled receptor-encoding cDNA was identified by low-stringency screening of a human erythroleukemia cell line cDNA library, using a p2y$_3$ receptor-encoding probe. On the basis of ATP-binding studies on transfected cells, the name p2y$_7$ was proposed for this receptor (Akbar *et al.*, 1996). However, p2y$_7$ is not a nucleotide receptor but a leukotriene B$_4$ (LTB$_4$) receptor (Yokomizo *et al.*, 1997). Several orphan sequences have tentatively been assigned to the P2Y receptor family, that is, the p2y$_5$, p2y$_9$, and p2y$_{10}$ receptors. No functional data are available to prove that these putative G protein-coupled receptors are true nucleotide-activated receptors.

## B. Nucleotide Selectivity

The precise functional role and nucleotide selectivity of natively expressed P2Y receptors have been difficult to establish because of the lack of specific, high-potency agonists and antagonists for most P2Y receptors, the lack of specific antibodies against extracellular domains, and the fact that many tissues express several P2Y receptors. Additional problems in defining the pharmacological profile of these receptors are as follows: (1) most tissues release nucleotides, which in turn autocrinally affect the activity of the receptor under study; (2) ectonucleotidases and ectonucleoside mono- and diphosphokinases actively metabolize and inter convert nucleotide agonists; and (3) many commercial preparations of nucleotides are impure. The problem of agonist cross-contamination can be minimized by monitoring the nucleotide purity, for example, by high-performance liquid chromatography (HPLC), and by adopting short time read-out assays for nucleotide-promoted receptor activation to reduce the impact of agonist metabolism. The availability of null cell lines in which P2Y receptors could be expressed and their coupling to second-messenger formation investigated in the

absence of interfering background have greatly facilitated the pharmacological studies that have defined the nucleotide selectivities of P2Y receptors.

Among naturally occurring nucleotides, ADP is the most potent agonist
acting on the $P2Y_1$ and $P2Y_{12}$ receptors. The $P2Y_2$ receptor is activated
equipotently by ATP and UTP, but not by nucleoside diphosphates. The
(human) $P2Y_4$ receptor is activated by UTP but not by ATP. UDP is the
most potent agonist for the $P2Y_6$ receptor, which is not activated by ATP or
UTP. The $P2Y_{11}$ receptor is activated by ATP but not by uridine
nucleotides. Finally, the $P2Y_{UDPG}$ receptor is activated by UDP-glucose
but not by UTP, UDP, ATP, or ADP (Table I).

There are no potent and specific competitive antagonists for most P2Y
receptors, and reliable radioligand-based assays are still not available for
receptor screening, and for mechanistic studies of agonist–receptor
interactions. Reactive Blue 2, suramin, trypan blue, and pyridoxal-
phosphate-6-azophenyl-2',4'-disulfonic acid (PPADS) have been widely
used to antagonize some P2Y receptor-mediated responses, but they are
poorly discriminating compounds that also act on P2X receptors and on
nonreceptor nucleotide-binding proteins (Burnstock and Williams, 2000;
Ralevic and Burnstock, 1998; Zhong et al., 2001) (Table I).

## C. G Protein Coupling and Second-Messenger Production

All heptahelical receptors, including the P2Y receptors, transduce
extracellular signals to downstream effectors by activating heterotrimeric
G protein complexes. Agonist binding to the receptor promotes guanine
nucleotide exchange ($G\alpha$ releases GDP and binds GTP), and dissociation of
the GTP-bound $G\alpha$ subunit from the $G\beta\gamma$ complex. Dissociated $G\alpha$-GTP
and $G\beta\gamma$ subunits interact and stimulate (or inhibit) target effectors,
including phospholipases, adenylyl cyclases, and ion channels, or promote
macromolecular assembly, thus eliciting second-messenger production and
functional changes. There are 4 major groups of $G\alpha$ proteins ($G_s$, $G_{i/o}$, $G_{q/11}$,
and $G_{12/13}$), which involve at least 17 distinct isoforms, each of which can
combine with a variety of $\beta$ and $\gamma$ isoforms (Neer, 1995).

Three transductional pathways are classically associated with P2Y
receptor occupation (Fig. 2): (1) activation of phospholipase C by $G_q$-
coupled P2Y receptors, (2) activation of adenylyl cyclase by $G_s$-coupled P2Y
receptors, and (3) inhibition of adenylyl cyclase by $G_i$-coupled P2Y
receptors. When expressed in 1321N1 cells, a human astrocytoma cell line
that does not express endogenous nucleotide receptors, the $P2Y_1$, $P2Y_2$,
$P2Y_4$, and $P2Y_6$ receptors are preferentially coupled to the pertussis toxin-
insensitive $G\alpha_q/\beta\gamma$ heterotrimer, promoting the activation of phospholipase

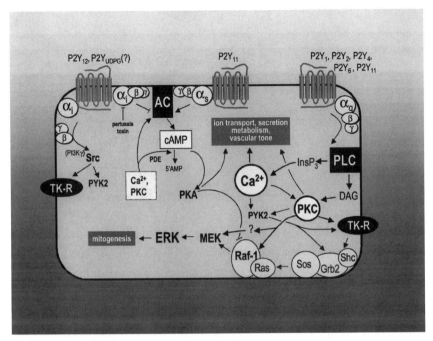

**FIGURE 2**  Schematic representation of the cellular responses regulated by G protein-coupled P2Y receptors. The indicated pathways should be considered as illustrative examples rather than as a complete list of all biochemical changes occurring downstream to P2Y receptor activation. Only a few responses regulated by $Ca^{2+}$, protein kinase A (PKA), protein kinase C (PKC), or MAPK are illustrated. AC, Adenylyl cyclase; PLC, phospholipase C; TK-R, tyrosine kinase receptor; $InsP_3$, inositol 1,4,5-triphosphate; DAG, diacylglycerol; cAMP, cyclic AMP; PDE, cAMP-phosphodiesterase. $\perp$, Inhibition.

C-$\beta$. Phospholipase C hydrolyzes phosphatidylinositol 4,5-biphosphate and generates two second messengers, inositol 1,4,5-triphosphate ($InsP_3$), which promotes the release of $Ca^{2+}$ from intracellular stores, and diacylglycerol (DAG), which activates protein kinase C. The $P2Y_{11}$ receptor also couples to $G_q$ and activates phospholipase C but, in addition, promotes the formation of cellular cyclic AMP, likely, although not formally proven, via activation of $G_s$ and adenylyl cyclase. The $P2Y_{12}$ receptor couples to $G_i$ and does not promote phospholipase C activation. Little is known about the signaling cascade triggered by the $P2Y_{UDPG}$ receptor, which, when transfected in HEK-293 cells, promoted no evident intracellular calcium mobilization in response to nucleotides, but robust intracellular calcium mobilization was observed after coexpressing the $P2Y_{UDPG}$ receptor with $G\alpha_{16}$, a $G\alpha$ subunit that promiscuously confers phospholipase C activation to cotransfected G protein-coupled receptors irrespective of whether these receptors are natively coupled to $G_q$, $G_i$, or $G_s$ (Chambers *et al.*, 2000). Thus, recombinant P2Y receptors expressed and characterized in heterologous systems appear to affect selectively a narrow spectrum of second-messenger production, consistent with the idea that specific intracellular domains influence the transductional interaction to G proteins.

In their natural environment, however, P2Y receptors may be less selective in their coupling to G proteins. For example, in 1321N1 astrocytoma cells stably expressing the $P2Y_6$ receptor, pertussis toxin had no effect on UDP-stimulated responses, but responses were inhibited $\sim$60% by pertussis toxin when the receptor was expressed in rat sympathetic neurons (Filippov *et al.*, 1999). There are many reports in the literature of native P2Y receptors in which some degree of inhibition by pertussis toxin (implicating $G_i/G_o$ coupling in native tissues) has no correspondence with the $G_q$-selective coupling manifested by recombinant receptors. For example, pertussis toxin inhibited nucleotide-stimulated phospholipase C and/or $Ca^{2+}$ responses in human neutrophils and eosinophils (Mohanty *et al.*, 2001; Walker *et al.*, 1991), and human 6CFSMEo$^-$ submucosal cells (Communi *et al.*, 1999a); bovine aortic and middle cerebral artery endothelial cells (Pirotton *et al.*, 1996; Purkiss *et al.*, 1994; Zhang *et al.*, 1997); rat renal mesangial (Pfeilschifter, 1990) and glioma cells (Lazarowski and Harden, 1994), astrocytes (Centemeri *et al.*, 1997; Jimenez *et al.*, 2000), pituicytes (Troadec *et al.*, 1999), and sympathetic neurons (Bofill-Cardona *et al.*, 2000); mouse pineal gland tumor cells (Suh *et al.*, 2001); and pig tracheal submucosal glands (Zhang and Roomans, 1999), rabbit visceral smooth muscle cells (Murthy and Makhlouf, 1998), Madin-Darby canine kidney (MDCK) cells (Yang *et al.*, 1997), and guinea pig cardiac endothelial cells (Yang *et al.*, 1996). In human erythroleukemia (HEL) cells, which endogenously express $G\alpha_{i-1}$, $G\alpha_{i-2}$, $G\alpha_q$, and $G\alpha_{16}$, ATP- or UTP-promoted $Ca^{2+}$ mobilization was inhibited $\sim$50% by pertussis toxin. However, $Ca^{2+}$ responses were totally suppressed in HEL cells transfected with $G\alpha_{16}$ antisense RNA, suggesting that the $P2Y_2$ receptor in these cells engages $G_{16}$ to mobilize $Ca^{2+}$, but also communicates with $G_i$ proteins (Baltensperger and Porzig, 1997). Thus the apparently selective signaling associated with recombinant P2Y receptors may reflect limitations in the expression system utilized.

## D. Regulation of Long-Term Events

Cell proliferation, differentiation, and apoptotic cell death, which not long ago were considered to be under the exclusive control of growth factor and cytokine receptors, are regulated also by G protein-coupled receptors, including P2Y receptors. Pioneering works by Burnstock, Neary,

Abbracchio, and others have indicated that extracellular nucleotides promote trophic responses in nerve tissues, ranging from induction of cell differentiation, mitogenesis, and morphogenetic changes to stimulation of synthesis or release of cytokines and neurotrophic factors (reviewed by Neary et al., 1996). More recently, P2Y receptors were shown to mediate proliferative events in astrocytes, vascular smooth muscle cells, endothelial cells, fibroblasts, and mesangial cells (Boarder, 1998), and to induce mitogenic (Harper et al., 1998) and apoptotic (Sellers et al., 2001) changes when they are overexpressed in heterologous cells.

A key event in P2Y receptor-mediated cell proliferation is the activation of the mitogen-activated protein kinase (MAPK) cascade. The MAPK cascade is regulated tightly through phosphorylation of regulatory domains, and removal of phosphates by specific phosphatases generally results in reduced MAPK activity. The efficient serial activation of the MAPK cascade components is also regulated by nonenzymatic protein–protein interactions involving scaffold and adapter proteins that spatially organize MAPK cascade components to discriminate and restrict the signal reception.

The MAPK family includes at least four groups of proline-directed serine/threonine kinases, (1) stress-activated protein kinase (SAPK), (2) p38 kinase, (3) BMK1/ERK5, and (4) extracellular signal-regulated protein kinase (ERK) (Widmann et al., 1999). ERK plays a central role in the MAPK cascade that controls cell proliferation and differentiation. Signaling through ERK involves sequential stimulation of several cytosolic protein kinases that eventually results in activation of regulatory molecules in the cytoplasm and in the nucleus. The first kinase in the MAPK/ERK cascade is Raf (MAPK kinase kinase or MEK kinase), which can be activated by polypeptide growth factor receptors through a sequence of events involving the formation of guanine nucleotide exchange complexes (Shc/Grb2-mediated recruitment of Sos) and activation of p21Ras. Activated Raf phosphorylates MEK, which in turn phosphorylates and activates ERK (Fig. 2). Activation of the MAPK/ERK cascade results in enhanced expression of transcription factors (e.g., c-Fos and c-Jun) and in the formation of nuclear activator protein 1 (AP-1) complexes, which in turn leads to gene expression and cell growth (Angel and Karin, 1991).

The ERK/MAPK cascade is activated shortly after P2Y receptor occupancy, and signals associated with P2Y receptor occupancy (e.g., dissociation of $G\beta\gamma$ from $G\alpha$-GTP, activation of protein kinase C, elevation of intracellular $Ca^{2+}$, and cyclic AMP formation) are known to affect ERK. Potential scenarios for regulation of MAPK cascade by P2Y receptors include the following: (1) $G_s$-coupled receptors evoke activation of cyclic AMP-dependent protein kinase A, which directly phosphorylates (and inactivates) Raf-1; (2) $G_i$-coupled receptors signal through ERK activation by the interaction of free G$\beta\gamma$ subunits with Src [in a phosphatidylinositol (PI) 3-kinase $\gamma$-dependent manner], which in turn phosphorylates a tyrosine kinase receptor, related adhesion focal tyrosine kinase (RAFTK or PYK2), or focal adhesion kinase, triggering a sequence of events similar to those used by receptor tyrosine kinases, which results in activation of Raf-1 by Ras; and (3) G$_q$-coupled receptors can activate ERK by protein kinase C-dependent mechanisms, which include direct phosphorylation of Raf by protein kinase C, activation of PYK2, or activation of receptor tyrosine kinases (Fig. 2).

The link between P2Y receptors and MAPK activation has not been elucidated in full, but there are several steps in the MAPK cascade that are known to be affected by P2Y receptor activation, either synergistically with or independently from growth factor receptors. In rat pheochromocytoma PC12 cells, the P2Y$_2$ receptor stimulates MAPK by a mechanism that involves activation of PYK2 by protein kinase C and Ca$^{2+}$, and the subsequent phosphorylation of the epithelial growth factor (EGF) receptor and recruitment of Grb2 (Soltoff, 1998). In astrocytes, nucleotides acting via the P2Y$_1$, P2Y$_2$, or P2Y$_4$ receptor activate a Ca$^{2+}$-independent form of protein kinase C (protein kinase C$\delta$) that in turn stimulates MEK, using a yet unknown MEK activator that is independent of Raf-1. Also in these cells, P2Y receptor activation can either activate Raf-1 synergistically with growth factor receptors, or inhibit the activation of Raf-1 by EGF, fibroblast growth factor (FGF), or platelet-derived growth factor (PDGF) without inhibiting MEK and ERK activation (Lenz et al., 2000, 2001). In human endothelial cells, UTP and ATP (presumably acting via the P2Y$_2$ receptor) stimulate MEK and MAPK by a Ca$^{2+}$- and protein kinase C-dependent mechanism that is independent of Raf but that requires integrin-modulated cell anchoring (Short et al., 2000).

The few examples described above of P2Y receptor–MAPK kinase cascade interactions are intended to be an introductory guide to this intricate signaling pathway. The reader is encouraged to consult excellent reviews covering in more detail various aspects of MAPK cascade signaling (Pearson et al., 2001) and the regulation of MAPK by P2Y receptors (Boarder and Hourani, 1998; Neary, 2000).

## III. THE P2Y RECEPTOR FAMILY

The preceding consideration of general characteristics of P2Y receptors is followed by a discussion of more specific aspects associated with molecular, pharmacological, and functional properties of each member of the P2Y receptor family.

## A. The P2Y₁ Receptor

The $P2Y_1$ receptor is perhaps the most studied P2Y receptor for which substantial progress in understanding its pharmacological and functional properties has been achieved. The $P2Y_1$ receptor was the first "purinergic" receptor unambiguously characterized as a G protein-coupled and phospholipase C-activating receptor, before its identification was confirmed through molecular approaches. Studies by Harden and colleagues, using the turkey erythrocyte membrane model for receptor-promoted phospholipase C activation, demonstrated that 2MeSATP, ATP, and other adenine nucleotides activate what at that time was called "the $P_{2Y}$-purinergic" receptor to increase the rate of exchange of GTP for GDP on the relevant guanine nucleotide regulatory protein, promoting inositol phosphate formation in these membrane preparations (Harden et al., 1988; Boyer et al., 1989). Several years elapsed until the $P2Y_1$ receptor was cloned from chicken (Webb et al., 1993) and subsequently from turkey (Filtz et al., 1994), human (Janssens et al., 1996), mouse and rat (Tokuyama et al., 1995), and cow (Henderson et al., 1995), and shown to have remarkable pharmacological and signaling similarities to the $P_{2Y}$-purinergic receptor originally described by Burnstock and Kennedy and to the turkey erythrocyte $P_{2Y}$-purinergic receptor characterized by Harden and colleagues. A recombinant human $P2Y_1$ receptor was purified to homogeneity and its coupling to a G protein was verified in vitro in proteoliposomes reconstituted in the presence of $G\alpha_q$ and $G\beta_1\gamma_2$. Activation of the $P2Y_1$ receptor with 2MeSADP promoted the hydrolysis of GTP in the reconstituted proteoliposomes and this action was potentiated by RGS2, a member of the regulator of G protein signaling family that stimulates the hydrolysis of the GTP bound to activated $G\alpha$ subunits (Dohlman and Thorner, 1997). Phosphorylation of RGS2 by protein kinase C resulted in loss of agonist-induced GTP hydrolysis, suggesting one mechanism of regulation of $P2Y_1$ receptor activity, that is, agonist-promoted protein kinase C activation decreases $G\alpha_q$-GTP hydrolysis by RGS (Cunningham et al., 2001).

The $P2Y_1$ receptor is activated exclusively by adenine nucleotides and displays higher selectivity for ADP over ATP. AMP is not an agonist on this receptor. The most potent $P2Y_1$ receptor agonist is 2MeSADP, which displays a median effective concentration ($EC_{50}$) value of $\sim10$ n$M$ for receptor-promoted phospholipace C activation. The human $P2Y_1$ receptor overexpressed in 1321N1 human astrocytoma cells is activated with an agonist potency order of 2MeSADP > 2MeSATP > ADP > ATP (Boyer et al., 1996; Schachter et al., 1996). Similar agonist selectivity was observed with avian, bovine, murine, and rat $P2Y_1$ receptors. Studies with human $P2Y_1$ receptor-transfected Jurkat T cells have suggested that this receptor is selective for ADP and ADP analogs and that ATP behaves as an antagonist (Leon et al., 1997; Hechler et al., 1998). Using highly purified ATP under conditions that minimize the action of ectonucleotidases that convert ATP to ADP, Palmer and colleagues demonstrated that ATP is a full agonist on human $P2Y_1$ receptor-expressing 1321N1 cells, but that the agonist activity of ATP decreased when the cells were subjected to agonist-induced desensitization; they concluded that ATP exhibits relatively low intrinsic activity at the $P2Y_1$ receptor and, consequently, the extent of the agonist activity of ATP depends on the extent of receptor reserve (Palmer et al., 1998). Adenosine-3'-phosphate-5'-phosphosulfate (A3P5PS) and deoxy-adenosine bisphosphate analogs such as $N$:(6)-methyl-2'-deoxyadenosine 3',5'-bisphosphate ($N^6$-MABP, also named MRS2179) are potent and specific $P2Y_1$ receptor antagonists (Boyer et al., 1996, 1998; Nandanan et al., 2000). $N^6$-MABP has also been used as a radioligand (in the form of [$^{33}$P] MRS2179) for the native $P2Y_1$ receptor of circulating platelets and for the recombinant $P2Y_1$ receptor expressed in 1321N1 human astrocytoma cells (Baurand et al., 2001). MRS2179 has an affinity constant ($K_d$) of $\sim$100 n$M$, consistent with the p$K_b$ value of 6.99 determined for the inhibition of ADP-induced phospholipase C activation.

On the basis of site-directed mutagenesis, molecular modeling, and the chemical synthesis of novel agonists and antagonists, the determinants of ligand recognition in $P2Y_1$ receptors were proposed to reside in positively charged and other conserved residues in TM3, TM6, and TM7, such as Arg-128, Arg-310, and Ser-314 (Hoffmann et al., 1999), and that arginine and lysine residues in the second and third extracellular loops are essential for high agonist potency (Jacobson et al., 1999).

The $P2Y_1$ receptor displays a relatively wide tissue distribution. Published data in the form of Northern blot analysis of human tissues indicate that the $P2Y_1$ receptor mRNA is highly expressed in prostate, ovary, and intestine, and at lower levels in heart, brain, lung, and placenta (Janssens et al., 1996). The $P2Y_1$ receptor mRNA is also expressed in vein endothelial cells (Leon et al., 1996) and human platelets (Jin et al., 1998a). In rat, the $P2Y_1$ receptor mRNA is most abundantly expressed in the heart and skeletal muscle, followed by the brain, lung, kidney, and liver (Tokuyama et al., 1995).

The $P2Y_1$ receptor plays a major role in platelet physiology. Pharmacological evidence had suggested the existence of different ADP-recognizing receptors in circulating platelets, which have now been identified as the $P2Y_1$, $P2Y_{12}$, and $P2X_1$ receptors (Hourani and Cusack, 1991; Hollopeter et al., 2001). Studies with $P2Y_1$ receptor-deficient (Fabre et al., 1999; Leon et al., 1999) and $G\alpha_q$-deficient mice (Offermanns et al., 1997; Ohlmann et al., 2000) have confirmed that the platelet $P2Y_1$ receptor obligatorily signals via $G_q$ to promote the activation of phospholipase C and the mobilization of

$[Ca^{2+}]$; to support platelet aggregation. *In vivo*, suppression of the $P2Y_1$ receptor gene increased bleeding time and decreased the incidence of thromboembolism in response to intravenous injection of collagen and ADP, indicating that the $P2Y_1$ receptor plays a central role in hemostasis. Activation of the vascular endothelial cell $P2Y_1$ receptor results in antithrombotic, relaxant responses mediated by the release of cyclooxygenase products (prostaglandin $I_2$, $PGI_2$) and NO from endothelial cells (Pirotton *et al.*, 1996; Boarder and Hourani, 1998). $P2Y_1$ receptor-mediated responses in nonvascular tissues include the propagation of calcium waves in dorsal spinal cord astrocytes (Fam *et al.*, 2000) and mouse airway epithelial cells (Homolya *et al.*, 2000), activation of N-type $Ca^{2+}$ channels in neurons (Filippov *et al.*, 2000), and elevation of intracellular $Ca^{2+}$ in rat hepatocytes (Dixon, 2000).

## B. The $P2Y_2$ Receptor

The observation by Dubyak and De Young (1985) that extracellular UTP promoted mobilization of intracellular $Ca^{2+}$ in Ehrlich ascites tumor cells triggered widespread interest in the biological actions of extracellular uridine nucleotides. Subsequent studies indicated that pyrimidine nucleotides affected a variety of cell functions including glucose transport, portal pressure, $K^+$ fluxes, and thromboxane release in perfused rat liver (Haussinger *et al.*, 1987, 1988), calcium mobilization in MDCK cells (Paulmichl and Lang, 1988), and human neutrophil chemotaxis and phosphoinositide hydrolysis in HL60 cells (Seifert *et al.*, 1989a,b; Stutchfield and Cockcroft, 1990). However, it was not clear at that time whether a combination of purine-selective and pyrimidine-selective receptors (as proposed by Seifert and Schultz, 1989) or a single ATP- and UTP-recognizing receptor accounted for the observed actions of these nucleotides. A G protein-coupled receptor with pharmacological similarities to what is now termed $P2Y_2$ receptor was first described by Cowen *et al.* (1990) and named $P_{2U}$-purinoceptor (Dubyak, 1991; O'Connor *et al.*, 1991).

In 1993, Lusting and colleagues reported the expression cloning of a nucleotide receptor from a neuroblastoma cell cDNA library, which was found by pharmacological approaches to correspond to the $P_{2U}$-purinoceptor (Lustig *et al.*, 1993). Subsequently, ortholog genes sharing 90% overall homology with the neuroblastoma cell $P_{2U}$-purinoceptor were cloned from human, rat, and canine cDNAs (Parr *et al.*, 1994; Rice *et al.*, 1995; Godecke *et al.*, 1996; Zambon *et al.*, 2000), and the receptor was renamed $P2Y_2$. In all species, ATP and UTP most potently activate the $P2Y_2$ receptor. Diadenosine polyphosphates ($Ap_nA$) are naturally occurring dinucleotides released by exocytotic mechanisms from a number of tissues and they interact with variable specificity with P2Y and P2X receptors (Ogilvie *et al.*, 1996; Pintor *et al.*, 2000). $Ap_4A$ is the most potent diadenosine nucleotide acting at the human $P2Y_2$ receptor stably expressed in 1321N1 cells, with a potency that is approximately one-fifth that of ATP and UTP. Although UDP and ADP were initially reported to be weak agonists on human astrocytoma cells expressing the human $P2Y_2$ receptor, subsequent studies showed that these effects were abolished by coincubating the samples with hexokinase (and glucose), which eliminates contaminating NTPs from NDP solutions (Nicholas *et al.*, 1996).

RNA analysis has confirmed the wide tissue distribution of $P2Y_2$ receptor transcripts. In humans, $P2Y_2$ receptor mRNA is abundantly detected by Northern blot in skeletal muscle, heart, liver, lung, placenta, and kidney as well as in primary cultures of nasal and kidney proximal tubule epithelial cells and in various epithelial cell lines (Communi *et al.*, 1996a; Parr *et al.*, 1994). In contrast, little $P2Y_2$ receptor mRNA is found in the brain (Parr *et al.*, 1994). In the rat, Northern blot analyses and RNAase protection assays indicate expression of $P2Y_2$ receptor transcripts in the liver, testis, esophagus, lung, trachea, heart, skeletal muscle, stomach, and kidney, but not in the brain. High levels of $P2Y_2$ receptor transcripts were also found in primary cultures of rat microvascular endothelial cells but messengers were low or absent in aortic smooth muscle cells and in freshly isolated cardiac myocytes (Godecke *et al.*, 1996). *In situ* hybridization revealed the expression of $P2Y_2$ receptor messengers in the epithelium of the esophagus and in the bronchial epithelium (Godecke *et al.*, 1996). Single-cell reverse transcriptase-polymerase chain reaction (RT-PCR) analysis indicates the presence of $P2Y_2$ receptor mRNA in freshly isolated astrocytes of the rat hyppocampus (Zhu and Kimelberg, 2001). $P2Y_2$ receptor transcripts were also detected by RT-PCR in primary cultures of rat brain capillary endothelial cells (Anwar *et al.*, 1999).

One problem in establishing a physiological role for the $P2Y_2$ receptor is that the effects of ATP and UTP observed in some human tissues could be due to the coexpression of ATP-selective receptors (e.g., $P2Y_{11}$ and P2X) and the UTP-selective $P2Y_4$ receptor. Pharmacological analyses of native $P2Y_2$ receptors have been hampered by the ubiquitous presence of ecto-ATPases and ectonucleoside diphosphokinase (Zimmermann, 2001), which, by acting in concert on locally released nucleotides and exogenous NTPs, transfer the $\gamma$-phosphate from one to another nucleotide (e.g., from exogenous UTP to endogenous ADP, thus resulting in ATP formation) (Lazarowski *et al.*, 2000). Further complications emanate from studies with animal models, in which other receptors may display agonist selectivities similar to the $P2Y_2$ receptor (e.g., mouse and rat $P2Y_4$ receptors). Two allelic variants of the human $P2Y_2$ receptor gene (Arg-334 to Cys-334), which cause no major pharmacological differences, have been identified (Janssens et al., 1999).

Efforts to establish detailed structure–activity relationships for adenine and uridine nucleotide analogs for the $P2Y_2$ receptor have been limited by the paucity of commercially available biologically active nucleotide derivatives. Besides ATP, UTP, and $Ap_4A$, only ATP$\gamma$S activates the $P2Y_2$ receptor, but with a potency that is one-tenth that of ATP. 2MeSATP and $\alpha,\beta$-MetATP are not $P2Y_2$ receptor agonists (Lazarowski et al., 1995). Several UTP analogs with variable $P2Y_2$ receptor agonist activities have been synthesized. Among them, $Up_4U$ and UTP$\gamma$S are the most potent and stable uridine nucleotide derivatives that activate the $P2Y_2$ receptor, with $EC_{50}$ values that closely approach UTP $EC_{50}$ values (Lazarowski et al., 1996; Pendergast et al., 2001). $Up_4U$ and UTP$\gamma$S also act on $P2Y_4$ receptors, but not on $P2Y_1$ or $P2Y_6$ receptors (Nicholas et al., 1996; Pendergast et al., 2001). Modifications in the pyrimidine ring results in less active compounds. In a study in which the activity of a series of ten 4-substituted pyrimidine nucleotide analogs was investigated, only 2 of them (4-thiouridine 5′-triphosphate and 4-thiohexyluridine 5′-triphosphate) displayed $EC_{50}$ values within an order of magnitude of the UTP $EC_{50}$ value (Shaver et al., 1997). No specific antagonist is known for the $P2Y_2$ receptor. Suramin is a weak, nonselective $P2Y_2$ receptor antagonist with a reported $pA_2$ value of 4.32 (Charlton et al., 1996). PPADS is ineffective. Site-directed mutagenesis studies of the mouse $P2Y_2$ receptor have suggested that positively charged residues in TM6 and TM7 are important for nucleotide binding (Erb et al., 1995).

A receptor with pharmacological characteristics of the $P2Y_2$ receptor has been described in many cell types and tissues. Airway epithelium, which expresses large amounts of $P2Y_2$ receptor mRNA, was one of the first tissues in which the nucleotide selectivity and second-messenger production of the natively expressed nucleotide receptor closely correlate with those of the stably expressed human $P2Y_2$ receptor (Brown et al., 1991; Mason et al., 1991; Lazarowski et al., 1995). The airway epithelial cell $P2Y_2$ receptor regulates several components of $Ca^{2+}$-dependent mucociliary clearance. Mucosal administration of ATP or UTP results in $Ca^{2+}$-stimulated chloride secretion that is independent of the cystic fibrosis transmembrane conductance regulator (CFTR), the cyclic AMP-regulated epithelial chloride channel that is defective in cystic fibrosis (Knowles et al., 1991; Mason et al., 1991; Donaldson and Boucher, 1998). The physiological importance of the $P2Y_2$ receptor in the control of airway epithelial cell responses has been confirmed in studies showing that nucleotide-promoted inositol phosphate formation, $Ca^{2+}$ mobilization, and ion secretion in airway epithelia were nearly abolished in the P2Y$_2$–/– mouse (Cressman et al., 1999; Homolya et al., 1999). In the human heart, the P2Y$_2$ receptor is involved in the regulation of contractile responses of coronary arteries (Malmsjo et al., 2000a). Functional studies also have implicated a physiological role for the P2Y$_2$ receptor in neutrophils and other blood cells, and in astrocytes and glial cells, endothelial cells, osteoblasts, fibroblasts, hepatocytes, smooth muscle cells, keratinocytes, and tumor cells (Pirotton et al., 1996; Ralevic and Burnstock, 1998).

Activation of most G protein-coupled receptors results in progressive loss of the capacity of the receptor to respond to agonist stimulation. Although P2Y$_2$ receptors expressed on airway epithelial and other cells are under the constant influence of constitutively released ATP and UTP (Lazarowski et al., 2000, 2001a), it is not known whether autocrine activation of the P2Y$_2$ receptor implies some degree of loss of receptor responsiveness. Essentially, little is known about the mechanisms involved in agonist-promoted P2Y$_2$ receptor desensitization. Using HA-tagged P2Y$_2$ receptor expressed in 1321N1 cells, Sromek and Harden (1998) demonstrated that agonist-induced desensitization of inositol phosphate responses occurred rapidly on receptor occupation, and that the functional loss of receptor activity was accompanied, although slowly and less extensively, by receptor internalization. Loss of cell surface immunoreactive HA–P2Y$_2$ receptors was reversed within 1 h of removal of agonist.

Studies of adenylyl cyclase $\beta_2$-adrenergic receptors indicate that receptor phosphorylation is at least one mechanism involved in agonist-promoted desensitization (Lohse, 1993; Freedman and Lefkowitz, 1996). Serine and threonine residues in the third intracellular loop and in the carboxyl-terminal domain of the P2Y$_2$ receptor are potential phosphorylation sites for G protein-coupled receptor kinases and protein kinase C (Fig. 1). Garrad and colleagues (1998) investigated the role of these residues in UTP-induced receptor desensitization and internalization, using a series of truncation mutants of the HA-tagged P2Y$_2$ receptor cDNA stably expressed in 1321N1 cells. In cells expressing the wild-type HA–P2Y$_2$ receptor, UTP promoted a rapid ($\sim$10 min) loss of approximately 80% of cell surface receptors and a similar degree of receptor sequestration was observed in cells expressing a truncation involving the most C-terminal sites for phosphorylation. In contrast, truncations involving additional phosphorylation sites in the C-terminal tail resulted in delayed sequestration of the P2Y$_2$ receptor during the UTP challenge. Truncation of 18 or more residues at the C terminus increased the concentration of UTP necessary to induce desensitization. These results suggest that phosphorylation at the C terminus might be important for receptor desensitization and internalization. Site-directed mutagenesis studies targeting the putative phosphorylation sites of the $P2Y_2$ receptor may be required to confirm the role of these residues in receptor downregulation. However, although activation of protein kinase C with phorbol esters resulted in strong receptor desensitization, that is, loss of UTP-elicited $Ca^{2+}$ mobilization, the cell surface density of the $P2Y_2$ receptor was not affected by this treatment. This apparent dissociation between protein kinase C-promoted receptor seques-tration and desensitization could reflect the fact that (1) receptor occupancy is necessary to allow phosphorylation of specific serines or threonines, (2) effector molecules located downstream from the receptor are phosphoryl-ated, or (3) protein kinase C is not involved in $P2Y_2$ receptor recycling. Thus, although models for receptor desensitization, internalization, delivery to lysosomes, and recycling to the cell surface have been established for many G protein-coupled receptors, it has not yet been determined how phosphorylation events affect the function of the $P2Y_2$ receptor. Interest-ingly, a direct, specific, interaction between the C terminus of the $P2Y_2$ receptor and p53-binding protein 2 (p53BP-2, a ubiquitous nonreceptor tyrosine kinase substrate) was revealed using the yeast two-hybrid system and *in vitro* binding assays with recombinant fusion proteins. The full-length $P2Y_2$ receptor expressed in human bronchial epithelial cells bound to fragments of p53BP-2 *in vitro*, but a truncated $P2Y_2$ receptor, lacking the entire C terminus, failed to bind p53BP-2, leading to the speculation that p53BP-2 may serve as a scaffolding/adapter protein to modulate $P2Y_2$ receptor signaling or trafficking (Szabo *et al.*, 2001).

## C. The $P2Y_4$ Receptor

After the cloning of the $P2Y_1$ and $P2Y_2$ receptors in 1993, it seemed that these receptors could account for most nucleotide-promoted second messenger-mediated responses reported at that time in the literature. However, highly selective uridine nucleotide responses were described that could not be explained by the pharmacological properties of known receptors. For example, UTP and/or UDP promoted activation of phospholipases C and phospholipase $A_2$ in rat C6-2B glioma cells (Lazarowski and Harden, 1994) and in rat macrophages (Liu *et al.*, 1995), and stimulated the neural release of catecholamines (Boehm *et al.*, 1995). These effects of uridine nucleotides, which were not mimicked by adenine nucleotides, suggested that pyrimidine nucleotide-selective receptors exist, and eventually led to the identification of the $P2Y_4$ and $P2Y_6$ receptors in 1995.

The $P2Y_4$ receptor was cloned from human genomic DNA and is 54% homologous to the human $P2Y_2$ receptor. The human $P2Y_4$ receptor was initially characterized as a UTP- and UDP-preferring phospholipase C-coupled receptor (Communi et al., 1995; Nguyen et al., 1995), and ATP exhibited a small and weak stimulatory effect (Communi et al., 1995; Charlton et al., 1996). It was shown later that ATP is not a true agonist but a low-potency antagonist for the human P2Y$_4$ receptor (Kennedy et al., 2000), and that the apparent ATP effect on phospholipase C resulted from conversion of exogenous ATP to UTP by an action of ectonucleoside diphosphokinase that phosphorylates endogenous UDP (Lazarowski et al., 1997a,b). In experiments in which highly pure nucleotides were used and nucleotide interconversion was carefully prevented, Nicholas and colleagues (1996) established that UDP has no agonist activity whatsoever on the P2Y$_4$ receptor.

The confusion initially generated in establishing the agonist selectivity of the P2Y$_4$ receptor was further complicated when the cloning of the rat and mouse P2Y$_4$ receptor revealed that, although sharing ~83% homology with the human receptor, they are not uridine nucleotide-selective proteins, but instead, ATP and ITP are also potent agonists (Bogdanov et al., 1998; Webb et al., 1998; Kennedy et al., 2000; Lazarowski et al., 2001b; Suarez-Huerta et al., 2001). Retrospectively, it is possible that some nucleotide actions in nonhuman tissues, previously assigned to the P$_{2U}$-purinergic/P2Y$_2$ receptor reflected, instead, involvement of the P2Y$_4$ receptor. The hydrolysis-resistant uridine nucleotide derivatives UTP$\gamma$S and Up$_4$U are potent P2Y$_4$ receptor agonists, but they do not discriminate between P2Y$_4$ and P2Y$_2$ receptors (Nicholas et al., 1996; Malmsjo et al., 2000a,b). The human P2Y$_4$ receptor is insensitive to suramin or PPADS (Charlton et al., 1996), while the rat and the mouse P2Y$_4$ receptors are noncompetitively antagonized by PPADS. The mouse P2Y$_4$ receptor is antagonized by high concentrations of Reactive Blue 2, but not by suramin (Suarez-Huerta et al., 2001).

The recombinant P2Y$_4$ receptor expressed in 1321N1 cells stimulates the formation of inositol phosphates, mostly by a pertussis toxin-insensitive pathway, although partial inhibition was observed when cells were challenged with UTP for 30 sec (Communi et al., 1996a). The P2Y$_4$ receptor studied in expression systems does not promote changes in the cellular levels of cyclic AMP.

The human P2Y$_4$ receptor is composed of 365 residues, only 23 of which differ from the rodent homologs. The remarkable differences in nucleotide preferences (UTP over ATP) displayed by the human (but not the rat or mouse) P2Y$_4$ receptor suggested that the agonist recognition site relies on a discrete set of amino acids capable of discriminating between urine and adenine rings. When examining the pharmacological properties of various rat/human chimeric constructs, Herold et al. (1999) observed that substitution of TM6 to the carboxyl terminus of the human P2Y$_4$ receptor with the corresponding region of the rat receptor yielded a chimeric receptor with wild-type rat characteristics. However, substitution of the amino terminus to TM3 of the human receptor with the corresponding rat $P2Y_4$ receptor region also yielded a chimeric receptor with wild-type rat $P2Y_4$ properties. These results suggest that multiple, noncontiguous motifs converge to the recognition site that discriminates between UTP and ATP.

Initially, the physiological role of the $P2Y_4$ receptor has been difficult to establish because of the unavailability of $P2Y_4$ receptor-selective high-potency agonists and antagonists, and by its low abundance and rather confined tissue distribution. Transcripts of the human $P2Y_4$ receptor were found by Northern blot only in placenta (Communi et al., 1995), although messages were detected by RT-PCR in human neutrophils and monocytes (Jin et al., 1998b), and in human coronary arteries (Malmsjo et al., 2000a). In the rat, $P2Y_4$ receptor mRNA could be detected by PCR (but not by Northern blot) in the brain, spinal cord, and in various peripheral tissues. And $P2Y_4$ receptor transcripts were revealed by in situ hybridization in rat pineal gland and the ventricular system (Webb et al., 1998). Additional RT-PCR studies indicate $P2Y_4$ receptor mRNA in rat aortic smooth muscle cells (Harper et al., 1998), primary cultures of rat brain capillary endothelial cells (Anwar et al., 1999), and in the murine liver, intestine, stomach, lung, and bladder, but not in the spleen, brain, and heart (Suarez-Huerta et al., 2001).

Association of the $P2Y_4$ receptor with well-established physiological responses is now emerging. Communi and colleagues (1999a) have investigated the expression of various P2Y receptors in human lung cells and found functional evidence of $P2Y_4$ receptor expression in 6CFSMEo$^-$ submucosal cells, where UTP but not ATP or UDP promoted the formation of inositol phosphates, consistent with the pharmacological profile of nucleotides acting on the human $P2Y_4$ receptor. In these cells, $P2Y_4$ mRNA could be detected by RT-PCR but not by Northern blot. A role for the $P2Y_4$ receptor in the control of ion transport in the gastrointestinal epithelia was revealed by studies with the $P2Y_2$–/– mouse. In the jejunum of wild-type mice, ATP and UTP equipotently induced changes in epithelial ion transport, and these effects were unaffected by $P2Y_2$ gene disruption (Cressman et al., 1999). In situ hybridization analysis confirmed the presence of $P2Y_4$ receptor transcripts in freshly excised mouse jejunum (Lazarowski et al., 2001b). Intestinal epithelial cells lack the $Ca^{2+}$-regulated $Cl^-$ channel that is typical of many other secretory epithelia (Clarke et al., 1994; Grubb, 1997) and activation of intestinal ion transport responses by UTP and ATP were missing in the CFTR-deficient mouse (Lazarowski et al., 2001b). Thus, $P2Y_4$ receptor regulation of ion transport responses in jejunum may involve $Ca^{2+}$-independent, CFTR-associated ion conductance. One speculation is that the $P2Y_4$ receptor of the mouse jejunal epithelium couples indirectly to the promotion of cyclic AMP, perhaps via $Ca^{2+}$-dependent activation of phospholipase $A_2$, prostaglandin (PGE) release, and activation of the adenylyl cyclase-coupled PGE receptors or, alternatively, activation of CFTR either by protein kinase C- or by $Ca^{2+}$-dependent elevation of cellular cyclic GMP levels. Irrespective of the mechanism involved, the $P2Y_4$ receptor appears to play a major role in the regulation of electrolyte transport in the murine gastrointestinal tract.

The $P2Y_4$ receptor may be involved in UTP-promoted responses reported in other tissues, but its role could be masked by a predominant $P2Y_2$ receptor. In the human vascular system, the antiproliferative action of UTP on smooth muscle cells could be explained only by activation of the $P2Y_4$ receptor (White et al., 2000). In addition, the $P2Y_4$ receptor may be responsible for the angiogenic action of UTP on guinea pig cultured endothelial cells (Satterwhite et al., 1999), and for the mitogen effect of UTP in aortic smooth muscle cells from hypertensive rats (Harper et al., 1998).

The human $P2Y_4$ receptor expressed in 1321N1 cells undergoes rapid desensitization during agonist challenge (Robaye et al., 1997). Consensus sequences for phosphorylation by protein kinase C and other kinases, identified in the third intracellular loop and the carboxyl-terminal tail of the $P2Y_4$ receptor, may be involved in nucleotide-promoted receptor desensitization and internalization. Applying mutational analysis, Brinson and Harden (2001) have identified residues in the carboxyl-terminal end that play major roles in agonist-promoted changes in the $P2Y_4$ receptor. In 1321N1 cells expressing the N-terminal HA-tagged wild-type $P2Y_4$ receptor, UTP promoted inositol phosphate accumulation that rapidly desensitized, reaching steady state within 10 min. This cessation of receptor-induced responses was accompanied by an ~50% loss of receptor immunoreactivity on the cell surface, which was recovered in an intracellular compartment. Cell surface receptor expression returned to control levels within 30 min of agonist removal, suggesting an efficient recycling of internalized $P2Y_4$ receptor. Because no changes in agonist-promoted receptor internalization were observed after downregulation of protein kinase C or after mutating the protein kinase C consensus site (RLRS) in the third intracellular loop of the $P2Y_4$ receptor, it was concluded that protein kinase C plays no major role in agonist-mediated $P2Y_4$ receptor downregulation. However, the 55-amino acid-long carboxyl-terminal region of the $P2Y_4$ receptor contains 11 serine/threonine residues that may be potentially phosphorylated not only by protein kinase C but also by members of the G protein-coupled receptor kinase family, and casein kinase. Simultaneous mutations Ser-333 and Ser-334 reduced the capacity of UTP to induce loss of surface $P2Y_4$ receptor. Furthermore, UTP-induced $P2Y_4$ receptor phosphorylation was totally lost in a truncation mutant that deleted all residues after position

332, but not after position 343. Together, these results indicated that phosphorylation of Ser-333 and Ser-334 is important for agonist-induced P2Y$_4$ receptor desensitization and internalization.

## D. The P2Y$_6$ Receptor

The P2Y$_6$ receptor is a highly selective uridine nucleotide receptor. In all species in which the P2Y$_6$ receptor has been cloned so far (i.e., human, rat, and mouse) it is potently activated by UDP, and ADP is a much weaker agonist. Specific receptors for pyrimidine nucleotides had been previously proposed to mediate the actions of UDP in several tissues, including C6-2B rat glioma cells and rat macrophages (Seifert and Schultz, 1989; Lazarowski and Harden, 1994; Liu *et al.*, 1995). However, the identification of the molecular species involved in these responses remained elusive even after the rat P2Y$_6$ receptor was cloned in 1995. Chang *et al.* (1995) reported the isolation from a rat aortic smooth muscle cell cDNA library of a novel nucleotide receptor that promoted the activation of phospholipase C with an agonist order potency of UTP > ADP > ATP. No effect on cyclic AMP accumulation was observed. The effect of UDP was not investigated in that study. This receptor, which was later named the P2Y$_6$ receptor, is 40, 44, and 38% homologous to the human P2Y$_4$, P2Y$_2$, and P2Y$_1$ receptors, respectively. Nicholas *et al.* (1996) have amplified the P2Y$_6$ receptor from a rat genomic DNA library, using PCR, and have stably expressed it in 1321N1 cells. When the nucleotide selectivity for inositol phosphate accumulation was assayed with highly purified nucleotides and conditions were carefully controlled to minimize nucleotide metabolism, UDP was revealed as the most potent P2Y$_6$ receptor agonist (EC$_{50}$ of 250 n$M$), while UTP and ADP acted only at high micromolar concentrations. Transcripts for the P2Y$_6$ receptor were identified in C6-2B cells (Nicholas *et al.*, 1996), strongly suggesting that the P2Y$_6$ receptor could account for the actions of UDP described previously on these cells (Lazarowski and Harden, 1994). The human (Communi *et al.*, 1996b) and the mouse (Lazarowski *et al.*, 2001b) P2Y$_6$ receptors display 83 and 91% overall homology to the rat P2Y$_6$ receptor, respectively. A chicken species 60% homologous to the rat P2Y$_6$ receptor was reported (Webb *et al.*, 1996) and characterized as a novel nucleoside diphosphate recognizing receptor activated preferentially by UDP (EC$_{50}$ of 0.12 $\mu M$) over ADP (EC$_{50}$ of 1.7 $\mu M$). This species was named the p2y$_3$ receptor, although it may represent the avian version of the mammalian P2Y$_6$ receptor (Li *et al.*, 1998).

Although UTP was reported as a full agonist at the cloned P2Y$_6$ receptors, UTP effects appear to be secondary to metabolic conversion to UDP, as suggested by kinetic differences between UDP and UTP on $Ca^{2+}$ responses in 1321N1 cells expressing the mouse $P2Y_6$ receptor. In these cells, UDP (100 n$M$) triggered full $Ca^{2+}$ mobilization within 2 min, while UTP (1 $\mu M$) induced a gradual $Ca^{2+}$ response, which reached a maximum only 20 min after its addition (Lazarowski et al., 2001b). 5Br-UDP is equipotent to UDP, and the stable uridine nucleotide derivatives UDP$\beta$S and Up$_3$U are also potent $P2Y_6$ receptor agonists that do not activate the $P2Y_1$, $P2Y_2$, or $P2Y_4$ receptor (Hou et al., 2001; Pendergast et al., 2001). Reactive Blue 2, PPADS, and suramin are weak and not selective $P2Y_6$ receptor antagonists (Robaye et al., 1997).

Northern blot analyses indicate a wide tissue distribution of $P2Y_6$ receptor messages. In humans, $P2Y_6$ receptor transcripts are abundantly expressed in placenta and spleen, and to a lesser extent in lung, thymus, small intestine, and blood leukocytes. Only a weak signal is observed in the kidney, heart, ovary, and colon. No $P2Y_6$ receptor message was detected in the brain (Communi et al., 1996b). In the rat, $P2Y_6$ receptor transcripts are abundantly expressed in the lung, stomach, intestine, spleen, mesentery, and aorta smooth muscle cells, and less abundantly in the heart and kidney. Again, no message could be detected in the brain by Northern blot (Chang et al., 1995). RT-PCR revealed $P2Y_6$ transcripts in brain capillary endothelial cells (Filippov et al., 1999), and in the rat neonatal whole heart (Webb et al., 1996).

The $P2Y_6$ receptor has been associated with nucleotide-promoted responses in several tissues. The most compelling evidence supporting a functional role for the $P2Y_6$ receptor derived from studies with the $P2Y_2-/-$ mouse model (Cressman et al., 1999). In freshly isolated gallbladder epithelium from wild-type mice, a robust $Cl^-$ secretory response to mucosal UDP was observed. UDP-promoted ion transport was retained intact in the $P2Y_2-/-$ mouse gallbladder epithelium, and was not mimicked by UTP or ATP. $P2Y_6$ receptor transcripts were revealed in this tissue, by RT-PCR. UDP-promoted responses in murine gallbladder epithelium also were unaffected after disrupting the CFTR gene, consistent with $P2Y_6$ receptor-promoted $Ca^{2+}$-activated $Cl^-$ conductance that is different from that of the cyclic AMP-regulated CFTR channel (Lazarowski et al., 2001b). $P2Y_6$ receptor-mediated responses were associated with the action of UDP on $Cl^-$ secretory responses in primary human nasal epithelium (Lazarowski et al., 1997c) and stimulation of ciliary activity in explants of human nasal epithelium (Morse et al., 2001). There is strong evidence of a physiological role for the $P2Y_6$ receptor in the rat cardiovascular system. $P2Y_6$ receptor transcripts are abundantly expressed in rat aorta smooth muscle cells where UDP and UDP$\beta$S promote mitogenic responses (Chang et al., 1995; Erlinge et al., 1998; Hou et al., 2002). In the isolated mesenteric artery, UDP and UDP$\beta$S stimulate potent contractile responses (Malmsjo et al., 2000b). The

$P2Y_6$ receptor may also be involved in the UDP-promoted $Ca^{2+}$ mobilization in rat astrocytes (Bolego *et al.*, 2001; Delicado *et al.*, 2001).

Unlike the $P2Y_2$ and $P2Y_4$ receptors, the $P2Y_6$ receptor expressed in 1321N1 cells desensitizes and internalizes slowly during agonist challenge, and fails to recycle after agonist removal (Robaye *et al.*, 1997; Brinson and Harden, 2001). The decrease in agonist-induced $P2Y_6$ receptor desensitization could be due to the fact that the $P2Y_6$ receptor lacks the consensus site for phosphorylation by protein kinase C in the third intracellular loop, and has only one potential site for phosphorylation in the carboxyl terminal. However, engineering the $P2Y_6$ receptor to express either a protein kinase C phosphorylation site in the third loop or the serine- and threonine-rich carboxyl-terminal tail of the $P2Y_4$ receptor did not suffice to confer UDP-induced loss of cell surface receptor (Brinson and Harden, 2001).

## E. The $P2Y_{11}$ Receptor

Communi *et al.* (1997) reported the cloning of the $P2Y_{11}$ receptor from human placenta cDNA and genomic DNA libraries. The human $P2Y_{11}$ receptor is 33% identical to the human $P2Y_1$ receptor, its closest homolog, and is activated by adenine but not uridine nucleotides to promote pertussis toxin-insensitive phosphoinositide breakdown, and cyclic AMP formation (Fig. 2). Although it is believed that the $P2Y_{11}$ receptor couples directly to $G_s$ and adenylyl cyclase to promote cyclic AMP formation, this pathway has not been formally proved, due in part to the fact that assaying agonist-promoted conversion of ATP to cyclic AMP in a cell-free system is hampered by the dual role of ATP, as $P2Y_{11}$ receptor agonist and adenylyl cyclase substrate. Moreover, Qi *et al.* (2001) have shown that nucleotide-promoted cellular cyclic AMP accumulation in $P2Y_{11}$ receptor-expressing 1321N1 astrocytoma or CHO cells was reduced up to 60–80% by downregulation of protein kinase C, and that chelating the intracellular $Ca^{2+}$ also inhibited, by 45%, the production of cyclic AMP in 1321N1 cells, but not in CHO cells. Thus, activation of protein kinase C and intracellular $Ca^{2+}$ mobilization as consequences of inositol lipid hydrolysis potentiated (in a cell type-dependent manner) the capacity of ATP to increase cyclic AMP levels via the $P2Y_{11}$ receptor.

Deoxy-ATP (dATP) and ATP are the most potent naturally occurring agonists at the human $P2Y_{11}$ receptor, but the receptor is more potently activated by 2',3'-benzoylbenzoyl-ATP (BzATP) and ATPγS and by the $P2Y_{12}$ receptor antagonist ARC67085 (2-propylthio-$\beta$,$\gamma$-dichloromethylene-D-ATP). The agonist potency order (ARC67085 > ATPγS $\approx$ BzATP > dATP > ATP > ADPβS > 2MeSATP) is the same for stimulation of inositol phosphate formation and for cyclic AMP accumulation, but agonist potencies for cyclic AMP formation are 10 to 15 times lower (i.e., $EC_{50}$ values are greater) than for inositol phosphate responses (Communi et al., 1999b; Qi et al., 2001), suggesting that receptor coupling to adenylyl cyclase is much weaker than coupling to phospholipase C. $ADP\beta S$, $AMP\alpha S$, and A3P5PS are partial agonists and inhibit ATP responses. Suramin is a relatively potent competitive antagonist for ATP-promoted inositol phosphate responses ($K_i = 0.8$ $\mu M$) and cyclic AMP responses ($IC_{50} = 16$ $\mu M$; $K_i$ value not determined) (Communi et al., 1999b). A canine cDNA encoding a receptor that is 70% identical to the human $P2Y_{11}$ receptor was cloned and functionally characterized as an adenine nucleotide-selective phospholipase C- and adenylyl cyclase-coupled receptor (Zambon et al., 2001). The canine $P2Y_{11}$ receptor, however, is activated by an agonist potency order of $ADP\beta S$ = 2MeSADP >> ADP > ATP, suggesting important structural differences between the human and canine putative orthologs.

The gene encoding the human $P2Y_{11}$ receptor is unusual among the P2Y receptor genes in that it contains a 1.9-kb intron between the exon encoding the six first amino acids of the receptor and the exon encoding the remaining seven-transmembrane region of the protein. In addition, the $P2Y_{11}$ gene forms an intergenic splicing with the adjacent *SSF1* gene, a rare phenomenon in normal mammalian cells that may have evolutionary implications (Communi et al., 2001).

Northern blot analysis indicates that $P2Y_{11}$ receptor transcripts are abundantly expressed in the spleen and weakly in the small intestine (Communi et al., 1997) and in the liver (Communi et al., 2001). No signal was detected in the heart, brain, lung, skeletal muscle, kidney, placenta, and pancreas (Communi et al., 1997). HL60 promyelocitic leukemia cells express an endogenous receptor with pharmacological characteristics of the $P2Y_{11}$ receptor (Conigrave et al., 1998). High levels of $P2Y_{11}$ receptor mRNA were found in these cells (Communi et al., 1997). Data from HL60 cells and other leukemia-derived cell lines suggest that the $P2Y_{11}$ receptor regulates a cyclic AMP-dependent pathway that controls granulocyte differentiation (Communi et al., 2000b; Wyden et al., 2000).

### F. The $P2Y_{12}$ Receptor

Pharmacological evidence indicates that two ADP-stimulated receptors are important for activation of circulating platelets. As discussed above, the $P2Y_1$ receptor is largely responsible for the increase in intracellular $Ca^{2+}$ concentrations in response to ADP necessary to support platelet activation. The second receptor involved in ADP-activated platelet aggregation (called $P2Y_{AC}$ or $P_{2T}$) has been predicted to inhibit the formation of cyclic AMP via activation of the $G_i$ family of G proteins (Hourani and Cusack, 1991; Gachet, 2000). The identity of the $G_i$-coupled ADP receptor of circulating platelets remained elusive for nearly 20 years, until the $P2Y_{12}$ receptor was molecularly and functionally characterized in two contexts, by Hollopeter *et al.* (2001) and by Zhang *et al.* (2001). Using an expression system in *Xenopus* oocytes that allows the detection of $G_i$-linked responses through an electrophysiological assay that measures inwardly activating $K^+$ channels, Hollopeter *et al.* (2001) reported the isolation of a clone from a rat platelet cDNA library that confers ADP-stimulated $K^+$ currents to injected oocytes. The human homolog of this clone, named the $P2Y_{12}$ receptor, was subsequently isolated and functionally characterized. As predicted for $G_i$-linked receptors, $P2Y_{12}$ receptor-mediated ADP actions were prevented by pertussis toxin. 2MeSADP is two orders of magnitude more potent than ADP (with half-maximal responses at 0.9 and 300 n$M$, respectively). 2MeSAMP and C1330-7, but not the $P2Y_1$ receptor antagonist A3P5P, inhibited ADP-induced $K^+$ currents in oocytes expressing the human $P2Y_{12}$ receptor. Expression of the $P2Y_{12}$ receptor in CHO cells resulted in ADP-mediated inhibition of forskolin-stimulated cyclic AMP formation (Hollopeter *et al.*, 2001). Zhang *et al.* (2001) identified ADP as the component in a spinal cord extract that most potently activated the orphan G protein-coupled receptor SP1999. SP1999 conferred $G_i$-linked ADP-dependent inhibition of forskolin-stimulated adenylyl cyclase to transfected CHO cells ($EC_{50}$ of 60 n$M$). Other nucleotides activated SP1999 with a rank order potency of 2MeSADP $\geq$ 2MeSATP > ADP > ADP$\beta$S > 2Cl-ATP > ATP$\gamma$S. As it turned out, SP1999 was identical to the $P2Y_{12}$ receptor. The $P2Y_{12}$ receptor is most closely related to the $P2Y_{UDPG}$ receptor (44% identity) but has little homology (19%) to the $P2Y_1$ receptor. ARC67085 and 2MeSAMP are potent antagonists for the $P2Y_{12}$ receptor. The thiol reagent *p*-chloromercuriphenylsulfonic acid (pCMBS) suppressed ADP-evoked responses in $P2Y_{12}$ receptor-expressing oocytes.

The $P2Y_{12}$ receptor mRNA is selectively and abundantly expressed in platelets and to a lesser extent in the brain, most likely in glial cells. The expression of $P2Y_{12}$ receptor protein on the surface of circulating platelets was confirmed with a polyclonal antiserum directed against the predicted amino terminus of the rat $P2Y_{12}$ receptor (Hollopeter *et al.*, 2001).

A mutation in the gene encoding the $P2Y_{12}$ receptor was identified in a patient with a mild bleeding disorder associated with impaired ADP-dependent platelet aggregation. ADP-stimulated $P2Y_1$ receptor-mediated platelet responses such as intracellular $Ca^{2+}$ mobilization and shape change were not affected in this patient (Hollopeter *et al.*, 2001).

The platelet $G_i$-coupled $P2Y_{12}$ receptor is essential for the full aggregation response to ADP and stabilization of platelet aggregates, and is the molecular target of the ADP-selective antiaggregating drugs ticlopidine and clopidogrel (Gachet, 2000; Bennett, 2001; Savi et al., 2001).

A receptor with pharmacological and signaling properties similar to those of the $P2Y_{12}$ receptor has been previously reported in C6 and C6-2B rat glioma cells (Valeins et al., 1992; Boyer et al., 1993). RT-PCR studies have confirmed the expression of $P2Y_{12}$ receptor transcripts in C6-2B cells (Jin et al., 2001).

## G. The $P2Y_{UDPG}$ Receptor

A G protein-coupled receptor for UDP-glucose ($P2Y_{UDPG}$) was identified by Chambers et al. (2000), using a reverse pharmacology strategy. By screening a large library of putative natural G protein-coupled receptor agonists, UDP-glucose and some closely related nucleotide-sugars were found to stimulate a yeast strain expressing the human K1AA0001 orphan receptor. The cells had been engineered to respond to agonist activation with expression of an inducible FUSI-lacZ reporter gene. When K1AA0001 was expressed in mammalian cells (HEK-293 cells) cotransfected with $G\alpha_{16}$ to promote coupling of intracellular $Ca^{2+}$, UDP-glucose and UDP-galactose elicited $Ca^{2+}$ responses with $EC_{50}$ values of 104 and 421 n$M$, respectively. UDP-glucuronic acid and UDP-$N$-acetylglucosamine were less potent agonists, but CDP-glucose, ADP-glucose, and other adenine, guanine, and cytidine nucleotide-sugars were inactive, as were UTP, UDP, ATP, and other nucleotides. Activation of $Ca^{2+}$ responses by UDP-glucose in HEK-293 cells transfected with K1AA0001 was not observed in the absence of $G\alpha_{16}$. However, UDP-glucose, but not ATP or UTP, promoted binding of GTP$\gamma$S to membranes from HEK-293 cells that were transfected with K1AA0001 alone, without recombinant G proteins. Nontransfected cells were unresponsive. The effect of UDP-glucose on GTP$\gamma$S binding was inhibited by pertussis toxin, suggesting that the expressed $P2Y_{UDPG}$ receptor coupled to native G proteins of the $G_{i/o}$ class. The UDP-glucose receptor shares 44% amino acid identity with the $P2Y_{12}$ receptor, but has little similarity to other P2Y receptors.

The $P2Y_{UDPG}$ receptor has a broad tissue distribution. Quantitative PCR indicated the highest levels of expression in placenta, adipose tissue, stomach, and intestine, and moderate levels in brain, spleen, lung, and heart (Chambers et al., 2000). Previously, few examples in the literature suggested pharmacological actions of UDP-glucose. UDP-glucose evoked depolarization of rat superior cervical ganglion (Connolly and Harrison, 1995), and promoted contractile responses in the rat phrenic diaphragm (Pastoris *et al.*, 1981). The involvement of the $P2Y_{UDPG}$ receptor in these responses has not been verified.

## IV. CONCLUDING COMMENTS

The biological significance of extracellular nucleotides has been supported by pharmacological studies of P2 receptors that are potently stimulated by ATP, UTP, ADP, and UDP; by direct demonstration of ATP and UTP release from various cell types and tissues under resting and stimulated conditions; and by the autocrine activation of P2 receptors by released nucleotides. The introduction of molecular biology in this area of study has led to the cloning and identification not only of P2Y receptors previously described by pharmacological criteria ($P2Y_1$, $P2Y_2$, $P2Y_6$, and $P2Y_{12}$), but also of receptors whose existence was not revealed by previous studies ($P2Y_4$, $P2Y_{11}$, and $P2Y_{UDPG}$). It is now clear that the P2Y receptor family is composed of adenine nucleotide-selective receptors ($P2Y_1$, $P2Y_{11}$, and $P2Y_{12}$), uridine nucleotide and nucleotide-sugar-selective receptors (human $P2Y_4$, $P2Y_6$, and $P2Y_{UDPG}$), and receptors that do not discriminate between purine and pyrimidine nucleotides ($P2Y_2$ and mouse/rat $P2Y_4$).

The availability of null systems in which recombinant P2Y receptors can be individually studied without the interference of endogenous background has revealed differences in selectivity in receptor $\rightarrow$ G protein $\rightarrow$ effector coupling, indicating that the $P2Y_1$, $P2Y_2$, $P2Y_4$, and $P2Y_6$ receptors are $G_q$ and phospholipase C linked, that the $P2Y_{11}$ couples to phospholipase C via $G_q$ and to adenylyl cyclase via $G_s$, and that the $P2Y_{12}$ receptor couples negatively to adenylyl cyclase via $G_i$ but does not activate phospholipase C. However, in their natural environment, P2Y receptors may be less discriminatory in their G protein coupling. It has been speculated that P2Y receptors may signal through phospholipase $A_2$ and phosphatidylcholine–phospholipase D in a phosphatidylinositide- and phospholipase C-independent manner (Neary *et al.*, 1999), and that they may also signal through nonenzymatic protein–protein interactions that may be dependent or independent of G protein activation (Erb *et al.*, 2001; Hall *et al.*, 1998; Szabo *et al.*, 2001).

The availability of the $P2Y_1$ and $P2Y_2$ knockout mice has opened new possibilities to explore the correspondence between the properties of cloned receptors with physiologically relevant functions. The availability of cloned receptors and the perspective of available receptor-deficient animal models should facilitate the development of selective and high-affinity P2Y receptor antagonists suitable for studying P2Y receptors in intact tissues, and for defining their role during short-term cell responses as well as in the complex and multisignaling pathways that transduce P2Y receptor occupancy into long-term responses such as differentiation, proliferation, and regulated cell death.

## References

Akbar, G. K. M., Dasari, V. R., Webb, T. E., Ayyanathan, K., Pillarisetti, K., Sandhu, A. K., Athwal, R. S., Daniel, J. L., Ashby, B., Barnard, E. A., and Kunapuli, S. P. (1996). Molecular cloning of a novel P2 purinoceptor from human erythroleukemia cells. *J. Biol. Chem.* **271,** 18363–18367.

Angel, P., and Karin, M. (1991). The role of Jun, Fos and the AP-1 complex in cell-proliferation and transformation. *Biochim. Biophys. Acta.* **1072,** 129–157.

Anwar, Z., Albert, J. L., Gubby, S. E., Boyle, J. P., Roberts, J. A., Webb, T. E., and Boarder, M. R. (1999). Regulation of cyclic AMP by extracellular ATP in cultured brain capillary endothelial cells. *Br. J. Pharmacol.* **128,** 465–471.

Baltensperger, K., and Porzig, H. (1997). The P2U purinoceptor obligatorily engages the heterotrimeric G protein G16 to mobilize intracellular $Ca^{2+}$ in human erythroleukemia cells. *J. Biol. Chem.* **272,** 10151–10159.

Baurand, A., Raboisson, P., Freund, M., Leon, C., Cazenave, J. P., Bourguignon, J. J., and Gachet, C. (2001). Inhibition of platelet function by administration of MRS2179, a P2Y1 receptor antagonist. *Eur. J. Pharmacol.* **412,** 213–221.

Bennett, J. S. (2001). Novel platelet inhibitors. *Annu. Rev. Med.* **52,** 161–184.

Boarder, M. R. (1998). Cyclic AMP and tyrosine kinase cascades in the regulation of cellular function by P2Y nucleotide receptors. *In* "The P2 Nucleotide Receptors" (J. Y. Turner, G. A. Weisman, and J. S. Fedan, eds.), pp. 413–426. Humana Press, Totowa, NJ.

Boarder, M. R., and Hourani, S. M. O. (1998). The regulation of vascular function by P2 receptors: Multiple sites and multiple receptors. *Trends Physiol. Sci.* **19,** 99–107.

Boehm, S., Huck, S., and Illes, P. (1995). UTP- and ATP-triggered transmitter release from rat sympathetic neurones via separate receptors. *Br. J. Pharmacol.* **116,** 2341–2343.

Bofill-Cardona, E., Vartian, N., Nanoff, C., Freissmuth, M., and Boehm, S. (2000). Two different signaling mechanisms involved in the excitation of rat sympathetic neurons by uridine nucleotides. *Mol. Pharmacol.* **57,** 1165–1172.

Bogdanov, Y. D., Dale, L., King, B. F., Whittock, N., and Burnstock, G. (1997). Early expression of a novel nucleotide receptor in the neural plate of *Xenopus* embryos. *J. Biol. Chem.* **272,** 12583–12590.

Bogdanov, Y. D., Wildman, S. S., Clements, M. P., King, B. F., and Burnstock, G. (1998). Molecular cloning and characterization of rat P2Y4 nucleotide receptor. *Br. J. Pharmacol.* **124,** 428–430.

Bolego, C., Centemeri, C., Abbracchio, M. P., Ceruti, S., Cattabeni, F., Jacobson, K. A., Puglisi, L., Rovati, G. E., Burnstock, G., and Nicosia, S. (2001). Two distinct P2Y receptors are involved in purine- and pyrimidine-evoked $Ca^{2+}$ elevation in mammalian brain astrocytic cultures. *Drug Dev. Res.* **52**(Suppl.), 122–132.

Boyer, J. L., Downes, C. P., and Harden, K. T. (1989). Kinetics of activation of phospholipase C by $P_{2Y}$ purinergic receptor agonists and guanine nucleotides. *J. Biol. Chem.* **264,** 884–890.

Boyer, J. L., Lazarowski, E. R., Chen, X.-H., and Harden, T. K. (1993). Identification of a $P_{2Y}$-purinergic receptor that inhibits adenylyl cyclase. *J. Pharm. Exp. Ther.* **267,** 1140–1146.

Boyer, J. L., Romero-Avila, T., Schachter, J. B., and Harden, T. K. (1996). Identification of competitive agonists of the P2Y1 receptor. *Mol. Pharmacol.* **50,** 1323–1329.

Boyer, J. L., Waldo, G. L., and Harden, T. K. (1997). Molecular cloning and expression of an avian G protein-coupled P2Y receptor. *Mol. Pharmacol.* **52**, 928–934.

Boyer, J. L., Mohanram, A., Camaioni, E., Jacobson, K. A., and Harden, T. K. (1998). Competitive and selective antagonism of P2Y1 receptors by $N^6$-methyl 2'-deoxyadenosine 3',5'-bisphosphate. *Br. J. Pharmacol.* **124**, 1–3.

Brinson, A. E., and Harden, T. K. (2001). Differential regulation of the uridine nucleotide-activated P2Y$_4$ and P2Y$_6$ receptors: SER-333 and SER-334 in the carboxyl terminus are involved in agonist-dependent phosphorylation desensitization and internalization of the P2Y$_4$ receptor. *J. Biol. Chem.* **276**, 11939–11948.

Brown, H. A., Lazarowski, E. R., Boucher, R. C., and Harden, T. K. (1991). Evidence that UTP and ATP regulate phospholipase C through a common extracellular 5'-nucleotide receptor in human airway epithelial cells. *Mol. Pharmacol.* **40**, 648–655.

Burnstock, G. (1972). Purinergic nerves. *Pharmacol. Rev.* **24**, 509–581.

Burnstock, G. (1978). A basis for distinguishing two types of purinergic receptor. *In* "Cell Membrane Receptors for Drugs and Hormones: A Multidisciplinary Approach" (L. B. Straub, ed.), pp. 107–118. Raven Press, New York.

Burnstock, G., and Kennedy, C. (1985). Is there a basis for distinguishing two types of P$_2$-purinoceptor? *Gen. Pharmacol.* **16**, 433–440.

Burnstock, G., and Williams, M. (2000). P2 purinergic receptors: Modulation of cell function and therapeutic potential. *J. Pharmacol. Exp. Ther.* **295**, 862–869.

Centemeri, C., Bolego, C., Abbracchio, M. P., Cattabeni, F., Puglisi, L., Burnstock, G., and Nicosia, S. (1997). Characterization of the Ca$^{2+}$ responses evoked by ATP and other nucleotides in mammalian brain astrocytes. *Br. J. Pharmacol.* **121**, 1700–1706.

Chambers, J. K., Macdonald, L. E., Sarau, H. M., Ames, R. S., Freeman, K., Foley, J. J., Zhu, Y., McLaughlin, M. M., Murdock, P., McMillan, L., Trill, J., Swift, A., Aiyar, N., Taylor, P., Vawter, L., Naheed, S., Szekeres, P., Hervieu, G., Scott, C., Watson, J. M., Murphy, A. J., Duzic, E., Klein, C., Bergsma, D. J., Wilson, S., and Livi, G. P. (2000). A G protein-coupled receptor for UDP-glucose. *J. Biol. Chem.* **275**, 10767–10771.

Chang, K., Hanaoka, K., Kumada, M., and Takuwa, Y. (1995). Molecular cloning and functional analysis of a novel P$_2$ nucleotide receptor. *J. Biol. Chem.* **270**, 26152–26158.

Charlton, S. J., Brown, C. A., Weisman, G. A., Turner, J. T., Erb, L., and Boarder, M. R. (1996). PPADS and suramin as antagonists at cloned P2Y- and P2U-purinoceptors. *Br. J. Pharmacol.* **118**, 704–710.

Clarke, L. L., Grubb, B. R., Yankaskas, J. R., Cotton, C. U., McKenzie, A., and Boucher, R. C. (1994). Relationship of a non-CFTR mediated chloride conductance to organ-level disease in cftr$^{-/-}$ mice. *Proc. Natl. Acad. Sci. USA* **91**, 479–483.

Communi, D., Pirotton, S., Parmentier, M., and Boeynaems, J.-M. (1995). Cloning and functional expression of a human uridine nucleotide receptor. *J. Biol. Chem.* **270**, 30849–30852.

Communi, D., Motte, S., Boeynaems, J. M., and Pirotton, S. (1996a). Pharmacological characterization of the human P2Y4 receptor. *Eur. J. Pharmacol.* **317**, 383–389.

Communi, D., Parmentier, M., and Boeynaems, J.-M. (1996b). Cloning, functional expression and tissue distribution of the human P2Y$_6$ receptor. *Biochem. Biophys. Res. Commun.* **222**, 303–308.

Communi, D., Govaerts, C., Parmentier, M., and Boeynaems, J. M. (1997). Cloning of a human purinergic P2Y receptor coupled to phospholipase C and adenylyl cyclase. *J. Biol. Chem.* **272**, 31969–31973.

Communi, D., Paindavoine, P., Place, G. A., Parmentier, M., and Boeynaems, J. M. (1999a). Expression of P2Y receptors in cell lines derived from the human lung. *Br. J. Pharmacol.* **127**, 562–568.

Communi, D., Robaye, B., and Boeynaems, J. M. (1999b). Pharmacological characterization of the human P2Y11 receptor. *Br. J. Pharmacol.* **128,** 1199–1206.

Communi, D., Janssens, R., Suarez-Huerta, N., Robaye, B., and Boeynaems, J. M. (2000a). Advances in signalling by extracellular nucleotides: The role and transduction mechanisms of P2Y receptors. *Cell Signal* **12,** 351–360.

Communi, D., Janssens, R., Robaye, B., Zeelis, N., and Boeynaems, J. M. (2000b). Rapid up-regulation of P2Y messengers during granulocytic differentiation of HL-60 cells. *FEBS Lett.* **475,** 39–42.

Communi, D., Suarez-Huerta, N., Dussossoy, D., Savi, P., and Boeynaems, J. M. (2001). Cotranscription and intergenic splicing of human P2Y11 and SSF1 genes. *J. Biol. Chem.* **276,** 16561–16566.

Conigrave, A. D., Lee, J. Y., van der, W. L., Jiang, L., Ward, P., Tasevski, V., Luttrell, B. M., and Morris, M. B. (1998). Pharmacological profile of a novel cyclic AMP-linked P2 receptor on undifferentiated HL-60 leukemia cells. *Br. J. Pharmacol.* **124,** 1580–1585.

Connolly, G. P., and Harrison, P. J. (1995). Structure–activity relationship of a pyrimidine receptor in the rat isolated superior cervical ganglion. *Br. J. Pharmacol.* **116,** 2764–2770.

Cowen, D. S., Sanders, M., and Dubyak, G. (1990). P2-purinergic receptors activate a guanine nucleotide-dependent phospholipase C in membranes from HL-60 cells. *Biochim. Biophys. Acta.* **1053,** 195–203.

Cressman, V. L., Lazarowski, E., Homolya, L., Boucher, R. C., Koller, B. H., and Grubb, B. R. (1999). Effect of loss of P2Y$_2$ receptor gene expression on nucleotide regulation of murine epithelial Cl$^-$ transport. *J. Biol. Chem.* **274,** 26461–26468.

Cunningham, M. L., Waldo, G. L., Hollinger, S., Hepler, J. R., and Harden, T. K. (2001). Protein kinase C phosphorylates RGS2 and modulates its capacity for negative regulation of G$_\alpha$ 11 signaling. *J. Biol. Chem.* **276,** 5438–5444.

Delicado, E. G., Jimenez, A. I., Castro, E., and Miras-Portugal, M. T. (2001). Cerebellar astrocytes coexpress different purinoceptors: Cross-talk between several transduction mechanisms. *Drug Dev. Res.* **52**(Suppl.), 114–121.

Dixon, C. J. (2000). Evidence that 2-methylthioATP and 2-methylthioADP are both agonists at the rat hepatocyte P2Y$_1$ receptor. *Br. J. Pharmacol.* **130,** 664–668.

Dohlman, H. G., and Thorner, J. (1997). RGS proteins and signaling by heterotrimeric G proteins. *J. Biol. Chem.* **272,** 3871–3874.

Donaldson, S. H., and Boucher, R. C. (1998). Therapeutic applications for nucleotides in lung disease. *In* "The P2 Nucleotide Receptors" (J. T. Turner, G. A. Weisman, and J. S. Fedan, eds.), pp. 413–424. Humana Press, Totowa, NJ.

Dranoff, J. A., O'Neill, A. F., Franco, A. M., Cai, S. Y., Connolly, G. C., Ballatori, N., Boyer, J. L., and Nathanson, M. H. (2000). A primitive ATP receptor from the little skate *Raja erinacea. J. Biol. Chem.* **275,** 30701–30706.

Dubyak, G. R. (1991). Signal transduction by P$_2$-purinergic receptors for extracellular ATP. *Am. J. Respir. Cell Mol. Biol.* **4,** 295–300.

Dubyak, G. R., and De Young, M. B. (1985). Intracellular Ca$^{2+}$ mobilization activated by extracellular ATP in Ehrlich ascites tumor cells. *J. Biol. Chem.* **260,** 10653–10661.

Erb, L., Garrad, R., Wang, Y., Quinn, T., Turner, J. T., and Weisman, G. A. (1995). Site-directed mutagenesis of P$_{2U}$ purinoceptors. Positively charged amino acids in transmembrane helices 6 and 7 affect agonist potency and specificity. *J. Biol. Chem.* **270,** 4185–4188.

Erb, L., Liu, J., Ockerhausen, J., Kong, Q., Garrad, R. C., Griffin, K., Neal, C., Krugh, B., Santiago-Perez, L. I., Gonzalez, F. A., Gresham, H. D., Turner, J. T., and Weisman, G. A. (2001). An RGD sequence in the P2Y$_2$ receptor interacts with $\alpha_v\beta_3$ integrins and is required for G$_o$-mediated signal transduction. *J. Cell Biol.* **153,** 491–501.

Erlinge, D., Hou, M., Webb, T. E., Barnard, E. A., and Moller, S. (1998). Phenotype changes of the vascular smooth muscle cell regulate P2 receptor expression as measured by quantitative RT-PCR. *Biochem. Biophys. Res. Commun.* **248**, 864–870.

Fabre, J. E., Nguyen, M., Latour, A., Keifer, J. A., Audoly, L. P., Coffman, T. M., and Koller, B. H. (1999). Decreased platelet aggregation, increased bleeding time and resistance to thromboembolism in P2Y1-deficient mice. *Nat. Med.* **5**, 1199–1202.

Fam, S. R., Gallagher, C. J., and Salter, M. W. (2000). P2Y$_1$ purinoceptor-mediated Ca$^{2+}$ signaling and Ca$^{2+}$ wave propagation in dorsal spinal cord astrocytes. *J. Neurosci.* **20**, 2800–2808.

Filippov, A. K., Webb, T. E., Barnard, E. A., and Brown, D. A. (1999). Dual coupling of heterologously-expressed rat P2Y6 nucleotide receptors to N-type Ca$^{2+}$ and M-type K$^+$ currents in rat sympathetic neurones. *Br. J. Pharmacol.* **126**, 1009–1017.

Filippov, A. K., Brown, D. A., and Barnard, E. A. (2000). The P2Y$_1$ receptor closes the N-type Ca$^{2+}$ channel in neurones, with both adenosine triphosphates and diphosphates as potent agonists. *Br. J. Pharmacol.* **129**, 1063–1066.

Filtz, T. M., Li, Q., Boyer, J. L., Nicholas, R. A., and Harden, T. K. (1994). Expression of a cloned P$_{2Y}$ purinergic receptor that couples to phospholipase C. *Mol. Pharmacol.* **46**, 8–14.

Freedman, N. J., and Lefkowitz, R. J. (1996). Desensitization of G protein-coupled receptors. *Recent Prog. Horm. Res.* **51**, 319–351.

Gachet, C. (2000). Platelet activation by ADP: The role of ADP antagonists. *Ann. Med.* **32**(Suppl. 1), 15–20.

Garrad, R. C., Otero, M. A., Erb, L., Theiss, P. M., Clarke, L. L., Gonzalez, F. A., Turner, J. T., and Weisman, G. A. (1998). Structural basis of agonist-induced desensitization and sequestration of the P2Y2 nucleotide receptor: Consequences of truncation of the C terminus. *J. Biol. Chem.* **273**, 29437–29444.

Godecke, S., Decking, U. K., Godecke, A., and Schrader, J. (1996). Cloning of the rat P2u receptor and its potential role in coronary vasodilation. *Am. J. Physiol.* **270**, C570–C577.

Grubb, B. R. (1997). Ion transport across the murine intestine in the absence and presence of CFTR. *Comp. Biochem. Physiol. A Physiol.* **118**, 277–282.

Hall, R. A., Ostedgaard, L. S., Premont, R. T., Blitzer, J. T., Rahman, N., Welsh, M. J., and Lefkowitz, R. J. (1998). A C-terminal motif found in the β$_2$-adrenergic receptor, P2Y1 receptor and cystic fibrosis transmembrane conductance regulator determines binding to the Na$^+$/H$^+$ exchanger regulatory factor family of PDZ proteins. *Proc. Natl. Acad. Sci. USA* **95**, 8496–8501.

Harden, T. K., Hawkins, P. T., Stephens, L., Boyer, J. L., and Downes, C. P. (1988). Phosphoinositide hydrolysis by guanosine 5'-[γ-thio]triphosphate-activated phospholipase C of turkey erythrocyte membranes. *Biochem. J.* **252**, 583–593.

Harden, T. K., Boyer, J. L., and Nicholas, R. A. (1995). P2-purinergic receptors: Subtype-associated signaling responses and structure. *Annu. Rev. Pharmacol. Toxicol.* **35**, 541–579.

Harper, S., Webb, T. E., Charlton, S. J., Ng, L. L., and Boarder, M. R. (1998). Evidence that P2Y4 nucleotide receptors are involved in the regulation of rat aortic smooth muscle cells by UTP and ATP. *Br. J. Pharmacol.* **124**, 703–710.

Haussinger, D., Stehle, T., and Gerok, W. (1987). Actions of extracellular UTP and ATP in perfused rat liver: A comparative study. *Eur. J. Biochem.* **167**, 65–71.

Haussinger, D., Busshardt, E., Stehle, T., Stoll, B., Wettstein, M., and Gerok, W. (1988). Stimulation of thromboxane release by extracellular UTP and ATP from perfused rat liver: Role of icosanoids in mediating the nucleotide responses. *Eur. J. Biochem.* **178**, 249–256.

Hechler, B., Vigne, P., Leon, C., Breittmayer, J. P., Gachet, C., and Frelin, C. (1998). ATP derivatives are antagonists of the P2Y1 receptor: Similarities to the platelet ADP receptor. *Mol. Pharmacol.* **53**, 727–733.

Henderson, D. J., Elliot, D. G., Smith, G. M., Webb, T. E., and Dainty, I. A. (1995). Cloning and characterisation of a bovine P2Y receptor. *Biochem. Biophys. Res. Commun.* **212,** 648–656.

Herold, C. L., Oi, A., Kennedy, C., Harden, T. K., and Nicholas, R. A. (1999). Analysis of chimeric human and rat P2Y4 receptors. *FASEB J.* **13**(Suppl. I), A464.

Hoffmann, C., Moro, S., Nicholas, R. A., Harden, T. K., and Jacobson, K. A. (1999). The role of amino acids in extracellular loops of the human P2Y1 receptor in surface expression and activation processes. *J. Biol. Chem.* **274,** 14639–14647.

Hollopeter, G., Jantzen, H. M., Vincent, D., Li, G., England, L., Ramakrishnan, V., Yang, R. B., Nurden, P., Nurden, A., Julius, D., and Conley, P. B. (2001). Identification of the platelet ADP receptor targeted by antithrombotic drugs. *Nature.* **409,** 202–207.

Homolya, L., Watt, W. C., Lazarowski, E. R., Koller, B. H., and Boucher, R. C. (1999). Nucleotide-regulated calcium signaling in lung fibroblasts and epithelial cells from normal and P2Y2 receptor–/– mice. *J. Biol. Chem.* **274,** 26454–26460.

Homolya, L., Steinberg, A. D., and Boucher, R. C. (2000). Cell to cell communication in response to mechanical stress via bilateral release of ATP and UTP in polarized epithelia. *J. Cell Biol.* **150,** 1349–1360.

Hou, M., Harden, T. K., Kuhn, C. M., Baldetorp, B., Lazarowski, E., Pendergast, W., Moeller, S., Edvinsson, L., and Erlinge, D. (2002). UDP is a growth factor for vascular smooth muscle cells by activation of $P2Y_6$ receptors. *Am. J. Physiol. Heart Circ. Physiol.* **282,** H784–H792.

Hourani, S. M. O., and Cusack, N. J. (1991). Pharmacological receptors on blood platelets. *Pharmacol. Rev.* **43,** 243–298.

Jacobson, K. A., Hoffmann, C., Kim, Y. C., Camaioni, E., Nandanan, E., Jang, S. Y., Guo, D. P., Ji, X. D., von Kugelgen, I., Moro, S., Ziganshin, A. U., Rychkov, A., King, B. F., Brown, S. G., Wildman, S. S., Burnstock, G., Boyer, J. L., Mohanram, A., and Harden, T. K. (1999). Molecular recognition in P2 receptors: Ligand development aided by molecular modeling and mutagenesis. *Prog. Brain Res.* **120,** 119–132.

Janssens, R., Communi, D., Pirotton, S., Samson, M., Parmentier, M., and Boeynaems, J.-M. (1996). Cloning and tissue distribution of the human $P2Y_1$ receptor. *Biochem. Biophys. Res. Commun.* **221,** 588–593.

Janssens, R., Paindavoine, P., Parmentier, M., and Boeynaems, J. M. (1999). Human P2Y2 receptor polymorphism: Identification and pharmacological characterization of two allelic variants. *Br. J. Pharmacol.* **127,** 709–716.

Jimenez, A. I., Castro, E., Communi, D., Boeynaems, J. M., Delicado, E. G., and Miras-Portugal, M. T. (2000). Coexpression of several types of metabotropic nucleotide receptors in single cerebellar astrocytes. *J. Neurochem.* **75,** 2071–2079.

Jin, J., Daniel, J. L., and Kunapuli, S. P. (1998). Molecular basis for ADP-induced platelet activation. II. The P2Y1 receptor mediates ADP-induced intracellular calcium mobilization and shape change in platelets. *J. Biol. Chem.* **273,** 2030–2034.

Jin, J., Dasari, V. R., Sistare, F. D., and Kunapuli, S. P. (1998). Distribution of P2Y receptor subtypes on haematopoietic cells. *Br. J. Pharmacol.* **123,** 789–794.

Jin, J., Tomlinson, W., Kirk, I. P., Kim, Y. B., Humphries, R. G., and Kunapuli, S. P. (2001). The C6-2B glioma cell $P2Y_{AC}$ receptor is pharmacologically and molecularly identical to the platelet $P2Y_{12}$ receptor. *Br. J. Pharmacol.* **133,** 521–528.

Kennedy, C., Qi, A. D., Herold, C. L., Harden, T. K., and Nicholas, R. A. (2000). ATP, an agonist at the rat P2Y4 receptor, is an antagonist at the human P2Y4 receptor. *Mol. Pharmacol.* **57,** 926–931.

King, B. F., Townsend-Nicholson, A., and Burnstock, G. (1998). Metabotropic receptors for ATP and UTP: Exploring the correspondence between native and recombinant nucleotide receptors. *Trends Pharmacol. Sci.* **19,** 506–514.

King, B. F., Burnstock, G., Boyer, J. L., Boeynaems, J.-M., Weisman, G. A., Kennedy, C., Jacobson, K. A., Humphries, R. G., Abbracchio, M. P., Gachet, C., and Miras-Portugal, M. T. (2000). "P2Y Receptors: The IUPHAR Compendium of Receptor Characterization of Classification 2000." International Union of Pharmacology, Irvine, CA.

Knowles, M. R., Clarke, L. L., and Boucher, R. C. (1991). Activation by extracellular nucleotides of chloride secretion in the airway epithelia of patients with cystic fibrosis. *N. Engl. J. Med.* **325,** 533–538.

Lazarowski, E. R., and Harden, T. K. (1994). Identification of a uridine nucleotide-selective G-protein-linked receptor that activates phospholipase C. *J. Biol. Chem.* **269,** 11830–11836.

Lazarowski, E. R., Watt, W. C., Stutts, M. J., Boucher, R. C., and Harden, T. K. (1995). Pharmacological selectivity of the cloned human P2U-receptor: Potent activation by diadenosine tetraphosphate. *Br. J. Pharmacol.* **116,** 1619–1627.

Lazarowski, E. R., Stutts, M. J., Watt, W. C., Brown, H. A., Boucher, R. C., and Harden, T. K. (1996). Enzymatic synthesis of $UTP_\gamma S$, a potent hydrolysis resistant agonist of $P_{2U}$-purinergic receptors. *Br. J. Pharmacol.* **117,** 203–209.

Lazarowski, E. R., Homolya, L., Boucher, R. C., and Harden, T. K. (1997a). Identification of an ectonucleoside diphosphokinase and its contribution to interconversion of P2 receptor agonists. *J. Biol. Chem.* **272,** 20402–20407.

Lazarowski, E. R., Homolya, L., Boucher, R. C., and Harden, T. K. (1997b). Direct demonstration of mechanically induced release of cellular UTP and its implication for uridine nucleotide receptor activation. *J. Biol. Chem.* **272,** 24348–24354.

Lazarowski, E. R., Paradiso, A. M., Watt, W. C., Harden, T. K., and Boucher, R. C. (1997c). UDP activates a mucosal-restricted receptor on human nasal epithelial cells that is distinct from the $P2Y_2$ receptor. *Proc. Natl. Acad. Sci. USA* **94,** 2599–2603.

Lazarowski, E. R., Boucher, R. C., and Harden, T. K. (2000). Constitutive release of ATP and evidence for major contribution of ecto- nucleotide pyrophosphatase and nucleoside diphosphokinase to extracellular nucleotide concentrations. *J. Biol. Chem.* **275,** 31061–31068.

Lazarowski, E. R., Boucher, R. C., and Harden, T. K. (2001a). Interplay of constitutively released nucleotides, nucleotide metabolism and activity of P2Y receptors. *Drug Dev. Res.* **53,** 66–71.

Lazarowski, E. R., Rochelle, L. G., O'Neal, W. K., Ribeiro, C. M., Grubb, B. R., Zhang, V., Harden, T. K., and Boucher, R. C. (2001b). Cloning and functional characterization of two murine uridine nucleotide receptors reveal a potential target for correcting ion transport deficiency in cystic fibrosis gallbladder. *J. Pharmacol. Exp. Ther.* **297,** 43–49.

Lenz, G., Gottfried, C., Luo, Z., Avruch, J., Rodnight, R., Nie, W. J., Kang, Y., and Neary, J. T. (2000). $P_{2Y}$ purinoceptor subtypes recruit different mek activators in astrocytes. *Br. J. Pharmacol.* **129,** 927–936.

Lenz, G., Goncalves, D., Luo, Z., Avruch, J., Rodnight, R., and Neary, J. T. (2001). Extracellular ATP stimulates an inhibitory pathway towards growth factor-induced cRaf-1 and MEKK activation in astrocyte cultures. *J. Neurochem.* **77,** 1001–1009.

Leon, C., Vial, C., Cazenave, J. P., and Gachet, C. (1996). Cloning and sequencing of a human cDNA encoding endothelial P2Y1 purinoceptor. *Gene* **171,** 295–297.

Leon, C., Hechler, B., Vial, C., Leray, C., Cazenave, J. P., and Gachet, C. (1997). The P2Y1 receptor is an ADP receptor antagonized by ATP and expressed in platelets and megakaryoblastic cells. *FEBS Lett.* **403,** 26–30.

Leon, C., Hechler, B., Freund, M., Eckly, A., Vial, C., Ohlmann, P., Dierich, A., Lemeur, M., Cazenave, J. P., and Gachet, C. (1999). Defective platelet aggregation and increased resistance to thrombosis in purinergic $P2Y_1$ receptor-null mice. *J. Clin. Invest.* **104,** 1731–1737.

Li, Q., Olesky, M., Palmer, R. K., Harden, T. K., and Nicholas, R. A. (1998). Evidence that the p2y3 receptor is the avian homologue of the mammalian P2Y6 receptor. *Mol. Pharmacol.* **54,** 541–546.

Liu, P., Wen, M., and Hayashi, J. (1995). Characterization of ATP receptor responsible for the activation of phospholipase A2 and stimulation of prostaglandin E2 production in thymic epithelial cells. *Biochem. J.* **308,** 399–404.

Lohse, M. J. (1993). Molecular mechanisms of membrane receptor desensitization. *Biochim. Biophys. Acta.* **1179,** 171–188.

Lustig, K. D., Shiau, A. K., Brake, A. J., and Julius, D. (1993). Expression cloning of an ATP receptor from mouse neuroblastoma cells. *Proc. Natl. Acad. Sci. USA* **90,** 5113–5117.

Malmsjo, M., Hou, M., Harden, T. K., Pendergast, W., Pantev, E., Edvinsson, L., and Erlinge, D. (2000a). Characterization of contractile P2 receptors in human coronary arteries by use of the stable pyrimidines uridine 5'-O-thiodiphosphate and uridine 5'-O-3-thiotriphosphate. *J. Pharmacol. Exp. Ther.* **293,** 755–760.

Malmsjo, M., Adner, M., Harden, T. K., Pendergast, W., Edvinsson, L., and Erlinge, D. (2000b). The stable pyrimidines UDP$\beta$S and UTP$\gamma$S discriminate between the P2 receptors that mediate vascular contraction and relaxation in the rat mesenteric artery. *Br. J. Pharmacol.* **131,** 51–56.

Mason, S. J., Paradiso, A. M., and Boucher, R. C. (1991). Regulation of transepithelial ion transport and intracellular calcium by extracellular adenosine triphosphate in human normal and cystic fibrosis airway epithelium. *Br. J. Pharmacol.* **103,** 1649–1656.

Mohanty, J. G., Raible, D. G., McDermott, L. J., Pelleg, A., and Schulman, E. S. (2001). Effects of purine and pyrimidine nucleotides on intracellular $Ca^{2+}$ in human eosinophils: Activation of purinergic P2Y receptors. *J. Allergy Clin. Immunol.* **107,** 849–855.

Morse, D. M., Smullen, J. L., and Davis, C. W. (2001). Differential effects of UTP, ATP, and adenosine of ciliary activity of human nasal epithelial cells. *Am. J. Physiol.* **280,** C1485–C1497.

Murthy, K. S., and Makhlouf, G. M. (1998). Coexpression of ligand-gated $P_{2X}$ and G protein-coupled $P_{2Y}$ receptors in smooth muscle: Preferential activation of P2Y receptors coupled to phospholipase C (PLC)-$\beta$1 via $G\alpha_{q/11}$. *J. Biol. Chem.* **273,** 4695–4704.

Nandanan, E., Jang, S. Y., Moro, S., Kim, H. O., Siddiqui, M. A., Russ, P., Marquez, V. E., Busson, R., Herdewijn, P., Harden, T. K., Boyer, J. L., and Jacobson, K. A. (2000). Synthesis, biological activity, and molecular modeling of ribose-modified deoxyadenosine bisphosphate analogues as P2Y$_1$ receptor ligands. *J. Med. Chem.* **43,** 829–842.

Neary, J. T. (2000). Trophic actions of extracellular ATP: Gene expression profiling by DNA array analysis. *J. Auton. Nerv. Syst.* **81,** 200–204.

Neary, J. T., Rathbone, M. P., Cattabeni, F., Abbracchio, M. P., and Burnstock, G. (1996). Trophic actions of extracellular nucleotides and nucleosides on glial and neuronal cells. *Trends Neurosci.* **19,** 13–18.

Neary, J. T., Kang, Y., Bu, Y., Yu, E., Akong, K., and Peters, C. M. (1999). Mitogenic signaling by ATP/P2Y purinergic receptors in astrocytes: Involvement of a calcium-independent protein kinase C, extracellular signal-regulated protein kinase pathway distinct from the phosphatidylinositol-specific phospholipase C/calcium pathway. *J. Neurosci.* **19,** 4211–4220.

Neer, E. J. (1995). Heterotrimeric G proteins: Organizers of transmembrane signals. *Cell* **80,** 249–257.

Nguyen, T., Erb, L., Weisman, G. A., Marchese, A., Heng, H. H. Q., Garrad, R. C., George, S. R., Turner, J. T., and O'Dowd, B. F. (1995). Cloning, expression, and chromosomal localization of the human uridine nucleotide receptor gene. *J. Biol. Chem.* **270,** 30845–30848.

Nicholas, R. A., Watt, W. C., Lazarowski, E. R., Li, Q., and Harden, T. K. (1996). Uridine nucleotide selectivity of three phospholipase C-activating $P_2$ receptors: Identification of a UTP-selective, a UTP-selective, and an ATP- and UTP-specific receptor. *Mol. Pharmacol.* **50,** 224–229.

O'Connor, S. E., Dainty, I. A., and Leff, P. (1991). Further subclassification of ATP receptors based on agonist studies. *Trends Pharmacol. Sci.* **12,** 137–141.

Offermanns, S., Toombs, C. F., Hu, Y. H., and Simon, M. I. (1997). Defective platelet activation in $G\alpha_q$-deficient mice. *Nature* **389,** 183–186.

Ogilvie, A., Blasius, R., Schulze-Lohoff, E., and Sterzel, R. B. (1996). Adenine dinucleotides: A novel class of signalling molecules. *J. Auton. Pharmacol.* **16,** 325–328.

Ohlmann, P., Eckly, A., Freund, M., Cazenave, J. P., Offermanns, S., and Gachet, C. (2000). ADP induces partial platelet aggregation without shape change and potentiates collagen-induced aggregation in the absence of $G\alpha_q$. *Blood* **96,** 2134–2139.

Palmer, R. K., Boyer, J. L., Schachter, J. B., Nicholas, R. A., and Harden, T. K. (1998). Agonist action of adenosine triphosphates at the human $P2Y_1$ receptor. *Mol. Pharmacol.* **54,** 1118–1123.

Parr, C. E., Sullivan, D. M., Paradiso, A. M., Lazarowski, E. R., Burch, L. H., Olsen, J. C., Erb, L., Weisman, G. A., Boucher, R. C., and Turner, J. T. (1994). Cloning and expression of a human $P_{2U}$ nucleotide receptor, a target for cystic fibrosis pharmacotherapy. *Proc. Natl. Acad. Sci. USA* **91,** 3275–3279.

Pastoris, O., Raimondo, S., Dossena, M., and Fulle, D. (1981). UDP-glucose effect on phrenic diaphragm preparation of the rat. *Farmaco Sci.* **36,** 721–728.

Paulmichl, M., and Lang, F. (1988). Enhancement of intracellular calcium concentration by extracellular ATP and UTP in Madin Darby canine kidney cells. *Biochem. Biophys. Res. Commun.* **156,** 1139–1143.

Pearson, G., Robinson, F., Beers, G. T., Xu, B., Karandikar, M., Berman, K., and Cobb, M. H. (2001). Mitogen-activated protein (MAP) kinase pathways: Regulation and physiological functions. *Endocr. Rev.* **22,** 153–183.

Pendergast, W., Yerxa, B. R., Douglass, J. G., III, Shaver, S.R., Dougherty, R. W., Redick, C. C., Sims, I. F., and Rideout, J. L. (2001). Synthesis and P2Y receptor activity of a series of uridine dinucleoside 5′-polyphosphates. *Bioorg. Med. Chem. Lett.* **11,** 157–160.

Pfeilschifter, J. (1990). Comparison of extracellular ATP and UTP signalling in rat renal mesangial cells. *Biochem. J.* **272,** 469–472.

Pintor, J., Diaz-Hernandez, M., Gualix, J., Gomez-Villafuertes, R., Hernando, F., and Miras-Portugal, M. T. (2000). Diadenosine polyphosphate receptors. from rat and guinea-pig brain to human nervous system. *Pharmacol. Ther.* **87,** 103–115.

Pirotton, S., Communi, D., Motte, S., Janssens, R., and Boeynaems, J. M. (1996). Endothelial P2-purinoceptors: Subtypes and signal transduction. *J. Auton. Pharmacol.* **16,** 353–356.

Purkiss, J. R., Wilkinson, G. F., and Boarder, M. R. (1994). Differential regulation of inositol 1,4,5-trisphosphate by co-existing P2Y-purinoceptors and nucleotide receptors on bovine aortic endothelial cells. *Br. J. Pharmacol.* **111,** 723–728.

Qi, A. D., Kennedy, C., Harden, T. K., and Nicholas, R. A. (2001). Differential coupling of the human $P2Y_{11}$ receptor to phospholipase C and adenylyl cyclase. *Br. J. Pharmacol.* **132,** 318–326.

Ralevic, V., and Burnstock, G. (1998). Receptors for purines and pyrimidines. *Pharmacol. Rev.* **50,** 413–492.

Rice, W. R., Burton, F. M., and Fiedeldey, D. T. (1995). Cloning and expression of the alveolar type II cell $P_{2U}$-purinergic receptor. *Am. J. Respir. Cell Mol. Biol.* **12,** 27–32.

Robaye, B., Boeynaems, J. M., and Communi, D. (1997). Slow desensitization of the human P2Y6 receptor. *Eur. J. Pharmacol.* **329,** 231–236.

Satterwhite, C. M., Farrelly, A. M., and Bradley, M. E. (1999). Chemotactic, mitogenic, and angiogenic actions of UTP on vascular endothelial cells. *Am. J. Physiol.* **276,** H1091–H1097.

Savi, P., Labouret, C., Delesque, N., Guette, F., Lupker, J., and Herbert, J. M. (2001). P2Y$_{12}$, a new platelet ADP receptor, target of clopidogrel. *Biochem. Biophys. Res. Commun.* **283,** 379–383.

Schachter, J. B., Li, Q., Boyer, J. L., Nicholas, R. A., and Harden, T. K. (1996). Second messenger cascade specificity and pharmacological selectivity of the human P2Y1-purinoceptor. *Br. J. Pharmacol.* **118,** 167–173.

Seifert, R., and Schultz, G. (1989). Involvement of pyrimidinoceptors in the regulation of cell functions by uridine and by uracil nucleotides. *Trends Pharmacol. Sci.* **10,** 365–369.

Seifert, R., Burde, R., and Schultz, G. (1989a). Activation of NADPH oxidase by purine and pyrimidine nucleotides involves G proteins and is potentiated by chemotactic peptides. *Biochem. J.* **259,** 813–819.

Seifert, R., Wenzel, K., Eckstein, F., and Schultz, G. (1989b). Purine and pyrimidine nucleotides potentiate activation of NADPH oxidase and degranulation by chemotactic peptides and induce aggregation of human neutrophils via G proteins. *Eur. J. Biochem.* **181,** 277–285.

Sellers, L. A., Simon, J., Lundahl, T. S., Cousens, D. J., Humphrey, P. P., and Barnard, E. A. (2001). Adenosine nucleotides acting at the human P2Y1 receptor stimulate mitogen-activated protein kinases and induce apoptosis. *J. Biol. Chem.* **276,** 16379–16390.

Shaver, S. R., Pendergast, W., Siddiqi, S. M., Yerxa, B. R., Croom, D. K., Dougherty, R. W., James, M. K., Jones, A. N., and Rideout, J. L. (1997). 4-Substituted uridine 5′-triphosphates as agonists of the P$_{2Y2}$ purinergic receptor. *Nucleotides Nucleosides* **16,** 1099–1102.

Short, S. M., Boyer, J. L., and Juliano, R. L. (2000). Integrins regulate the linkage between upstream and downstream events in G protein-coupled receptor signaling to mitogen-activated protein kinase. *J. Biol. Chem.* **275,** 12970–12977.

Soltoff, S. P. (1998). Related adhesion focal tyrosine kinase and the epidermal growth factor receptor mediate the stimulation of mitogen-activated protein kinase by the G-protein-coupled P$_{2Y2}$ receptor: Phorbol ester or [Ca$^{2+}$]$_i$ elevation can substitute for receptor activation. *J. Biol. Chem.* **273,** 23110–23117.

Sromek, S. M., and Harden, T. K. (1998). Agonist-induced internalization of the P2Y$_2$ receptor. *Mol. Pharmacol.* **54,** 485–494.

Stutchfield, J., and Cockcroft, S. (1990). Undifferentiated HL60 cells respond to extracellular ATP and UTP by stimulating phospholipase C activation and exocytosis. *FEBS Lett.* **262,** 256–258.

Suarez-Huerta, N., Pouillon, V., Boeynaems, J., and Robaye, B. (2001). Molecular cloning and characterization of the mouse P2Y$_4$ nucleotide receptor. *Eur. J. Pharmacol.* **416,** 197–202.

Suh, B. C., Kim, J. S., Namgung, U., Han, S., and Kim, K. T. (2001). Selective inhibition of $\beta_2$-adrenergic receptor-mediated cAMP generation by activation of the P2Y$_2$ receptor in mouse pineal gland tumor cells. *J. Neurochem.* **77,** 1475–1485.

Szabo, K., Stutts, M. J., Boucher, R. C., and Milgram, S. L. (2001). The COOH-terminus of the human P2Y2 receptor binds p53-binding protein 2 (53BP2). *Mol. Biol. Cell* **11,** 105a.

Tokuyama, Y., Hara, M., Jones, E. M. C., Fan, Z., and Bell, G. I. (1995). Cloning of rat and mouse P$_{2Y}$ purinoceptors. *Biochem. Biophys. Res. Commun.* **211,** 211–218.

Troadec, J. D., Thirion, S., Petturiti, D., Bohn, M. T., and Poujeol, P. (1999). ATP acting on P2Y receptors triggers calcium mobilization in primary cultures of rat neurohypophysial astrocytes (pituicytes). *Pflugers Arch.* **437,** 745–753.

Valeins, H., Merle, M., and Labouesse, J. (1992). Pre-steady state study of $\beta$-adrenergic and purinergic receptor interaction in C6 cell membranes: Undelayed balance between positive and negative coupling to adenylyl cyclase. *Mol. Pharmacol.* **42**, 33–41.

van der Wyden, L., Rakyan, V., Luttrell, B. M., Morris, M. B., and Conigrave, A. D. (2000). Extracellular ATP couples to cAMP generation and granulocytic differentiation in human NB4 promyelocytic leukaemia cells. *Immunol. Cell Biol.* **78**, 467–473.

Vanhoutte, P. M., Humphrey, P. P., and Spedding, M. (1996). X. International Union of Pharmacology recommendations for nomenclature of new receptor subtypes. *Pharmacol. Rev.* **48**, 1–2.

Walker, B. A., Hagenlocker, B. E., Douglas, V. K., Tarapchak, S. J., and Ward, P. A. (1991). Nucleotide responses of human neutrophils. *Lab. Invest.* **64**, 105–112.

Webb, T. E., Simon, J., Krishek, B. J., Bateson, A. N., Smart, T. G., King, B. F., Burnstock, G., and Barnard, E. A. (1993). Cloning and functional expression of a brain G-protein-coupled ATP receptor. *FEBS Lett.* **324**, 219–225.

Webb, T. E., Boluyt, M. O., and Barnard, E. A. (1996). Molecular biology of P2Y purinoceptors: Expression in rat heart. *J. Auton. Pharmacol.* **16**, 303–307.

Webb, T. E., Henderson, D. J., Roberts, J. A., and Barnard, E. A. (1998). Molecular cloning and characterization of the rat P2Y4 receptor. *J. Neurochem.* **71**, 1348–1357.

White, P. J., Kumari, R., Porter, K. E., London, N. J., Ng, L. L., and Boarder, M. R. (2000). Antiproliferative effect of UTP on human arterial and venous smooth muscle cells. *Am. J. Physiol. Heart Circ. Physiol.* **279**, H2735–H2742.

Widmann, C., Gibson, S., Jarpe, M. B., and Johnson, G. L. (1999). Mitogen-activated protein kinase: Conservation of a three-kinase module from yeast to human. *Physiol. Rev.* **79**, 143–180.

Yang, C. M., Tsai, Y. J., Pan, S. L., Tsai, C. T., Wu, W. B., Chiu, C. T., Luo, S. F., and Ou, J. T. (1997). Purinoceptor-stimulated phosphoinositide hydrolysis in Madin-Darby canine kidney (MDCK) cells. *Naunyn Schmiedebergs Arch. Pharmacol.* **356**, 1–7.

Yang, S., Buxton, I. L., Probert, C. B., Talbot, J. N., and Bradley, M. E. (1996). Evidence for a discrete UTP receptor in cardiac endothelial cells. *Br. J. Pharmacol.* **117**, 1572–1578.

Yokomizo, T., Izumi, T., Chang, K., Takuwa, Y., and Shimizu, T. (1997). A G-protein-coupled receptor for leukotriene B4 that mediates chemotaxis. *Nature* **387**, 620–624.

Zambon, A. C., Hughes, R. J., Meszaros, J. G., Wu, J. J., Torres, B., Brunton, L. L., and Insel, P. A. (2000). $P2Y_2$ receptor of MDCK cells: Cloning, expression, and cell-specific signaling. *Am. J. Physiol.* **279**, F1045–F1052.

Zambon, A. C., Brunton, L. L., Barrett, K. E., Hughes, R. J., Torres, B., and Insel, P. A. (2001). Cloning, expression, signaling mechanisms, and membrane targeting of $P2Y_{11}$ receptors in Madin Darby canine kidney cells. *Mol. Pharmacol.* **60**, 26–35.

Zhang, A. L., and Roomans, G. M. (1999). Regulation of intracellular calcium by extracellular nucleotides in pig tracheal submucosal gland cells. *Respir. Physiol.* **118**, 237–246.

Zhang, F. L., Luo, L., Gustafson, E., Lachowicz, J., Smith, M., Qiao, X., Liu, Y. H., Chen, G., Pramanik, B., Laz, T. M., Palmer, K., Bayne, M., and Monsma, F. J., Jr. (2001). ADP is the cognate ligand for the orphan G protein-coupled receptor SP1999. *J. Biol. Chem.* **276**, 8608–8615.

Zhang, H., Weir, B. K., Marton, L. S., Lee, K. S., and Macdonald, R. L. (1997). P2 purinoceptors in cultured bovine middle cerebral artery endothelial cells. *J. Cardiovasc. Pharmacol.* **30**, 767–774.

Zhong, Y., Dunn, P. M., and Burnstock, G. (2001). Multiple P2X receptors on guinea-pig pelvic ganglion neurons exhibit novel pharmacological properties. *Br. J. Pharmacol.* **132**, 221–233.

Zhu, Y., and Kimelberg, H. K. (2001). Developmental expression of metabotropic $P2Y_1$ and $P2Y_2$ receptors in freshly isolated astrocytes from rat hippocampus. *J. Neurochem.* **77,** 530–541.

Zimmermann, H. (2001). Ectonucleotidases: Some recent development and note on nomenclature. *Drug Dev. Res.* **52,** 44–56.

# CHAPTER 4

# Extracellular ATP-Gated P2X Purinergic Receptor Channels

**Amanda Taylor Boyce and Erik M. Schwiebert**
Departments of Cell Biology and Physiology and Biophysics, and Gregory Fleming James
Cystic Fibrosis Research Center, University of Alabama at Birmingham, Birmingham,
Alabama 35294

## I. SUMMARY

It was known for decades that application of ATP as an extracellular agonist on a cell or tissue led to "permeabilization" of the cell plasma membrane. There were many interpretations of this phenomenon, although a role in exacerbation of cell injury was a common conclusion. The receptor or receptors thought to underlie this effect were classified as the P2Z receptors in the early nomenclature. However, there were a handful of dogged researchers who hypothesized that extracellular ATP plays a physiological role in promoting ionic permeability in the plasma membrane, by activating ion channels. Channel activation in many experimental paradigms was so rapid that electrophysiologists hypothesized further that ATP might bind to and gate an ion channel directly. These hypotheses and experiments were the genesis of the P2X purinergic receptor channel field.

At least seven P2X receptor channel subtypes (P2XRs, as we refer to them) have now been identified and characterized. Other tissue-specific and developmentally regulated P2XRs are still emerging, as are P2XRs from lower model organisms such as zebrafish. P2XRs are calcium-permeable nonselective cation channels that, like the glutamate, $N$-methyl-D-aspartate (NMDA), and kainate receptor channels at excitatory synapses, are gated by an extracellular agonist, show divalent and monovalent permeation, and divalent cation block. P2XRs are thought to trigger signaling by mediating calcium influx from extracellular stores and by changing the plasma membrane potential; however, other signaling mechanisms may be governed by ATP-gated receptor channels. P2XRs, as well as the P2Y G-protein-coupled receptors, also stimulate other ion channels in native cells, complicating their definition in specific cell models. This is complicated further by the fact that P2XRs form homomultimers or heteromultimers like their relatives in the two transmembrane-spanning ion channel superfamily. Heteromeric P2XRs often show a mixed phenotype with regard to pharmacology, desensitization, and biophysics.

This review serves as a comprehensive treatment of all aspects of P2XR cell physiology. The biophysics of P2XRs are presented only briefly, since a companion chapter on P2XR biophysics is contributed in this volume by Egan, Voigt, and their co-workers. It is hoped that these reviews will kindle new interest or strengthen current interest in the study of P2X receptor channels.

## II. EXTRACELLULAR ATP-ACTIVATED ION CHANNELS

B. Bean was an early pioneer in the P2XR field. In many different cell and tissue systems, Bean and co-workers were applying ATP and related agonists to the extracellular surface of tissues, cells, and membrane patches and observing instantaneous activation of ion channel current, as determined by patch-clamp and voltage-clamp electrophysiology. The puzzling feature of this early work was the variability in the recordings. In seminal papers and reviews about this original electrophysiological research on native cells (Bean, 1990, 1992; Bean and Friel, 1990; Bean *et al.*, 1990), Bean explains that single-channel conductances of as little as 1 picosiemen (pS) to as much as 100 pS could be observed. The different channels also had different biophysical current signatures, with differing kinetics, dependence on voltage, and rate of inactivation (later determined to be desensitization). The permeability of the channels to monovalent and divalent ions also varied. This was exceedingly frustrating, because the variability in the recordings often made this research difficult to publish. Moreover, this issue of ATP permeabilization of cells always loomed in the background as a troubling issue. In hindsight, a subset of these channels was indeed P2XRs of varying types and, possibly, different multimeric mixtures. Other channels recorded immediately after ATP stimulation may have been separate ion channels that P2XRs and, possibly, P2Y G protein-coupled receptors were stimulating via rapid signal transduction mechanisms. These hypothesized cellular mechanisms of extracellular ATP-stimulated ion channels are shown in Fig. 1. Needless to say, only a handful of researchers remained excited about these ATP-stimulated ion channels and pressed on.

There were physiological reasons for believing that extracellular ATP-gated ion channels existed. Burnstock proposed the existence of so-called "purinergic nerves" in the gut and in the nervous system that were neither cholinergic or adrenergic (Burnstock, 1972). By inference, the ATP that was released from the synapse had to bind to a receptor channel to propagate the

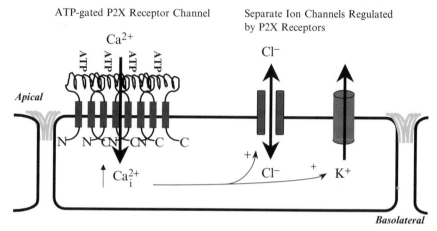

**FIGURE 1**   P2X ATP-gated ion channels versus ATP-regulated ion channels through P2X receptors. Simple schematic illustrating that care must be taken in evaluating ATP-activated ion channels that could be bona fide P2XRs or separate ion channels regulated by P2XRs and dependent on cytosolic calcium.

action potential. Moreover, the ATP that was applied to whole cell patch-clamp and voltage-clamp preparations activated ion channels as rapidly as it was observed to activate the acetylcholine receptor channel in neurons and muscle. There was also a relative potency of different nucleotide agonists in the rapid activation of these ion channels, suggesting that this was a receptor-mediated effect that had physiological meaning. This work was a classic example of the adage that no data should be completely discounted or discarded because it cannot be fully explained at the time of experimentation.

## III. MOLECULAR CLONING

### A. P2X₁ and P2X₂ via Expression Cloning in Xenopus Oocytes

P2Y G protein-coupled receptors were cloned and sequenced several years prior to the emergence of P2X receptor molecular biology. Because of this, it was thought for several years that the ionic channels activated by extracellular ATP were regulated by G protein-dependent signaling mechanisms triggered by the P2Y G protein-coupled receptors, then referred to as P2Y (later renamed $P2Y_1$), P2U (later renamed $P2Y_2$), and P2Z (later determined to be the $P2X_7$ receptor channel). This was also the period in the study of ion channel regulation in which the relative roles of the $\alpha$ subunit versus the $\beta\gamma$ dimer of heterotrimeric G proteins were being debated.

It took some undaunted laboratories led by D. Julius and by A. North, G. Buell, and A. Surprenant to make the enormous commitment to expression cloning in *Xenopus* oocytes. Working simultaneously and publishing concurrently in 1994, Julius and co-workers isolated and cloned a cDNA originally derived from a cDNA library made from the PC-12 pheochromocytoma cell line (Brake *et al.*, 1994), while Buell and co-workers isolated and cloned a related but distinct cDNA from a rat vas deferens smooth muscle cDNA library (Valera *et al.*, 1994). The latter was named P2X$_1$ in a unifying nomenclature, while the Julius cDNA was classified as P2X$_2$. The nomenclature, P2X, was begun, because this ATP-gated channel was also a receptor that was previously unappreciated, an "X-factor" to the purinergic receptor field. Figure 2 shows a schematic of a P2XR gene and its transcription and translation to a P2XR channel protein. The curious finding from the molecular and functional characterization of these receptor channels was that they were most similar to inwardly rectifying K$^+$ channels

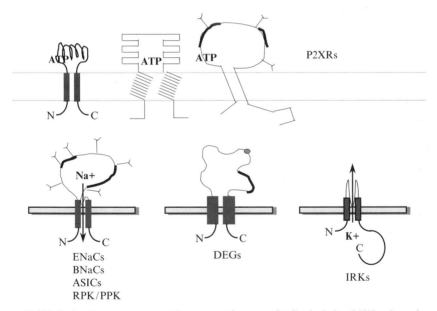

**FIGURE 2** The two transmembrane-spanning superfamily includes P2XR channels. Various schematics of P2XR topology are shown *(top)* versus ENaCs, BNaCs, and ASICs from mammalian cells, pickpocket (PPK) and ripped pocket (RPK) from *Drosophila*, the *degenerin* (DEG) channels from *Drosophila,* and inwardly rectifying K$^+$ channels (IRKs) from mammalian cells and lower organisms. It is important to note that the huge family of IRK channels lacks the large extracellular domain present in the other families depicted.

($K_{IR}$s), amiloride-sensitive $Na^+$ channels (ENaCs), and the *Caenorhabditis elegans* degenerins (DEGs), channels normally found in nonexcitable cells (see below) (Benos and Stanton, 1999; Bianchi and Driscoll, 2002; Kellenberger and Schild, 2002). Their predicted structure, which has held up to experimentation, revealed intracellular N and C termini, two transmembrane-spanning $\alpha$ helices, and a large extracellular region or domain with N-linked glycosylation sites and two cysteine-rich regions. In fact, the extracellular domain accounts for approximately 70% of the molecular mass of P2XR proteins. Figure 2 illustrates the structural features of P2XRs and how they relate to the other two transmembrane-spanning cation channel families. In these 2 initial clones and other members that emerged in later cloning, 10 cysteines are conserved in the same sites in all P2X receptor channel subtypes. This is eerily similar to the ENaC and DEG families and implies a similar three-dimensional structure that begs to be elucidated. This extracellular domain makes P2XRs diverge somewhat from the two transmembrane-spanning, inwardly rectifying $K^+$ channels. These revelations were somewhat surprising, because it was hypothesized by some that these P2X receptor channels would resemble voltage-dependent ion channels because of their known expression in excitable cells.

## B. P2XR: Exclusive Membership in the Two Transmembrane-Spanning Cation Channel Superfamily

Once these genes were isolated and the protein structure predicted, it became clear that P2XRs were unique members of a large superfamily of two transmembrane-spanning ion channels that includes the many inwardly rectifying $K^+$ channels ($K_{IR}$s), the mechanosensitive *Drosophila pickpocket* and *ripped pocket* channels and the *C. elegans degenerin* (DEG) channels, and the amiloride-sensitive epithelial $Na^+$ channels (ENaCs) and more newly cloned relatives (the "acid-sensing ion channels," the ASICs, and "brain ENaC channels," the BNaCs) (Benos and Stanton, 1999; Bianchi and Driscoll, 2002; Kellenberger and Schild, 2002). The structures of the P2XRs and the ENaCs are most similar, despite the fact that there is little or no sequence homology. P2XRs and the ASICs and BNaCs also share a similar overall structure. Although there is little or no sequence homology, the similar structure of the P2XR family and the ENaC family also suggests similar cell biology with regard to multimerization (see below) and, possibly, gating by an external ligand (although there have been no natural ligands identified for the ENaC family). The P2XR and ENaC families differ from the inwardly rectifying $K^+$ channels with regard to the large extracellular domain that is present in P2XRs and ENaCs, but is lacking in $K_{IR}$ channels.

## C. P2X₃ via Homology Cloning in Mammals and Zebrafish

The original P2XR cDNAs were followed quickly by homology cloning strategies that yielded many other members of the P2X gene family. Seguela *et al.* performed homology cloning, using a degenerate RT-PCR fragment isolated from rat brain RNA, and cloned a full-length cDNA from a rat brain cDNA library (Seguela *et al.*, 1996). This cDNA, $P2X_3$, was rapidly desensitizing (like $P2X_1$ but unlike $P2X_2$), potentiated by $Zn^{2+}$, and reversibly blocked by suramin, Reactive Blue 2, and pyridoxalphosphate-6-axophenyl-2′,4′-disulfonic acid (PPADS). Its expression was enriched in Purkinje cells and the granule cells of the cerebellum and in $CA_3$ pyramidal cells of the hippocampus. This was intriguing, because zinc-enriched mossy fibers innervate this region. This is perhaps the widespread neuronal specific subtype of the P2X family that is involved in fast excitatory synaptic transmission. More recently, a homolog of the $P2X_3$ receptor channel was isolated from zebrafish, *Danio rerio* (Egan *et al.*, 2000). Three groups published identification, characterization, and expression of $zP2X_3$ in 2000 (Egan *et al.*, 2000; Boue-Grabot *et al.*, 2000a; Norton *et al.*, 2000). One surprise is that similar homologs in other model organisms such as bacteria, yeast, *Drosophila*, and *C. elegans* have not emerged, despite the full sequencing of the genomes of these models (Fig. 3, See Color Insert). It is likely that more P2XR homologs and functions will emerge from these models. Not surprisingly, splice variants of the original members of the P2X superfamily have also been identified (Lynch *et al.*, 1999; Hardy *et al.*, 2000). P2XRs are prone to alternative splicing. This area of P2XR research is quite complex and is still emerging.

## D. Homology Cloning of Mammalian P2X₄, P2X₅, and P2X₆

Three independent laboratories isolated P2X4 by homology cloning in 1996 (Buell *et al.*, 1996; Soto *et al.*, 1996a,b; Wang *et al.*, 1996). Buell and co-workers showed that P2X4 was expressed in serous epithelial cells from salivary gland as well as in brain and sensory ganglia, the first demonstration of P2XR expression outside of smooth muscle cells or neurons. Another unique aspect of this new subtype was its insensitivity to PPADS. Stuhmer and colleagues cloned a similar cDNA from rat brain. In their hands, the P2XR channel was also insensitive to suramin. $P2X_4$ also displays high permeability to $Ca^{2+}$ over other cations and was modulated by $Zn^{2+}$ (as was $P2X_3$). Stuhmer's groups also showed a broader distribution of expression into different tissues. Like $P2X_2$, $P2X_4$ desensitizes poorly (unlike $P2X_1$ and $P2X_3$). Seino and co-workers cloned

**(a)   N-Terminus**

```
P2X1 MARRL  QDELS   AF x FF   EYDTP   RMVLV   RNK
P2X2 MVRRL  ARG xC  WSA FW    DYETP   KVIVV   RNR
P2X3 Mx x x x  x x NxC  ISDFF   TYETT   KSVVV   KSW
P2X4 MAGCx  CSVLG   SFxLF   EYDTP   RIVLI   RSR
P2X5 MGQAA  WKGFV   xLSLF   DYKTA   KFVVA   KSK
P2X6 MASAV  AAALV   SWG FL  DYKTE   KYVMT   RNC
P2X7 MPACC  x x xxS WNDVF   QYETN   KVTRI   QSV
     1           10              20              28
```

**(b)   M1 alpha-Helix**

```
P2X1 K KVGV   IF RL I  QLVVL  VYV IG  W xVFV  YEK GYQ
P2X2 R RL GF  VHRMV   QLL IL  LY FVx  WYVFI   VQK SYQ
P2X3 WT I GI  IN RAV   QLL II  SY FVG  W xVFL  HEK AYQ
P2X4 R KVGL   MNRAV   QLL IL  AYV IG  WxVFV   WEKGYQ
P2X5 K KVGL   LYRVL   QL I IL L YLLI  WxVFL   I KK SYQ
P2X6 CWVGI   SQR LL  QLGVV VYV IG  WxALL   AKK GYQ
P2X7 V NYGT   IKW IL  HMTVF  S YV Sx  x FALM  SDK LYQ
     28 30            40            50            58
```

**(c)   Extracellular Domain**

```
P2X1 YEK GYQT S  S x DLI  SSVSV  KLK GL  AVTQx  xxxxxxx  LQGLG  PQVW
P2X2 VQK SYQD S  E TGPE  SSI IT  KVKG I  TMSEx  xxxxxxx  xxxxx  DKVW
P2X3 HEK AYQV R  DTAI E  SSVVT  KVKGF  GRYAx  xxxxxxx  xxxxx  NRVM
P2X4 WEK GYQE T  D xSVV  SSVTT  KAKGV  AVTNx  xxxxxxx  TSQLG  FR IW
P2X5 I KK SYQD I  D TSLQ  SAVVT  KVKGV  AYTNx  xxxxxxx  TTMLG  ERLW
P2X6 AKK GYQEW  DMDPQ  ISV I T  KLKGV  SVTQx  xxxxxxx  VKELE  KRLW
P2X7 SDK  LYQR K  E P xLI  SSVHT  KVKGV  AEVTE  NVTEGG  VTKLV  HGI F
     53        60          70          80          90      95
```

```
P2X1 DVADY  VFPAH  G DSSF  VVMTN  FIVTP  QQTQG  HCAEN  PE xGx  G IC
P2X2 DVEEY  VKPPE  GGSVV  SII TR  IEVTP  SQT LG  TCPE S  MRVHS  STC
P2X3 DVSDY  VTP PQ  G TSVF  V II TK  MIVTE  NQMQG  FCPEN  EE KYR  xxC
P2X4 DVADY  VI PAQ  E ENSL  FIMTN  MIVTV  NQT QS  TCPE I  PD KTx  S IC
P2X5 DVADF  VI PSQ  G ENVF  FVVTN  LIVTP  NQR QS  TCAER  EG I PD  GEC
P2X6 DVADF  VRPSQ  G ENVF  FLVTN  FLVTP  AQV QG  RCPEH  PS VPL  ANC
P2X7 DTADY  TLPLQ  G NSxF  FVMTN  YLKSE  GQE QK  LCPEY  PS RGK  QxC
     96     100        110         120         130    135
```

**FIGURE 3**   *(Continued)*

**(c) Extracellular Domain**

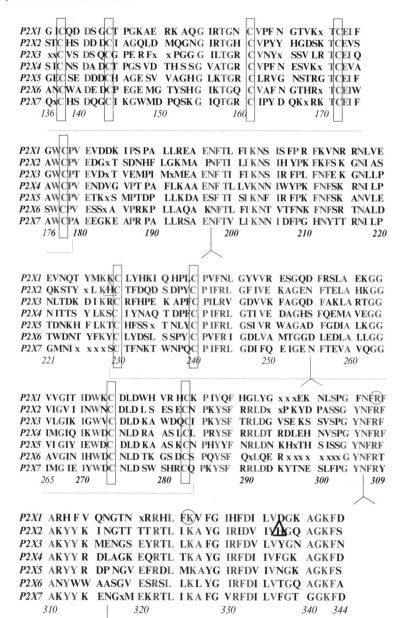

```
P2X1  G ICQD DS GCT PGKAE  RK AQG  IRTGN  C VPF N  GTVKx  TCEI F
P2X2  STCHS DD DCI AGQLD  MQGNG  IRTGH  C VPYY  HGDSK  TCEVS
P2X3  xxCVS DS QCG PE R Fx  x PGG G  ILTGR  C VNYx  SSV LR  TCEI Q
P2X4  S ICNS DA DCT PGS VD  TH S SG  VATGR  C VPF N  ESVKx  TCEVA
P2X5  GECSE DDDCH AGE SV  VAGHG  LKTGR  C LRVG  NSTRG  TCEI F
P2X6  ANCWA DE DCP EGE MG  TYSHG  IKTGQ  C VAF N  GTHRx  TCEIW
P2X7  QxCHS DQGCI KGWMD  PQSK G  IQTGR  C IPY D  QKxRK  TCEI F
      136    140          150          160          170

P2X1  GWCPV EVDDK IPS PA  LLREA  ENFTL  FI KNS  IS FP R  FKVNR  RNLVE
P2X2  AWCPV EDGxT SDNHF  LGKMA  PNFTI  LI KNS  IH YPK  FKFS K  GNI AS
P2X3  GWCPT EVDxT VEMPI  MxMEA  ENF TI  FI KNS  IR FPL  FNFE K  GNLLP
P2X4  AWCPV ENDVG VPT PA  FLKAA  ENF TL  LVKNN  IWYPK  FNFSK  RNI LP
P2X5  AWCPV ETKxS MPTDP  LLKDA  ESF TI  SI KNF  IR FPK  FNFSK  ANVLE
P2X6  SWCPV ESSxA VPRKP  LLAQA  KNFTL  FI KNT  VTFNK  FNFSR  TNALD
P2X7  AWCPA EEGKE APR PA  LLRSA  ENFTV  LI KNN  I DFPG  HNYTT  RNI LP
      176    180         190          200          210          220

P2X1  EVNQT YMKKC LYHKI  Q HPLC PVFNL  GYVVR  ESGQD  FRSLA  EKGG
P2X2  QKSTY x L KHC TFDQD  S DPYC P IFRL  GF IVE  KAGEN  FTELA  HKGG
P2X3  NLTDK D I KRC RFHPE  K APFC PILRV  GDVVK  FAGQD  FAKLA  RTGG
P2X4  N ITTS Y LKSC I YNAQ  T DPFC P IFRL  GTI VE  DAGHS  FQEMA  VEGG
P2X5  TDNKH F LKTC HFSS x  T NLYC P IFRL  GSI VR  WAGAD  FGDIA  LKGG
P2X6  TWDNT YFKYC LYDSL  S SPY C PVFR I  GDLVA  MTGGD  LEDLA  LLGG
P2X7  GMNI x  x x x SC TFNKT  WNPQC P IFRL  GDI FQ  E IGE N  FTEVA  VQGG
      221          230          240          250          260

P2X1  VVGIT IDWKC DLDWH  VR HCK  P IYQF  HGLYG  x x xEK  NLSPG  FNFRF
P2X2  VIGV I INWNC DLD L S  ES ECN  PKYSF  RRLDx  xP KYD  PASSG  YNFRF
P2X3  VLGIK IGWVC DLD KA  WDQCI  PKYSF  TRLDG  VSE KS  SVSPG  YNFRF
P2X4  IMGIQ IKWDC NLD RA  AS LCL  PRYSF  RRLDT  RDLEH  NVSPG  YNFRF
P2X5  VI GIY IEWDC DLD KA  AS KCN  PHYYF  NRLDN  KHxTH  S ISSG  YNFRF
P2X6  AVGIN IHWDC NLD TK  GS DCS  PQYSF  QxLQE  R x x x x  x xxx G  YNFRT
P2X7  IMG IE IYWDC NLD SW  SHRCQ  PKYSF  RRLDD  KYTNE  SLFPG  YNFRY
      265    270          280          290          300          309

P2X1  ARH F V QNGTN  xRRHL  FKV FG  IHFDI  LVDGK  AGKFD
P2X2  AKYY K I NGTT  TT RTL  I KA YG  IRIDV  I VHGQ  AGKFS
P2X3  AKYY K MENGS  EYRTL  LKA FG  IRFDV  LVYGN  AGKFN
P2X4  AKYY R DLAGK  EQRTL  TKA YG  IRFDI  IVFGK  AGKFD
P2X5  ARYY R DP NGV  EFRDL  MKA YG  IRFDV  IVNGK  AGKFS
P2X6  ANYWW AASGV  ESRSL  LKL YG  IRFDI  LVTGQ  AGKFA
P2X7  AKYY K ENGxM  EKRTL  I KA FG  VRFDI  LVFGT  GGKFD
      310          320          330          340    344
```

**FIGURE 3** *(Continued)*

## (d) M2 alpha-Helix

```
P2X1  I IPTM  TTIGS  GI G IF   GVA TV   LCD L L  LLHI L  PK
P2X2  LIPT I  INLAT  ALT SI    GVG SF   LCD W I  LLT FM  NK
P2X3  I IPT I  ISSVA  AFT SV   GVG TV   LCD I I  LLN FL  KG
P2X4  I IPTM  INVGS  GLA LL    GVA TV   LCD V I  VLY CM KK
P2X5  I IPTV  INI GS  GLA LM   GAG AF   FCD LV  L IY L I  RK
P2X6  I IPTA  ITVGT  GAAWL     GMVTF    LCD L L  LLY VD  RE
P2X7  LIQLV  VYIGS  TLS YF     GLA TV   CI D L I  I NT YA  ST
      345      350            360               370        376
```

## (e) C-Terminus

```
P2X1  xxxxxxxxxxxxxxxx xxx RH  YYKQK  KFKYA  EDMGP  GEGEH
P2X2  xxxxxxxxxxxxxxxx xxx NK  LYSHK  KFDKV  RTPKH  P SSRW
P2X3  xxxxxxxxxxxxxxxx xxx AD  HYKAR  KFEE V  TETTL  KGTAS
P2X4  xxxxxxxxxxxxxxxx xxx KY  YYRDK  KYKYV  EDYEQ  GLSG E
P2X5  xxxxxxxxxxxxxxxx xxx SE  FYRDK  KFEKV  RGQKE  DANVE
P2X6  xxxxxxxxxxxxxxxx xxx AG  FYWRT  KYEEA  RA PKA  TTNS A
P2X7  CCRSR  VYPSC  KCCEP  CAV NE  YYYRK  KCEP I  VE PKP  TLKYV
      377  380           390            400           410     416
```

```
P2X1  DPVAT  SST LG  LQENM RTS stop
P2X2  PVTLA  LVLGQ  I P PP P  SHY SQ  DQP PS  PPSGE  GPTLG  EGAEL
P2X3  TNPVF  VFA SD  TV EKQ  STDSG  AYS IG  H stop
P2X4  MNQ stop
P2X5  VEANE  NEMEQ  PE DEP  LERVR  QDEQS  QELAQ SGRKQ  NSNSQ
P2X6  stop
P2X7  SFVDE  PHIWM  VDQQL  LGKSL  QDVKG QEVPR  PQTDF  LELSR
      417  420          430          440          450
```

## (f) C-Terminus

```
P2X2  PLAVQ  SPRPC  S I SAL  TEQVV  DTLGQ  HMGQR   PPVP E  PSQQD
P2X5  VLLEP  ARFGL  RENAI  VNVKQ SQ I LH  PVKT stop
P2X7  LS LSL  HHSPP  I PGQP EEMQL LQ I EA  V P RSR   DSPDW  CQCGN
      457  460      470          480          490
```

```
P2X2  STSTD  PKGLA  QL stop
P2X7  CLPSQ LP ENR  RALEE  LCCRR KPGQC ITTSE LFSKI  VLSRE  ALQLL
      497  500           510          520          530          540
```

```
P2X7  LLYQE  PLLAL  EGEAI NSKLR  HCAYR  SYATW  RFVSQ  DMADF AILPS
      542          550          560          570          580
```

```
P2X7  CCRWK  IRKEF  PKTQG  QYSGF KYPT stop
      587  590      600          610
```

FIGURE 3   ( For legend, see opposite page.)

$P2X_4$ from a rat pancreatic islet cDNA library. They showed expression in endocrine tissues and in hormone-secreting cells within those tissues. They showed partial block by suramin, Reactive Blue 2, and DIDS and a lack of effect of $\alpha\beta$-methylene-ATP, an agonist potent for the previously cloned P2XRs. Despite similarities in nucleotide and amino acid sequence, differences in cation permeability, sensitivity to agonists and antagonists, and degree of desensitization of the receptor channel after binding agonist began to distinguish the P2XR subtypes. Later work again documented splice variants in mouse $mP2X_4$ and $mP2X_{4a}$ (lacking a 27-amino acid region in the extracellular domain) (Townsend-Nicholson et al., 1999). A different splice variant was found for human $P2X_4$ in smooth muscle and brain (Dhulipala et al., 1998). Six different $P2X_4$ isoforms were cloned from Xenopus laevis oocytes (Juranka et al., 2001). This makes expression and study of P2XRs in Xenopus oocytes problematic, to say the least.

Stuhmer and co-workers also cloned a fifth P2XR, the $P2X_5$ receptor, from a heart cDNA library (Garcia-Guzman et al., 1996). $P2X_5$ was expressed primarily in rat heart; however, in situ hybridization also revealed expression in brain, spinal cord, and adrenal gland. Functional expression

---

**FIGURE 3** (A–E) Amino acid sequence alignment of the P2XR genes and channel proteins. A scan of the literature revealed that, although sequence alignments were published, particular motifs as well as partially and fully conserved residues were not highlighted in any particular review. We also looked for sequences or motifs that had a relationship to the ENaC family or to other cation channel families. Amino acid residues in red are conserved across all subtypes, while amino acid residues in blue are conserved in at least five of the seven subtypes. An x indicates a gap in the alignment for a particular subtype or subtypes. Boxed residues indicate sites of cation binding near and within the pore region. Boldface boxes indicate agreement between two laboratories. Circled residues indicate key residues in ATP recognition. A rectangle spanning alignment that encloses a cysteine residue marks each of the 10 conserved cysteines in the extracellular domain. A triangle-enclosed histidine residue shows a site for proton binding, while the boldface triangle-enclosed histidine is a $Zn^{2+}$-binding residue. The Y symbol along the alignment shows an N-linked glycosylation site. Two motifs were identified in the M2 $\alpha$ helix and are boxed. One motif was L/FCD, which was found near the cytoplasmic end of this transmembrane segment in every subtype except P2X7. A negatively charged amino acid near a cysteine within an $\alpha$ helix is not something to be overlooked. Central to the M2 $\alpha$ helix is a GVG motif. This is eerily similar to the GYG motif defined by MacKinnon and co-workers in the Streptomyces lividans inwardly rectifying $K^+$ channel. This motif was shared by $P2X_1$ through $P2X_5$, but is partially lost in $P2X_6$ and $P2X_7$. The interesting feature is that it lies in a transmembrane $\alpha$ helix in P2XRs, while it is in a P loop in IRKs and other six transmembrane-spanning cation channels. In the C terminus, little is conserved among the subtypes except a tyrosine and a lysine early in the shared sequence near the cytosolic end of the M2 $\alpha$ helix. One other motif, PPPPXXY, was observed in $P2X_2$. This motif is also found in $\alpha$-ENaC and is mutated in Liddle syndrome, causing a hyperactive ENaC channel. (See Color Insert.)

revealed a $P2X_5$ receptor channel that desensitizes poorly, was activated by $\alpha\beta$-methylene-ATP, and was inhibited by suramin and PPADS. These properties differed greatly from those of the $P2X_4$ protein cloned and characterized in the same year by this laboratory. Rather, $P2X_5$ resembled $P2X_2$ in its biophysical and pharmacological properties. Stuhmer's group also cloned $P2X_6$ from rat brain during the same year. Expression of $P2X_6$ was restricted to regions of the brain; however, heterologous expression of $P2X_6$ alone did not show ATP-gated channel activity. They concluded that $P2X_6$ may need to coassemble in a multimeric P2XR channel complex to contribute to ATP-gated receptor channel activity.

Buell and co-workers also cloned $P2X_5$ and $P2X_6$ concomitantly (Collo et al., 1996). In this study, they also performed an extremely thorough analysis of expression of the first six P2XRs in various tissues. An enormous table in the Collo et al. article (1996) shows the regional expression in cells and nuclei of the brain concomitantly; the reader is encouraged to review this extensive information. In addition to extensive analysis of brain tissues, they showed that $P2X_1$ through $P2X_6$ were expressed in sensory ganglia, whereas all but $P2X_3$ were expressed in spinal cord. However, new evidence was shown in these in situ hybridization studies for $P2X_4$ expression in salivary gland and bronchial epithelium. This piece of evidence spawned our vigorous interest in epithelial P2XRs. While $P2X_1$ was exclusively found in smooth muscle, $P2X_2$ alone was found in adrenal medulla (Collo et al., 1996). P2XR subtype expression is revisited below. These new family members desensitized poorly and were relatively insensitive to $\alpha\beta$-methylene-ATP, suramin, and PPADS. These data were, however, contrary to work from Stuhmer's laboratory (Garcia-Guzman et al., 1996). This provided a first clue, among others described above, that differences in properties could be explained by heteromultimerization of expressed P2XRs with endogenous P2XRs, a caveat that confounds the functional and biophysical characterization of these receptor channels. Given that Xenopus oocytes and even HEK-293 cells after confluence express endogenous P2XRs (Worthington et al., 1999), caution is necessary to validate P2XR properties.

### E. $P2X_7$: The Elusive P2Z Receptor

Nevertheless, a gap remained in the molecular biology. The P2Z receptor, known to be abundant in hematopoietic cells, remained elusive. Surprenant et al. found a P2XR by homology cloning in rat brain that had a significantly longer carboxy terminus (Surprenant et al., 1996). When expressed, this new P2XR subtype showed normal P2X receptor cation channel function with lower micromolar concentrations of agonist but also the ability to form larger membrane pores that allow larger molecules such as fluorescent dyes to permeate. The latter function required stimulation with higher millimolar ATP concentrations. This newest subtype, $P2X_7$, was proposed and later confirmed to be the P2 receptor that was previously known as the permeabilizing P2 receptor, P2Z. This important molecular biological result explained the long-standing ability of extracellular ATP to permeabilize cells.

## F. Tissue-Specific and Developmentally Regulated P2XRs: $P2X_8$ and $P2X_M$

Two additional P2XR subtypes have been found that may be unique: $P2X_8$ in chick (Bo et al., 2000) and $P2X_8$ in tadpole (Jensik et al., 2001). In these two separate studies, the chick and tadpole $P2X_8$ homologs are expressed abundantly throughout early stages of development, but this expression fades and is lost in the adult. In the chick, $P2X_8$ was cloned from embryonic skeletal muscle, and its functional properties are similar to those of $P2X_1$ and $P2X_3$ with regard to desensitization and pharmacology. On embryonic day 10, $P2X_8$ was found mainly in skeletal muscle, brain, and heart, with some expression in gizzard and retina. On embryonic days 3 and 4, $cP2X_8$ was expressed in myotome, neurotube, notochord, and stomach. In the tadpole, $P2X_8$ was found in skin, heart, eye, brain, and skeletal muscle. Although most properties were similar to the chick $P2X_8$ receptor and to $P2X_1$ and $P2X_3$, this laboratory found that $\alpha\beta$-methylene-ATP was a poor agonist for tadpole $P2X_8$, a finding that differed from chick $P2X_8$ and the desensitizing $P2X_1$ and $P2X_3$. Thus, tadpole $P2X_8$ might be considered a unique P2X receptor subtype.

An intriguing cancer study found yet another P2XR subtype. Urano et al., while screening the human genome for p53-binding sites on the 5' flanking promoters of open reading frames, isolated a novel gene inducible by wild-type p53 (Urano et al., 1997). The gene encoded a 431-amino acid polypeptide with significant homology to the P2XR family. This protein, like the original two P2XRs, shows similarities to *RP-2*, a gene activated in thymocytes undergoing apoptosis. Northern blot analysis revealed expression almost exclusive to skeletal muscle, and thus it was named $P2X_M$. Further analysis revealed a correlation of a loss in $P2X_M$ with occurrence of soft tissue tumors (Nawa et al., 1999). Mouse $P2X_M$ showed similar properties, with some expression in lung (unlike human $P2X_M$) (Nawa et al., 1998). To our knowledge, the biophysical, pharmacological, and large pore properties of $P2X_M$ have not been investigated.

## G. Genomic Organization of P2XR Genes and P2XR Knockouts

Shown in Fig. 4 is an example of the genomic organization of P2XR genes. Several genomic clones of P2XRs have been studied for their exon/intron structure. The human $P2X_1$ gene was localized to chromosome 17, with a major mRNA transcript of 2.9 kb and a predicted protein of 399 amino acids of the polypeptide structure and topology outlined above (Longhurst *et al.*, 1996). This transcript was found by Northern blot to be abundantly expressed in human bladder and moderately expressed in leukocytes, pancreas, prostate, small intestine, colon, testis, and ovary. $P2X_1$ was also expressed in fetal liver but not in adult liver. More recently, the mouse $P2X_1$ gene, localized to chromosome 11, spans 10 kb of chromosomal DNA and has 12 exons (Liang *et al.*, 2001). All splice sites conformed to the GT-AG motif, and exon–intron boundaries conformed to genomic clones of other P2XR family members. A single transcriptional start site was found 233 bp upstream of a single, major methionine translation initiation site, arguing for a short 5′ promoter with tight transcription regulation. The

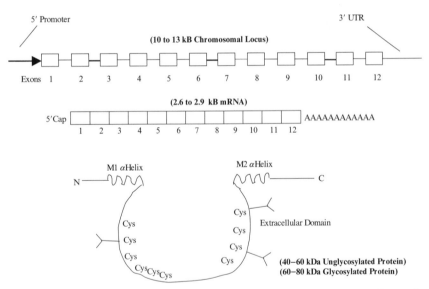

**FIGURE 4**   P2XR genomic organization, messenger RNA generation, and P2XR protein translation. Not all of the P2XR genes have been cloned at the genomic DNA level. However, those that have display a similar genomic structure. With the exception of $P2X_5$ (13 exons), the P2XR gene locus has 12 exons that are transcribed into a 2.6 to 2.9-kb messenger RNA. This mRNA encodes a protein of 40–60 kDa (the variability comes with the length of the C terminus) in its non glycosylated form, and a form approximately 20 kDa larger when glycosylated at at least three N-linked glycosylation sites.

mouse $P2X_3$ gene was the first genomic clone to be characterized. Localized to chromosome 2p by fluorescence *in situ* hybridization (FISH), this gene also was determined to have 12 exons and similar exon–intron boundary motifs (Souslova *et al.*, 1997). The encoded protein is 397 amino acids and has 99% homology to rat $P2X_3$. Like mouse $P2X_1$, transcription start sites are only 162–168 bp upstream of the methionine start codon. The 5′ promoter was also noted to lack any TATA or CCAAT boxes for classic transcriptional control. Voigt and co-workers elucidated the mouse $P2X_5$ gene. It was localized to chromosome 11 by FISH and it contains 13 exons within 13 kb of the chromosomal DNA (Cox *et al.*, 2001). The major transcript is 2.6 kb in length and encodes a 455-amino acid protein, 95% homologous with rat $P2X_5$. Mouse $P2X_5$ was expressed in most tissues, but most abundantly in heart and kidney. Transcriptional start sites are only 30–70 bp upstream of the start methionine. Chicken $P2X_5$ was similar in genomic structure (Ruppelt *et al.*, 2001). Finally, mouse $P2X_M$ also has 12 exons and spans 10 kb of chromosomal DNA (Nawa *et al.*, 1998).

Transgenic knockout of P2XR genes has begun in an effort to understand their physiological roles more thoroughly. Sometimes, these studies produce more questions than answers. However, emerging data have come forth. Mice lacking $P2X_1$ receptor channels expressed one major phenotype, reduced fertility due to lack of ejection of sperm from the testes into the vas deferens (Mulryan *et al.*, 2000). The interpretation was that $P2X_1$ receptors are the main purinergic receptor on testes and vas deferens vascular smooth muscle cells. The sympathetic-driven contractile response is known to be mediated in large part by P2X receptors; however, this study showed that $P2X_1$ was the principal postsynaptic receptor channel involved. Fertility was reduced by 90% in homozygous knockout mice, but not in wild-type or heterozygous mice, not because of defects in spermatogenesis, but because of defects in ejection of sperm through the vas deferens into the ejaculate (Mulryan *et al.*, 2000). Any other phenotypes in the mice were subtle and not detected as yet in this study.

Burnstock and co-workers have generated a $P2X_3$ knockout mouse (Vlaskovska *et al.*, 2001). With this mouse, this laboratory has shown that two different populations of sensory neurons are impaired in their function due to a lack of $P2X_3$ (Vlaskovska *et al.*, 2001; Zhong *et al.*, 2001). The suburothelial plexus of the mouse bladder showed abundant staining for $P2X_3$. Although ATP was released similarly in response to bladder distension in $P2X_3^{+/+}$ versus $P2X_3^{-/-}$ mice, the bladder afferent nerve activity was attenuated in response to distension (Vlaskovska *et al.*, 2001). The loss of this activity in the $P2X_3^{-/-}$ mice could be rescued by exogenous ATP analogs or ATP itself, while normal activity in the $P2X_3^{+/+}$ afferents was blocked by trinitrophenyl-ATP, PPADS, and capsaicin (Vlaskovska *et al.*, 2001). These data suggest that $P2X_3$ is a major neural sensory receptor on a population of bladder nerve fibers that transduce bladder distension into an electrical signal. Specialized neurons of the dorsal root ganglia respond to ATP as a principal neurotransmitter by a mechanism that is transduced by different P2XR subtypes: $P2X_2$, $P2X_2/P2X_3$ heteromultimers, and $P2X_3$. Burnstock and co-workers also examined the activity of dorsal root ganglion (DRG) and nodose neurons (specialized neurons of the DRG involved in pain perception) in $P2X_3^{-/-}$ mice (Zhong *et al.*, 2001). Because $P2X_3$ has a distinctive, rapidly desensitizing response to ATP, patch-clamp electrophysiology of the neurons was employed to look for differences between $P2X_3^{+/+}$ and $P2X_3^{-/-}$ DRG neurons failed to desensitize after ATP stimulation, and both subsets of neurons failed to respond to $\alpha\beta$-methylene-ATP. The responses that were persistent in the $P2X_3^{-/-}$ neurons were consistent with intact, endogenous expression of $P2X_2$. These studies showed that, although DRG and nodose neurons remained responsive to ATP, the nature of the neuronal sensory response to the ATP neurotransmitter in the spinal cord is significantly altered and more long-lived.

## IV. DETERMINATION OF PATTERNS AND LEVELS OF EXPRESSION IN SPECIFIC CELLS AND TISSUES

P2XR subtypes are expressed abundantly in brain and spinal cord, and the most comprehensive analysis of their relative expression was from Collo *et al.* (1996). The functional role of P2XRs in different excitable cell preparations is addressed in more detail below. In this review, we choose to focus on the expression of P2XRs in other tissues. This is an emerging area of research and is not as appreciated. Although the P2XR as an excitatory postsynaptic receptor channel is easy to envision and characterize, the physiological roles of P2XRs on nonexcitable cells have not been elucidated fully.

### A. Sensory Organs

Although sensory organs do contain neurons that convey sensory inputs to higher neural centers, evidence has emerged that P2XRs are expressed in other more specialized, nonexcitable cell types in inner ear, retina, and taste bud. Multiple P2XRs have been localized to spiral and vestibular ganglia of the inner ear, the cochlear nucleus, and primary auditory neurons (Xiang *et al.*, 1999; Brandle *et al.*, 1999; Salih *et al.*, 1999). In addition, $P2X_2$

receptors or splice variants of $P2X_2$ (Chen et al., 2000) have been found biochemically, molecularly, and functionally in Deiters' cells of the cochlea (Chen and Bobbin, 1998), the organ of Corti within the cochlea (Parker et al., 1998), outer hair cells (Brandle et al., 1999), and epithelial cells lining the stria vascularis, Reissner's membrane, and the tectorial membrane (Xiang et al., 1999). $P2X_2$ staining was also intense in the hair cell stereocilia, indicating a key role in sound transduction (Housley et al., 1999).

In the retina, P2XRs are also abundant. Postnatally, $P2X_2$ and $P2X_4$ were found in the rat retina and choroid, while $P2X_3$ and $P2X_5$ were found only in rat retina (Brandle et al., 1998a). No polymerase chain reaction (PCR) product for $P2X_1$ or $P2X_6$ was amplified from these tissues of the eye. $P2X_7$ was studied in a later article by Brandle et al. (1998b). These data showed $P2X_7$ expression only in the retina. In particular, different populations of amacrine cells and ganglion cells were positive within the inner nuclear layer and the ganglion cell layer of the retina. Electrophysiology of postnatal rat retinal ganglion and amacrine cells revealed ATP-gated channels with properties of P2XRs only in retinal ganglion cells (Taschenberger et al., 1999). ATP, ADP, and $\alpha\beta$-methylene-ATP (but not adenosine) activated an inwardly rectifying current that had a high permeability to $Ca^{2+}$ versus $Na^+$ or $Cs^+$, a single-channel conductance estimated by noise analysis to be 2.3 pS, and was antagonized variably by suramin. Because of heterogeneity in the relative effects of ADP, $\alpha\beta$-methylene-ATP, and suramin, the possibility that multiple P2XRs were expressed in retinal ganglia was formed. Single-cell reverse transcriptase (RT)-PCR by Wheeler-Schilling's laboratory of retinal ganglion cells revealed expression of $P2X_3$, $P2X_4$, $P2X_5$, and $P2X_7$ (Wheeler-Schilling et al., 2001). Immunolabeling of the ganglion and inner cell layers revealed staining for $P2X_3$ and $P2X_4$. In Müller cells, another specialized neuron within the retinal cell layers that has properties of astroglia, Jabs et al. found evidence of expression of $P2X_3$, $P2X_4$, and $P2X_5$ but not of $P2X_7$ (Jabs et al., 2000). Similar results were found for bipolar cells of the retina (Wheeler-Schilling et al., 2000). Abundant expression of multiple P2XRs in the many cell layers of the retina argues convincingly for a key physiological role for P2XRs in photo-transduction. Unlike the inner ear, specialized epithelial cells in the eye were not examined for P2XR expression.

In addition to transduction of pain, light, and sound, P2XRs were identified in circumvallate fungiform papillae of the rat tongue. Nerve fibers innervating the taste buds expressed $P2X_3$ abundantly, as did other nerve fibers within the tongue, whereas $P2X_2$ was present but staining was less intense (Bo et al., 1999). As for the eye research described above, the specialized sensory epithelia of the taste bud were not examined.

## B. Vascular Smooth Muscle, Skeletal Muscle, and Heart

The role of $P2X_1$ in the smooth muscle cells of the testes and vas deferens has already been well documented with regard to cloning of $P2X_1$ and the $P2X_1$ knockout mouse. The urinary bladder appears to be another important tissue regulated by local ATP release and P2XRs. In detrusor, ureteral, and bladder blood vessel smooth muscle, $P2X_1$ immunoreactivity was observed routinely and markedly on the smooth muscle cell membrane. A large degree of $P2X_2$ staining and some $P2X_5$ and $P2X_6$ staining were observed in submembrane compartments of smooth muscle cells. The fine capillary network supplying the bladder tissue stained intensely with $P2X_4$, while the urothelial membrane showed high $P2X_5$ immunoreactivity. The $P2X_4$ result is suggestive of endothelial cell immunoreactivity (see below). Only nuclear membrane staining was observed with $P2X_7$ (Lee et al., 2000). A separate study confirmed the high $P2X_1$ expression in urinary bladder smooth muscle and in smooth muscle surrounding ureter and blood vessels; however, Elneil et al. found $P2X_3$ on the urothelium (Elneil et al., 2001). In studies of kidney collecting duct (see below), $P2X_3$ and $P2X_4$ with some $P2X_5$ expression was documented, suggesting that urothelium and renal tubular epithelium may share in abundant and functional relevant P2XR expression.

While P2XR expression remains and is pronounced in smooth muscle, P2XR expression in skeletal muscle is notable during early stages of development, while expression fades during the postnatal period and in adult tissue. An exception is the p53-induced gene encoding $P2X_M$, whose expression protects against development of soft tissue tumors, especially those in skeletal muscle. Burnstock's group studied expression of P2XRs in rat skeletal muscle in embryos and pups (Meyer et al., 1999; Ryten et al., 2001). In an initial study, $P2X_5$ and $P2X_6$ were found during early developmental stages and were lost in later development and in the adult. In a follow-up study, $P2X_5$ expression appeared first and transiently and was followed by transient expression of $P2X_6$. However, in a finding somewhat contradictory to the previous study, sustained expression of $P2X_2$ was found that persisted albeit weakly in the adult. The previous study concluded that no other P2XR subunit was present in skeletal muscle. Because P2XR expression correlated with myotube formation and acetylcholine receptor redistribution to neuromuscular junction, P2XRs may play a key developmental role in skeletal muscle. $P2X_4$ was found in chick cardiac muscle myocytes. A positive ionotropic effect of ATP on heart muscle may be mediated by this P2XR subtype and allow ATP-induced calcium influx that was blocked by antisense oligonucleotides to $P2X_4$ (Hu et al., 2002). In addition, glycosylated (58 kDa) and nonglycosylated (44 kDa) forms of the cP2X$_4$ receptor were found in cardiac myocyte membranes, a similar phenotype to the epithelial P2X$_4$ receptor (see below). Only the 58-kDa glycosylated form of cP2X$_4$ could be biotinylated. These results provide a new tissue where P2XR expression and function may be critical.

## C. Endothelia

Work has shown the presence of P2XRs in human vascular endothelial cells. Two separate studies by Yamamoto *et al.* showed that P2X$_4$ mediated $Ca_i^{2+}$ influx induced by shear stress or exogenous ATP agonists (Yamamoto *et al.*, 2000a,b). By Northern blot analysis, they showed abundant expression of P2X$_4$ in human endothelial cells from umbilical vein, aorta, pulmonary artery, and skin microvessels (Yamamoto *et al.*, 2000b). Other subtypes including P2X$_1$, P2X$_3$, P2X$_5$, and P2X$_7$ were also found, but were much less abundant than P2X$_4$ (Yamamoto *et al.*, 2000b). Shinozuka *et al.* showed that ATP release was critical for the modulation of endothelial cell volume regulation (Shinozuka *et al.*, 2001). Using immunohistochemistry and confocal microscopy, Ray *et al.* found expression of P2X$_1$–P2X$_4$, P2X$_7$, and P2Y$_2$ in the endothelial cell layer of arteries and veins (Ray *et al.*, 2002). P2X$_5$ and P2X$_6$ were less abundant. P2X$_4$ expression differed dramatically between artery and vein; expression was 14.6-fold higher in saphenous vein versus radial and internal mammary arteries, where expression was judged to be low. This difference in expression was argued to have implications for the choice of veins versus arteries in coronary artery bypass grafts (Ray *et al.*, 2002).

Schwiebert *et al.* performed an integrated study of primary human endothelial cells grown as monolayers, where they studied ATP release, ATP receptor expression, and ATP-regulated $Ca_i^{2+}$ signaling (Schwiebert *et al.*, 2001). They showed that ATP release was directed across the apical membrane exclusively. ATP release occurred constitutively under basal conditions and was potentiated by ionomycin or thapsigargin, two different $Ca_i^{2+}$-mobilizing agents, and hypotonic challenge. In contrast, hypertonicity inhibited ATP release. ATP release was inhibited at 4°C versus 25°C, suggesting exocytosis as a mechanism of ATP release. Diisothiocyanatos-tilbene-disulfonic acid (DIDS), the anion channel and transporter inhibitor, also inhibited ATP release markedly under basal and hypotonic conditions, suggesting a transport mechanism for ATP release. Specific RT-PCR confirmed the expression of the P2Y$_2$ receptor; however, degenerate RT-PCR and differential DNA sequencing of white bacterial colonies bearing the various P2XR PCR product inserts in plasmids revealed abundant expression of P2X$_4$ and P2X$_5$ and rare expression of P2X$_1$, P2X$_7$, and a partial sequence homolog found in three different endothelial cell samples. Immunoblotting of human umbilical vein endothelial cell (HUVEC) membrane protein also revealed abundant expression of $P2X_4$ and $P2X_5$. $Ca_i^{2+}$ imaging with Fura-2/AM revealed that endogenous ATP maintained basal $Ca_i^{2+}$ at constant levels, because a purinergic scavenger cocktail and a global P2 purinergic receptor antagonist, suramin, lowered basal $Ca_i^{2+}$ in endothelial cells. This argued for an autocrine purinergic signaling cascade that may govern basal endothelial cell function. Ramirez and Kunze have found similar results in bovine aortic endothelium (Ramirez and Kunze, 2002). Specific RT-PCR revealed expression of $P2X_4$, $P2X_5$, and $P2X_7$. Whole cell and outside-out patch-clamp recording showed that 2-methylthio-ATP (2-MeS-ATP) stimulated a slowly desensitizing, inwardly rectifying cation current that had a single-channel conductance of 36 pS and was insensitive to the various P2XR antagonists. These properties were most consistent with $P2X_4$. Benzoylbenzoyl-ATP (BzBzATP) activated a 9-pS cation channel that did not desensitize and was partially blocked by suramin and Reactive Blue 2. Interestingly, despite the expression of $P2X_7$, a lack of large pore formation with chronic application of agonist was noted in this endothelial cell preparation. The challenge will be to characterize fully the physiological roles of P2XRs in endothelial cells.

## D. Epithelia

In the context of expression profiling in excitable tissues, hints in these initial studies led to the characterization of P2XR expression and/or function in epithelial cells *in vitro* and *in vivo*. This is the focus of a separate section of this review (see below) and is a primary focus of research by the authors of this review (Taylor *et al.*, 1999).

## E. Bone

An emerging role for purinergic receptors in the cells that constitute skeletal bone is being appreciated. Bone remodeling within microenvironments is regulated by autocrine and paracrine factors and by local mechanical forces that are poorly understood. Hoebertz *et al.* have performed a comprehensive study on P2X and P2Y receptor expression in bone cells (Hoebertz *et al.*, 2000). $P2X_2$ was found to be abundant in osteoclasts, osteoblasts, and chondrocytes. $P2X_5$ and $P2Y_2$ were expressed on osteoblasts and chondrocytes, while $P2X_4$ and $P2X_7$ were present on osteoclasts. $P2X_2$ expression on osteoclasts was of particular interest,

## (a) N-Terminus

```
P2X1  MARRL  QDELS  AF x FF   EYDTP   RMVLV  RNK
P2X2  MVRRL  ARG xC WSAFW    DYETP   KVIVV  RNR
P2X3  Mxx x x  x x NxC  ISDFF  TYETT   KSVVV  KSW
P2X4  MAGCx  CSVLG  SFxLF    EYDTP   RIVLI  RSR
P2X5  MGQAA  WKGFV  xLSLF    DYKTA   KFVVA  KSK
P2X6  MASAV  AAALV  SWG FL   DYKTE   KYVMT  RNC
P2X7  MPACC  x x xxS WNDVF   QYETN   KVTRI  QSV
       1         10          20         28
```

## (b) M1 alpha-Helix

```
P2X1  K KVGV   IFRL I   QLVVL  VYV IG  W xVEV  YEK GYQ
P2X2  R RL GF   VHRMV   QLL IL  LY FVx  WYVFI   VQK SYQ
P2X3  WT I GI   INRAV   QLL II  SY FVG  W xVFL  HEK AYQ
P2X4  R KVGL   MNRAV   QLL IL  AYV IG  WxVFV   WEK GYQ
P2X5  K KVGL   LYRVL   QL I IL  L YLLI  WxVFL   I KK SYQ
P2X6  CWVGI   SQR LL   QLGVV  VYV IG  WxALL   AKK GYQ
P2X7  V NYGT   IKW IL   HMTVF  SYV Sx  x FALM  SDK LYQ
       28 30              40         50        58
```

## (c) Extracellular Domain

```
P2X1  YEK GYQT S  S x DLI  SSVSV  (KLKQL)  AVTQx  xxxxxx  LQGLG  PQVW
P2X2  VQK SYQD S  E TGPE  (SSIDT   KVKG I  TMSEx  xxxxxx  xxxxx  DKVW
P2X3  HEK AYQV R  DTAI E  SSVVT   KVKGF  GRYAx  xxxxxx  xxxxx  NRVM
P2X4  WEK GYQE T  D xSVV  SSVTT   KAKGV  AVTNx  xxxxxx  TSQLG  FR IW
P2X5  I KK SYQD I  D TSLQ  SAVVT   KVKGV  AYTNx  xxxxxx  TTMLG  ERLW
P2X6  AKK GYQEW  DMDPQ  ISV I T  KLKGV  SVTQx  xxxxxx  VKELE  KRLW
P2X7  SDK LYQR K  E P xLI  SSVHT   KVKGV  AEVTE  NVTEGG  VTKLV  HGI F
       53        60          70         80         90     95
```

```
P2X1  DVADY  VFPAH  G DSSF  VVMTN  FIVTP  QQTQG  H CAEN  PE xGx  G IC
P2X2  DVEEY  VKPPE  GGSVV  SI I TR  IEVTP  SQT LG  T CPE S  MRVHS  ST C
P2X3  DVSDY  VTP PQ  G TSVF  V I I TK  MIVTE  NQMQG  F CPEN  EE KYR  xx C
P2X4  DVADY  VI PAQ  E ENSL  FIMTN  MIVTV  NQT QS  T CPE I  PD KTx  S I C
P2X5  DVADF  VI PSQ  G ENVF  FVVTN  LIVTP  NQR QS  T CAER  EG I PD  GE C
P2X6  DVADF  VRPSQ  G ENVF  FLVTN  FLVTP  AQV QG  R CPEH  PS VPL  AN C
P2X7  DTADY  TLPLQ  G NSxF  FVMTN  YLKSE  GQE QK  L CPEY  PS RGK  Qx C
       96    100         110        120         130     135
```

## (c)  Extracellular Domain

```
P2X1  G I CQD DS GCT PGKAE  RK AQG  IRTGN  C VPF N  GTVKx  TCEI F
P2X2  STC HS DD DCI AGQLD  MQGNG IRTGH  C VPYY  HGDSK  TCEVS
P2X3  xxC VS DS QCG PE R Fx  x PGG G  ILTGR  C VNYx  SSV LR  TCEI Q
P2X4  S IC NS DA DCT PGS VD  TH S SG VATGR  C VPF N  ESVKx  TCEVA
P2X5  GEC SE DDDCH AGE SV  VAGHG LKTGR  C LRVG  NSTRG  TCEI F
P2X6  ANC WA DE DCP EGE MG  TYSHG IKTGQ  C VAF N  GTHRx  TCEIW
P2X7  QxC HS DQGCI KGWMD PQSK G  IQTGR  C IPY D  QKxRK  TCEI F
      136    140         150              160              170
```

```
P2X1  GW CPV EVDDK  IPS PA  LLREA  ENFTL  FI KNS  IS FP R  FKVNR  RNLVE
P2X2  AW CPV EDGxT  SDNHF  LGKMA  PNFTI  LI KNS  IH YPK  FKFS K  GNI AS
P2X3  GW CPT EVDxT  VEMPI  MxMEA  ENF TI  FI KNS  IR FPL  FNFE K  GNLLP
P2X4  AW CPV ENDVG  VPT PA  FLKAA  ENF TL  LVKNN  IWYPK  FNFSK  RNI LP
P2X5  AW CPV ETKxS  MPTDP  LLKDA  ESF TI  SI KNF  IR FPK  FNFSK  ANVLE
P2X6  SW CPV ESSxA  VPRKP  LLAQA  KNFTL  FI KNT  VTFNK  FNFSR  TNALD
P2X7  AW CPA EEGKE  APR PA  LLRSA  ENFTV  LI KNN  I DFPG  HNYTT  RNI LP
      176    180        190            200           210           220
```

```
P2X1  EVNQT  YMKKC  LYHKI  Q HPLC PVFNL  GYVVR  ESGQD  FRSLA  EKGG
P2X2  QKSTY  x L KHC TFDQD  S DPYC P IFRL  GF IVE  KAGEN  FTELA  HKGG
P2X3  NLTDK  D I KRC RFHPE  K APFC PILRV  GDVVK  FAGQD  FAKLA  RTGG
P2X4  N ITTS  Y LKSC HFSS x  T NLYC P IFRL  GTI VE  DAGHS  FQEMA  VEGG
P2X5  TDNKH  F LKTC HFSS x  T NLYC P IFRL  GSI VR  WAGAD  FGDIA  LKGG
P2X6  TWDNT  YFKYC LYDSL  S SPYC PVFR I  GDLVA  MTGGD  LEDLA  LLGG
P2X7  GMNI x  xxxSC TFNKT  WNPQC P IFRL  GDI FQ  E IGE N  FTEVA  VQGG
      221          230          240          250          260
```

```
P2X1  VVGIT  IDWKC DLDWH  VR HCK P IYQF  HGLYG  x x xEK  NLSPG  FNFRF
P2X2  VIGV I  INWNC DLD LS  ES ECN PKYSF  RRLDx  xP KYD PASSG  YNFRF
P2X3  VLGIK  IGWVC DLD KA  WDQCI PKYSF  TRLDG  VSE KS  SVSPG  YNFRF
P2X4  IMGIQ  IKWDC NLD RA  AS LCL PRYSF  RRLDT  RDLEH  NVSPG  YNFRF
P2X5  VI GIY  IEWDC DLD KA  AS KCN PHYYF  NRLDN  KHxTH  SISSG  YNFRF
P2X6  AVGIN  IHWDC NLD TK  GS DCS PQYSF  QxLQE  R x xx x  x xxx G YNFRT
P2X7  IMG IE  IYWDC NLD SW  SHR CQ PK YSF  RRLDD  KYTNE  SLFPG  YNFRY
      265    270          280          290          300          309
```

```
P2X1  ARH F V  QNGTN  xRRHL  FKV FG  IHFDI  LV DGK  AGKFD
P2X2  AKYY K  I NGTT  TT RTL  I KA YG  IRIDV  IVHGQ  AGKFS
P2X3  AKYY K  MENGS  EYRTL  LKA FG  IRFDV  LVYGN  AGKFN
P2X4  AKYY R  DLAGK  EQRTL  TKA YG  IRFDI  IVFGK  AGKFD
P2X5  ARYY R  DP NGV  EFRDL  MKAYG  IRFDV  IVNGK  AGKFS
P2X6  ANYWW  AASGV  ESRSL  LKL YG  IRFDI  LVTGQ  AGKFA
P2X7  AKYY K  ENGxM  EKRTL  I KA FG  VRFDI  LVFGT  GGKFD
      310          320          330          340    344
```

## (d) M2 alpha-Helix

```
P2X1  I IPTM  TTIGS  GI G IF  GVATV  LCDL L  LLHI L  PK
P2X2  LIPT I  INLAT  ALT SI  GVG SF  LCDW I  LLT FM  NK
P2X3  I IPT I  ISSVA  AFT SV  GVG TV  LCDI I  LLN FL  KG
P2X4  I IPTM  INVGS  GLA LL  GVA TV  LCDV I  VLY CM  KK
P2X5  I IPTV  INI GS  GLA LM  GAG AF  FCDLV  L IY LI  RK
P2X6  I IPTA  ITVGT  GAAWL  GMVTF  LCDL L  LLY VD  RE
P2X7  LIQLV  VYIGS  TLS YF  GLA TV  CI DL I  I NT YA  ST
       345    350         360         370      376
```

## (e) C-Terminus

```
P2X1  xxxxxxxxxxxxxxxxx xxx RH  YYKQK  KFKYA  EDMGP  GEGEH
P2X2  xxxxxxxxxxxxxxxxx xxx NK  L YSHK  KFDKV  RTPKH  P SSRW
P2X3  xxxxxxxxxxxxxxx xxx AD  HYKAR  KFEE V  TETTL  KGTAS
P2X4  xxxxxxxxxxxxxxxxx xxx KY  YYRDK  KYKYV  EDYEQ  GLSG E
P2X5  xxxxxxxxxxxxxxxxx xxx SE  FYRDK  KFEKV  RGQKE  DANVE
P2X6  xxxxxxxxxxxxxxxxx xxx AG  FYWRT  KYEEA  RA PKA  TTNS A
P2X7  CCRSR  VYPSC  KCCEP  CAV NE  YYYRK  KCEP I  VE PKP  TLKYV
       377 380         390         400          410       416
```

```
P2X1  DPVAT  SST LG  LQENM RTS stop
P2X2  PVTLA  LVLGQ  I P PP P  SHYSQ  DQP PS  PPSGE  GPTLG  EGAEL
P2X3  TNPVF  VFA SD  TV EKQ  STDSG  AYS IG  H stop
P2X4  MNQ stop
P2X5  VEANE  NEMEQ  PE DEP  LERVR  QDEQS  QELAQ SGRKQ  NSNSQ
P2X6  stop
P2X7  SFVDE  PHIWM  VDQQL  LGKSL  QDVKG  QEVPR  PQTDF  LELSR
       417 420         430          440         450
```

## (f) C-Terminus

```
P2X2  PLAVQ  SPRPC  S I SAL  TEQVV  DTLGQ  HMGQR  PPVP E  PSQQD
P2X5  VLLEP  ARFGL  RENAI  VNVKQ SQ I LH  PVKT stop
P2X7  LS LSL  HHSPP  I PGQP  EEMQL LQ I EA  V P RSR  DSPDW  CQCGN
       457 460         470         480          490
```

```
P2X2  STSTD  PKGLA  QL stop
P2X7  CLPSQ  LP ENR  RALEE  LCCRR  KPGQC  ITTSE  LFSKI  VLSRE  ALQLL
       497 500          510         520         530          540
```

```
P2X7  LLYQE  PLLAL  EGEAI  NSKLR  HCAYR  SYATW  RFVSQ  DMADF  AILPS
       542         550         560         570         580
```

```
P2X7  CCRWK  IRKEF  PKTQG  QYSGF  KYPT stop
       587 590         600         610
```

**CHAPTER 4, FIGURE 3** (A–E) Amino acid sequence alignment of the P2XR genes and channel proteins. A scan of the literature revealed that, although sequence alignments were published, particular motifs as well as partially and fully conserved residues were not highlighted in any particular review. We also looked for sequences or motifs that had a relationship to the ENaC family or to other cation channel families. Amino acid residues in red are conserved across all subtypes, while amino acid residues in blue are conserved in at least five of the seven subtypes. An $x$ indicates a gap in the alignment for a particular subtype or subtypes. Boxed residues indicate sites of cation binding near and within the pore region. Boldface boxes indicate agreement between two laboratories. Circled residues indicate key residues in ATP recognition. A rectangle spanning alignment that encloses a cysteine residue marks each of the 10 conserved cysteines in the extracellular domain. A triangle-enclosed histidine residue shows a site for proton binding, while the boldface triangle-enclosed histidine is a $Zn^{2+}$-binding residue. The $Y$ symbol along the alignment shows an N-linked glycosylation site. Two motifs were identified in the M2 $\alpha$ helix and are boxed. One motif was L/FCD, which was found near the cytoplasmic end of this transmembrane segment in every subtype except P2X7. A negatively charged amino acid near a cysteine within an $\alpha$ helix is not something to be overlooked. Central to the M2 $\alpha$ helix is a GVG motif. This is eerily similar to the GYG motif defined by MacKinnon and co-workers in the *Streptomyces lividans* inwardly rectifying $K^+$ channel. This motif was shared by $P2X_1$ through $P2X_5$, but is partially lost in $P2X_6$ and $P2X_7$. The interesting feature is that it lies in a tramsmembrane $\alpha$ helix in P2XRs, while it is in a P loop in IRKs and other six transmembrane-spanning cation channels. In the C terminus, little is conserved among the subtypes except a tyrosine and a lysine early in the shared sequence near the cytosolic end of the M2 $\alpha$ helix. One other motif, PPPPXXY, was observed in $P2X_2$. This motif is also found in $\alpha$-ENaC and is mutated in Liddle syndrome, causing a hyperactive ENaC channel.

because activation of this receptor channel is potentiated by low extracellular pH, a microenvironment found at sites of resorption in skeletal bone. These results point to a key role for release of nucleotide agonists and P2 receptors in bone remodeling. An added observation in a separate paper by Hoebertz *et al.* (2001) showed that ADP and the ADP analog, 2-MeS-ADP, potentiated bone resorption greatly at the sites of ruffled and pitted membrane of osteoclasts. Two additional studies in 2001 have shown expression and function of the $P2X_7$ receptor in osteoblasts (Gartland *et al.*, 2001) and osteoclasts (Naemsch *et al.*, 2001).

## F. Endocrine Organs

Burnstock and co-workers have led the field in documenting P2X receptor expression in endocrine tissues. Initial work documented abundant $P2X_2$ receptor expression in the nuclei and neurons of the hypothalamus (Xiang *et al.*, 1998). Further immunohistochemical work found abundant $P2X_2$ protein expression in all aspects of neurons emanating from paraventricular and supraoptic nuclei of the hypothalamus leading down into the posterior pituitary that are involved intimately in vasopressin and oxytocin secretion (Loesch *et al.*, 1999). Subpopulations of neurons that were $P2X_2$ positive and $P2X_2$ negative in hypothalamus and neurohypophysis were noted. Burnstock and co-workers have also done P2XR expression studies in thyroid and adrenal gland (Glass and Burnstock, 2001; Afework and Burnstock, 1999, 2000; Liu *et al.*, 1999). In addition to expression of multiple P2XRs in vascular smooth muscle and endothelia of the rat thyroid and adrenal glands, thyroid follicular cells expressed $P2X_3$, $P2X_4$, and $P2X_5$. Different cells of the adrenal gland are enriched with P2XRs. Original functional work on chromaffin cells suggested that $P2X_2$ was the major P2XR present in epinephrine-secreting cells. However, biochemical and molecular work that followed showed that $P2X_5$, $P2X_6$, and $P2X_7$ were most prominent in this cell type. Innervating nerve fibers and neurons showed expression of $P2X_1$–$P2X_3$. Cells in the adrenal cortex stained positively for $P2X_4$–$P2X_7$. $P2X_4$ was confined to the zona reticularis, while staining followed this order of intensity: $P2X_6 > P2X_7 > P2X_5$. The abundant and redundant expression of P2XRs in adrenal gland suggests an important role of purinergic signaling in the local control of adrenal gland function. Petit *et al.* showed roles for P2Y and P2X receptors in stimulating insulin secretion from pancreatic beta cells in the absence or presence of the requisite extracellular glucose stimulus (Petit *et al.*, 1998). Because ADP$\beta$S and $\alpha\beta$-methylene-ATP were similarly effective and stimulate particular P2Y and P2X receptor subtypes, this argues for expression of $P2Y_1$ and

$P2X_1$, $P2X_2$, $P2X_3$, or $P2X_2/P2X_3$ heterodimers in the pancreatic beta cell. Burnstock's group also studied expression of all P2XR subtypes in rat testis (Glass *et al.*, 2001). Immunohistochemistry and Western blotting showed the expression of $P2X_1$, $P2X_2$, $P2X_3$, $P2X_5$, and $P2X_7$ in rat testis. $P2X_1$ was present exclusively in blood vessels in vascular smooth muscle as documented above in the context of the $P2X_1$ knockout mouse. $P2X_2$, $P2X_3$, and $P2X_5$ were found in the germ cells in different stages of the seminiferous epithelium, whereas $P2X_7$ was present in low levels in all stages. $P2X_2$ and $P2X_3$ tracked together during seminiferous epithelial maturation. Sertoli cells showed expression of $P2X_2$ and $P2X_3$ during spermatogenesis; however, Leydig cells were altogether devoid of P2X receptor expression. Taken together, expression of multiple P2XR subtypes in endocrine organs argues for tight, local control of endocrine tissue function by autocrine purinergic signaling.

## V. BIOPHYSICAL PROPERTIES OF P2XRS IN NATIVE AND HETEROLOGOUS CELLS

Egan and Voigt (see their chapter on P2XRs in this volume) focus on the biophysics of P2XRs. This section summarizes the key aspects of the extracellular ATP-gated P2X nonselective cation channels for the non-electrophysiologist. For the patch-clamp electrophysiologists, please refer to the chapter by Egan and Voigt in this volume. Figure 3 also depicts particular amino acid residues within the P2XR subtype amino acid sequence alignment that have been implicated by site-directed mutagenesis in cation binding and permeability, ATP recognition, cysteine-rich domains, and other phenotypes.

P2XRs are nonselective cation channels that open when external ATP binds to the outer face of the channel protein. ATP binding opens the pore of the channel, which is poorly selective for cations. In other words, divalent as well as monovalent cations can bind to and permeate the P2XR channel pore. Ding and Sachs studied this elegantly for the P2X2 channel protein, expressed in HEK-293 cells devoid of endogenous P2XRs when maintained as isolated cells (Ding and Sachs, 1999). Outside-out patches were performed so that the ATP agonist had free access to the extracellular domain of the receptor channel protein. While monovalent cations were permeant in the sequence $K^+ > Rb^+ > Cs^+ > Na^+ > Li^+$, divalent cations also entered the pore. However, divalent cations such as $Mn^{2+}$, $Mg^{2+}$, $Ca^{2+}$, and $Ba^{2+}$ also blocked the channel in a manner characterized as "fast block," reducing the amplitude of the single-channel currents. Importantly, organic cations such as $NMDG^+$, $Tris^+$, $TMA^+$, and $TEA^+$ were virtually impermeant. The channels were "flickery," in that they opened in brief bursts. The single-channel conductance was 49 pS in the presence of $K^+$ and 35 pS in the presence of $Na^+$. P2XR $Na^+$ currents were partially blocked by $Mg^{2+}$ and $Ca^{2+}$. A subsequent article showed that divalent cations also speeded inactivation of the $P2X_2$ channels (Ding and Sachs, 2000).

Interestingly, Wong, Burnstock, and Gibb also studied P2XRs recorded from rat hippocampal granule cells within brain slices (Wong et al., 2000). Previous electrophysiological and expression analysis suggested the presence of $P2X_4$ and $P2X_6$ in one study and of $P2X_1$, $P2X_2$, $P2X_4$, and $P2X_6$ in another study. A channel with flickery behavior consistent with the work described above by Ding and Sachs was found in three patches that had a conductance of 32 pS and an extremely short mean open time and burst duration. In lay terms, the channel barely opens on binding agonist and spends most of its time in the closed state. A greater conductance P2XR was observed in 19 of 98 patches with a conductance of 56 pS, but with similar short open time and burst duration. A high concentration of ATP (1 m$M$) was required to open these channels. The short openings related to an open probability (the time the channel is open divided by the total time of the recording) of less than 0.1. The more specific $P2X_1$ and $P2X_3$ agonist $\alpha\beta$-methylene-ATP (40 $\mu M$) produced a similar activation of the greater conductance channel, suggesting that $P2X_1$ contributed the larger channel. One troubling result was the lack of marked inhibitory effects by suramin (40 $\mu M$). This result is consistent with $P2X_4$ and $P2X_6$ receptor channels, which are refractory to suramin. These results suggest that heteromultimeric P2XR channels that contained both $P2X_1$ and $P2X_4$ and/or $P2X_6$ might have been conferring the channels recorded in these neurons.

Close examination of these two different ATP-activated channels revealed different current signatures. The 32-pS channel was quite flickery, like those described by Ding and Sachs; however, the greater conductance channel displayed slower kinetics with much longer open times. These results, generated in native cells, must take into account the possibility that the second channel may be a separate ion channel regulated by P2XRs (perhaps a channel gated by cytosolic calcium that elevated in the cell due to entry through the P2XRs into the cell).

Site-directed mutagenesis has been used to probe many regions critical to the activity of P2XR channels. With regard to cation permeation through the pore, Egan, Voigt, and co-workers used $P2X_2$ receptor channel expression in HEK-293 cells and substituted wild-type amino acids residues in a 28-residue stretch that included the predicted first $\alpha$-helical transmembrane domain with cysteines (Haines et al., 2001a,b). Using cysteine substitution (so-called "cysteine scanning mutagenesis"), it is possible to probe these residues chemically with $Ag^+$ ions or with sulfhydryl-reacting reagents such as methanethiosulfonates. Twenty-three of the 28 mutants were functional with regard to sensitivity to the ligand, ATP. Five mutants that corresponded to residues H33, R34, I50, K53, and S54 reacted with $Ag^+$ to competitively block ATP-gated cation channel activity, suggesting that these amino acids were critical for cation binding as they traversed the pore of the P2XR channel. These results suggested that residues at both ends of the first $\alpha$-helical transmembrane domain contribute to cation binding and movement through the pore. In more radical mutagenesis studies, Haines *et al.* replaced the first $\alpha$-helical transmembrane domain of $P2X_2$ (insensitive to $\alpha\beta$-methylene-ATP) with the same domain from $P2X_1$ or $P2X_3$ that are sensitive to $\alpha\beta$-methylene-ATP. This swap of the first transmembrane (TM) segment conferred sensitivity to $\alpha\beta$-methylene-ATP in the $P2X_2$ receptor. Mutagenesis of the C-terminal or extracellular portion of this first $\alpha$ helix revealed a role in binding of the nucleotide agonist and, by inference, the gating or opening of the pore within P2XRs. Migita *et al.* also probed the second $\alpha$-helical transmembrane domain for residues key in divalent cation binding and conduction through P2XRs (Migita *et al.*, 2001). $Ca^{2+}$ permeability through P2X2 expressed in HEK-293 cells was thought to be mediated by amino acids with negatively charged side chains, D315 and D349. This had no effect on divalent cation permeation. Rather, the size of side chains that were exposed to the pore of the channel was critical. Mutation of polar threonines at positions 336 and 339 as well as of serine at position 340 affected cation permation profoundly. The largest changes were induced by replacement of these residues with a tyrosine, whose bulky side chain prevented $Ca^{2+}$ movement altogether. Taken together, these three studies by Egan, Voigt, and co-workers showed that the two transmembrane $\alpha$ helices cooperate in conduction of monovalent and divalent cations through the pore and also are involved in ATP binding and gating of the pore as well.

North and colleagues also studied $P2X_2$ channels in HEK-293 cells but probed a larger stretch of amino acids around the first transmembrane segment (Jiang *et al.*, 2001). Their cysteine-scanning mutagenesis revealed that D15, P19, V23, V24, G30, Q37, and F44 were involved in cation binding and permeation. V48 reacted with all forms of methanethiosulfonates, suggesting an extracellular location. In their hands, an R34C mutant passed no current, and their work suggested that K53 and S54, implicated by Egan and co-workers, were extracellular. These results suggest that interpretation of cysteine-scanning mutagenesis is a tricky business. It is possible that amino acids that lie outside of the membrane may "dip" or "loop" into and interact with the $\alpha$ helices to affect permeation. Lessons from the KcsA two transmembrane-spanning $K^+$ channel crystallized by MacKinnon and co-workers revealed this structure feature of these channels.

In addition to the evidence generated by Egan, Voigt, and co-workers regarding ATP gating of the channel by binding to the extracellular face of the first transmembrane segment (Haines et al., 2001a), North and co-workers also addressed this issue in the $P2X_2$ receptor expressed in HEK-293 cells (Jiang et al., 2000). Alanine substitution mutagenesis of 30 polar residues in the presumptive extracellular domain revealed two positively charged lysine residues (L69 and L71) near the same region proximal to transmembrane domain 1 implicated by Egan and Voigt. Substitution with cysteines for these lysines and other flanking amino acid residues showed similar effects. These experiments implicated S65 and I67 as well in ATP binding and gating of the $P2X_2$ channel. In particular, I67 was identified to be critical to the ATP-binding site (Jiang et al., 2000). Perhaps, anionic ATP is attracted to and bound by these lysines and the flanking serine and isoleucine. On binding, ATP may mask these closely coupled lysines with positively charged side chains, removing steric hindrance and allowing cations to move from the external channel vestibule into the pore.

Ennion et al. studied the extracellular domain of the $P2X_1$ receptor, searching for residues critical for ATP binding (Ennion et al., 2000). This study and the one described above concurred. They, too, targeted conserved lysines and arginines, postulating that positively charged side chains may attract and/or bind anionic ATP. Many were irrelevant; however, like the $P2X_2$ receptor studies above, K68 and K70 external to transmembrane domain 1 and R292 and K309 external to transmembrane domain 2 were key to ATP recognition. Taken together, these two studies on different P2XR subtypes show that the ATP-binding pocket is indeed at the extracellular face of the cation pore close to the channel vestibule.

$P2X_1$ and $P2X_3$ receptor channels desensitize rapidly on binding ATP and opening, while the other major subtypes ($P2X_2$ and $P2X_4$ through $P2X_7$) inactivate slowly or not at all. Stojilkovic and co-workers swapped a stretch of six amino acids of $P2X_3$ and $P2X_4$ with the corresponding stretch of R371 through P376 (Koshimizu et al., 1999). $P2X_4$ sequence substitution slowed receptor desensitization, while $P2X_1$ and $P2X_3$ sequence substitution caused rapid desensitization. Coexpression of $P2X_2$ and $P2X_3$ caused a phenotype that had delayed desensitization, a phenotype intermediate to $P2X_2$ and $P2X_3$ expressed alone. Thus, this highly charged stretch of amino acids in the C terminus plays a significant role in desensitization. Seguela and colleagues identified a protein kinase C phosphorylation site, TX(K/R), at amino acids 18 through 20 in the N terminus of the $P2X_2$ receptor, as critical for regulation of receptor desensitization (Boue-Grabot et al., 2000b). Normally a slowly or poorly desensitizing P2X subtype, mutation of the N terminus of $P2X_2$ (T18A, T18N, or K20T) showed rapid desensitization. Phosphorylation of this site within the N terminus was verified biochemically. Examination of the N-terminal sequences of all P2X subtypes reveals that $P2X_1$ and $P2X_3$, rapidly desensitizing subtypes, lack this threonine at position 18, while the other more slowly desensitizing subtypes possess this conserved protein kinase C (PKC) site. Still another laboratory generated data suggesting that the hydrophobic transmembrane segments were critical in desensitization (Werner *et al.*, 1996). Chimeric swapping of the TM domains from $P2X_1$ and $P2X_3$ with $P2X_2$ affected desensitization markedly. Taken together, these results suggest that multiple, if not all, regions of the P2X receptor accessible to the cytosol are involved in receptor desensitization following ATP binding and channel opening.

## VI. HOMO- AND HETEROMULTIMERIZATION OF P2XRS

As members of the two transmembrane-spanning ion channel superfamily (see above), P2X receptors were long suspected to form homomultimers or, in some cell models, heteromultimers of mixed P2XR subtypes. Torres, Egan, and Voigt explored this elegantly with parallel biochemical and physiological methods in the HEK-293 cell heterologous cell system with multiple studies (Haines *et al.*, 1999; Torres *et al.*, 1999a). The most elegant of these studies was the examination of the ability of all P2XR subtypes to oligomerize or multimerize (Torres *et al.*, 1999a). Coimmunoprepitation of epitope-tagged P2XRs was undertaken to probe the ability of P2XRs to form homomultimers and heteromultimers. Hemagglutinin (HA) and FLAG tags added to the C terminus of each P2XR subtype did not affect cation channel function or other properties such as desensitization. The exception in this work was $P2X_6$, where even the wild-type construct failed to generate cation currents. Again, with the exception of $P2X_6$, all other P2XRs tested ($P2X_1$ through $P2X_5$ and $P2X_7$) were capable of forming homomultimers. The authors maintain that the inability of $P2X_6$ to form oligomers may explain why it fails to form functional channels. The stoichiometry of these multimers is unknown; however, dimers, trimers, and tetramers have been postulated (Torres *et al.*, 1999a). With regard to heteromultimers, all P2XRs were capable of forming multimeric complexes with other subunits, with the notable exception of $P2X_7$. A helpful table in the article by Torres *et al.* (1999a) summarizes which P2XR subtypes oligomerize with specific other P2XR subtypes. $P2X_1$ and $P2X_2$ coassemble with themselves as homomultimers and with $P2X_3$, $P2X_5$, and $P2X_6$, but not with $P2X_4$ or $P2X_7$. $P2X_3$ coassembles with $P2X_1$, $P2X_2$, and $P2X_5$, but not with $P2X_4$, $P2X_6$, or $P2X_7$. $P2X_4$ was expressed at high levels, making it difficult to assess in this analysis. However, when the conditions were made favorable, $P2X_4$ was found to oligomerize with itself and with $P2X_5$ and

$P2X_6$. This result was helpful to our laboratory, which found that $P2X_4$ and $P2X_5$ were most abundantly expressed in human vascular endothelial cells and in human and rodent airway, gastrointestinal, and kidney epithelial cell models. $P2X_5$ coassembled with every other P2XR subtype, except $P2X_7$. $P2X_6$ coassembles with $P2X_1$, $P2X_2$, and $P2X_4$, but fails to interact with $P2X_3$, itself, or $P2X_7$. These seminal results help explain some of the functional studies described below that show physiological evidence for heteromultimeric assembly of unique pairs of P2XR subunits.

## A. $P2X_2/P2X_3$ Channels

Surprenant and colleagues coexpressed $P2X_2$ and $P2X_3$ (Lewis et al., 1995), taking advantage of the fact that $P2X_2$ was a poorly desensitizing receptor channel and $P2X_3$ was a rapidly desensitizing receptor channel. Assessing desensitization of the ATP-gated cation currents, they found that coexpression of equal amounts of $P2X_2$ and $P2X_3$ revealed an ATP-gated cation current that had a mixed desensitization phenotype. While $P2X_2$ did not desensitize within the time frame measured and $P2X_3$ desensitized fully, $P2X_2/P2X_3$ currents desensitized slowly with a time course that lay in between that of each P2XR subtype expressed alone. The rationale for this study was reconstitution of ATP-gated currents normally observed in rat dorsal root ganglion sensory neurons. $P2X_2/P2X_3$ currents most closely resembled these native currents, while other combinations failed to do so. Egan, Voigt, and co-workers performed similar experiments; however, they also verified the coassembly biochemically (Torres et al., 1999a). Work by Surprenant, North, and colleagues showed further evidence for $P2X_2/P2X_3$ channels using three different chemical antagonists (Spelta et al., 2002).

## B. $P2X_1/P2X_5$ Channels

Work has also been performed to identify heteromultimeric $P2X_1/P2X_5$ channels. *In situ* hybridization and immunocytochemical work revealed an overlap in the expression of these two subtypes in spinal cord motor neurons (Le et al., 1999). Egan, Voigt, and co-workers performed similar studies and obtained a similar conclusion, supporting the notion that this heteromultimeric $P2X_1/P2X_5$ had unique biophysical, pharmacological, and desensitization properties (Torres et al., 1998b; Haines et al., 1999). Like the $P2X_2/P2X_3$ channels, $P2X_1/P2X_5$ had a slowly desensitizing phenotype that was intermediate to the rapidly desensitizing $P2X_1$ and nondesensitizing $P2X_5$ gated by ATP. Torres et al. showed that 30 $\mu M$ $\alpha\beta$-methylene-ATP stimulated $P2X_1$ but failed to activate $P2X_5$. Interestingly, the $EC_{50}$ for $\alpha\beta$-methylene-ATP at the $P2X_1$ receptor was 5 $\mu M$, while it was lower for the $P2X_1/P2X_5$ channels (1.6 $\mu M$). In a somewhat different data set, Le $et\ al.$ showed that $\alpha\beta$-methylene-ATP ($EC_{50}$ of 1.1 $\mu M$) and trinitrophenyl-ATP (TNP-ATP, $IC_{50}$ of 64 n$M$) were an effective agonist and antagonist, respectively, for the $P2X_1/P2X_5$ channels, but not effective at all for $P2X_1$ or $P2X_5$ homomeric channels. Both Torres $et\ al.$ and Le $et\ al.$ showed coimmunoprecipitation in their expression systems with differentially epitope-tagged $P2X_1$ and $P2X_5$. Haines $et\ al.$ did additional pharmacology on $P2X_1$ channels, $P2X_5$ channels, and $P2X_1/P2X_5$ channels. Only ATP and 2-methylthio-ATP were full agonists for the $P2X_1/P2X_5$ channels. Suramin and PPADS were antagonists of equal potency for $P2X_1/P2X_5$ channels versus $P2X_1$ channels, while TNP-ATP was more effective at $P2X_1/P2X_5$ heteromeric channels. Surprenant $et\ al.$ followed this work on $P2X_1/P2X_5$ and also performed characterization in HEK-293 cells. However, they compared this work with native currents recorded presumably from vascular smooth muscle cells in the submucosa of arterioles from guinea pig (Surprenant $et\ al.$, 2000). Their characterization of $P2X_1/P2X_5$ receptor channels revealed that the ATP response curve had a threshold of 1 n$M$, the lowest effective agonist dose found for any flavor of P2X receptor channel. They also found that TNP-ATP was a weak agonist as well as a noncompetitive antagonist for this heteromeric channel. Furthermore, in contrast to individual homomeric P2XR subtypes that are stimulated by a change in the $pH_o$ away from physiologic in one direction versus the other, $P2X_1/P2X_5$ channels were inhibited by both alkaline and acidic extracellular pH changes. The nanomolar sensitivity of this heteromer revealed a steady endogenous release of ATP from HEK-293 cells that could be calibrated to show a basal constitutive release of 3–300 n$M$ ATP. Noradrenaline potentiated ATP-gated $P2X_1/P2X_5$ currents significantly. This and other properties described above were found in the excitatory junctional potentials recorded from vascular smooth muscle cells on submucosal arterioles, suggesting that purinergic nerves may influence vascular tone in different vascular beds by stimulating these highly ATP-sensitive heteromeric $P2X_1/P2X_5$ channels.

## C. Additional Evidence for Heteromeric P2XR Channels

$P2X_6$ receptor channels do not assemble into functional homomeric channels; however, they can coassemble with other P2XR subtypes. Seguela and co-workers tested the hypothesis that $P2X_6$ might influence the properties of another P2XR subtype, $P2X_4$, if coassembled with it

(Le *et al.*, 1998). $P2X_4$ alone was slowly desensitizing and weakly responsive to either the agonist $\alpha\beta$-methylene-ATP or the antagonists PPADS and suramin. In contrast, however, heteromeric $P2X_4/P2X_6$ channels are activated by low micromolar $\alpha\beta$-methylene-ATP ($EC_{50}$ of 12 $\mu M$) and are blocked effectively by suramin. In addition, the dye Reactive Blue 2 also inhibited the $P2X_4/P2X_6$ channels; however, it potentiated $P2X_4$ homomeric channels modestly. This work argued for a heteromeric $P2X_4/P2X_6$ channel with unique pharmacological properties that might be found in native neurons in the central nervous system, where they are naturally coexpressed. Similar work was done with $P2X_6$ when coassembled with $P2X_2$. King *et al.* (2000) showed that reduction in potency of ATP was seen in $P2X_2/P2X_6$ channels versus $P2X_2$ homomeric channels. The range was also narrower with regard to effects of extracellular pH, and the effect of ATP was curiously biphasic in the presence of the coagonist $ZnCl_2$. In both studies, the rate of desensitization was unaffected. What is unclear is why $P2X_6$ does not form a functional channel in and of itself; Egan and Voigt showed that it cannot multimerize, a finding that was confirmed by Seguela and co-workers. However, it is possible that the $P2X_6$ subtype lies within the cell rather than on the cell surface. When coexpressed with another subtype, $P2X_6$ may be trafficked to the cell surface along with the other subtype. Nevertheless, these studies provide further proof for heteromultimerization, which complicates the study of P2XRs in native cells (see below). Another confounding question concerns the stoichiometry of the homomeric or heteromeric P2XR channel complexes. This has been a conundrum for other two transmembrane-spanning cation channel subfamilies such as the ENaCs. For two transmembrane-spanning potassium channels, it has been shown that they likely form tetramers. However, this is still an open and unsolved issue for this ion channel superfamily that needs to be solved. It is also important to note that the stoichiometry may differ depending on the subtype composition.

## VII. BIOCHEMISTRY OF P2XR CHANNELS IN NATIVE AND HETEROLOGOUS CELLS

### A. Membrane Topology and Glycosylation

In 1998, Newbolt *et al.* performed an elegant study defining the membrane topology that was previously predicted for the P2XR family (Newbolt *et al.*, 1998). On the basis of homology with other two transmembrane-spanning cation channel subfamilies, it was predicted that the N and C termini would be intracellular, that the channel would have two transmembrane $\alpha$ helices, and that the vast majority of the protein would be extracellular. Importantly, there are also three asparagines that could act as potential N-linked glycosylation sites (residues N182, N239, and N298 with the wild-type $P2X_2$ receptor). Newbolt *et al.* found that mutation of the asparagines to serines was tolerated at one position, but not when two or three sites were mutated (Newbolt *et al.*, 1998). This argued that glycosylation was essential for function. This finding was confirmed by Torres, Egan, and Voigt in a parallel study (Torres *et al.*, 1998a) and by Rettinger, Aschrafi, and Schamlzing for the rat $P2X_1$ receptor (Rettinger *et al.*, 2000a). To assess topology, artificial N-linked glycosylation sites of the conserved motif were inserted in various regions of the $P2X_2$ protein to determine whether they were functionally glycosylated. Motifs inserted into the N and C termini were not used for glycosylation; however, sites inserted into the presumed extracellular side of the transmembrane $\alpha$ helices were used, suggesting that they were exposed to the extracellular face of the plasma membrane. In the $P2X_2$ receptor, the calculated molecular mass is between 52 and 53 kDa. However, immunoblotting revealed a broad band at 65 kDa. This was shown to be due to glycosylation that enhanced the molecular mass of the protein, because tunicamycin treatment of the cells or endoglycosidase H treatment of the protein lysates shifted the detected band from 65 to 52 kDa. Mutation of one or two asparagines reduced the weight of the $P2X_2$ protein to an intermediate value, while mutation of all three mimicked the effect of inhibition of glycosylation or stripping of carbohydrate residues from the protein.

## B. Biochemistry of P2XRs in Native Cells

A study has shed new light of P2XR biochemistry in native cells and tissue. Hu *et al.* studied the protein biochemistry of the chick $P2X_4$ receptor in heart and in cultured embryonic ventricular myocytes (Hu *et al.*, 2002). Before this study, most of the protein work on P2XRs was performed on P2XR subunits expressed in heterologous systems. Not only did Hu *et al.* show compelling evidence of a role for $P2X_4$ in calcium influx from extracellular stores and in contractile amplitude of the chick heart, they also showed that native $P2X_4$ receptors were resistant to various rather harsh detergents under various biochemical conditions. Initially, they verified that the $P2X_4$ receptor was found in two forms: a nonglycosylated 44-kDa form and a glycosylated 58-kDa form. Tunicamycin shifted the 58-kDa form to the 44-kDa form, as shown above in heterologous expression systems for $P2X_1$ and $P2X_2$. In this study, the nonglycosylated form was resistant to solubilization in 1% Triton X-114 in the absence or presence of 1% deoxycholic acid or 0.5 $M$ urea or in 1% Nonidet P-40 or 1 $M$ urea. Only the glycosylated form was biotinylated and reacted with streptavidin, suggesting that glycosylation was necessary for normal trafficking. Interestingly, the glycosylated form was soluble in all of the detergent combinations listed above. Molecular mass forms larger than 58 kDa are also evident in many of the lanes, approximately 100–150 kDa. Our laboratory has confirmed these results in human epithelial and endothelial cell models for the human P2X$_4$ receptor, a tight band that represents the nonglycosylated form at 40–45 kDa and a broader band at 60–65 kDa (Schwiebert *et al.*, 2001, 2002). Similar results have also been obtained in non-cystic fibrosis (non-CF) and CF human airway epithelial cells (A. T. Boyce and E. M. Schwiebert, unpublished observations). Our laboratory has also seen higher molecular weight forms in immunoblotting that are competed away by the peptide immunogen, suggesting higher order glycosylation and/ or a detergent-resistant multimer. The prominent carbohydrate addition on this large extracellular loop, along with disulfide bonds between conserved cysteines (see below) and the reaction with the extracellular domain of the ATP agonist, heavy metals, protons, and cations, provide a complex and dynamic protein worth further investigation in native cells.

## C. Essential Nature of Conserved Cysteines

As introduced above, the P2XR extracellular domain has 2 cysteine-rich regions, within which each P2XR has 10 conserved cysteine residues in identical locations. This argues for a complex, three-dimensional structure that begs to be crystallized. ENaCs have 14 conserved cysteine residues in their extracellular domains. Work suggests that ENaCs have Kunitz-like domains that are susceptible to protease cleavage, which is argued to be a major mode of activation for these channels (Bridges *et al.*, 2001). Further proof for this is that protease inhibitors attenuate ENaC activity. The effect of proteases on P2XRs has not been tested.

The role of the cysteines in forming intra- and interchain disulfide bonds and creating trafficking-competent P2X$_1$ receptor channels was addressed by Ennion and Evans (2002). An important finding in this study was the inability to label wild-type receptors with MTSEA-biotin, suggesting that all 10 cysteine residues are engaged in disulfide bonds. When most single cysteines were mutated, only modest effects were observed on ATP potency at the receptor channel. However, exceptions were C261A and C270A, where the peak current amplitudes were reduced by almost 100%. This, however, was determined to be an effect on trafficking. The other single-cysteine mutants did allow MTSEA-biotin labeling, suggesting that companion cysteines were not free to react with the methanethiosulfonate compound. On the basis of their work, they proposed the following pairs of cysteines that were disulfide linked: C117–C165, C126–C149, and C132–C159 in the first cysteine-rich region and C217–C227 and C261–C270 in the second cysteine-rich region. Channel function is not affected significantly by these bonds; however, trafficking is severely disrupted by the elimination of the C261–C270 bond or by the C117–C165 bond together with another bond.

## D. Pharmacology and Chemistry of P2XR Channel Activity

Therefore, the extracellular domain of the P2X receptor channel is a rich background for complex chemistry. As it contains more than half of the molecular mass of the receptor channel protein, like its relatives in the ENaC/ degenerin family, this protein can be thought of more as a receptor than a channel. In fact, its extracellular domain is so large and complex, perhaps P2XRs can be thought of as sensors rather than receptors. This chemistry begins with the binding of its gating ligand, ATP. Agonists and antagonists have been presented in the context of work discussed above. However, work has identified additional antagonists for P2XR subtypes, including TNP-ATP (Burgard et al., 2000; Lewis et al., 1998), NF-279, a suramin analog (Rettinger et al., 2000b; Klapperstuck et al., 2000), and pyridoxal-5′-phosphate-6-(2′-naphthylazo-6′-nitro-4′,8′-disulfonate) (PPNDS) (Lambrecht et al., 2000). Despite these efforts, these antagonists affect mainly $P2X_1$, $P2X_2$, and $P2X_3$ or mixtures thereof. The lack of agonists that are selective between P2Y and P2X receptor families, and the lack of agonists and antagonists that define particular P2XR subtypes (in particular, $P2X_4$, $P2X_5$, and $P2X_6$), are a major roadblock in the study of P2 receptors and P2X receptors that needs to be overcome.

Not only does the P2XR extracellular domain have sites for agonist and antagonist binding, N-linked glycosylation, cysteine bonding, and, possibly, protease cleavage, protons and heavy metals react within this extracellular domain as well. Stoop, Surprenant, and North explored the sensitivity of four P2XR subtypes to changes in extracellular pH ($pH_o$) (Stoop et al., 1997). $P2X_3$ homomultimers are only mildly inhibited by acidic pH ($pH_0$ of 6.3 vs. 7.3) and are unaffected by an alkaline $pH_0$ of 8.3. In contrast, $P2X_2$ homomultimers and $P2X_2/P2X_3$ heteromultimers show profound stimulation by acidic $pH_o$ and almost complete inhibition at alkaline $pH_o$. The authors noted that the data pointed to a single site being modulated by protons. For $P2X_1$ and $P2X_4$, opposite effects were observed with acidic versus alkaline $pH_o$. Acidic $pH_o$ profoundly inhibited $P2X_4$ currents in particular, while alkaline $pH_o$ had little effect. Site-directed mutagenesis of the extracellular domain of the rat $P2X_2$ receptor revealed that acidic pH potentiated ATP stimulation more than 4-fold in wild-type channels and in eight mutant channels in which extracellular histidines were mutated (Clyne et al., 2002). Only one histidine mutant, H319A, attenuated the effect of acidic pH down to 1.4-fold. Substitution of a lysine instead of an alanine reduced the $EC_{50}$ for ATP 40-fold. Extracellular $Zn^{2+}$ is known to potentiate and inhibit the responses of various P2XR subtypes to ATP. The maximal effective dose is only 20 $\mu M$, which falls within physiological limits for this trace element. $Zn^{2+}$ potentiates ATP gating of $P2X_2$, $P2X_3$, $P2X_4$, $P2X_2/P2X_3$, $P2X_2/P2X_6$, and $P2X_4/P2X_6$, while it inhibits the effect of ATP on $P2X_1$ and $P2X_7$. $ZnCl_2$ potentiated the response to ATP in wild-type and seven histidine mutants by 8-fold; however, in two mutants, H120A and H213A, $Zn^{2+}$ had no effect (Clyne et al., 2002). Because these two $Zn^{2+}$-reactive histidine residues are rather distant from one another, two different sites may be required for the effect of $Zn^{2+}$ or the two histidines may interact with each other in a three-dimensional model of the extracellular loop. Taken together, these results suggest that multiple but different histidine residues, the $pK_a$ of which falls within the range tested in these studies, bind protons and heavy metals to modulate the gating by the physiologic agonist, ATP. Again, Fig. 3 highlights the histidine residues implicated in $H^+$ and $Zn^{2+}$ binding.

## VIII. EPITHELIAL P2XRS

Collo et al., in the context of in situ hybridization for P2XR subtypes in neural and other tissues, found staining in bronchial epithelium for $P2X_4$ and $P2X_6$. To our knowledge, this was the first of many hints that extracellular ATP-gated P2X receptor channels may be expressed by epithelial cells. This observation both puzzled and intrigued our laboratory. Why would an epithelial cell express P2X receptors, when P2Y G protein-coupled receptors have already been found in countless epithelial cell models? However, perhaps P2XRs subserved a unique physiological role or roles for the epithelial cell, heretofore unappreciated.

### A. Why Would an Epithelial Cell Express a P2XR?

In 1998, concomitant with the early stages of our work with degenerate RT-PCR primers to the P2XR gene family, two studies provided additional indications of the epithelial expression of P2XRs. Filipovic et al. showed functional evidence for a P2X receptor channel in LLC-PK1 cells, a porcine kidney epithelial cell line with distal nephron properties (Filipovic *et al.*, 1998). Degenerate RT-PCR amplication of mRNA/cDNA from this cell line revealed a sequence that was most similar to $P2X_1$, although it was unique when compared with the known cloned members of the P2XR gene family. Silberberg and co-workers also showed functional evidence for an extracellular ATP-gated cation channel in freshly isolated rabbit ciliated epithelial cells (Korngreen *et al.*, 1998). In a subsequent study, they discovered this ATP-gated channel on the cilia and found it to be critical for ciliary beat and inhibited by extracellular $Na^+$ (Ma *et al.*, 1999). This latter observation was critical to P2X-specific cytosolic calcium measurements that our laboratory has performed in airway epithelial cells (see below).

## B. Multiple P2XRs in a Single Epithelial Cell

Because of the description of difficult early work identifying and cloning P2XRs and the heterogeneity in the ATP-gated cation currents observed in patch-clamp recording, we decided to begin with molecular biology to define whether P2XRs were expressed in epithelial cells and, if so, which P2XR subtypes. With degenerate primers designed to an alignment of rat $P2X_1$, $P2X_2$, and $P2X_3$, a PCR product of the expected size was amplified from every epithelial and endothelial total RNA/cDNA sample. Because this experiment could amplify $P2X_1$, $P2X_2$, $P2X_3$, or any other subtype, we subcloned and sequenced multiple white colonies per cell line or primary culture to gauge which subtypes were expressed and in what amounts, abundantly or rarely. Four articles from our laboratory describe this work (Taylor *et al.*, 1999; McCoy *et al.*, 1999; Schwiebert *et al.*, 2001, 2002). These articles show results from specific human cell lines and/or primary cultures as well as some mouse cell lines. In short, a given epithelial cell model may express as many as four different P2XR subtypes. In the context of this work, the most abundant subtypes were identified, lower abundance P2XR subtypes were also detected, and novel P2X-like sequences that represented sequences with only partial homology to a cloned P2XR subtype were also identified in lung epithelial cells, cholangiocytes, and in human vascular endothelial cells. This work was performed before subtype-specific antibodies were developed and well characterized; however, they allowed us to determine that P2XRs were expressed in epithelial and endothelial cells; that homomultimers and heteromultimers might be expressed in epithelial cells; and what biophysical, pharmacological, and cell biological properties might be displayed by epithelial P2XRs.

Table I summarizes the pooled results for non-CF and CF airway epithelial cell models, gastrointestinal (GI) epithelial cell models, normal and polycystic kidney disease (PKD) renal epithelial cell models, and human vascular endothelial cells. More than 500 sequences were read in these analyses. The data are presented in this way to show several key points. First, $P2X_4$ is found in every category (Table I). It appears to be the major P2XR subtype expressed by epithelial and endothelial barrier cells. Biochemical work with a $P2X_4$-specific antibody verifies these findings (Schwiebert *et al.*, 2001, 2002). Biotinylation of $P2X_4$ shows expression on the apical and basolateral membrane of human airway epithelial cell monolayers, while $P2X_5$ immunocytochemistry reveals only apical localization (A. T. Boyce and E. M. Schwiebert, unpublished observations). Functional work described below also is consistent with a major

**TABLE I**

P2XR Subtype Sequences Found by Degenerate RT-PCR of the P2XR Family in Primary and Immortalized Epithelial Cells from Lung and Airway Epithelia, Gastrointestinal Epithelia, Kidney Epithelia, and Vascular Endothelia[a]

| "Pooled" cell type | $P2X_1$ | $P2X_2$ | $P2X_3$ | $P2X_4$ | $P2X_5$ | $P2X_7$ | $P2X_L$ | $P2X_H$ | $P2X_{Endo}$ | Total |
|---|---|---|---|---|---|---|---|---|---|---|
| Normal human kidney | 0 | 0 | 0 | **46** | 42 | 5 | NA | NA | NA | 93 |
| Human ADPKD | 0 | 0 | 0 | **14** | 12 | 0 | NA | NA | NA | 26 |
| Normal mouse kidney | 4 | 1 | 22 | **19** | 0 | 0 | NA | NA | NA | 46 |
| Mouse ARPKD | 0 | 1 | 11 | **5** | 0 | 0 | NA | NA | NA | 17 |
| Normal human airway | 0 | 0 | 19 | **33** | 28 | 0 | 5 | NA | NA | 85 |
| CF human airway | 0 | 21 | 2 | **8** | 22 | 14 | 1 | NA | NA | 68 |
| Human GI | 11 | 23 | 1 | **24** | 33 | 0 | NA | NA | NA | 92 |
| Rat GI | 0 | 2 | 2 | **12** | 0 | 0 | NA | 2 | NA | 18 |
| Human vascular endothelial primaries | 1 | 0 | 0 | **39** | 27 | 3 | NA | NA | 6 | 76 |
| | 16 | 48 | 57 | **190** | 164 | 22 | 6 | 2 | 6 | 521 |
| Primary cultures | 5 | 2 | 19 | **107** | 85 | 9 | 1 | 2 | 6 | |
| Cell lines | 11 | **46** | **38** | 83 | **79** | 13 | 5 | 0 | 0 | |

*Abbreviations:* ADPKD, Autosomal dominant polycystic kidney disease; ARPKD, autosomal recessive PKD; GCF, cystic fibrosis; GI, gastrointestinal; NA, not available.[a]

[a]$P2X_6$ is not included, because it was never detected in any of our samples. There could be two reasons for this: lack of expression or lack of primer recognition. More than 500 sequences were read in our laboratory and the DNA sequencing was done ourselves (**use of** a CORE facility would have bankrupted our laboratory). $P2X_4$ is found in every category, $P2X_4$ and $P2X_5$ are the most abundant sequences, and the distribution of mainly $P2X_4$ and $P2X_5$ is tighter in primary cultures than in cell lines.

contribution from $P2X_4$ receptors and, possibly, coassembly with $P2X_5$ receptors. Second, human normal and PKD kidney epithelial cell models show this abundant expression of $P2X_4$ and $P2X_5$ (Table I). However, mouse normal and PKD kidney epithelial cell models, predominantly derived from collecting duct, show a shift to a mixture of primarily $P2X_4$ and $P2X_3$ (Table I). Third, in normal human airway epithelial cell models, abundant expression of $P2X_4$ and $P2X_5$, with significant expression of $P2X_3$, is apparent. However, CF human airway epithelial cell models show a larger spread in expression, with virtually every subtype expressed (Table I). The significance of this is unknown at present. Fourth, a similar broad distribution of subtypes is also observed in GI epithelial cell models. This may represent the fact that intestinal, pancreatic, and hepatic cell models are pooled in this category. Fifth, in another barrier cell, the endothelial cell, $P2X_4$ and $P2X_5$ are again abundant, with rare expression of $P2X_1$, $P2X_7$, and $P2X_{Endo}$ (sequences with partial but not complete homology to known P2XR subtypes). $P2X_L$ and $P2X_H$ refer to lung and hepatic sequences of the same ilk (Table I). Finally, compartmentalizing the data between primary cultures and cell lines revealed a tighter distribution in primary cultures that centered on $P2X_4$ and $P2X_5$, while the distribution was broader for the cell lines. Having said that, to our surprise, many of the cell lines maintain expression of predominantly $P2X_4$ and $P2X_5$ at the protein level (A. T. Boyce and E. M. Schwiebert, unpublished observations). Taken together, these results show that epithelial cells from a variety of tissues express P2XRs abundantly.

## C. P2XRs Are Ion Channels and Regulators of Separate Ion Channels in Epithelial Cells

As described in detail above, P2X receptors or sensors are also extracellular ATP-gated calcium-permeable nonselective cation channels. However, the cause of much of the heterogeneity in ATP-activated ion channels may result from the fact that the rise in $Ca^{2+}$ mediated by P2XRs may activate other ion channels in a rapid manner. In epithelial cells, in the context of our laboratory's interests (Taylor et al., 1999), putative P2XR-regulated ion channels include $Ca^{2+}$ activated chloride and potassium channels. At the same time, $Ca^{2+}$ inhibits epithelial ENaC $Na^+$ channels (Ling et al., 1992). These possibilities are included in an airway epithelial cell model adapted from one published previously (Taylor et al., 1999), in which we have begun to understand how P2XRs may regulate epithelial cell functions (Fig. 5). In this schematic, we show a prominent role of P2XRs in stimulating ciliary beat through mediating $Ca^{2+}$ influx directly into the cilium. In what may also be of therapeutic benefit in cystic fibrosis (CF), P2XRs stimulate chloride secretion when activated in either the apical or basolateral membrane. Apical activation may involve $Ca^{2+}$-dependent signaling and activation of $Ca^{2+}$-dependent chloride channels directly. Basolateral activation may involve P2XRs themselves acting as potassium channels or by activation of $Ca^{2+}$-dependent potassium channels. As such, P2XRs may be critical in regulating transepithelial NaCl and water transport. Moreover, because KCl efflux is critical in regulatory volume decrease following hypotonic cell swelling, P2XRs may bind ATP released by the epithelial cells themselves and transduce that autocrine signal into efficient cell volume regulation. At the same time, P2XRs may inhibit ENaC $Na^+$ channels to quell any regulatory volume increase mechanisms,

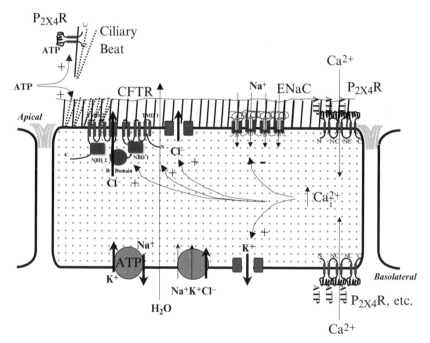

**FIGURE 5**  Simplified epithelial cell model of P2XR channel function and P2XR regulation of additional epithelial ion channels. P2XRs on ciliated epithelial cells stimulate ciliary beat, potentiate cell volume regulation, and stimulate chloride, potassium, and fluid secretion. P2XRs may also inhibit sodium absorption. It is thought that P2XRs accomplish this by mediating a sustained influx of calcium from extracellular stores. Calcium may act directly on the ion channels or the cilium or it may stimulate calcium-dependent protein kinases in a more elaborate signaling cascade.

critical in recovery from the opposite osmotic stress, hypertonicity. The physiological role of ATP release, P2X receptors, P2Y receptors, and P2 receptor-triggered signal transduction cascades in airway epithelial cell biology is reviewed in another chapter (see Braunstein and Schwiebert in this volume).

### D. Dissection of the P2XR-Specific Current from a Separate Ion Channel That the P2XR Activates in a Native Epithelial Cell

We have been able to define conditions under which to study P2XR-specific currents in epithelial cells, while eliminating contaminating ion channels that might be opened by P2XRs. Fura-2 imaging of cytosolic $Ca^{2+}$ has allowed us to dissect out P2X- from P2Y-mediated increases in intracellular $Ca^{2+}$ in airway epithelial cells (Zsembery et al., 2003). Early on, we found it difficult to detect P2X-mediated increases in cytosolic $Ca^{2+}$. Only in early primary cultures of renal epithelial cells were we able to see P2XR-mediated increases (Schwiebert et al., 2002), as measured with the agonist BzBz-ATP. On the other hand, P2Y-mediated responses with UTP, UDP, or ADP were quite reproducible. We felt that, perhaps, we were losing the expression of a key P2XR subtype on passage of the primary cultures that conferred $Ca^{2+}$ permeability or that P2XRs were somehow regulating the function of store-operated or voltage-dependent $Ca^{2+}$ channels. However, expression of the key P2XR subtypes was robust in primary cultures and in cell lines, as assessed by molecular and biochemical means.

Then, by analogy to store-operated or the TRP (transient receptor potential) $Ca^{2+}$ entry channels, we reasoned that extracellular $Na^+$ might compete with extracellular $Ca^{2+}$, because $Na^+$ was in 100-fold excess in physiological solutions. In an IB3-1 CF human airway epithelial cell line, substitution of $Na^+$ with $N$-methyl-D-glucamine (NMDG) unmasked a BzBzATP-triggered increase in $Ca_i^{2+}$ that was sustained and dependent on extracellular $Ca^{2+}$ (Zsembery et al., 2003). Increasing extracellular $Ca^{2+}$ from 1 to 5 m$M$ augmented the response. The presence of $Mg^{2+}$ also partially blocked the response, suggesting that P2XR channels act much like glutamate receptor channels functionally (Zsembery and Schwiebert, unpublished observation). Because this sustained increase in $Ca_i^{2+}$ induced by BzBzATP and ATP itself was not mimicked by $\alpha\beta$-methylene-ATP, because acidic $pH_o$ blocked the response, and because alkaline $pH_o$ and extracellular $Zn^{2+}$ potentiated the response to ATP, we surmise that P2X$_4$ may be the major P2XR subtype that conferred this sustained increase in $Ca_i^{2+}$-derived from extracellular stores (Zsembery et al., 2002). We have also defined conditions within the Fura-2 fluorescence assay of the IB3-1 CF cells to perform high-throughput screening experiments to identify higher affinity agonists for P2XRs that may have therapuetic potential for cystic fibrosis.

From this molecular, biochemical, and fluorescence imaging of $Ca_i^{2+}$, we now have a clear idea of how to design patch-clamp experiments to study the biophysical properties of epithelial P2XRs. First, the agonist, ATP or BzBzATP, must be in close proximity to the extracellular face of the channel. In single-channel recordings in the cell-attached or inside-out "excised" modes, the agonist must be included in the pipette solution or perfused into the pipette by a pipette perfusion system. In outside-out patches or in whole cell patch-clamp recordings, the agonist can be added to the bath. Second, in native epithelial cells, $Cl^-$ and $K^+$ must be removed from the solution, because P2XRs and P2YRs stimulate calcium-dependent $Cl^-$ and $K^+$ channels. Third, $Na^+$ and $Mg^{2+}$ must be removed, because they inhibit $Ca^{2+}$ permeability through the channels. Fourth, an external alkaline pH in the bath solution as well as inclusion of 20 $\mu M$ $ZnCl_2$ will augment the ATP or BzBzATP gating of the P2XR channels. Figure 6 shows that we have designed patch-clamp recordings using BzBzATP as the agonist in an extracellular $BaSO_4$ solution that also has a pH of 8.0 and 20 $\mu M$ $ZnCl_2$. Although the senior author of this chapter was trained as a patch-clamp electrophysiologist, we maintain that, if we began with patch-clamping to define P2XR expression in epithelial cells 5 years ago, we would be at a loss to define epithelial P2XR expression and function. Future studies under these defined conditions will be performed on the plasma membrane of isolated human and mouse airway and kidney epithelial cells, the apical membrane of human and mouse airway and kidney epithelial cells grown as polarized monolayers, and in HEK-293 cells transfected with a single P2XR or mixtures of P2XRs thought to be expressed in the epithelial cells studied in parallel. Figure 6 illustrates this experimental design, currently being utilized by our laboratory.

## IX. PHYSIOLOGICAL ROLES FOR P2XRS IN ENDOTHELIA, VASCULAR SMOOTH MUSCLE, AND SPECIALIZED NEURONS

### A. Endothelial Cells and Vascular Smooth Muscle

P2XRs are emerging as a critical player in vascular physiology. Not only have multiple P2XRs been defined in vascular cells such as those in endothelium and vascular smooth muscle, but P2XRs have also been found on erthyrocytes (Light *et al.*, 1999, 2001). In vascular smooth muscle, the best example of the functional role of P2XRs comes from the $P2X_1$-deficient mouse. In this knockout mouse, vascular smooth muscle in the testes fails to

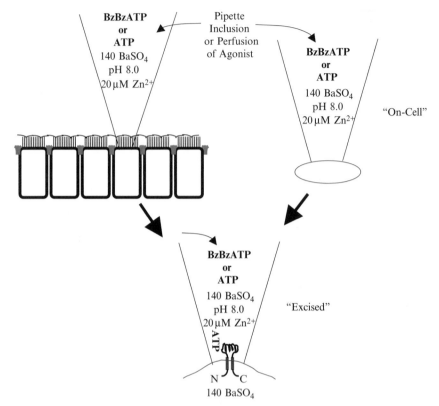

FIGURE 6 Experimental design of "on-cell" patch-clamp experiments with pipette perfusion or pipette inclusion of agonist in the apical membrane of a polarized epithelium and in the plasma membrane of a transfected, heterologous HEK-293 cell. See text for elaborate rationale and description of why this design was established. 140 $BaSO_4$ refers to a millimolar concentration.

contract and sperm are not ejected into the vas deferens. This is a unique cause of sterility. $P2X_1$ immunocytochemistry yields gratifying results in testes vascular smooth muscle, where it is so abundant that the entire tissue is ablaze with $P2X_1$ staining (Mulryan et al., 2000). Through $P2X_1$ and, likely, additional P2XR subtypes, a sustained increase in $Ca_i^{2+}$ derived from extracellular stores would stimulate contraction mediated by $Ca^{2+}$ directly as well as sustained $Ca^{2+}$ stimulation of $Ca^{2+}$-binding proteins such as calmodulin and $Ca^{2+}$-dependent protein kinases.

In the endothelium, multiple P2XRs have been found (Yamamoto et al., 2000a,b; Shinozuka et al., 2001; Ray et al., 2002; Schwiebert et al., 2002; Ramirez and Kunze, 2002). It is less clear in the endothelium how P2XRs may affect function. They could change the membrane potential of the endothelial cell membrane as a $Ca^{2+}$- and $Na^+$-permeable channel, or they could induce a sustained increase in $Ca_i^{2+}$ derived from extracellular stores. If it is the latter, this could cause a profound increase in nitric oxide (NO) production as well as its many derivatives (in particular, peroxynitrite, $ONOO^-$). This could cause a relaxation in vascular smooth muscle. In fact, Sprague, Lonigro, and colleagues have established a paradigm in vascular purinergic biology in which erythrocytes, deformed when squeezing through the smallest blood vessels, release ATP (see Chapter 8 in this volume by Sprague *et al.*) This could have an autocrine effect on the red blood cell (RBC), given that they express P2X receptor channels. They also hypothesize that ATP would then bind to P2Y or P2X receptors in the endothelium in a paracrine manner, where it would cause NO synthesis and release. NO would then diffuse to neighboring vascular smooth muscle cells and cause them to relax. This paradigm is highlighted and illustrated in Fig. 7. Because all three relevant vascular cell types express P2X receptors, the source of the ATP released in this system and the relative activation of P2X receptors in the different cell types would determine reflexation or contraction of the blood vessel. Nevertheless, this is an intriguing paradigm for purinergic signaling (see Chapter 8 in this volume by Sprague *et al.*), as is a parallel body of literature implicating adenosine as a local metabolite that influences vascular tone (Katori and Berne, 1966).

## B. Neurons

It is clear that P2XRs act as neuronal and synaptic receptor channels to initiate and/or propagate action potentials as well as to modify synaptic

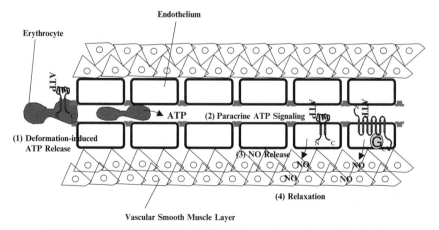

**FIGURE 7**  Physiological roles of P2XRs in vasculature. See text for details.

transmission and action potential propagation. This is highlighted as part of Burnstock's chapter, Purinergic Signaling in the Nervous System (this volume). This section seeks to highlight the many mechanisms by which P2XRs affect synaptic transmission and action potential formation and propagation.

The existence of purinergic nerves is well documented. Specialized enteric neurons that innervate the GI tract, nodose nociceptive sensory neurons in the dorsal root ganglia of the spinal cord, and sensory ganglia activated by outer hair cells in the inner ear are all well-documented examples. However, it is not unreasonable to assume that P2XRs may be expressed in all neurons as one or more subtypes. $P2X_2$ receptors are expressed in autonomic ganglia functionally and biochemically (Liu and Adams, 2001). Peripheral neurons of neural crest origin and placodal origin express mRNAs for multiple P2XRs (Dunn et al., 2001). In particular, dorsal root ganglia express $P2X_3$, nodose nociceptive neurons express $P2X_2$ and $P2X_3$, and sympathetic ganglia express $P2X_2$ only or $P2X_2$ and $P2X_3$, depending on the species. Rubio and Soto showed elegant staining of cerebellar Purkinje cell dendritic spines for $P2X_2$, $P2X_4$, and $P2X_6$, while no staining was apparent on climbing fibers or basket cells (Rubio and Soto, 2001). In hippocampal $CA_1$ pyramidal cells, postsynaptic membranes apposed to the Schaffer collaterals also stained for $P2X_2$, $P2X_4$, and $P2X_6$. Double labeling of P2XRs and $\alpha$-amino-3-hydroxy-5-methyl-4-isoxazole propionic acid (AMPA) glutamate receptors revealed that the P2XR-expressing neurons were also glutamatergic. Kanjhan et al. performed a detailed study on P2X2 expression in the rat central nervous system (Kanjhan et al., 1999). In short, P2X2 was found in virtually every region of the CNS, in the spinal cord, and in the dorsal and ventral horns in ganglia. P2XRs are also in sensory nerves of sensory organs such as the eye and the ear. Shibuya et al. showed expression of multiple P2XRs ($P2X_3$, in particular) in rat supraoptic neurons (Shibuya et al., 1999). Housley and co-workers have shown P2X receptor expression on the outer hair cells of the cochlea on the sensory cilia. Once activated by sound, the outer hair cells activate both purinergic and glutamatergic sensory ganglions that also express P2XRs (Housley, 2000; Munoz et al., 1999). Taken together, the abundant and broad expression of P2XRs in all neural tissues argues for key physiological roles in neurotransmission and, by association, processes such as sound and light transduction, long-term potentiation (LTP), facilitation, depression, and nociception as well as autonomic control over peripheral tissues.

P2XRs may act via several mechanisms to affect neurotransmission. First, the classic role of P2XRs in these specialized purinergic neurons is to act as the postsynaptic receptor for ATP, which is released from the presynaptic nerve terminal as the major neurotransmitter (Fig. 8). Because P2XRs are $Ca^{2+}$- and $Na^{+}$-permeable channels gated by extracellular ATP, ATP and

P2XRs would act as excitatory neurotransmitter and receptor channel, respectively, positively propagating an action potential. In fact, the biophysical properties of P2XRs are quite similar to those of the glutamate receptor channel family. Second, P2XRs can also modify the effect of another neurotransmitter at a given synapse (Fig. 8). Jo and Schlichter detected corelease of ATP with $\gamma$-aminobutyric acid (GABA) in neurons of the dorsal horn that affected postsynaptic neurons (Jo and Schlichter, 1999). ATP stimulation of an excitatory P2X receptor channel concomitant with GABA stimulation of an inhibitory receptor channel complicated the overall sensory output from that postsynaptic neuron. In fact, ATP is coreleased with many different neurotransmitters in addition to GABA, including acetylcholine and epinephrine. In this study, these authors also showed that adenosine modulated the GABA response, presumably generated by metabolism of ATP. In contrast, Sokolova et al. showed that there was negative cross-talk between anionic GABA channels and cationic P2XR channels in rat dorsal root ganglion neurons (Sokolova et al., 2001). Hugel and Schlichter showed, however, that presynaptic P2X receptor channels can facilitate inhibitory GABA-ergic transmission (Hugel and Schlichter, 2000). They also showed that postsynaptic ATP could increase the frequency of miniature inhibitory postsynaptic currents mediated by GABA or glycine. It may be imagined that presynaptic facilitation may be due to evoking a sustained increase in $Ca_i^{2+}$ derived from extracellular stores, which would trigger exocytosis of a given neurotransmitter, stimulatory or inhibitory. The facilitation of postsynaptic potentials is less clear; however, cytosolic $Ca^{2+}$ increase could also play a positive role here.

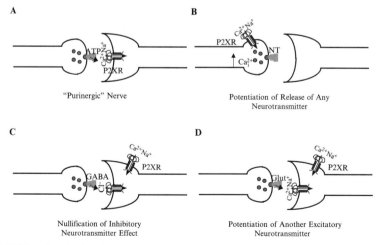

**FIGURE 8**   Physiological roles of P2XRs in specialized neurons. See text for details.

Indeed, Boehm showed that ATP evoked noradrenaline release from sympathetic ganglia via a $Ca^{2+}$-dependent mechanism (Boehm, 1999). In contrast, Boehm showed that P2Y receptors played an inhibitory role in noradrenaline release. The difference was temporal. Fast activation of P2XRs was stimulatory, whereas slower activation of metabotropic P2Y receptors was attenuating. Third, P2XRs can augment the effect of excitatory neurotransmitters that also propagate action potentials in postsynaptic neurons via rapid $Ca^{2+}$ and $Na^+$ influx (Fig. 8). Pankratov et al. showed that glutamatergic transmission at Schaffer collaterals of the $CA_1$ area of the hippocampus were augmented by ATP and $Zn^{2+}$ and were blocked by P2 receptor antagonists suramin and PPADS (Pankratov et al., 1998). A similar result was found for glutamate-evoked excitatory postsynaptic currents that were potentiated by ATP-gated P2XRs in dorsal horn neurons of the rat spinal cord (Nakatsuka and Gu, 2001) and in trigeminal mesencephalic nucleus in the brainstem (Khakh and Henderson, 1998). In fact, fast synaptic transmission mediated by P2X receptors in $CA_3$ pyramidal cells of hippocampal slices was found to be driven solely by ATP, was unaffected by GABA or enkephalin, and was potentiated only mildly by glutamate. This result argued for a classic role of ATP and P2XRs in neurotransmission independent of other neurotransmitters in central nervous system neurons (Mori et al., 2001). As alluded to above, ATP and P2XRs heretofore were thought to act in classic neurotransmission only in specialized neurons or purinergic nerves. Fourth, P2XRs could affect neuronal function via increases in $Ca_i^{2+}$ derived from extracellular stores (Fig. 8). Evidence from Stojilkovic and co-workers in hypothalamic neurons as well as from Kim and co-workers in PC-12 cells showed that P2XRs acted as $Ca^{2+}$ channels themselves but also facilitated the activation of voltage-dependent and voltage-independent $Ca^{2+}$ entry channels in these excitable cell models by also allowing $Na^+$ influx and a subsequent membrane depolarization (Koshimizu et al., 2000; Hur et al., 2001). In PC-12 cells, Landreth and co-workers showed that ATP activation of P2X receptors led to activation of the mitogen-activated protein kinases (MAP kinases) ERK1 and ERK2, via a mechanism dependent on extracellular $Ca^{2+}$ (Swanson et al., 1998). ATP but not adenosine, 5'-AMP, ADP, UTP, or $\alpha\beta$-methylene-ATP induced kinase activation, while suramin and Reactive Blue 2 inhibited ATP induction. An upstream regulator of the MAP kinases, the calcium-activated tyrosine kinase Pyk-2, was also activated rapidly by ATP. This may be due to the sustained increase in $Ca_i^{2+}$ derived from extracellular stores and mediated by P2XRs. Taken together, these results show that P2XRs may affect neuronal signaling and function at multiple key levels. These possible functions of P2XRs in neurotransmission are illustrated in Fig. 8.

## X. FUTURE DIRECTIONS

It has not been long since the cloning of the first P2XR subtypes; however, the amount of literature on P2XRs continues to increase at a lightning pace. In preparing this review, the literature had to be screened weekly, because new literature was emerging that quickly. We apologize if we missed the newest work.

It is difficult to predict where this work might go. An advance in studying the movement and redistribution of P2XRs in real time, by tagging them with green fluorescent protein (GFF), has allowed the visualization of $P2X_2$ aggregation within nerves in seconds (Khakh et al., 2001). Perhaps tagging of different P2XR subtypes with different fluorescent protein tags [e.g., with red and blue fluorescent proteins (RFP and BFP, respectively)] could allow the probing of homo- and heteromultimerization. The stoichiometry of P2XR multimers still needs to be defined. Only then will we have a clearer picture of how P2XRs function as receptors and channels. Having said that, crystallization of a P2XR, prokaryotic or eukaryotic, is imminent. At the least, solving the structure of the complicated and elegant extracellular domain, which reacts with so many different ligands, is mandatory. One glaring hole in the study of P2XRs is the lack of specific pharmacological tools. This needs to be remedied if we are to be more specific about the mixtures of different P2XRs in a given cell model. Subtype-specific in antibodies are now emerging and will help this work. In addition to the zebrafish $P2X_3$ homolog, it is quite possible that P2XR homologs are indeed present in lower organisms and, even, in prokaryotes. Identification of such homologs, deleting such genes, and elucidating their function in C. elegans, Escherichia coli, or Drosophila might be helpful. Mammalian single or double knockouts of P2XRs will also be informative.

As we leap ahead to these new advances, a few caveats need to be kept clearly in mind. First, investigators must be careful in the biophysical characterization of P2XRs in native cells. As P2XRs could activate separate and distinct ion channels via $Ca_i^{2+}$, the potential for recording a contaminating ion channel that is, in fact, not a P2XR channel is quite high. Second, in native cells, molecular biology and biochemistry must coincide with electrophysiology to account for all of the P2XRs that may be expressed in a given cell model. Only then can precise electrophysiology be performed. Finally, verification in HEK-293 cells or oocytes expressing exogenous P2XRs should be performed in parallel, provided that care is taken with these heterologous cell systems so as not to cause expression of endogenous P2XRs.

Nevertheless, our laboratory finds P2XRs to be ideal, multifaceted membrane proteins to study in our favorite native cell models, epithelial and endothelial cells. P2XRs act as receptors for ligands and those receptor–ligand interactions are modified by many extracellular factors. In this light, we actually view P2XRs more as sensors than as ATP receptors or cation channels. P2XRs also act as ion channels and can also regulate other ion channels via increases in $Ca_i^{2+}$. And finally, P2XRs can affect signaling in a profound manner: not just via a sustained increase in $Ca_i^{2+}$ derived from extracellular stores but also via protein kinases, the activation of which is triggered by such a sustained $Ca^{2+}$ increase. To put it simply, what is not to love about P2XRs!

## Acknowledgments

Amanda T. Boyce was formerly Amanda L. Taylor. E.M.S. thanks Amanda for initiating and launching an exciting and productive project concerning epithelial P2X receptor channels. Work on epithelial P2XRs by E.M.S. and A.T.B. was supported by an NIH grant (ROI HL63934).

## References

Afework, M., and Burnstock, G. (1999). Distribution of P2X receptors in the rat adrenal gland. *Cell Tissue Res.* **298**, 449–456.

Afework, M., and Burnstock, G. (2000). Localization of P2X receptors in the guinea pig adrenal gland. *Cells Tissues Organs* **167**, 297–302.

Bean, B. P. (1990). ATP-activated channels in rat and bullfrog sensory neurons: Concentration dependence and kinetics. *J. Neurosci.* **10**, 1–10.

Bean, B. P. (1992). Pharmacology and electrophysiology of ATP-activated ion channels. *Trends Pharmacol. Sci.* **13**, 87–90.

Bean, B. P., and Friel, D. D. (1990). ATP-activated channels in excitable cells. *Ion Channels.* **2**, 169–203.

Bean, B. P., Williams, C. A., and Ceeler, P. W. (1990). ATP-activated channels in rat and bullfrog sensory neurons: Current–voltage relation and single-channel behavior. *J. Neurosci.* **10**, 11–19.

Benos, D. J., and Stanton, B. A. (1999). Functional domains within the degenerin/epithelial sodium channel (Deg/ENaC) superfamily of ion channels. *J. Physiol.* **520**, 631–644.

Bianchi, L., and Driscoll, M. (2002). Protons at the gate: DEG/ENaC ion channels help us feel and remember. *Neuron.* **34**, 337–340.

Bo, X., Alavi, A., Xiang, Z., Oglesby, I., Ford, A., and Burnstock, G. (1999). Localization of ATP-gated P2X2 and P2X3 receptor immunoreactive nerves in rat taste buds. *Neuroreport* **10**, 1107–1111.

Bo, X., Schoepfer, R., and Burnstock, G. (2000). Molecular cloning and characterization of a novel ATP P2X receptor subtype from embryonic chick skeletal muscle. *J. Biol. Chem.* **275**, 14401–14407.

Boehm, S. (1999). ATP stimulates sympathetic transmitter release via presynaptic purinoceptors. *J. Neurosci.* **19**, 737–746.

Boue-Grabot, E., Akimenko, M. A., and Seguela, P. (2000a). Unique functional properties of a sensory neuronal P2X ATP-gated channel from zebrafish. *J. Neurochem.* **75**, 1600–1607.

Boue-Grabot, E., Archambault, V., and Seguela, P. (2000b). A protein kinase C site highly conserved in P2X subunits confer the desensitization kinetics of P2X$_2$ ATP-gated channels. *J. Biol. Chem.* **275**, 10190–10195.

Brake, A. J., Wagenbach, M. J., and Julius, D. (1994). New structural motif for ligand-gated ion channels defined by an ionotropic ATP receptor. *Nature* **371**, 519–523.

Brandle, U., Guenther, E., Irrle, C., and Wheeler-Schilling, T. H. (1998a). Gene expression of the P2X receptors in the rat retina. *Brain Res. Mol. Brain Res.* **59**, 269–272.

Brandle, U., Kohler, K., and Wheeler-Schilling, T. H. (1998b). Expression of the P2X7-receptor subunit in neurons of the rat retina. *Brain Res. Mol. Brain Res.* **62**, 106–109.

Brandle, U., Zenner, H. P., and Ruppersberg, J. P. (1999). Gene expression of P2X-receptors in the developing inner ear of the rat. *Neurosci. Lett.* **273**, 105–108.

Bridges, R. J., Newton, B. B., Pilewski, J. M., Devor, D. C., Poll, C. T., and Hall, R. L. (2001). Sodium transport in normal and CF human bronchial epithelial cells is inhibited by BAY 39-9437. *Am. J. Physiol. Lung Cell Mol. Physiol.* **281**, L16–L23.

Buell, G., Lewis, C., Collo, G., North, R. A., and Surprenant, A. (1996). An antagonist-insensitive P2X receptor expressed in epithelia and brain. *EMBO J.* **15**, 55–62.

Burgard, E. C., Niforatos, W., van Biesen, T., Lynch, K. J., Kage, K. L., Touma, E., Kowaluk, E. A., and Jarvis, M. F. (2000). Competitive antagonism of recombinant P2X2/P2X3 receptors by 2′,3′-$O$-(2,4,6-trinitrophenyl) adenosine 5′-triphosphate (TNP-ATP). *Mol. Pharmacol.* **58**, 1502–1510.

Burnstock, G. (1972). Purinergic nerves. *Pharmacol. Rev.* **24**, 509–581.

Chen, C., and Bobbin, B. P. (1998). P2X receptors in cochlear Deiters' cells. *Br. J. Pharmacol.* **124**, 337–344.

Chen, C., Parker, M. S., Barnes, A. P., Deininger, P., and Bobbin, R. P. (2000). Functional expression of three P2X$_2$ receptor splice variants from guinea pig cochlea. *J. Neurophysiol.* **83**, 1502–1509.

Clyne, J. D., LaPointe, L. D., and Hume, R. I. (2002). The role of histidine residues in modulation of the rat P2X$_2$ purinoceptor by zinc and pH. *J. Physiol.* **539**, 347–359.

Collo, G., North, R. A., Kawashima, E., Merlo-Pich, E., Neidhart, S., Surprenant, A., and Buell, G. (1996). Cloning of P2X5 and P2X6 receptors and the distribution and properties of an extended family of ATP-gated ion channels. *J. Neurosci.* **16**, 2495–2507.

Cox, J. A., Barmina, O., and Voigt, M. M. (2001). Gene structure, chromosomal localization, cDNA cloning and expression of the mouse ATP-gated ionotropic receptor P2X5 subunit. *Gene* **270**, 145–152.

Dhulipala, P. D., Wang, Y. X., and Kotlikoff, M. I. (1998). The human P2X4 receptor gene is alternatively spliced. *Gene* **207**, 259–266.

Ding, S., and Sachs, F. (1999). Ion permeation and block of P2X$_2$ purinoceptors: Single channel recordings. *J. Membr. Biol.* **172**, 215–223.

Ding, S., and Sachs, F. (2000). Inactivation of P2X$_2$ purinoceptors by divalent cations. *J. Physiol.* **522**, 199–214.

Dunn, P. M., Zhong, Y., and Burnstock, G. (2001). P2X receptors in peripheral neurons. *Prog. Neurobiol.* **65**, 107–134.

Egan, T. M., Cox, J. A., and Voigt, M. M. (2000). Molecular cloning and functional characterization of the zebrafish ATP-gated ionotropic receptor P2X$_3$ subunit. *FEBS Lett.* **475**, 287–290.

Elneil, S., Skepper, J. N., Kidd, E. J., Williamson, J. G., and Ferguson, D. R. (2001). Distribution of P2X$_1$ and P2X$_3$ receptors in the rat and human urinary bladder. *Pharmacology* **63**, 120–128.

Ennion, S., and Evans, R. J. (2002). Conserved cysteine residues in the extracellular loop of the human P2X$_1$ receptor form disulfide bonds and are involved in receptor trafficking to the cell surface. *Mol. Pharmacol.* **61**, 303–311.

Ennion, S., Hagan, S., and Evans, R. J. (2000). The role of positively charged amino acids in ATP recognition of human P2X$_1$ receptors. *J. Biol. Chem.* **275**, 29361–29367.

Filipovic, D. M., Adebanjo, O. A., Zaidi, M., and Reeves, W. B. (1998). Functional and molecular evidence for P2X receptors in LLC-PK1 cells. *Am. J. Physiol. Renal Physiol.* **274,** F1070–F1077.

Garcia-Guzman, M., Soto, F., Laube, B., and Stuhmer, W. (1996). Molecular cloning and functional expression of a novel rat heart P2X purinoceptor. *FEBS Lett.* **388,** 123–127.

Gartland, A., Hipskind, R. A., Gallagher, J. A., and Bowler, W. B. (2001). Expression of a P2X7 receptor by a subpopulation of human osteoblasts. *J. Bone Miner. Res.* **16,** 846–856.

Glass, R., and Burnstock, G. (2001). Immunohistochemical identification of cells expressing ATP-gated cation channels (P2X receptors) in the adult rat thyroid. *J. Anat.* **198,** 569–579.

Glass, R., Bardini, M., Robson, T., and Burnstock, G. (2001). Expression of nucleotide P2X receptor subtypes during spermatogenesis in the adult rat testis. *Cells Tissues Organs* **169,** 377–387.

Haines, W. R., Torres, G. E., Voigt, M. M., and Egan, T. M. (1999). Properties of the novel ATP-gated ionotropic receptor composed of the $P2X_1$ and $P2X_5$ isoforms. *Mol. Pharmacol.* **56,** 720–727.

Haines, W. R., Migita, K., Cox, J. A., Egan, T. M., and Voigt, M. M. (2001a). The first transmembrane domain of the P2X receptor subunit participates in the agonist-induced gating of the channel. *J. Biol. Chem.* **276,** 32793–32798.

Haines, W. R., Voigt, M. M., Migita, K., Torres, G. E., and Egan, T. M. (2001b). On the contribution of the first transmembrane domain to whole-cell current through an ATP-gated ionotropic P2X receptor. *J. Neurosci.* **21,** 5885–5892.

Hardy, L. A., Harvey, I. J., Chambers, P., and Gillespie, J. I. (2000). A putative alternatively spliced variant of the $P2X_1$ purinoreceptor in human bladder. *Exp. Physiol.* **85,** 461–463.

Hoebertz, A., Townsend-Nicholson, A., Glass, R., Burnstock, G., and Arnett, T. R. (2000). Expression of P2 receptors in bone and cultured bone cells. *Bone* **27,** 503–510.

Hoebertz, A., Meghji, S., Burnstock, G., and Arnett, T. R. (2001). Extracellular ADP is a powerful osteolytic agent: Evidence for signaling through the $P2Y_1$ receptor on bone cells. *FASEB J.* **15,** 1139–1148.

Housley, G. D. (2000). Physiological effects of extracellular nucleotides in the inner ear. *Clin. Exp. Physiol. Pharmacol.* **27,** 575–580.

Housley, G. D., Kanjhan, R., Raybould, N. P., Greenwood, D., Salih, S. G., Jarlebark, L., Burton, L. D., Setz, V. C., Cannell, M. B., Soeller, C., Christie, D. L., Usami, S., Matsubara, A., Yoshie, H., Ryan, A. F., and Thorne, P. R. (1999). Expression of the $P2X_2$ receptor subunit of the ATP-gated ion channel in the cochlea: Implications for sound transduction and auditory neurotransmission. *J. Neurosci.* **19,** 8377–8388.

Hu, B., Senkler, C., Yang, A., Soto, F., and Liang, B. T. (2002). $P2X_4$ receptor is a glycosylated cardiac receptor mediating a positive inotropic response to ATP. *J. Biol. Chem.* **277,** 15752–15757.

Hugel, S., and Schlichter, R. (2000). Presynaptic P2X receptors facilitate inhibitory GABAergic transmission between cultured rat spinal cord dorsal horn neurons. *J. Neurosci.* **20,** 2121–2130.

Hur, E. M., Park, T. J., and Kim, K.-T. (2001). Coupling of L-type voltage-sensitive calcium channels to P2X purinoceptors in PC-12 cells. *Am. J. Physiol. Cell Physiol.* **280,** C1121–C1129.

Jabs, R., Guenther, E., Marquordt, K., and Wheeler-Schilling, T. H. (2000). Evidence for $P2X_3$, $P2X_4$, $P2X_5$ but not for $P2X_7$ containing purinergic receptors in Muller cells of the rat retina. *Brain Res. Mol. Brain Res.* **76,** 205–210.

Jensik, P. J., Holbird, D., Collard, M. W., and Cox, T. C. (2001). Cloning and characterization of a functional P2X receptor from larval bullfrog skin. *Am. J. Physiol. Cell Physiol.* **281**, C954–C962.

Jiang, L. H., Rassendren, F., Surprenant, A., and North, R. A. (2000). Identification of amino acid residues contributing to the ATP-binding site of a purinergic P2X receptor. *J. Biol. Chem.* **275**, 34190–34196.

Jiang, L. H., Rassendren, F., Spelta, V., Surprenant, A., and North, R. A. (2001). Amino acid residues involved in gating identified in the first membrane-spanning domain of the rat P2X$_2$ receptor. *J. Biol. Chem.* **276**, 14902–14908.

Jo, Y. H., and Schlichter, R. (1999). Synaptic corelease of ATP and GABA in cultured spinal neurons. *Nat. Neurosci.* **2**, 241–245.

Juranka, P. F., Haghighi, A. P., Gaertner, T., Cooper, E., and Morris, C. E. (2001). Molecular cloning and functional expression of *Xenopus laevis* oocyte ATP-activated P2X4 channels. *Biochim. Biophys. Acta.* **1512**, 111–124.

Kanjhan, R., Housley, G. D., Burton, L. D., Christie, D. L., Kippenberger, A., Thorne, P. R., Luo, L., and Ryan, A. F. (1999). Distribution of the P2X2 receptor subunit of the ATP-gated ion channels in the rat central nervous system. *J. Comp. Neurol.* **407**, 11–32.

Katori, M., and Berne, R. M. (1966). Release of adenosine from anoxic hearts: Relationship to coronary flow. *Circ. Res.* **19**, 420–425.

Kellenberger, S., and Schild, L. (2002). Epithelial sodium channel/degenerin family of ion channels: A variety of functions for a shared structure. *Physiol. Rev.* **82**, 735–767.

Khakh, B. S., and Henderson, G. (1998). ATP receptor-mediated enhancement of fast excitatory neurotransmitter release in the brain. *Mol. Pharmacol.* **54**, 372–378.

Khakh, B. S., Smith, W. B., Chiu, C. S., Ju, D., Davidson, D., and Lester, H. A. (2001). Activation-dependent changes in receptor distribution and density morphology in hippocampal neurons expressing P2X2-green fluorescent protein receptors. *Proc. Natl. Acad. Sci. USA* **98**, 5288–5293.

King, B. F., Townsend-Nicholson, A., Wildman, S. S., Thomas, T., Spyer, K. M., and Burnstock, G. (2000). Coexpression of rat P2XR and P2X6 subunits in *Xenopus* oocytes. *J. Neurosci.* **20**, 4871–4877.

Klapperstuck, M., Buttner, C., Nickel, P., Schmalzing, G., Lambrecht, G., and Markwardt, F. (2000). Antagonism by the suramin analogue NF279 on human P2X$_1$ and P2X$_7$ receptors. *Eur. J. Pharmacol.* **387**, 245–252.

Korngreen, A., Ma, W., Priel, Z., and Silberberg, S. D. (1998). Extracellular ATP directly gates a cation-selective channel in rabbit airway ciliated epithelial cells. *J. Physiol.* **508**, 703–720.

Koshimizu, T., Koshimizu, M., and Stojilkovic, S. S. (1999). Contributions of the C-terminal domain to the control of P2X receptor desensitization. *J. Biol. Chem.* **274**, 37651–37657.

Koshimizu, T. A., Van Goor, F., Tomic, M., Wong, A. O., Tanoue, A., Tsujimoto, G., and Stojilkovic, S. S. (2000). Characterization of calcium signaling by purinergic receptor channels expressed in excitable cells. *Mol. Pharmacol.* **58**, 936–945.

Lambrecht, G., Rettinger, J., Baumert, H. G., Czeche, S., Damer, S., Ganso, M., Hildebrandt, C., Niebel, B., Spatz-Kumbel, G., Schmalzing, G., and Mutschler, E. (2000). The novel pyridoxal-5′-phosphate derivative PPNDS potently antagonizes activation of P2X$_1$ receptors. *Eur. J. Pharmacol.* **387**, R19–R21.

Le, K. T., Babinski, K., and Seguela, P. (1998). Central P2X4 and P2X6 channel subunits coassemble into a heteromeric ATP receptor. *J. Neurosci.* **18**, 7152–7159.

Le, K. T., Boue-Grabot, E., Archambault, V., and Seguela, P. (1999). Functional and biochemical evidence for heteromeric ATP-gated channels composed of P2X1 and P2X5 subunits. *J. Biol. Chem.* **274**, 15415–15419.

Lee, H. Y., Bardini, M., and Burnstock, G. (2000). Distribution of P2X receptors in the urinary bladder and the ureter of the rat. *J. Urol.* **163**, 2002–2007.

Lee, J. H., Chiba, T., and Marcus, D. C. (2001) P2X2 receptor mediates stimulation of parasensory cation absorption by cochlear outer sulcus cells and vestibular transitional cells. *J. Neurosci.* **21**, 9168–9174.

Lewis, C., Neidhart, S., Holy, C., North, R. A., Buell, G., and Surprenant, A. (1995). Coexpression of P2X2 and P2X3 receptor subunits can account for ATP-gated currents in sensor neurons. *Nature* **377**, 432–435.

Lewis, C. J., Surprenant, A., and Evans, R. J. (1998). 2′,3′-O-(2,4,6-trinitrophenyl) adenosine 5′-triphosphate (TNP-ATP)—a nanomolar affinity antagonist at rat mesenteric artery P2X receptor ion channels. *Br. J. Pharmacol.* **124**, 1463–1466.

Liang, S. X., Jenkins, N. A., Gilbert, D. J., Copeland, N. G., and Phillips, W. D. (2001). Structure and chromosome location of the mouse P2X$_1$ purinoceptor gene (*P2rx1*). *Cytogenet. Cell Genet.* **92**, 333–336.

Light, D. B., Capes, T. L., Gronau, R. T., and Adler, M. R. (1999). Extracellular ATP stimulates volume decrease in *Necturus* red blood cells. *Am. J. Physiol. Cell Physiol.* **277**, C480–C491.

Light, D. B., Dhalstrom, P. K., Gronau, R. T., and Baumann, N. L. (2001). Extracellular ATP activates a P2 receptor in *Necturus* erythrocytes during hypotonic swelling. *J. Membr. Biol.* **182**, 193–202.

Ling, B. N., Kokko, K. E., and Eaton, D. C. (1992). Inhibition of apical sodium channels in rabbit cortical collecting tubules by basolateral prostaglandin E2 is modulated by protein kinase C. *J. Clin. Invest.* **90**, 1328–1334.

Liu, M., Dunn, P. M., King, B. F., and Burnstock, G. (1999). Rat chromaffin cells lack P2X receptors while those of the guinea-pig express a P2X receptor with novel pharmacology. *Br. J. Pharmacol.* **128**, 61–68.

Loesch, A., Miah, S., and Burnstock, G. (1999). Ultrastructural localisation of ATP-gated P2X2 receptor immunoreactivity in the rat hypothalamo–neurohypophysial system. *J. Neurocytol.* **28**, 495–504.

Longhurst, P. A., Schwegel, T., Folander, K., and Swanson, R. (1996). The human P2X1 receptor: Molecular cloning, tissue distribution, and localization to chromosome 17. *Biochim. Biophys. Acta* **1308**, 185–188.

Lynch, K. J., Touma, E., Niforatos, W., Kage, K. L., Burgard, E. C., van Biesen, T., Kowaluk, E. A., and Jarvis, M. F. (1999). Molecular and functional characterization of human P2X$_2$ receptors. *Mol. Pharmacol.* **56**, 1171–1181.

Ma, W., Korngreen, A., Uzlaner, N., Priel, Z., and Silberberg, S. D. (1999). Extracellular sodium regulates airway ciliary motility by inhibiting a P2X receptor. *Nature* **400**, 894–897.

McCoy, D. E., Taylor, A. L., Kudlow, B. A., Karlson, K., Slattery, M. J., Schwiebert, L. M., Schwiebert, E. M., and Stanton, B. A. (1999). Nucleotides regulate NaCl transport across mIMCD-K2 cells via P2X and P2Y purinergic receptors. *Am. J. Physiol. Renal Physiol.* **277**, F552–F559.

Meyer, M. P., Groschel-Stewart, U., Robson, T., and Burnstock, G. (1999). Expression of two ATP-gated ion channels, P2X5 and P2X6, in developing chick skeletal muscle. *Dev. Dyn.* **216**, 442–449.

Migita, K., Haines, W. R., Voigt, M. M., and Egan, T. M. (2001). Polar residues of the second transmembrane domain influence cation permeability of the ATP-gated P2X2 receptor. *J. Biol. Chem.* **276**, 30934–30941.

Mori, M., Heuss, C., Gahwiler, B. H., and Gerber, U. (2001). Fast synaptic transmission mediated by P2X receptors in CA3 pyramidal cells of rat hippocampal slice cultures. *J. Physiol.* **535**, 115–123.

Mulryan, K., Gitterman, D. P., Lewis, C. J., Vial, C., Leckie, B. J., Cobb, A. L., Brown, J. E., Conley, E. C., Buell, G., Pritchard, C. A., and Evans, R. J. (2000). Reduced vas deferens contraction and male infertility in mice lacking P2X1 receptors. *Nature* **403**, 86–89.

Munoz, D. J., Thorne, P. R., and Housley, G. D. (1999). P2X receptor-mediated changes in cochlear potentials arising from exogenous adenosine 5′ triphosphate in endolymph. *Hear. Res.* **138**, 56–64.

Naemsch, L. N., Dixon, S. J., and Sims, S. M. (2001). Activity-dependent development of P2X7 current and $Ca^{2+}$ entry in rabbit osteoclasts. *J. Biol. Chem.* **276**, 39107–39114.

Nakatsuka, T., and Gu, J. G. (2001). ATP P2X receptor-mediated enhancement of glutamate release evoked EPSCs in dorsal horn neurons of the rat spinal cord. *J. Neurosci.* **21**, 6522–6531.

Nawa, G., Urano, T., Tokino, T., Ochi, T., and Miyoshi, Y. (1998). Cloning and characterization of the murine P2XM receptor gene. *J. Hum. Genet.* **43**, 262–267.

Nawa, G., Miyoshi, Y., Yoshikawa, H., Ochi, T., and Nakamura, Y. (1999). Frequent loss of expression or aberrant alternative splicing of P2XM, a p53-inducible gene, in soft-tissue tumours. *Br. J. Cancer* **80**, 1185–1189.

Newbolt, A., Stoop, R., Virginio, C., Surprenant, A., North, R. A., Buell, G., and Rassendren, F. (1998). Membrane topology of an ATP-gated ion channel (P2X receptor). *J. Biol. Chem.* **273**, 15177–15182.

Norton, W. H., Rohr, K. B., and Burnstock, G. (2000). Embryonic expression of a P2X$_3$ receptor encoding gene in zebrafish. *Mech. Dev.* **99**, 149–152.

Pankratov, Y., Castro, E., Miras-Portugal, M. T., and Krishtal, O. (1998). A purinergic component of the excitatory postsynaptic currents mediated by P2X receptors in the CA1 neurons of the rat hippocampus. *Eur. J. Neurosci.* **10**, 3898–3902.

Parker, M. S., Larroque, M. L., Camphell, J. M., Bobbin, R. P., and Deininger, P. L. (1998). Novel variant of the P2X2 ATP receptor from the guinea pig organ of Corti. *Hear. Res.* **121**, 62–70.

Petit, P., Hillaire-Buys, D., Manteghetti, M., Debrus, S., Chapal, J., and Loubatieres-Mariani, M. M. (1998). Evidence for two different types of P2 receptors stimulating insulin secretion from pancreatic B cell. *Br. J. Pharmacol.* **125**, 1368–1374.

Ramirez, A. N., and Kunze, D. L. (2002). P2X purinergic receptor channel expression and function in bovine aortic endothelium. *Am. J. Physiol. Heart Circ. Physiol.* **282**, H2106–H2116.

Ray, F. R., Huang, W., Slater, M., and Barden, J. A. (2002). Purinergic receptor distribution in endothelial cells in blood vessels: A basis for selection of coronary artery grafts. *Atherosclerosis* **162**, 55–61.

Rettinger, J., Aschrafi, A., and Schmalzing, G. (2000a). Roles of individual N-glycans for ATP potency and expression of the rat P2X1 receptor. *J. Biol. Chem.* **275**, 33542–33547.

Rettinger, J., Schmalzing, G., Damer, S., Muller, G., Nickel, P., and Lambrecht, G. (2000b). The suramin analogue NF279 is a novel and potent antagonist selective for the P2X$_1$ receptor. *Neuropharmacology* **39**, 2044–2053.

Rubio, M. E., and Soto, F. (2001). Distinct localization of P2X receptors at excitatory postsynaptic specializations. *J. Neurosci.* **21**, 641–653.

Ruppelt, A., Ma, W., Borchardt, K., Silberberg, S. D., and Soto, F. (2001). Genomic structure, developmental distribution and functional properties of the chicken P2X$_5$ receptor. *J. Neurochem.* **77**, 1256–1265.

Ryten, M., Hoebertz, A., and Burnstock, G. (2001). Sequential expression of three receptor subtypes for extracellular ATP in developing rat skeletal muscle. *Dev. Dyn.* **221**, 331–341.

Salih, S. G., Housley, G. D., Raybould, N. P., and Thorne, P. R. (1999). ATP-gated ion channel expression in primary auditory neurones. *Neuroreport.* **10**, 2579–2586.

Schwiebert, E. M., Wallace, D. P., Braunstein, G. M., King, S. R., Peti-Peterdi, J., Hanaoka, K., Guggino, W. B., Guay-Woodford, L. M., Darwin Bell, P., Sullivan, L. P., Grantham, J. J., and Taylor, A. L. (2002). Autocrine extracellular purinergic signaling in epithelial cells derived from polycystic kidneys. *Am. J. Physiol. Renal Physiol.* **282**, F763–F775.

Schwiebert, L. M., Rice, W. C., Kudlow, B. A., Taylor, A. L., and Schwiebert, E. M. (2001). Extracellular ATP signaling and P2X nucleotide receptors in monolayers of primary human vascular endothelial cells. *Am. J. Physiol. Cell Physiol.* **282**, C289–C301.

Seguela, P., Haghighi, A., Soghomonian, J. J., and Cooper, E. (1996). A novel neuronal P2x ATP receptor ion channel with widespread distribution in the brain. *J. Neurosci.* **16**, 448–455.

Shibuya, I., Tanaka, K., Hattori, Y., Vezono, Y., Harayama, N., Noguchi, J., Veta, Y., Izumi, F., and Yamashita, H. (1999). Evidence that multiple P2X purinoceptors are functionally expressed in rat supraoptic neurones. *J. Physiol.* **514**, 351–367.

Shinozuka, K., Tanaka, N., Kawasaki, K., Mizuno, H., Kubota, Y., Nakamura, K., Hashimoto, M., and Kunitomo, M. (2001). Participation of ATP in cell volume regulation in the endothelium after hypotonic stress. *Clin. Exp. Pharmacol. Physiol.* **28**, 799–803.

Sokolova, E., Nistri, A., and Giniatullin, R. (2001). Negative cross-talk between anionic $GABA_A$ and cationic P2X ionotropic receptors of rat dorsal root ganglion neurons. *J. Neurosci.* **21**, 4958–4968.

Soto, F., Garcia-Guzman, M., Karschin, C., and Stuhmer, W. (1996a). Cloning and tissue distribution of a novel P2X receptor from rat brain. *Biochem. Biophys. Res. Commun.* **223**, 456–460.

Soto, F., Garcia-Guzman, M., Gomez-Hernandez, J. M., Hollmann, M., Karschin, C., and Stuhmer, W. (1996b). P2X4: An ATP-activated ionotropic receptor cloned from rat brain. *Proc. Natl. Acad. Sci. USA* **93**, 3684–3688.

Souslova, V., Ravenall, S., Fox, M., Wells, D., Wood, J. N., and Akopian, A. N. (1997). Structure and chromosomal mapping of the mouse P2X3 gene. *Gene* **195**, 101–111.

Spelta, V., Jiang, L. H., Surprenant, A., and North, R. A. (2002). Kinetics of antagonist actions at rat P2X2/P2X3 heteromeric receptors. *Br. J. Pharmacol.* **135**, 1524–1530.

Stoop, R., Supranant, A., and North, R. A. (1997). Different sensitivities to pH of ATP-induced currents at four cloned P2X receptors. *J. Neurophysiol.* **78**, 1837–1840.

Surprenant, A., Rassendren, F., Kawashima, E., North, R. A., and Buell, G. (1996). The cytolytic P2Z receptor for extracellular ATP identified as a P2X receptor (P2X7). *Science* **272**, 735–738.

Surprenant, A., Schneider, D. A., Wilson, H. L., Galligan, J. J., and North, R. A. (2000). Functional properties of heteromeric P2X1/P2X5 receptors expressed in HEK cells and excitatory junction potentials in guinea pig submucosal arterioles. *J. Auton. Nerv. Syst.* **81**, 249–263.

Swanson, K. D., Reigh, C., and Landreth, G. E. (1998). ATP-stimulated activation of the mitogen-activated protein kinase through ionotropic P2X2 purinoceptors in PC12 cells: Difference in purinoceptor sensitivity in two PC12 cell lines. *J. Biol. Chem.* **273**, 19965–19971.

Taschenberger, H., Juttner, R., and Grantyn, R. (1999). $Ca^{2+}$-permeable P2X receptor channels in cultured rat retinal ganglion cells. *J. Neurosci.* **19**, 3353–3366.

Taylor, A. L., Schwiebert, L. M., Smith, J. J., King, C., Jones, J. R., Sorscher, E. J., and Schwiebert, E. M. (1999). Epithelial P2X purinergic receptor channel expression and function. *J. Clin. Invest.* **104**, 875–884.

Torres, G. E., Egan, T. M., and Voight, M. M. (1998a). N-linked glycosylation is essential for the functional expression of the recombinant P2X2 receptor. *Biochemistry* **37**, 14845–14851.

Torres, G. E., Haines, W. R., Egan, T. M., and Voigt, M. M. (1998b). Co-expression of P2X1 and P2X5 receptor subunits reveals a novel ATP-gated ion channel. *Mol. Pharmacol.* **54**, 989–993.

Torres, G. E., Egan, T. M., and Voight, M. M. (1999a). Hetero-oligomeric assembly of P2X receptor subunits. *J. Biol. Chem.* **274**, 6653–6659.

Torres, G. E., Egan, T. M., and Voigt, M. M. (1999b). Identification of a domain involved in ATP-gated ionotropic receptor subunit assembly. *J. Biol. Chem.* **274**, 22359–22365.

Townsend-Nicholson, A., King, B. F., Wildman, S. S., and Burnstock, G. (1999). Molecular cloning, functional characterization and possible cooperativity between the murine P2X4 and P2X4a receptors. *Brain. Res. Mol. Brain Res.* **64**, 246–254.

Urano, T., Nishimori, H., Han, H., Furuhata, T., Kimura, Y., Nakamura, Y., and Tokino, T. (1997). Cloning of P2XM, a novel human P2X receptor gene regulated by p53. *Cancer Res.* **57**, 3281–3287.

Valera, S., Hussy, N., Evans, R. J., Adami, N., North, R. A., Surprenant, A., and Buell, G. (1994). A new class of ligand-gated ion channel defined by P2x receptor for extracellular ATP. *Nature* **371**, 516–519.

Vlaskovska, M., Kasakov, L., Rong, W., Bodin, P., Bardini, M., Cockayne, D. A., Ford, A. P., and Burnstock, G. (2001). P2X3 knock-out mice reveal a major sensory role for urothelial released ATP. *J. Neurosci.* **21**, 5670–5677.

Wang, C. Z., Namba, N., Gonoi, T., Inagaki, N., and Seino, S. (1996). Cloning and pharmacological characterization of a fourth P2X receptor subtype widely expressed in brain and peripheral tissues including various endocrine tissues. *Biochem. Biophys. Res. Commun.* **220**, 196–202.

Werner, P., Seward, E. P., Buell, G. N., and North, R. A. (1996). Domains of P2X receptors involved in desensitization. *Proc. Natl. Acad. Sci. USA* **93**, 15485–15490.

Wheeler-Schilling, T. H., Marquordt, K., Kohler, K., Jabs, R., and Guenther, E. (2000). Expression of purinergic receptors in bipolar cells of the rat retina. *Brain Res. Mol. Brain Res.* **76**, 415–418.

Wheeler-Schilling, T. H., Marquordt, K., Kohler, K., Guenther, E., and Jabs, R. (2001). Identification of purinergic receptors in retinal ganglion cells. *Brain Res. Mol. Brain Res.* **92**, 177–180.

Wong, A. Y., Burnstock, G., and Gibb, A. J. (2000). Single channel properties of P2X ATP receptors in outside-out patches from rat hippocampal granule cells. *J. Physiol.* **527**, 529–547.

Worthington, R. A., Dutton, J. L., Poronnik, P., Bennett, M. R., and Barden, J. A. (1999). Localisation of P2X receptors in human salivary gland epithelial cells and human embryonic kidney cells by SDS–PAGE gel electrophoresis/Western blotting and immunofluoresence. *Electrophoresis* **20**, 2065–2070.

Xiang, Z., Bo, X., Oglesby, I., Ford, A., and Burnstock, G. (1998). Localization of ATP-gated P2X2 receptor immunoreactivity in the rat hypothalamus. *Brain Res.* **813**, 390–397.

Xiang, Z., Bo, X., and Burnstock, G. (1999). P2X receptor immunoreactivity in the rat cochlea, vestibular ganglion and cochlear nucleus. *Hear. Res.* **128**, 190–196.

Yamamoto, K., Korenaga, R., Kamiya, A., and Ando, J. (2000). Fluid shear stress activates $Ca^{2+}$ influx into human endothelial cells via P2X4 purinoceptors. *Circ. Res.* **87**, 385–391.

Yamamoto, K., Korenaga, R., Kamiya, A., Qi, Z., Sokabe, M., and Ando, J. (2000). $P2X_4$ receptors mediate ATP-induced calcium influx in human vascular endothelial cells. *Am. J. Physiol. Heart Circ. Physiol.* **279**, H285–H292.

Zhong, Y., Dunn, P. M., Bardini, M., Ford, A. P., Cockayne, D. A., and Burnstock, G. (2001). Changes in P2X receptor responses of sensory neurons from P2X3-deficient mice. *Eur. J. Neurosci.* **14,** 1784–1792.

Zsembery, A., and Schwiebert, E. M. (2002). Epithelial P2X purinergic receptor channels function as an extracellular sensor for ATP, cations, protons, and heavy metals. (Submitted.).

Zsembery, A., Boyce, A. T., *et al.* (2002). Dissection of P2Y and P2X nucleotide receptor-triggered increases in cytosolic calcium in human cystic fibrosis airway epithelial cells. (Submitted.).

Zsembery, A., Boyce, A. T., Liang, L., Peti-Peterdi, J., Bell, P. D., and Schwiebert, E. M. (2003). Sustained calcium entry through P2X nucleotide receptor channels in human airway epithelial cells. *J. Biol. Chem.* Epub A.

# CHAPTER 5

# Molecular and Cell Biology of Adenosine Receptors

**B. R. Cobb**[*†] **and J. P. Clancy**[††]
Departments of [*]Human Genetics and [†]Pediatrics, University of Alabama at Birmingham, Birmingham, Alabama 35205 and [‡]the Gregory Fleming James Cystic Fibrosis Research Center, University of Alabama at Birmingham, Birmingham, Alabama 35294

## I. INTRODUCTION

The purine nucleoside adenosine (ADO) has been of interest to researchers and clinicians for nearly 70 years due to its physiological role in cell signaling as well as its potential and observed therapeutic applications (Drury and Szent-Gyorgyi 1929). The therapeutic potential was first considered when ADO, unlike adenine, was found to be nontoxic and rapidly metabolized ($t_{1/2}$ = 3–6 s) when administered intravenously to humans. As early as 1929, the physiological effects of ADO were beginning to be defined, primarily due to its effects on the cardiovascular system, including coronary vasodilation and systemic hypotension (Drury and Szent-Gyorgyi, 1929). In modern medicine, ADO is the treatment of choice for certain supraventricular tachyarrhythmias, and can be useful to aid in the diagnosis of other tachycardias due to its profound, but short-lived, effects on atrioventricular nodal conduction (Belardinelli *et al.*, 1995). ADO exerts its effects through receptor signaling, and the development of ADO receptor agonists is currently an important line of research that may have relevance to many physiological processes and medical conditions. This includes cardiovascular functions, degenerative central nervous system (CNS) diseases, schizophrenia, memory and behavior, rheumatic and other inflammatory disorders, and epithelial ion transport.

ADO is a unique signaling molecule in that it utilizes both extracellular and intracellular signaling pathways to produce widespread effects in many different cells and organ systems (Berne, 1963; Williams, 1987; Ramkumar *et al.*, 1988). Found in virtually all living cells, ADO produces effects by activating specific cell surface $P_1$ purinergic receptors in a concentration-dependent, tissue and subtype-specific manner (Berne, 1963; van Calker *et al.*, 1979; Londos *et al.*, 1980). ADO receptors are members of the G-protein-coupled receptor (GPCR) superfamily with seven characteristic membrane-spanning domains, extracellular tertiary structure that deter-mines agonist selectivity, and intracellular structure that contributes to coupling with many G-proteins (van Calker *et al.*, 1979; Londos *et al.*, 1980; Bruns *et al.*, 1986). The purpose of this chapter is to discuss specific aspects of $P_1$ purinergic receptor molecular and cell biology, including pharmaco-logical and structural characteristics, functional significance, and the modulatory role these receptors can have on signaling as part of both homeostatic and pathophysiological processes.

## II. MOLECULAR PHARMACOLOGY

### A. Identification and Characterization of Subtypes

ADO primarily interacts with receptors that are distinct from those that interact with other nucleotides, such as adenosine triphosphate (ATP). Using biochemical, physiological, and pharmacological methods, two classes of purinergic receptors, $P_1$ (ADO selective) and $P_2$ (ATP selective), have been identified based on the following findings: (1) methylxanthines competitively antagonize ADO-mediated cell processes but not those mediated by ATP, (2) ADO but not ATP frequently increases intracellular cyclic adenosine monophosphate (cAMP) levels, and (3) ATP, adenosine diphosphate (ADP), AMP, and ADO have different relative potencies at ADO-selective receptors compared with ATP-selective receptors. $P_1$ purinergic receptors are characterized by their ADO selectivity over its phosphorylated nucleotides, whereas $P_2$ purinergic receptors are selective for ATP and related compounds over ADO (Burnstock 1978). For reviews on $P_2$ receptors, the reader is referred to two chapters on $P_{2X}$ receptors and a single chapter on $P_{2Y}$ receptors within this volume.

With the help of molecular cloning techniques, human $A_1$, $A_{2A}$, $A_{2B}$, and $A_3$ $P_1$ purinergic receptors have been identified with ~45% sequence homology exhibited between the four receptor subtypes (Libert et al., 1989; Cunha et al., 1994). With this information, it has become possible to begin characterizing each subtype and developing selective agonists and antagonists. With the aid of molecular cloning technology, researchers have also begun to better understand receptor activation and desensitization, as well as other regulatory processes including cell signaling, and the role that these receptors play in health and disease. For more extensive detail on structure–function relationships between $P_1$ purinergic receptors and ligands, the reader is directed to several excellent, focused reviews (Olah et al., 1995; Palmer and Stiles, 1995, 1997; Kaiser and Quinn, 1999; Olah and Stiles, 1992, 1995, 2000).

### B. $A_1$ Adenosine Receptors

The $A_1$ receptor subtype is a 326 amino acid protein of approximately 36,000 Da that has been cloned from a wide variety of species (Table I). Heterologous mammalian expression systems have shown that agonists bind to $A_1$ receptors with an order of potency of (−)-RG 14719 > 2-chloro-$N^6$-cyclopentyladenosine (CCPA) > $N^6$-cyclopentyladenosine (CPA) > (−)-$N^6$-($R$-phenylisopropyl)adenosine (R-PIA)

**154**                                                              Cobb and Clancy

**TABLE I**

Characterization of $P_1$ Purinergic Receptors

| Receptor (subtype) | Size (aa = amino acids) | Species (cloned) | Chromosome location |
|---|---|---|---|
| $A_1$ | 326 aa, 36 kDa | Human, bovine, dog, rabbit, chick, guinea pig, rat, mouse | 1q32.1 |
| $A_{2A}$ | 412 aa, 44 kDa | Human, dog, guinea pig, rat, mouse | 22q11.2 |
| $A_{2B}$ | 332 aa, 36 kDa | Human, chicken, rat, mouse | 17p11.2–12 |
| $A_3$ | 318 aa, 36 kDa | Human, sheep, dog, rabbit, rat, mouse | 1p13.3 |

> 5′-$N$-ethylcarboxamidoadenosine (NECA) > (+)-$N^6$-phenylisopropyla-denosine (S-PIA) > NECA > 2-[4-(2-carboxyethyl)phenethyl] aminoade-nosine-5′-$N$-ethylcarboxamide (CGS 21680) that (1) is distinct from other $P_1$ purinergic receptors and (2) possesses a higher (up to 10-fold) affinity for both agonists and antagonists than other ($A_2$ or $A_3$) $P_1$ receptors (Table II) (Mahan et al., 1991; Libert et al., 1992; Olah et al., 1992; Klotz et al., 1998). For example, the $K_d$ value for [$^3$H]CCPA in Chinese hamster ovary (CHO) cells transiently expressing human $A_1$ receptors is extremely low (0.6 n$M$). In general, $N^6$-substituted ADO analogs are selective for $A_1$ receptors, with CCPA being the most selective identified thus far. (−)-RG 14718 is considered the most potent $A_1$ receptor agonist, with high selectivity (2000-fold) for $A_1$ over $A_{2A}$ receptors. NECA is a relatively nonselective and potent agonist for both $A_1$ and $A_2$ receptors (Table III).

Many of the $A_1$ receptor antagonists are xanthine based, including commonly known drugs such as caffeine and theophylline. 1,3-Dipropyl-8-cyclopentylxanthine (DPCPX, $K_d$ value = 3.9 n$M$) and the nonxanthine, nonselective compound CGS 15943 (9-chloro-2-(2-furyl)[1,2,4]triazolo[1,5-c]quinazolin-5-amine, $K_i$ value = 3.5 n$M$) appear to be the most potent $A_1$ receptor antagonists in radioligand competitive binding studies (Klotz et al., 1998; Ito et al., 1999). $A_1$ receptors are structurally distinct from other $P_1$ purinergic receptors as well as other GPCRs, possessing several unique characteristics that are presently of uncertain functional significance. For example, the third intracellular loop has a short sequence of 34 amino acids that is roughly one-sixth the size of other GPCRs (such as muscarinic and adrenergic receptors) capable of inhibiting adenylate cyclase (Olah et al., 1992). There is also a potential fatty acid binding site on the C-terminal tail,

**TABLE II**

Expression of $P_1$ Purinergic Receptor Subtypes in Various Tissues[a]

| $A_1$ | $A_{2A}$ | $A_{2B}$ | $A_3$ |
|---|---|---|---|
| Brain, widespread (1–5) | Heart (6) | Cecum, colon (6,7) | Neurons various (8) |
| Spinal cord (3) | Kidney (6) | Bladder (6,7) | Brain (2) |
| Smooth muscle (9) | Platelets (10) | Brain, widespread (1,2) | Sperm, widespread (6) |
| Adipocytes (11) | Brain, striatum (12) | Neurons, various (6) | Mast cells (6) |
| Heart (13) | Smooth muscle (9) | Mast cells (14) | Lung (abundant) (6,15) |
| Kidney (16) | Neutrophils (17) | Airway cells (18,19) | Kidney (6) |
| Thyroid (20) | Olfactory tubercule (21) | Neutrophils (17) | Heart (6) |
| Neutrophils (22,23) | Neurons, various (6) | Fibroblasts (24) | Neutrophils (25) |
| Macrophages (22,23) | Lung (6) | Macrophages (26) | Macrophages (27) |
| Neurons, various (6) | Liver (6) | | Testes, widespread (28) |
| Retina (29) | | | Lung, abundant (15) |
| Bladder (6) | | | |
| Testis (3,28,30) | | | |

[a]*References:* (1) Daly and Butts-Lamb (1983); (2) Dunwiddie and Masino (2001); (3) Reppert and Weaver (1991); (4) Mahan and McVittie (1991); (5) Rivkees and Price (1995); (6) Dixon and Gubitz (1996); (7) Stehle and Rivkees (1992); (8) Ribeiro and Sebastiao (1986); (9) Ramkumar and Barrington (1990); (10) Wilken and Tawfik-Schlieper (1990); (11) Londos and Cooper (1980); (12) Bruns and Lu (1986); (13) Lohse and Elger (1988); (14) Feoktistov and Biaggioni (1997); (15) Walker and Jacobson (1997); (16) Arend and Handler (1989); (17) Fredholm and Zhang (1996); (18) Pauwels and Joos (1995); (19) Clancy and Ruiz (1999); (20) Okajima and Sato (1989); (21) Kaelin-Lang and Lauterburg (1999); (22) Salmon and Cronstein (1990); (23) Cronstein (1994); (24) Bruns (1980); (25) Forsythe and Ennis (1999); (26) Xaus and Mirabet (1999); (27) McWhinney and Dudley (1996); (28) Rivkees (1994); (29) Blazynski and Perez (1991); (30) Bhat *et al.* (1991).

in addition to relatively few ($n = 5$) consensus sequences for phosphorylation (see Fig. 1).

## C. $A_{2A}$ Receptors

$A_{2A}$ receptors are $\sim$45,000-Da proteins composed of 410–412 amino acids, and have been cloned from multiple mammalian species, including human, mouse, dog, guinea pig, rat, and canine (Ongini and Fredholm,

**TABLE III**

Prominent $P_1$ Purinergic Receptor Agonists/Antagonists and Relative Affinities[a]

| Receptor subtype | Agonist selectivity | Antagonists | Affinity (for adenosine) |
|---|---|---|---|
| $A_1$ | CCPA > CPA > R-PIA > ADO > NECA > S-PIA > CGS 21680 | DPCPX, CGS 15493, CPX | High affinity (15–100 n$M$)* |
| $A_{2A}$ | CGS 21680 > ADO > NECA > R-PIA~S-PIA | DPMDX, KF17837 | Highest affinity for CGS 21680 (~150 n$M$)* |
| $A_{2B}$ | NECA > ADO ≥ R-PIA = IB- MECA >> CGS 21680 | Enprofylline, XAC | Low affinity (0.5–20 $\mu M$)* |
| $A_3$ | 2Cl-IB-MECA > DBXRM >> ADO | MRE 300F20, MRS 1220 | Low affinity (1–10 $\mu M$)* |

[a]MRS 1220, 9-chloro-2-(2-furyl)-5-phenylacetylamino[1,2,4]triazolo[1,5-c]quinazoline; CGS 15943, 9-chloro-2-(2-furyl)[1,2,4]triazolo[1,5-c]quinazolin-5-amine; KF17837, (E)-1,3-Dipropyl-8-3,4-dimethoxystyryl-7-methylxanthine; CPX, 8-cyclopentyl-1,3-dipropylxanthine; DBXRM, 1,3-dibutylxanthine-7-riboside-5'-N-methylcarboxamide; MRE 3008F20, 5N-(4-methoxyphenylcarbamoyl)amino-8-propyl-2-(2-furyl)pyrazolo[4,3-e]-1,2,4-triazolo[1,5-c]pyrimidine; XAC, 8-[4-[[[[(2-aminoethyl)amino]-carbonyl]methyl]oxy]phenyl]-1,3-dipropylxanthine (Williams, 1987; Dunwiddie et al., 2001). For all other abbreviations, see text.

1996; Fredholm et al., 2000). The carboxy-terminal tail of the $A_{2A}$ receptors is comparatively larger than the $A_1$ subtype yet the functional significance of this extended sequence is unknown. $A_{2A}$ receptors are pharmacologically distinct from other subtypes primarily because of their high affinity binding to [$^3$H]CGS 21680 and [$^{125}$I]2-[[4-[2-[2-phenylmethly-carbonylamino]ethylaminocarbonyl]ethyl]phenyl]ethylamino-5'-N-ethyl-carboxamidoadenosine ([$^{125}$I]PAPA-APEC), both of which are much more selective for $A_{2A}$ receptors than [$^3$H]NECA and the PIA stereoisomers. Typically, $A_2$ subtype selectivity is accomplished by modifying the 2-position of ADO. Recent synthetic ligands generated using comparative molecular field analysis (CoMFA) have enhanced selectivity for $A_{2A}$ receptors (Rieger et al., 2001). This technology uses physicochemical modeling to design agonists with reduced $A_1$ receptor and enhanced $A_{2A}$ receptor selectivity.

Several potential $A_2$ receptor antagonists with 8-styryl modification of 1,2,7-alkylxanthines have been synthesized. 1,3-Dipropyl-7-methyl-8-(3,5-dimethoxystyryl)xanthine (DPMDX) was found to be a potent ($K_i$ = 24 n$M$) adenosine antagonist that is 110-fold more selective for $A_2$ receptors over $A_1$ receptors (Jacobson et al., 1993).

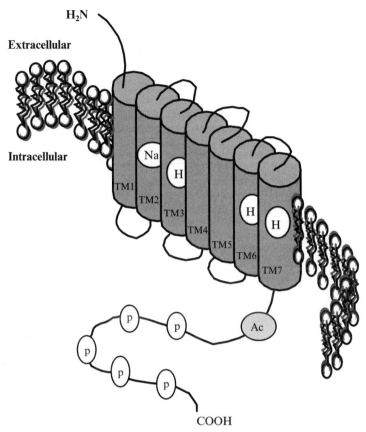

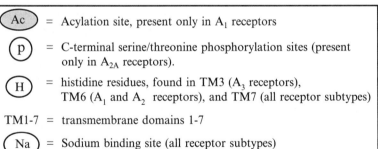

| | | |
|---|---|---|
| Ac | = | Acylation site, present only in $A_1$ receptors |
| p | = | C-terminal serine/threonine phosphorylation sites (present only in $A_{2A}$ receptors). |
| H | = | histidine residues, found in TM3 ($A_3$ receptors), TM6 ($A_1$ and $A_2$ receptors), and TM7 (all receptor subtypes) |
| TM1-7 | = | transmembrane domains 1-7 |
| Na | = | Sodium binding site (all receptor subtypes) |

**FIGURE 1** Structural comparisons between $P_1$ purinergic receptors. All $P_1$ purinergic receptors have seven transmembrane domains, characteristic of G-protein-coupled receptors. Consistent among subtypes are three extracellular loops and three intracellular loops that vary in size. The main differences between receptor subtypes includes phosphorylation sites, fatty acylation sites, histidine residues (that may be involved in ligand binding or desensitization), and sodium binding sites.

## D. $A_{2B}$ Receptors

$A_{2B}$ receptors, first evaluated in human fibroblasts (Bruns, 1980), have ~75 fewer amino acids than $A_{2A}$ receptors (332 amino acids total), with a molecular weight of approximately 35,000 Da. These receptors have also been cloned from many different species and tissue types including humans, and are expressed throughout the brain (Daly et al., 1983; Fredholm et al., 2000). $A_{2B}$ receptor characterization has required unique approaches to distinguish it from other $P_1$ receptors. $A_1$ and $A_{2A}$ receptors have historically been characterized in terms of agonist selectivity by performing radioligand binding studies using receptor clones heterologously expressed in eukaryotic cells. To characterize $A_{2B}$ receptors, however, demonstration of direct adenylate cyclase activity has been required in order to determine isotype-specific activity. This is due to a lack of known selective agonists that distinguish $A_{2B}$ receptors from other $P_1$ receptors. NECA nonselectively binds $A_{2B}$ receptors ($EC_{50} = 2\ \mu M$), while activating other subtypes ($A_1$ and $A_{2A}$) with greater (n$M$) affinity. The $A_{2B}$ receptor is characterized by a potency profile of NECA > R-PIA = $N^6$-(3-iodobenzyl)-adenosine-5'-$N$-methyluronamide (IB-MECA) > CGS 21680, but the receptor cannot be distinguished from other $P_1$ purinergic receptor subtypes based on this potency series alone. Interestingly, the $A_{2B}$ subtype has a relatively low affinity for [$^3$H]NECA, whereas the $A_{2A}$ subtype binds [$^3$H]NECA with much higher affinity (Bruns et al., 1986). CGS 21680 is a relatively ineffective agonist for $A_{2B}$ receptors (but very selective for $A_{2A}$ receptors), whereas R-PIA and IB-MECA in micromolar concentrations are nonselective for $A_{2B}$ versus $A_{2A}$ receptors (Jarvis and Williams, 1988; Feoktistov and Biaggioni, 1993; Brackett and Daly, 1994; Alexander et al., 1996). Identification of $A_{2B}$ receptors, therefore, has involved demonstration of direct adenylate cyclase activation in a xanthine-sensitive, NECA-inducible manner that is poorly responsive to CGS 21680 or PAP-APEC stimulation.

Although highly selective antagonists are preferable for subtype identification, they have yet to be developed for $A_{2B}$ receptors. However, some xanthine derivatives are moderately potent $A_2$ receptor antagonists, similar to that of other $P_1$ purinergic receptor subtypes (Feoktistov and Biaggioni, 1993; Brackett and Daly, 1994). Enprofylline, for example, was found to have a dissociation constant of approximately 7 $\mu M$, and appears to be a selective albeit not particularly potent $A_{2B}$ receptor antagonist. Other more potent yet less selective antagonists have been characterized and are reviewed by Feoktistov and Biaggioni (1997). More potent and specific $A_{2B}$ receptor ligands are needed to aid in the characterization of this receptor, to distinguish it from other $P_1$ purinergic receptors, and to clarify its physiological role.

## E. $A_3$ Receptors

Using CHO cells stably expressing a unique clone with sequence similarity to the $A_1$ and $A_{2A}$ receptors (isolated from a rat brain cDNA library), Zhou and colleagues discovered a $P_1$ purinergic receptor with a pharmacological profile distinct from $A_1$ and $A_2$ subtypes (Zhou *et al.*, 1992). Additionally, the cloned receptor was found to be relatively insensitive to antagonism by most xanthine derivatives, whereas NECA stimulation resulted in inhibition of forskolin-stimulated cAMP accumulation through a pertussis toxin-sensitive G-protein. This novel receptor subtype, composed of 337 amino acids, was designated $A_3$ and has subsequently been cloned from human, rat, dog, rabbit, and sheep (Linden, 1994; Murrison *et al.*, 1996; Sajjadi *et al.*, 1996; Hill *et al.*, 1997).

Selective activation of $A_3$ receptors requires modification of ADO in the $N^6$ and $5'$ positions. An example is IB-MECA, a potent agonist with moderate selectivity (Jacobson *et al.*, 1997). 2-Chloro-$N^6$-(3-iodobenzyl)-adenosine-$5'$-$N$-methyluronamide (2Cl-IB-MECA), however, is 2500- and 1400-fold more selective for the rat $A_3$ receptor subtype compared with the $A_1$ or $A_{2A}$ receptor, respectively (Jacobson, 1998). Additional synthetic ligands selective for the $A_3$ receptor subtype have been developed over the past 3 years (van Muijlwijk-Koezen *et al.*, 1998a,b; Varani *et al.*, 1998; van Tilburg *et al.*, 1999; Von Lubitz *et al.*, 1999). Differences in mRNA and protein expression among species homologs may represent distinct $A_3$ receptor subtypes rather than species variation of a single $A_3$ receptor. Additional genomic, proteomic, and pharmacological studies will be needed to better characterize the $A_3$ subtype.

## III. ADENOSINE RECEPTOR–EFFECTOR COUPLING AND INTRACELLULAR SIGNALING

The best studied intracellular signaling pathway that ADO receptors influence is the adenylate cyclase system (Sattin and Rall, 1970). $A_{2A}$ and $A_{2B}$ receptors are thought to couple to $G_S$ and activate adenylate cyclase, which catalyzes the conversion of ATP to $5'$-cyclic adenosine monophosphate (cAMP). $A_1$ and $A_3$ receptors, in contrast, couple to $G_i$ and, in general, inhibit adenylate cyclase and cAMP production. Studies over the past decade have broadened our view of the signaling pathways influenced by $P_1$ purinergic receptors, with many of the receptor subtypes modulating second messenger systems independent of cAMP (see Table IV). In general, the specific effector systems that couple to $P_1$ receptor subtypes vary in a tissue-specific manner that is dependent on both the level of receptor

TABLE IV

Effector Coupling Systems

| Receptor subtype | G-Protein | Additional coupling systems | Ion channels |
|---|---|---|---|
| A$_1$ | G$_{i\alpha1,2,3}$ and G$_{zero}$ | PKA($-$), PLC, PKC, IP$_3$, MAP kinase, PLA$_2$ | Ca$^{2+}$, K$^+$, Cl$^-$ channels |
| A$_{2A}$ | G$_{s\alpha}$ and G$_{olf}$ | PKA, PKC | N-/P-type Ca$^{2+}$ channel |
| A$_{2B}$ | G$_{s\alpha}$ and G$_0$ | MAP kinase (via G$_{q/11}$), PLC, PKA, PLA$_2$ | Ca$^{2+}$-regulated Cl$^-$ channel |
| A$_3$ | G$_{i\alpha2,3}$ and G$_{zero}$ | PLC, PLD, PLA$_2$ | Ca$^{2+}$ (undefined) channel |

subtype expression and the diversity of the effector coupling systems innate to the specific tissue of interest. This view might help to explain some of the apparent discrepancies found when comparing different methodologies that use different model systems.

## A. A$_1$ Receptors and Their Effector Coupling Systems

It has been demonstrated in reconstituted systems that A$_1$ receptors can interact with the G$_{i1,2,3}$ and G$_0$ and inhibit adenylate cyclase (Freissmuth et al., 1991). A$_1$ receptors are somewhat promiscuous, with the capacity to couple to a variety of effector systems other than adenylate cyclase, including phospholipase C (PLC), protein kinase C (PKC), and the inositol triphosphate (IP$_3$) signaling system, and several ion channels including Ca$^{2+}$, K$^+$, and Cl$^-$ channels (Linden, 1991; Biber et al., 1997; Abebe and Mustafa, 1998; Dickenson and Hill, 1998a; Pilitsis and Kimelberg, 1998; Santos et al., 1998; Walker et al., 1998).

Characterizing A$_1$ receptors in terms of their signaling systems has largely been made possible by using A$_1$ receptor expression systems in CHO cells. Using this approach, stimulation with A$_1$ receptor agonists has been shown to result in the following effects: (1) activation of the inositol phosphate signaling cascade (with increases in IP$_3$ and the release of stored Ca$^{2+}$), (2) the activation of the mitogen-activated protein (MAP) kinase signaling pathway, and (3) direct coupling to PLC in a G$_{3\alpha/3\beta\gamma}$-dependent manner (Murthy and Makhlouf, 1995; Olah and Stiles, 1995; Palmer and Stiles, 1995; Dickenson et al., 1998; Dickenson and Hill, 1998a,b).

The effector systems that couple to A$_1$ receptors also have been studied extensively in neurons. A$_1$ receptor stimulation can produce excitatory neurotransmission by modulating K$^+$ and Ca$^{2+}$ conductances (Picano and

Abbracchio, 2000). For example, ADO inhibits voltage-gated $Ca^{2+}$ influx via $A_1$ receptor activation. Preventing $Ca^{2+}$ accumulation within neurons prevents the release of neurotransmitters (mainly glutamate), which can dually inhibit both presynaptic excitatory effects and (by default) downstream postsynaptic excitatory effects.

Recent studies have also focused on the role of $A_1$ receptors and PKC signaling cascades. ADO release, acting through $A_1$ receptors, is capable of positively regulating insulin and leptin in isolated rat adipocytes utilizing a PLC/PKC-dependent pathway (Cheng et al., 2000). ADO stimulates insulin-dependent leptin release through PLC/PKC activity.

## B. $A_{2A}$ Receptors and Their Effector Coupling Systems

$A_{2A}$ receptors are thought to couple predominantly to adenylate cyclase. This has been demonstrated in a variety of tissues and cell types, including platelets, striatum, basal ganglia, vasculature, and smooth muscle cells. Unlike $A_1$ receptors, few alternative effector systems have been shown to link this receptor subtype to PKA-independent signaling pathways. However, $A_{2A}$ receptors have been shown to regulate additional G-proteins that are not coupled to adenylate cyclase. In striatal cholinergic neurons, $A_{2A}$ receptor subtype stimulation has been shown to lead to two parallel signaling pathways that utilize different G-proteins (Gubitz et al., 1996). One of the pathways (that activates adenylate cyclase) involves $G_s$ coupling and PKA-dependent activation of P-type $Ca^{2+}$ channels. The other pathway involves a cholera-toxin-insensitive G-protein that activates a PKC-dependent N-type $Ca^{2+}$ channel. Both pathways are sensitive to an $A_{2A}$ receptor antagonist (KF17837), are not additive, and PKA signaling is capable of inhibiting the second one. Finally, evidence suggests that $A_{2A}$ receptors are also able to couple to $G_{olf}$ (named for its identification in olfactory epithelium), a receptor subtype that is highly expressed in neurons of the striatum in the brain, in excess of $G_s$ expression (Kull et al., 2000).

## C. $A_{2B}$ Receptors and Their Effector Coupling Systems

$A_{2B}$ receptors couple to $G_s$ proteins and activate adenylate cyclase, and this has been demonstrated in various systems including, the brain, fibroblasts, and airway cells. Additionally, $A_{2B}$ receptors couple to various intracellular signal transduction pathways outside the adenylate cyclase system. For example, $A_{2B}$ receptors have been shown to couple to MAP kinase pathways via $G_{q/11}$ proteins (Feoktistov and Biaggioni, 1995;

Auchampach *et al.*, 1997; Feoktistov *et al.*, 1998). In mast cells, $A_{2B}$ receptors also couple to $G_{q/11}$ and activate phophatidylinositol-specific PLC, resulting in the release of diacylglycerol (DAG) and $IP_3$ (Marquardt and Walker, 1994; Marquardt *et al.*, 1994; Feoktistov and Biaggioni, 1995; Linden *et al.*, 1999). DAG, in turn, stimulates PKC, whereas $IP_3$ mobilizes $Ca^{2+}$ from intracellular stores.

Other studies support a role for $A_{2B}$ receptors in PLC signaling. Activation of $A_{2B}$ receptors expressed in *Xenopus* oocytes induces a $Ca^{2+}$-regulated $Cl^-$ conductance, thought to be modulated by PLC (Yakel *et al.*, 1993). In human mast HMC-1 cells, $A_{2B}$ receptors couple to $G_q$ and activate $\beta$-PLC, resulting in both $Ca^{2+}$ mobilization in a phosphatidylinositol-specific manner, and the release of interleukin-8 (IL-8) (Feoktistov and Biaggioni, 1995). In these studies, $A_{2B}$ receptor stimulation led to $Ca^{2+}$ mobilization that was insensitive to both cholera and pertussis toxin. Additionally, IL-8 release was sensitive to the $A_{2B}$ receptor antagonist, enprofylline.

In human erythroleukemia (HEL) cells, $A_{2B}$ receptor coupling to $G_s$ also influences intracellular $Ca^{2+}$ levels by potentiating $Ca^{2+}$ influx in a cAMP-independent manner (Feoktistov *et al.*, 1994). Stimulation of $A_{2B}$ receptors can also potentiate $Ca^{2+}$ influx through a $Ca^{2+}$ channel, activating PKA in neurons that innervate the guinea pig hippocampus (Mogul *et al.*, 1993). These studies, along with others, reinforce the role for the receptor subtype in coupling mechanisms outside of adenylate cyclase.

## D. $A_3$ Receptors and Their Effector Coupling Systems

$A_3$ receptors are thought to predominantly couple to either $G_{i\alpha2}$ or $G_{i\alpha3}$ in a pertussis toxin-sensitive manner, producing inhibitory effects on adenylate cyclase and cAMP production. In schaffer collateral-CA1 synapses and in a rat mast cell line, $A_3$ receptors are also able to stimulate PLC (Palmer and Stiles, 1995) and PLD (Ali *et al.*, 1996), respectively. In addition, $A_3$ receptors expressed in cultured airway smooth muscle cells appear to be capable of activating phospholipase $A_2$ (PLA$_2$) and releasing arachidonic acid from cell membranes (Michoud *et al.*, 1999). In this system, ATP and 5-hydroxytryptamine were shown to stimulate $Ca^{2+}$ release, which could be enhanced by a specific $A_3$ receptor agonist (IB-MECA). The enhancing effects of IB-MECA were shown to be sensitive to AACOCF$_3$ (a PLA$_2$-specific inhibitor) suggesting a role for cPLA$_2$ and arachidonic acid signaling.

Although additional signaling mechanisms are not well documented for this less promiscuous subtype, activation of $A_3$ receptors in human monocytes has been shown to inhibit NADPH oxidase activity independent of cAMP elevation or alterations in $Ca^{2+}$ levels. These results suggest that

$A_3$ receptors might utilize additional signaling pathways not yet described (Broussas *et al.*, 1999).

## E. Receptor Desensitization

Upon prolonged agonist exposure, GPCRs are known to undergo a loss of sensitivity to agonists (Grady *et al.*, 1997). Several studies support desensitization of $P_1$ receptors (Chern *et al.*, 1993; Palmer *et al.*, 1995; Palmer and Stiles, 1995; Ciruela *et al.*, 1997; Peters *et al.*, 1998). Two phases characterize this desensitization, and are based on short-term versus long-term agonist exposure. Short-term exposure results in uncoupling of $P_1$ receptors from G-proteins through receptor phosphorylation (of serine and tyrosine residues by protein kinase A and C), whereas long-term agonist exposure results in receptor downregulation, including receptor internalization and/or reduced receptor synthesis. In CHO cells transiently expressing the $A_1$ or the $A_3$ receptor subtype, different uncoupling rates have been demonstrated following sustained agonist exposure. The differences appear to be in part due to the ability of $A_3$ receptors (but not $A_1$ receptors) to undergo rapid phosphorylation (and subsequent internalization) through the activation of G-protein-coupled receptor kinases (GRK) (Ferguson *et al.*, 2000). $A_{2A}$ and $A_{2B}$ receptors have also been shown to undergo short-term desensitization mediated by phosphorylation through a specific GRK (GRK2). For these $A_2$ receptors, $P_1$ receptor phosphorylation was dependent on both agonist concentration and the level of GRK expression (Mundell *et al.*, 1998).

$A_{2B}$ receptor stimulation with NECA induces pulmonary vasodilation in isolated lung and pulmonary artery smooth muscle cells, followed by desensitization that requires $G_{s\alpha}$-adenylate cyclase signaling (Haynes *et al.*, 1999). $A_{2B}$ receptors found on the apical and basolateral surfaces of polarized T84 epithelial monolayers differentially desensitize following agonist stimulation (Sitaraman *et al.*, 2000). Prolonged stimulation of $A_{2B}$ receptors on the basolateral surface produces desensitization of $A_{2B}$ receptors on the apical membrane, but not the converse. The nature of this cross-cellular desensitization, however, is not clear.

## IV. ADENOSINE RECEPTORS IN HEALTH AND DISEASE

The intracellular and extracellular concentration of ADO is regulated by many factors, including (1) its production by both *de novo* and salvage pathways for purine nucleotide biosynthesis, (2) the amount of ADO released from cells by facilitated diffusion, (3) ADO reuptake, (4) the activity of degradation pathways, and (5) metabolic conversion following removal of phosphate groups from ATP and related compounds through $5'$-ectonucleotidase activity. The cellular effects that ADO influences are determined by the tissue levels and the regional $P_1$ purinergic receptor subtypes that are expressed. Adenosines' capacity to modulate multiple biological processes in numerous cell types, including inflammatory cells, mast cells, smooth muscle cells in bronchi and vasculature, intestinal and airway epithelial cells, neurons, and platelets makes it a versatile, quintessential signaling molecule. Although concentrations of ADO are tightly regulated, alterations in the levels of ADO can have pathophysiological ramifications, in that ADO serves as a cardioprotector, chemoprotector, immunomodulator, ion transport regulator, and neuromodulator.

## A. ADO Receptors and Protective Effects in the Cardiovascular System

The role of ADO in the cardiovascular system has been well studied but continues to be an active area of research. Through $P_1$ purinergic receptor signaling, ADO acts as a potent modulator of many cardiovascular functions, including blood pressure, vascular tone, and heart rate, and may be particularly useful in understanding the pathogenesis of myocardial infarction in animal models. During ischemic challenge, ADO is released to the interstitium, where it is believed to exert a cardioprotective effect and reduce myocardial infarct extension. In addition, ischemia followed by reperfusion can extend vascular injury, leading to contractile dysfunction, apoptosis, and cellular necrosis. Ischemia-reperfusion injury is characterized by neutrophilic invasion, alterations in ionic pump activity, edema, and generation of superoxide radicals. ADO appears to modulate many of the components that contribute to cardiac injury. Although $P_1$ purinergic receptors have been known to be expressed in the heart for some time, the specific mechanisms by which ADO exerts its cardioprotective effects have only recently been investigated. The nature of the cardioprotective effect of ADO depends on whether it is used as pretreatment, during ischemia, or with reperfusion. Cardioprotection by ADO is thought to be mediated by $A_1$ receptor activation and opening of an ATP-sensitive potassium channel ($K_{ATP}$) when used prior to or during ischemia (Baxter and Yellon, 1999; Nojiri et al., 1999). Pretreatment with ADO during ischemic preconditioning (following 5 min of ischemia prior to receiving longer, more damaging ischemia) has been shown in several animal models to reduce myocardial infarct size and augment postischemic recovery (McCully et al., 2001). ADO receptor activation during ischemia can also result in reduced infarct size, whereas enhanced postischemic recovery requires $P_1$ purinergic stimulation during ischemia and reperfusion.

Neutrophil recruitment is characteristic of ischemic vascular injury. Each receptor subtype seems to harness its defense through distinct signaling pathways. During reperfusion, $A_2$ receptor activation may inhibit neutrophil attachment to the endothelium and, therefore, reduce inflammation and myocardial injury. Stimulating $A_2$ receptors before reperfusion inhibits neutrophil function, and does not appear to involve $K_{ATP}$-channel activation (Zhao et al., 1997). Activation of $A_3$ receptors has also been shown to inhibit neutrophil adhesion to the endothelium, as well as attenuate postischemic contractile dysfunction through PKC-dependent activation of $K_{ATP}$ channels (Liang, 1998; Tracey et al., 1998; Thourani et al., 1999). Following coronary artery occlusion in rabbits, infarct size can be reduced by activating $K_{ATP}$ channels. ADO can accomplish this by stimulation of $A_1$ receptors and nitric oxide synthetase (NOS)-dependent signaling, or $A_3$ receptors and NOS-independent signaling (Takano et al., 2001).

The cardioprotective effects of ADO are not limited to ischemia-induced vascular injury. ADO accumulates in the systemic circulation in patients with both ischemic and nonischemic chronic heart failure (CHF) (Funaya et al., 1997). In a recent study, patients with CHF who were treated with either dilazep or dipyridamole (which inhibit ADO transport into the cell) had increases in plasma ADO levels, associated with improved maximal oxygen consumption and ejection fraction (Kitakaze et al., 1998). These studies suggest that there may be therapeutic implications for the management of cardiac and noncardiac vascular disorders by influencing ADO metabolism or by using $P_1$ receptor ligands.

## B. ADO Receptors and Cancer

Several observations have demonstrated that ADO stimulation of specific $P_1$ purinergic receptors modulates the growth and differentiation of both normal and cancer cells. At micromolar concentrations, ADO stimulates proliferation of bone marrow cells, thymocytes, and swiss mouse 3T3 cells. Treating HT29 cells (a colonic adenocarcinoma cell line) with adenosine deaminase (ADA, which converts ADO into inosine) or with an $A_1$ receptor antagonist decreases cell growth rates (Lelievre et al., 1998). At higher concentrations ($>100$ $\mu M$), ADO can induce apoptosis in both normal and transformed cells in vitro. This has been demonstrated in lckNT transgenic mice that overexpress ecto-5'-nucleotidase (5'-NTase) in cortical thymocytes. In these mice, 5'-NTase activity was increased 100-fold without alterations in the plasma concentration of ADO, whereas ADA activity was shown to prevent accumulation of ADO and maintain it at low concentrations (Resta *et al.*, 1997). Treatment with an ADA inhibitor (2'-deoxycoformycin) resulted in a 30-fold increase in ADO concentration, levels that are capable of inducing apoptosis.

Released ADO has been shown to abrogate muscle cell tumor growth, and in a rat lymphoma cell line, $A_3$ receptor stimulation has been shown to have antiproliferative effects by inducing cell cycle arrest during $G_0/G_i$ (Fishman *et al.*, 2000). *In vivo*, a combined effect of chemotherapy and $A_3$ receptor stimulation may theoretically enhance antiproliferative characteristics.

In hypoxic tissues, ADO modulates angiogenesis. In a recent study, antagonists of $A_2$ receptors were shown to mediate hypoxia-induced antiangiogenic properties in CD45 positive lymphocytes isolated from peritoneal ascitic fluid of ovarian cancer patients (Barcz *et al.*, 2000). Additionally, cell motility imperative to tumor progression was found to be enhanced in ADO-stimulated melanoma cells, which was inhibited by a $P_1$ receptor antagonist (Woodhouse *et al.*, 1998). In another study, ADO released from hypoxic regions within solid tumors (in the presence of an inhibitor of ADA) inhibited the adhesion of anti-CD3-activated killer lymphocytes to syngeneic MCA-38 colon adenocarcinoma cells by up to 60%. This effect appeared to be through stimulation of $A_3$ receptors on the effector cells (MacKenzie *et al.*, 1994).

These studies and others highlight the role that ADO might play in cellular differentiation and tumor development. ADO has been shown to have both inhibitory and stimulatory effects that are consistent with signaling through specific $P_1$ purinergic receptor subtypes. A better understanding of the effects of ADO on tumor growth (particularly in regions of hypoxia), angiogenesis, and metastasis may lead to new targets for future cancer therapy.

## C. $P_1$ Purinergic Receptors as Regulators of Inflammation

$P_1$ purinergic receptors mediate both pro- and antiinflammatory effects in various organ systems and white blood cell (WBC) types. This section will highlight examples of both pro- and antiinflammatory actions mediated through $P_1$ purinoreceptors with an emphasis on mucosal inflammation.

Evidence suggests that ADO can function as a potent antiinflammatory agent. Extracellular ADO can be produced during cell stress and following ecto-5'-NTase breakdown of ATP and other ADO nucleosides, with accumulation in inflammatory exudates. ADO release is promoted by neutrophil-dependent production of oxygen radicals, in addition to clinically used antiinflammatory agents (Cronstein, 1995; Gadangi *et al.*,

1996). Methotrexate and sulfasalazine are part of the treatment of chronic inflammatory conditions such as rheumatoid arthritis and Crohn's disease. The antiinflammatory effects of these agents appear to be mediated by enhanced ADO release with subsequent inhibition of neutrophil activity (discussed below) (Cronstein et al., 1993; Gadangi et al., 1996; Morabito et al., 1998; Cronstein et al., 1999). $A_{2B}$ receptors may also play a protective role in arthritis. Activation of this receptor subtype in cultured synoviocytes has been shown to inhibit gene expression of matrix metalloproteinases, which are involved in the pathological degradation of bone and cartilage seen in rheumatoid arthritis (Boyle et al., 1996).

The most potent antiinflammatory effects of ADO come from studies involving $A_{2A}$ receptors. Activation of this receptor subtype during an inflammatory response results in inhibition of neutrophil adhesion and a reduction in the generation of superoxide radicals (Cronstein et al., 1983, 1986, 1990, 1992; Bullough et al., 1995). In macrophages, $A_{2A}$ receptor activation also inhibits superoxide anion generation, tumor necrosis factor (TNF)-$\alpha$, IL-12 secretion, and the proliferation of T cells and promotes apoptosis of activated T cells (Leonard et al., 1987; Le Vraux et al., 1993; Prabhakar et al., 1995; Genestier et al., 1998; Link et al., 2000). These findings demonstrating the toxic effects of ADO on T cells may be important to better understanding T cell dysfunction in immune disorders such as severe combined immune deficiency (SCID), a genetically inherited disorder that involves defective ADA function.

$P_1$ purinergic receptors not only regulate the function of inflammatory cells, but can also effect gene expression. For example, $A_{2B}$ receptors play a crucial role in cycloxygenase-2 gene regulation and the synthesis of prostaglandin $E_2$ in microglial cells, as well as inducing IL-6 expression (Fiebich et al., 1996). $A_{2B}$ receptors, therefore, may play a role in the inflammatory responses seen in neurodegenerative diseases such as Alzheimer's disease, where IL-6 may contribute to disease pathogenesis (Bauer et al., 1991; Wood et al., 1993; Huberman et al., 1995; Forloni et al., 1997). $A_3$ receptor agonists have also been shown to regulate gene expression, inhibiting transcriptional activation of macrophage inflammatory protein $1\alpha$ (MIP-$1\alpha$), a protein that promotes neutrophil chemotaxis (Szabo et al., 1998).

Several studies also suggest a role for ADO in asthma pathogenesis. Evidence for this role was first recognized when it was shown that ADO and AMP inhalation produced dose-dependent bronchoconstriction in asthmatics with allergic triggers, but not in normal subjects or subjects with other airway diseases (CF, bronchectasis) (Cushley et al., 1983; Holgate et al., 1991). Excessive amounts of ADO have also been found in bronchial lavage fluid of asthmatic individuals (Driver et al., 1993). It appears that $A_1$ receptors are involved in asthma pathogenesis since they are overexpressed in allergic rabbits and rats, and $A_1$ receptors are upregulated in bronchial smooth muscle tissue exposed to human asthmatic serum (Driver *et al.*, 1993; Ali *et al.*, 1994; Pauwels and Joos, 1995; el-Hashim *et al.*, 1996; Hakonarson and Grunstein, 1998). ADO also induces mast cell degranulation through stimulation of $A_{2B}$ receptors, leading to histamine and IL-8 release (Cronstein *et al.*, 1992; Feoktistov *et al.*, 1998). $A_{2B}$ receptor-specific antagonists might, therefore, one day serve a role in asthma therapy, blocking mast cell degranulation.

ADO may also "fine tune" WBC activity, depending on local ADO concentrations and differential $P_1$ purinergic receptor affinities. For example, $A_1$, $A_{2A}$, $A_{2B}$, and $A_3$ receptors are all expressed in neutrophils. However, each receptor subtype plays a different role in regulating neutrophil-induced inflammatory responses. $A_1$ receptors stimulate proinflammatory functions such as chemotaxis, superoxide radical formation, and neutrophil adherence (Rose *et al.*, 1988; Cronstein *et al.*, 1983, 1985, 1990, 1992; Krump and Borgeat, 1999). The stimulation of $A_2$ receptors, however, inhibits many inflammatory neutrophil functions, including superoxide formation, inflammatory mediator release, phagocytosis, and neutrophil adherence. Different concentrations of ADO, therefore, may mediate pro- or antiinflammatory responses, supporting a model in which ambient production of ADO in tissues modulates inflammation. Peripheral human neutrophil suspensions also spontaneously release ADO, which, upon binding to $A_2$ receptors, inhibits important proinflammatory functions such as $LTB_4$ production (Krump *et al.*, 1997).

ADO has also been shown to modulate important inflammatory functions in peripheral macrophages, including the modulation of ADO receptor expression by various inflammatory mediators. For example, Xaus and colleagues have demonstrated that interferon (INF)-$\gamma$ upregulates $A_{2B}$ receptor expression in murine bone marrow macrophages (Xaus *et al.*, 1999). ADO stimulation also inhibits several proinflammatory functions in peripheral macrophages and macrophage cell lines, including IFN-$\gamma$-induced expression of MHC class II genes, NOS, LPS-induced IL-10, TNF-$\alpha$ and NO production, macrophage colony-stimulating factor (M-CSF)-induced macrophage proliferation, MIP-$\alpha$ expression, and immunostimulated IL-12, IL-6, and NO production (Hasko *et al.*, 1996; Sajjadi *et al.*, 1996; Hon *et al.*, 1997; Szabo *et al.*, 1998; Xaus *et al.*, 1999).

## D. Effects of $P_1$ Purinergic Receptor Agonists on Ion Channel Activity in Intestinal and Airway Epithelia

Many studies have investigated the role of $P_1$ purinergic receptors in both intestinal and airway epithelium. Mucosal stimulation of canine airway monolayers with ADO results in $Cl^-$ secretion that is sensitive to $P_1$ receptor blockade (Pratt *et al.*, 1986). Lazarowski and colleagues extended these findings by demonstrating that human airway epithelial monolayers secrete $Cl^-$ in response to ADO and ADO analogs with a rank order of potency supporting $A_2$ receptor subtype stimulation (Lazarowski *et al.*, 1992). More recently, Furukawa and co-workers found that ADO and its analogs stimulate $Cl^-$ secretion in primary cultures of gerbil middle ear epithelium (Furukawa *et al.*, 1998). In each of these investigations, ADO led to the production of cAMP.

In the intestinal epithelium, migration of white blood cells, including eosinophils and neutrophils, is thought to play a role in transepithelial ion transport characteristic of inflammatory disorders such as infectious diarrhea and allergic colitis. Activated neutrophils release a secretory factor thought to be an ATP metabolite that is capable of activating electrogenic $Cl^-$ secretion across T-84 monolayers (Matthews *et al.*, 1993). Similarly, activated eosinophils also invoke $Cl^-$ secretion across T-84 colonic cell monolayers, and the nature of this response is dependent on the conversion of $5'$-adenosine monophosphate to ADO by apically localized ecto-$5'$-nucleotidase activity (Resnick *et al.*, 1993). Later studies concluded that $A_{2B}$ receptors were responsible for activated $Cl^-$ secretion in these experiments, signaling predominantly through cAMP and the cystic fibrosis transmembrane conductance regulator (CFTR) (Strohmeier *et al.*, 1995).

Evidence for a cAMP-independent pathway involved in $A_2$ receptor activation of $Cl^-$ secretion has been suggested by Barrett and colleagues (Barrett and Bigby, 1993). Their studies showed that ADO stimulates arachidonic acid (AA) release from T-84 cells and that exogenous AA augmented $Cl^-$ secretion. An inhibitor of cytosolic phospholipase $A_2$ ($cPLA_2$), an enzyme that releases AA from cell membrane phospholipids, was found to attenuate ADO-stimulated $Cl^-$ secretion. This study provided evidence that additional, cAMP-independent signaling mechanisms such as arachidonic acid signaling might influence ADO stimulated $Cl^-$ secretion in intestinal epithelia.

Recent studies have also begun to more fully investigate the relationship between $P_1$ receptors and CFTR. Defective or absent CFTR activity leads to abnormal airway epithelial $Cl^-$ and $Na^+$ transport in a variety of tissues where it is expressed, which, in turn, is felt to alter the composition of the epithelial surface liquid and/or mucus (Schwiebert *et al.*, 1999; Wine, 1999). CFTR is a PKA-regulated $Cl^-$ channel, which also regulates a variety of other channels [including the epithelial sodium channel (ENaC), the outwardly rectified chloride channel (ORCC), the ROMK potassium channel, and possibly an ATP release pathway]. Using rabbits exposed to $SO_2$ as a model for bronchitis, Iwase and co-workers found that injury to the rabbit airway induced the upregulation of CFTR transcription, which was associated with the appearance of a new ATP-activated $Cl^-$ current (Iwase *et al.*, 1997). This current was dependent on conversion of ATP to ADO by surface $5'$-ectonucleotidase activity. More recently, ADO, in addition to its phosphorylated nucleosides (AMP, ADP, ATP), activated wild-type CFTR-dependent halide efflux, which was mediated by $A_{2B}$ receptors (Clancy *et al.*, 1999). Additionally, clinically relevant, mutant CFTR molecules available at the cell membrane could be activated using $A_{2B}$ receptor stimulation *in vitro*. Collectively, these studies provide a rationale for investigating $A_{2B}$ and its role in regulating CFTR and ion transport *in vivo*.

New evidence suggests that ADO receptors might also be involved in signaling transducisomes that reside at the cell surface, anchored through protein–protein interactions. CFTR is thought to be preferentially regulated by membrane-bound PKAII, which is anchored at specific sites to A kinase anchoring proteins (AKAPs) (Hausken *et al.*, 1994; Gray *et al.*, 1998; Short *et al.*, 1998; Colledge and Scott, 1999; Mohler *et al.*, 1999; Moyer *et al.*, 1999, 2000; Wang *et al.*, 1998, 2000; Milewski *et al.*, 2001). CFTR has been shown to bind a scaffolding protein, EBP50 (ezrin-binding protein), that in turn binds to ezrin, an AKAP (Dransfield *et al.*, 1997). The role of ezrin in CFTR regulation is believed to involve linking the protein to the actin cytoskeleton, and possibly participating in membrane targeting. In Calu3 cells (an airway epithelial serous cell line that expresses $A_2$ receptors and CFTR), a synthetic peptide, Ht-31, was used to block the interaction between PKAII and ezrin in an attempt to determine how $A_2$ receptors regulate CFTR (Huang *et al.*, 2000). Ht-31 blocked ADO-stimulated activation of CFTR in whole cell voltage-clamp studies, suggesting a role for this AKAP in $A_2$ receptor-coupled activation of CFTR. Additionally, forskolin and ATP were able to activate CFTR when applied to the cytoplasmic surface of excised membrane patches. Together, these elegant studies provide evidence that $A_2$ receptors and CFTR may be spatially compartmentalized and allow efficient regulation of CFTR activity by ADO.

### E. $P_1$ Purinergic Receptors and the Role of ADO in the CNS: Studies from an $A_{2A}$ Receptor Knockout

Accumulating evidence points toward ADO as an important homeostatic modulator and, in general, a neuroprotector of the CNS. With the exception of $A_{2A}$ receptors that are expressed primarily in the striatum, olfactory tubercle, and nucleus accumbens, $P_1$ receptors are widely expressed in the brain based on studies involving ligand binding, *in situ* hybridization, and RT-PCR (Dixon *et al.*, 1996). Manifestation of $P_1$ purinergic receptor activity in the CNS include regulation of behavior, sleep and arousal, locomotor activities, and pathological conditions such as CNS ischemia and possibly influencing neurodegenerative diseases including Parkinson's disease. Throughout the CNS, ADO is felt to frequently "fine tune" the primary components of synaptic transmission, and this modulation is accomplished through $P_1$ purinergic receptors. For a more complete discussion of ADO and $P_1$ receptor activity in the CNS, the reader is directed to several excellent recent reviews (Richardson et al., 1997; Sweeney, 1997; Kaiser and Quinn, 1999; Moreau and Huber, 1999; Svenningsson et al., 1999; Cunha, 2001; Dunwiddie and Masino, 2001).

Similar to the heart, ADO protects against ischemia-induced neuronal injury in the CNS. ADO influences pre- and postsynaptic neurotransmission by modulating the release and actions of specific neurotransmitters and neuropeptide modulators, including vasoactive intestinal peptide (VIP) and calcitonin gene-related peptide (CGRP) (Ribeiro, 1999; Sebastiao et al., 2000). Recently, a role for ADO receptors as regulators of other receptors in the nervous system such as $N$-methyl-D-aspartate (NMDA) receptors, metabotropic glutamate receptors, and nicotinic autofacilitary receptors has been identified. $A_1$ receptor stimulation also protects against methamphet-amine-induced neurotoxicity to nigrostriatal dopaminergic neurons in animal models (Delle Donne and Sonsalla, 1994). It has been suggested that ADO activation of $A_1$ receptors plays a role in glutamate-mediated neuronal injury following ischemia. During ischemic stress ADO is released, and has been shown to be associated with attenuated glutamate neurotoxicity and delayed onset of paraplegia in a New England white rabbit paraplegic model (Nakamichi, 1998). Although the nature of this effect is unclear, it appears that ADO may act through $A_1$ receptors to reduce the effect of excitatory amino acids (such as glutamate) by altering NMDA receptors.

Recent studies completed in $A_{2A}$ receptor knockout mice have provided insight into the role of this receptor in the brain in vivo. Caffeine, a well-established $A_{2A}$ receptor antagonist, is known to produce both exploratory behavior as well as alertness, both of which were found to be depressed in $A_{2A}$ receptor knockout mice (Ledent et al., 1997). Caffeine also has been shown to enhance anxiety levels and act as an analgesic. Consistent with these observations, $A_{2A}$ receptor knockouts were found to be more aggressive and anxious, and to have notably reduced pain responses compared to wild-type mice. Finally, knockout mice had increased blood pressure and heart rate as well as more efficient platelet aggregation compared to wild-type mice. These studies suggest that the effects of ADO on vasodilation and inhibition of platelet aggregation are likely to be $A_{2A}$ receptor specific.

$A_{2A}$ receptor knockout mice have also provided evidence for involve-ment of these receptors in the pathogenesis of Parkinson's disease, a neurodegenerative disease that results from depletion of dopaminergic neurons in the nigrostriatal pathway of the brain. Clinical manifestations of Parkinson's disease are thought to result from alterations in the dopaminergic pathways within the basal ganglia and the GABA-releasing spiny neurons of the striatum. ADO activation of $A_{2A}$ receptors is thought to antagonize dopaminergic signaling by decreasing the binding affinity of dopamine to $D_2$ dopamine receptors (Ferre et al., 1993, 1998). Because Parkinson's disease is characterized by depletion of dopamine, $A_{2A}$ receptor antagonists could theoretically be therapeutic targets for developing drugs to treat patients with Parkinson's disease. In a $D_2$ receptor transgenic knockout mouse model of Parkinson's disease (Baik et al., 1995), an $A_{2A}$ receptor antagonist was found to rescue motor deficits observed in the knockouts in a dopamine-independent manner. In a separate study, caffeine was found to have less locomotor effects in $D_2$ receptor knockout mice (versus wild-type mice), suggesting a functional coupling between $D_2$ receptors and $A_{2A}$ receptors (Zahniser et al., 2000). $P_1$ purinergic receptor signaling might also be relevant to additional neurodegenerative disorders such as Alzheimer's disease and Huntington's disease (Sebastiao and Ribeiro, 1996; Haas and Selbach, 2000).

## V. CONCLUDING REMARKS

ADO, by interacting with $P_1$ purinergic receptor subtypes, acts as a potent regulatory autocrine that "fine tunes" many physiological processes in different tissues and organ systems. Each receptor subtype is wired into a distinct and complex network of signaling pathways that allows ADO to have both stimulatory and inhibitory effects necessary for proper homeostatic control. Alterations in this regulatory relationship may be associated with many different pathological processes including progressive neurological disorders, inflammatory disorders, defective epithelial ion transport, and cardiovascular disorders that involve arrhythmias and ischemia. A better understanding of the cell biology of $P_1$ purinergic receptors and signaling may provide novel therapeutic targets for these and other pathological conditions.

**References**
Abebe, W., and Mustafa, S. J. (1998). A1 adenosine receptor-mediated Ins(1,4,5)P3 generation in allergic rabbit airway smooth muscle. Am. J. Physiol. **275**(5 pt. 1), L990–997.
Alexander, S. P., Cooper, J., et al. (1996). Characterization of the human brain putative A2B adenosine receptor expressed in Chinese hamster ovary (CHO.A2B4) cells. Br. J. Pharmacol. **119**(6), 1286–1290.

Ali, H., Choi, O. H., *et al.* (1996). Sustained activation of phospholipase D via adenosine A3 receptors is associated with enhancement of antigen- and Ca(2+)-ionophore-induced secretion in a rat mast cell line. *J. Pharmacol. Exp. Ther.* **276**(2), 837–845.

Ali, S., Mustafa, S. J., *et al.* (1994). Adenosine-induced bronchoconstriction and contraction of airway smooth muscle from allergic rabbits with late-phase airway obstruction: Evidence for an inducible adenosine A1 receptor. *J. Pharmacol. Exp. Ther.* **268**(3), 1328–1334.

Auchampach, J. A., Jin, X., *et al.* (1997). Canine mast cell adenosine receptors: Cloning and expression of the A3 receptor and evidence that degranulation is mediated by the A2B receptor. *Mol. Pharmacol.* **52**(5), 846–860.

Baik, J. H., Picetti, R., *et al.* (1995). Parkinsonian-like locomotor impairment in mice lacking dopamine D2 receptors. *Nature* **377**(6548), 424–428.

Barcz, E., Sommer, E., *et al.* (2000). Adenosine receptor antagonism causes inhibition of angiogenic activity of human ovarian cancer cells. *Oncol. Rep.* **7**(6), 1285–1291.

Barrett, K. E., and Bigby, T. D. (1993). Involvement of arachidonic acid in the chloride secretory response of intestinal epithelial cells. *Am. J. Physiol.* **264**(2 pt. 1), C446–452.

Bauer, J., Strauss, S., *et al.* (1991). IL-6-mediated events in Alzheimer's disease pathology. *Immunol. Today* **12**(11), 422.

Baxter, G. F., and Yellon, D. M. (1999). ATP-sensitive K+ channels mediate the delayed cardioprotective effect of adenosine A1 receptor activation. *J. Mol. Cell Cardiol.* **31**(5), 981–989.

Belardinelli, L., Shryock, J. C., *et al.* (1995). Ionic basis of the electrophysiological actions of adenosine on cardiomyocytes. *FASEB J.* **9**(5), 359–365.

Berne, R. M. (1963). Cardiac nucleotides in hypoxia: Possible role in the regulation of coronary blood flow. *Am. J. Physiol.* **204**(2), 317–322.

Biber, K., Klotz, K. N., *et al.* (1997). Adenosine A1 receptor-mediated activation of phospholipase C in cultured astrocytes depends on the level of receptor expression. *J. Neurosci.* **17**(13), 4956–4964.

Boyle, D. L., Sajjadi, F. G., *et al.* (1996). Inhibition of synoviocyte collagenase gene expression by adenosine receptor stimulation. *Arthritis Rheum.* **39**(6), 923–930.

Brackett, L. E., and Daly, J. W. (1994). Functional characterization of the A2b adenosine receptor in NIH 3T3 fibroblasts. *Biochem. Pharmacol.* **47**(5), 801–814.

Broussas, M., Cornillet-Lefebvre, P., *et al.* (1999). Inhibition of fMLP-triggered respiratory burst of human monocytes by adenosine: Involvement of A3 adenosine receptor. *J. Leukocyte Biol.* **66**(3), 495–501.

Bruns, R. F. (1980). Adenosine receptor activation in human fibroblasts: Nucleoside agonists and antagonists. *Can. J. Physiol. Pharmacol.* **58**(6), 673–691.

Bruns, R. F., Lu, G. H., *et al.* (1986). Characterization of the A2 adenosine receptor labeled by [3H]NECA in rat striatal membranes. *Mol. Pharmacol.* **29**(4), 331–346.

Bullough, D. A., Magill, M. J., *et al.* (1995). Adenosine activates A2 receptors to inhibit neutrophil adhesion and injury to isolated cardiac myocytes. *J. Immunol.* **155**(5), 2579–2586.

Burnstock, G. (1978). "A Basis for Distinguishing Two Types of Purinergic Receptors." Raven Press, New York.

Cheng, J. T., Liu, I. M., *et al.* (2000). Role of adenosine in insulin-stimulated release of leptin from isolated white adipocytes of Wistar rats. *Diabetes* **49**(1), 20–24.

Chern, Y., Lai, H. L., *et al.* (1993). Multiple mechanisms for desensitization of A2a adenosine receptor-mediated cAMP elevation in rat pheochromocytoma PC12 cells. *Mol. Pharmacol.* **44**(5), 950–958.

Ciruela, F., Saura, C., *et al.* (1997). Ligand-induced phosphorylation, clustering, and desensitization of A1 adenosine receptors. *Mol. Pharmacol.* **52**(5), 788–797.

Clancy, J. P., Ruiz, F. E., *et al.* (1999). Adenosine and its nucleotides activate wild-type and R117H CFTR through an A2B receptor-coupled pathway. *Am. J. Physiol.* **276**(2 pt. 1), C361–369.

Colledge, M., and Scott, J. D. (1999). AKAPs: From structure to function. *Trends Cell Biol.* **9**(6), 216–221.

Cronstein, B. N. (1995). The antirheumatic agents sulphasalazine and methotrexate share an anti-inflammatory mechanism. *Br. J. Rheumatol.* **34**(Suppl. 2), 30–32.

Cronstein, B. N., Kramer, S. B., *et al.* (1983). Adenosine: A physiological modulator of superoxide anion generation by human neutrophils. *J. Exp. Med.* **158**(4), 1160–1177.

Cronstein, B. N., Kramer, S. B., *et al.* (1985). Adenosine modulates the generation of superoxide anion by stimulated human neutrophils via interaction with a specific cell surface receptor. *Ann. N.Y. Acad. Sci.* **451**, 291–301.

Cronstein, B. N., Levin, R. I., *et al.* (1986). Adenosine: An endogenous inhibitor of neutrophil-mediated injury to endothelial cells. *J. Clin. Invest.* **78**(3), 760–770.

Cronstein, B. N., Daguma, L., *et al.* (1990). The adenosine/neutrophil paradox resolved: Human neutrophils possess both A1 and A2 receptors that promote chemotaxis and inhibit O2 generation, respectively. *J. Clin. Invest.* **85**(4), 1150–1157.

Cronstein, B. N., Levin, R. I., *et al.* (1992). Neutrophil adherence to endothelium is enhanced via adenosine A1 receptors and inhibited via adenosine A2 receptors. *J. Immunol.* **148**(7), 2201–2206.

Cronstein, B. N., Naime, D., *et al.* (1993). The antiinflammatory mechanism of methotrexate. Increased adenosine release at inflamed sites diminishes leukocyte accumulation in an *in vivo* model of inflammation. *J. Clin. Invest.* **92**(6), 2675–2682.

Cronstein, B. N., Montesinos, M. C., *et al.* (1999). Salicylates and sulfasalazine, but not glucocorticoids, inhibit leukocyte accumulation by an adenosine-dependent mechanism that is independent of inhibition of prostaglandin synthesis and p105 of NFkappaB. *Proc. Natl. Acad. Sci. USA* **96**(11), 6377–6381.

Cunha, R. A. (2001). Adenosine as a neuromodulator and as a homeostatic regulator in the nervous system: Different roles, different sources and different receptors. *Neurochem. Int.* **38**(2), 107–125.

Cunha, R. A., Johansson, B., *et al.* (1994). Evidence for functionally important adenosine A2a receptors in the rat hippocampus. *Brain Res.* **649**(1–2), 208–216.

Cushley, M. J., Tattersfield, A. E., *et al.* (1983). Inhaled adenosine and guanosine on airway resistance in normal and asthmatic subjects. *Br. J. Clin. Pharmacol.* **15**(2), 161–165.

Daly, J. W., Butts-Lamb, P., *et al.* (1983). Subclasses of adenosine receptors in the central nervous system: Interaction with caffeine and related methylxanthines. *Cell. Mol. Neurobiol.* **3**(1), 69–80.

Delle Donne, K. T., and Sonsalla, P. K. (1994). Protection against methamphetamine-induced neurotoxicity to neostriatal dopaminergic neurons by adenosine receptor activation. *J. Pharmacol. Exp. Ther.* **271**(3), 1320–1326.

Dickenson, J. M., and Hill, S. J. (1998a). Involvement of G-protein betagamma subunits in coupling the adenosine A1 receptor to phospholipase C in transfected CHO cells. *Eur. J. Pharmacol.* **355**(1), 85–93.

Dickenson, J. M., and Hill, S. J. (1998b). Potentiation of adenosine A1 receptor-mediated inositol phospholipid hydrolysis by tyrosine kinase inhibitors in CHO cells. *Br. J. Pharmacol.* **125**(5), 1049–1057.

Dickenson, J. M., Blank, J. L., *et al.* (1998). Human adenosine A1 receptor and P2Y2-purinoceptor-mediated activation of the mitogen-activated protein kinase cascade in transfected CHO cells. *Br. J. Pharmacol.* **124**(7), 1491–1499.

Dixon, A. K., Gubitz, A. K., *et al.* (1996). Tissue distribution of adenosine receptor mRNAs in the rat. *Br. J. Pharmacol.* **118**(6), 1461–1468.

Dransfield, D. T., Bradford, A. J., *et al.* (1997). Ezrin is a cyclic AMP-dependent protein kinase anchoring protein. *EMBO J.* **16**(1), 35–43.

Driver, A. G., Kukoly, C. A., *et al.* (1993). Adenosine in bronchoalveolar lavage fluid in asthma. *Am. Rev. Respir. Dis.* **148**(1), 91–97.

Drury, A. N., and Szent-Gyorgyi, A. (1929). The physiological activity of adenine compounds with special reference to their effects upon the mammalian heart. *J. Physiol.* **68**, 213–237.

Dunwiddie, T. V., and Masino, S. A. (2001). The role and regulation of adenosine in the central nervous system. *Annu. Rev. Neurosci.* **24**, 31–55.

el-Hashim, A., D'Agostino, B., *et al.* (1996). Characterization of adenosine receptors involved in adenosine-induced bronchoconstriction in allergic rabbits. *Br. J. Pharmacol.* **119**(6), 1262–1268.

Feoktistov, I., and Biaggioni, I. (1993). Characterization of adenosine receptors in human erythroleukemia cells and platelets: Further evidence for heterogeneity of adenosine A2 receptor subtypes. *Mol. Pharmacol.* **43**(6), 909–914.

Feoktistov, I., and Biaggioni, I. (1995). Adenosine A2b receptors evoke interleukin-8 secretion in human mast cells. An enprofylline-sensitive mechanism with implications for asthma. *J. Clin. Invest.* **96**(4), 1979–1986.

Feoktistov, I., and Biaggioni, I. (1997). Adenosine A2B receptors. *Pharmacol. Rev.* **49**(4), 381–402.

Feoktistov, I., Murray, J. J., *et al.* (1994). Positive modulation of intracellular Ca2+ levels by adenosine A2b receptors, prostacyclin, and prostaglandin E1 via a cholera toxin-sensitive mechanism in human erythroleukemia cells. *Mol. Pharmacol.* **45**(6), 1160–1167.

Feoktistov, I., Polosa, R., *et al.* (1998). Adenosine A2B receptors: A novel therapeutic target in asthma? *Trends Pharmacol. Sci.* **19**(4), 148–153.

Ferguson, G., Watterson, K. R., *et al.* (2000). Subtype-specific kinetics of inhibitory adenosine receptor internalization are determined by sensitivity to phosphorylation by G protein-coupled receptor kinases. *Mol. Pharmacol.* **57**(3), 546–552.

Ferre, S., Snaprud, P., *et al.* (1993). Opposing actions of an adenosine A2 receptor agonist and a GTP analogue on the regulation of dopamine D2 receptors in rat neostriatal membranes. *Eur. J. Pharmacol.* **244**(3), 311–315.

Ferre, S., Torvinen, M., *et al.* (1998). Adenosine A1 receptor-mediated modulation of dopamine D1 receptors in stably cotransfected fibroblast cells. *J. Biol. Chem.* **273**(8), 4718–4724.

Fiebich, B. L., Biber, K., *et al.* (1996). Adenosine A2b receptors mediate an increase in interleukin (IL)-6 mRNA and IL-6 protein synthesis in human astroglioma cells. *J. Neurochem.* **66**(4), 1426–1431.

Fishman, P., Bar-Yehuda, S., *et al.* (2000). Adenosine acts as an inhibitor of lymphoma cell growth: A major role for the A3 adenosine receptor. *Eur. J. Cancer* **36**(11), 1452–1458.

Forloni, G., Mangiarotti, F., *et al.* (1997). Beta-amyloid fragment potentiates IL-6 and TNF-alpha secretion by LPS in astrocytes but not in microglia. *Cytokine* **9**(10), 759–762.

Fredholm, B. B., Arslan, G., *et al.* (2000). Structure and function of adenosine receptors and their genes. *Naunyn Schmiedebergs Arch. Pharmacol.* **362**(4–5), 364–374.

Freissmuth, M., Schutz, W., *et al.* (1991). Interactions of the bovine brain A1-adenosine receptor with recombinant G protein alpha-subunits. Selectivity for rGi alpha-3. *J. Biol. Chem.* **266**(27), 17778–17783.

Funaya, H., Kitakaze, M., *et al.* (1997). Plasma adenosine levels increase in patients with chronic heart failure. *Circulation* **95**(6), 1363–1365.

Furukawa, M., Ikeda, K., *et al.* (1998). A2 adenosine receptors in Mongolian gerbil middle ear epithelium and their regulation of Cl⁻ secretion. *Acta Physiol. Scand.* **163**(1), 103–112.

Gadangi, P., Longaker, M., *et al.* (1996). The anti-inflammatory mechanism of sulfasalazine is related to adenosine release at inflamed sites. *J. Immunol.* **156**(5), 1937–1941.

Genestier, L., Paillot, R., *et al.* (1998). Immunosuppressive properties of methotrexate: Apoptosis and clonal deletion of activated peripheral T cells. *J. Clin. Invest.* **102**(2), 322–328.

Grady, E. F., Bohm, S. K., *et al.* (1997). Turning off the signal: Mechanisms that attenuate signaling by G protein-coupled receptors. *Am. J. Physiol.* **273**(3 pt. 1), G586–601.

Gray, P. C., Scott, J. D., *et al.* (1998). Regulation of ion channels by cAMP-dependent protein kinase and A-kinase anchoring proteins. *Curr. Opin. Neurobiol.* **8**(3), 330–334.

Gubitz, A. K., Widdowson, L., *et al.* (1996). Dual signalling by the adenosine A2a receptor involves activation of both N- and P-type calcium channels by different G proteins and protein kinases in the same striatal nerve terminals. *J. Neurochem.* **67**(1), 374–381.

Haas, H. L., and Selbach, O. (2000). Functions of neuronal adenosine receptors. *Naunyn Schmiedebergs Arch. Pharmacol.* **362**(4–5), 375–381.

Hakonarson, H., and Grunstein, M. M. (1998). Regulation of second messengers associated with airway smooth muscle contraction and relaxation. *Am. J. Respir. Crit. Care Med.* **158**(5 pt. 3), S115–122.

Hasko, G., Szabo, C., *et al.* (1996). Adenosine receptor agonists differentially regulate IL-10, TNF-alpha, and nitric oxide production in RAW 264.7 macrophages and in endotoxemic mice. *J. Immunol.* **157**(10), 4634–4640.

Hausken, Z. E., Coghlan, V. M., *et al.* (1994). Type II regulatory subunit (RII) of the cAMP-dependent protein kinase interaction with A-kinase anchor proteins requires isoleucines 3 and 5. *J. Biol. Chem.* **269**(39), 24245–24251.

Haynes, J. Jr., Obiako, B., *et al.* (1999). 5-(N-Ethylcarboxamido)adenosine desensitizes the A2b-adenosine receptor in lung circulation. *Am. J. Physiol.* **276**(6 pt. 2), H1877–1883.

Hill, R. J., Oleynek, J. J., *et al.* (1997). Cloning, expression and pharmacological characterization of rabbit adenosine A1 and A3 receptors. *J. Pharmacol. Exp. Ther.* **280**(1), 122–128.

Holgate, S. T., Church, M. K., and Polosa, R. (1991). Adenosine: A positive modulator of airway inflammation in asthma. *Ann. N. Y. Acad. Sci.* **629**, 227–236.

Hon, W. M., Moochhala, S., *et al.* (1997). Adenosine and its receptor agonists potentiate nitric oxide synthase expression induced by lipopolysaccharide in RAW 264.7 murine macrophages. *Life Sci.* **60**(16), 1327–1335.

Huang, P., Trotter, K., *et al.* (2000). PKA holoenzyme is functionally coupled to CFTR by AKAPs. *Am. J. Physiol. Cell Physiol.* **278**(2), C417–422.

Huberman, M., Sredni, B., *et al.* (1995). IL-2 and IL-6 secretion in dementia: Correlation with type and severity of disease. *J. Neurol. Sci.* **130**(2), 161–164.

Ito, H., Maemoto, T., *et al.* (1999). Pyrazolopyridine derivatives act as competitive antagonists of brain adenosine A1 receptors: [35S]GTPgammaS binding studies. *Eur. J. Pharmacol.* **365**(2–3), 309–315.

Iwase, N., Sasaki, T., *et al.* (1997). Signature current of SO2-induced bronchitis in rabbit. *J. Clin. Invest.* **99**(7), 1651–1661.

Jacobson, K. A. (1998). Adenosine A3 receptors: Novel ligands and paradoxical effects. *Trends Pharmacol. Sci.* **19**(5), 184–191.

Jacobson, K. A., Gallo-Rodriguez, C., *et al.* (1993). Structure-activity relationships of 8-styrylxanthines as A2-selective adenosine antagonists. *J. Med. Chem.* **36**(10), 1333–1342.

Jacobson, K. A., Park, K. S., *et al.* (1997). Pharmacological characterization of novel A3 adenosine receptor-selective antagonists. *Neuropharmacology* **36**(9), 1157–1165.

Jarvis, M. F., and Williams, M. (1988). Differences in adenosine A-1 and A-2 receptor density revealed by autoradiography in methylxanthine-sensitive and insensitive mice. *Pharmacol. Biochem. Behav.* **30**(3), 707–714.

Kaiser, S. M., and Quinn, R. J. (1999). Adenosine receptors as potential therapeutic targets. *Drug Discov. Today* **4**(12), 542–551.

Kitakaze, M., Minamino, T., *et al.* (1998). Elevation of plasma adenosine levels may attenuate the severity of chronic heart failure. *Cardiovasc. Drugs Ther.* **12**(3), 307–309.

Klotz, K. N., Hessling, J., *et al.* (1998). Comparative pharmacology of human adenosine receptor subtypes—characterization of stably transfected receptors in CHO cells. *Naunyn Schmiedebergs Arch. Pharmacol.* **357**(1), 1–9.

Krump, E., and Borgeat, P. (1999). Adenosine. An endogenous inhibitor of arachidonic acid release and leukotriene biosynthesis in human neutrophils. *Adv. Exp. Med. Biol.* **447**, 107–115.

Krump, E., Picard, S., *et al.* (1997). Suppression of leukotriene B4 biosynthesis by endogenous adenosine in ligand-activated human neutrophils. *J. Exp. Med.* **186**(8), 1401–1406.

Kull, B., Svenningsson, P., *et al.* (2000). Adenosine A(2A) receptors are colocalized with and activate g(olf) in rat striatum. *Mol. Pharmacol.* **58**(4), 771–777.

Lazarowski, E. R., Mason, S. J., *et al.* (1992). Adenosine receptors on human airway epithelia and their relationship to chloride secretion. *Br. J. Pharmacol.* **106**, 774–782.

Ledent, C., Vaugeois, J. M., *et al.* (1997). Aggressiveness, hypoalgesia and high blood pressure in mice lacking the adenosine A2a receptor. *Nature* **388**(6643), 674–678.

Lelievre, V., Muller, J. M., *et al.* (1998). Adenosine modulates cell proliferation in human colonic adenocarcinoma. I. Possible involvement of adenosine A1 receptor subtypes in HT29 cells. *Eur. J. Pharmacol.* **341**(2–3), 289–297.

Leonard, E. J., Shenai, A., *et al.* (1987). Dynamics of chemotactic peptide-induced superoxide generation by human monocytes. *Inflammation* **11**(2), 229–240.

Le Vraux, V., Chen, Y. L., *et al.* (1993). Inhibition of human monocyte TNF production by adenosine receptor agonists. *Life Sci.* **52**(24), 1917–1924.

Liang, B. T. (1998). Protein kinase C-dependent activation of KATP channel enhances adenosine-induced cardioprotection. *Biochem. J.* **336**(pt. 2), 337–343.

Libert, F., Parmentier, M., *et al.* (1989). Selective amplification and cloning of four new members of the G protein-coupled receptor family. *Science* **244**(4904), 569–572.

Libert, F., Van Sande, J., *et al.* (1992). Cloning and functional characterization of a human A1 adenosine receptor. *Biochem. Biophys. Res. Commun.* **187**(2), 919–926.

Linden, J. (1991). Structure and function of A1 adenosine receptors. *FASEB J.* **5**(12), 2668–2676.

Linden, J. (1994). Cloned adenosine A3 receptors: Pharmacological properties, species differences and receptor functions. *Trends Pharmacol. Sci.* **15**(8), 298–306.

Linden, J., Thai, T., *et al.* (1999). Characterization of human A(2B) adenosine receptors: Radioligand binding, western blotting, and coupling to G(q) in human embryonic kidney 293 cells and HMC-1 mast cells. *Mol. Pharmacol.* **56**(4), 705–713.

Link, A. A., Kino, T., *et al.* (2000). Ligand-activation of the adenosine A2a receptors inhibits. IL-12 production by human monocytes. *J. Immunol.* **164**(1), 436–442.

Londos, C., Cooper, D. M., *et al.* (1980). Subclasses of external adenosine receptors. *Proc. Natl. Acad. Sci. USA* **77**(5), 2551–2554.

MacKenzie, W. M., Hoskin, D. W., *et al.* (1994). Adenosine inhibits the adhesion of anti-CD3-activated killer lymphocytes to adenocarcinoma cells through an A3 receptor. *Cancer Res.* **54**(13), 3521–3526.

Mahan, L. C., McVittie, L. D., *et al.* (1991). Cloning and expression of an A1 adenosine receptor from rat brain. *Mol. Pharmacol.* **40**(1), 1–7.

Marquardt, D. L., and Walker, L. L. (1994). Inhibition of protein kinase A fails to alter mast cell adenosine responsiveness. *Agents Actions* **43**(1–2), 7–12.

Marquardt, D. L., Walker, L. L., *et al.* (1994). Cloning of two adenosine receptor subtypes from mouse bone marrow-derived mast cells. *J. Immunol.* **152**(9), 4508–4515.

Matthews, J. B., Awtrey, C. S., *et al.* (1993). Na(+)-K(+)-2Cl− cotransport and Cl− secretion evoked by heat-stable enterotoxin is microfilament dependent in T84 cells. *Am. J. Physiol.* **265**(2 pt. 1), G370–378.

McCully, J. D., Toyoda, Y., *et al.* (2001). Adenosine-enhanced ischemic preconditioning: Adenosine receptor involvement during ischemia and reperfusion. *Am. J. Physiol. Heart Circ. Physiol.* **280**(2), H591–602.

McWhinney, C. D., Dudley, M. W., Bowlin, T. L., Peet, N. P., Schook, L., Bradshaw, M., De, M., Borcherding, D. R., and Edwards, C. K., III. (1996). Activation of adenosine A3 receptors on macrophages inhibits tumor necrosis factor, alpha. *Eur. J. Pharmacol.* **310**, 209–213.

Michoud, M. C., Tao, F. C., *et al.* (1999). Mechanisms of the potentiation by adenosine of adenosine triphosphate-induced calcium release in tracheal smooth-muscle cells. *Am. J. Respir. Cell Mol. Biol.* **21**(1), 30–36.

Milewski, M. I., Mickle, J. E., *et al.* (2001). A PDZ-binding motif is essential but not sufficient to localize the C terminus of CFTR to the apical membrane. *J. Cell Sci.* **114**(pt. 4), 719–726.

Mogul, D. J., Adams, M. E., *et al.* (1993). Differential activation of adenosine receptors decreases N-type but potentiates P-type Ca2+ current in hippocampal CA3 neurons. *Neuron* **10**(2), 327–334.

Mohler, P. J., Kreda, S. M., *et al.* (1999). Yes-associated protein 65 localizes p62(c-Yes) to the apical compartment of airway epithelia by association with EBP50. *J. Cell Biol.* **147**(4), 879–890.

Morabito, L., Montesinos, M. C., *et al.* (1998). Methotrexate and sulfasalazine promote adenosine release by a mechanism that requires ecto-5′-nucleotidase-mediated conversion of adenine nucleotides. *J. Clin. Invest.* **101**(2), 295–300.

Moreau, J. L., and Huber, G. (1999). Central adenosine A(2A) receptors: An overview. *Brain Res. Rev.* **31**(1), 65–82.

Moyer, B. D., Denton, J., *et al.* (1999). A PDZ-interacting domain in CFTR is an apical membrane polarization signal. *J. Clin. Invest.* **104**(10), 1353–1361.

Moyer, B. D., Duhaime, M., *et al.* (2000). The PDZ-interacting domain of cystic fibrosis transmembrane conductance regulator is required for functional expression in the apical plasma membrane. *J. Biol. Chem.* **275**(35), 27069–27074.

Mundell, S. J., Luty, J. S., *et al.* (1998). Enhanced expression of G protein-coupled receptor kinase 2 selectively increases the sensitivity of A2A adenosine receptors to agonist-induced desensitization. *Br. J. Pharmacol.* **125**(2), 347–356.

Murrison, E. M., Goodson, S. J., *et al.* (1996). Cloning and characterisation of the human adenosine A3 receptor gene. *FEBS Lett.* **384**(3), 243–246.

Murthy, K. S., and Makhlouf, G. M. (1995). Adenosine A1 receptor-mediated activation of phospholipase C-beta 3 in intestinal muscle: Dual requirement for alpha and beta gamma subunits of Gi3. *Mol. Pharmacol.* **47**(6), 1172–1179.

Nakamichi, T. (1998). [Glutamate neurotoxicity during spinal cord ischemia—the neuroprotective effects of adenosine]. *Jpn. J. Thorac. Cardiovasc. Surg.* **46**(4), 354–360.

Nojiri, M., Tanonaka, K., *et al.* (1999). Involvement of adenosine receptor, potassium channel and protein kinase C in hypoxic preconditioning of isolated cardiomyocytes of adult rat. *Jpn. J. Pharmacol.* **80**(1), 15–23.

Olah, M. E., and Stiles, G. L. (1992). Adenosine receptors. *Annu. Rev. Physiol.* **54**, 211–225.

Olah, M. E., and Stiles, G. L. (1995). Adenosine receptor subtypes: Characterization and therapeutic regulation. *Annu. Rev. Pharmacol. Toxicol.* **35**, 581–606.

Olah, M. E., and Stiles, G. L. (2000). The role of receptor structure in determining adenosine receptor activity. *Pharmacol. Ther.* **85**(2), 55–75.

Olah, M. E., Ren, H., *et al.* (1992). Cloning, expression, and characterization of the unique bovine A1 adenosine receptor. Studies on the ligand binding site by site-directed mutagenesis. *J. Biol. Chem.* **267**(15), 10764–10770.

Olah, M. E., Ren, H., *et al.* (1995). Adenosine receptors: Protein and gene structure. *Arch. Int. Pharmacodyn. Ther.* **329**(1), 135–150.

Ongini, E., and Fredholm, B. B. (1996). Pharmacology of adenosine A2A receptors. *Trends Pharmacol. Sci.* **17**(10), 364–372.

Palmer, T. M., and Stiles, G. L. (1995). Adenosine receptors. *Neuropharmacology* **34**(7), 683–694.

Palmer, T. M., and Stiles, G. L. (1997). Structure-function analysis of inhibitory adenosine receptor regulation. *Neuropharmacology* **36**(9), 1141–1147.

Palmer, T. M., Benovic, J. L., *et al.* (1995). Agonist-dependent phosphorylation and desensitization of the rat A3 adenosine receptor. Evidence for a G-protein-coupled receptor kinase-mediated mechanism. *J. Biol. Chem.* **270**(49), 29607–29613.

Pauwels, R. A., and Joos, G. F. (1995). Characterization of the adenosine receptors in the airways. *Arch. Int. Pharmacodyn. Ther.* **329**(1), 151–160.

Picano, E., and Abbracchio, M. P. (2000). Adenosine, the imperfect endogenous anti-ischemic cardio-neuroprotector. *Brain. Res. Bull.* **52**(2), 75–82.

Pilitsis, J. G., and Kimelberg, H. K. (1998). Adenosine receptor mediated stimulation of intracellular calcium in acutely isolated astrocytes. *Brain Res.* **798**(1–2), 294–303.

Prabhakar, U., Brooks, D. P., *et al.* (1995). Inhibition of LPS-induced TNF alpha production in human monocytes by adenosine (A2) receptor selective agonists. *Int. J. Immunopharmacol.* **17**(3), 221–224.

Pratt, A. D., Clancy, G., and Welsh, M. J. (1986). Mucosal adenosine stimulates chloride secretion in canine tracheal epithelium. *Am. J. Physiol.* **251**, C167–174.

Ramkumar, V., Pierson, G., *et al.* (1988). Adenosine receptors: Clinical implications and biochemical mechanisms. *Prog. Drug Res.* **32**, 195–247.

Resnick, M. B., Colgan, S. P., *et al.* (1993). Activated eosinophils evoke chloride secretion in model intestinal epithelia primarily via regulated release of 5'-AMP. *J. Immunol.* **151**(10), 5716–5723.

Resta, R., Hooker, S. W., *et al.* (1997). Insights into thymic purine metabolism and adenosine deaminase deficiency revealed by transgenic mice overexpressing ecto-5'-nucleotidase (CD73). *J. Clin. Invest.* **99**(4), 676–683.

Ribeiro, J. A. (1999). Adenosine A2A receptor interactions with receptors for other neurotransmitters and neuromodulators. *Eur. J. Pharmacol.* **375**(1–3), 101–113.

Richardson, P. J., Kase, H., *et al.* (1997). Adenosine A2A receptor antagonists as new agents for the treatment of Parkinson's disease. *Trends Pharmacol. Sci.* **18**(9), 338–344.

Rieger, J. M., Brown, M. L., *et al.* (2001). Design, synthesis, and evaluation of novel A2A adenosine receptor agonists. *J. Med. Chem.* **44**(4), 531–539.

Rose, F. R., Hirschhorn, R., *et al.* (1988). Adenosine promotes neutrophil chemotaxis. *J. Exp. Med.* **167**(3), 1186–1194.

Sajjadi, F. G., Boyle, D. L., *et al.* (1996). cDNA cloning and characterization of A3i, an alternatively spliced rat A3 adenosine receptor variant. *FEBS Lett.* **382**(1–2), 125–129.

Santos, P. F., Santos, M. S., *et al.* (1998). Modulation of [3H]acetylcholine release from cultured amacrine-like neurons by adenosine A1 receptors. *J. Neurochem.* **71**(3), 1086–1094.

Sattin, A., and Rall, T. W. (1970). The effect of adenosine and adenine nucleotides on the cyclic adenosine 3′,5′-phosphate content of guinea pig cerebral cortex slices. *Mol. Pharmacol.* **6**(1), 13–23.

Schwiebert, E. M., Benos, D. J., *et al.* (1999). CFTR is a conductance regulator as well as a chloride channel. *Physiol. Rev.* **79**(1 Suppl), S145–166.

Sebastiao, A. M., and Ribeiro, J. A. (1996). Adenosine A2 receptor-mediated excitatory actions on the nervous system. *Prog. Neurobiol.* **48**(3), 167–189.

Sebastiao, A. M., Macedo, M. P., *et al.* (2000). Tonic activation of A(2A) adenosine receptors unmasks, and of A(1) receptors prevents, a facilitatory action of calcitonin gene-related peptide in the rat hippocampus. *Br. J. Pharmacol.* **129**(2), 374–380.

Short, D. B., Trotter, K. W., *et al.* (1998). An apical PDZ protein anchors the cystic fibrosis transmembrane conductance regulator to the cytoskeleton. *J. Biol. Chem.* **273**, 19797–19801.

Sitaraman, S. V., Si-Tahar, M., *et al.* (2000). Polarity of A2b adenosine receptor expression determines characteristics of receptor desensitization. *Am. J. Physiol. Cell Physiol.* **278**(6), C1230–1236.

Strohmeier, G. R., Reppert, S. M., *et al.* (1995). The A2b adenosine receptor mediates cAMP responses to adenosine receptor agonists in human intestinal epithelia. *J. Biol. Chem.* **270**(5), 2387–2394.

Svenningsson, P., Le Moine, C., *et al.* (1999). Distribution, biochemistry and function of striatal adenosine A2A receptors. *Prog. Neurobiol.* **59**(4), 355–396.

Sweeney, M. I. (1997). Neuroprotective effects of adenosine in cerebral ischemia: Window of opportunity. *Neurosci. Biobehav. Rev.* **21**(2), 207–217.

Szabo, C., Scott, G. S., *et al.* (1998). Suppression of macrophage inflammatory protein (MIP)-1alpha production and collagen-induced arthritis by adenosine receptor agonists. *Br. J. Pharmacol.* **125**(2), 379–387.

Takano, H., Bolli, R., *et al.* (2001). A(1) or A(3) adenosine receptors induce late preconditioning against infarction in conscious rabbits by different mechanisms. *Circ. Res.* **88**(5), 520–528.

Thourani, V. H., Nakamura, M., *et al.* (1999). Adenosine A(3)-receptor stimulation attenuates postischemic dysfunction through K(ATP) channels. *Am. J. Physiol.* **277**(1 pt. 2), H228–235.

Tracey, W. R., Magee, W., *et al.* (1998). Selective activation of adenosine A3 receptors with N6-(3-chlorobenzyl)-5′-N-methylcarboxamidoadenosine (CB-MECA) provides cardioprotection via KATP channel activation. *Cardiovasc. Res.* **40**(1), 138–145.

van Calker, D., Muller, M., *et al.* (1979). Adenosine regulates via two different types of receptors, the accumulation of cyclic AMP in cultured brain cells. *J. Neurochem.* **33**(5), 999–1005.

van Muijlwijk-Koezen, J. E., Timmerman, H., *et al.* (1998a). A novel class of adenosine A3 receptor ligands. 1. 3-(2-Pyridinyl)isoquinoline derivatives. *J. Med. Chem.* **41**(21), 3987–3993.

van Muijlwijk-Koezen, J. E., Timmerman, H., *et al.* (1998b). A novel class of adenosine A3 receptor ligands. 2. Structure affinity profile of a series of isoquinoline and quinazoline compounds. *J. Med. Chem.* **41**(21), 3994–4000.

van Tilburg, E. W., von Frijtag Drabbe Kunzel, J., *et al.* (1999). N6,5′-Disubstituted adenosine derivatives as partial agonists for the human adenosine A3 receptor. *J. Med. Chem.* **42**(8), 1393–1400.

Varani, K., Cacciari, B., *et al.* (1998). Binding affinity of adenosine receptor agonists and antagonists at human cloned A3 adenosine receptors. *Life Sci.* **63**(5), 81–87.

Von Lubitz, D. K., Lin, R. C., *et al.* (1999). Chronic administration of adenosine A3 receptor agonist and cerebral ischemia: Neuronal and glial effects. *Eur. J. Pharmacol.* **367**(2–3), 157–163.

Walker, E. M., Bispham, J. R., *et al.* (1998). Nonselective effects of the putative phospholipase C inhibitor, U73122, on adenosine A1 receptor-mediated signal transduction events in Chinese hamster ovary cells. *Biochem. Pharmacol.* **56**(11), 1455–1462.

Wang, S., Raab, R. W., *et al.* (1998). Peptide binding consensus of the NHE-RF-PDZ1 domain matches the C-terminal sequence of cystic fibrosis transmembrane conductance regulator (CFTR). *FEBS Lett.* **427**(1), 103–108.

Wang, S., Yue, H., *et al.* (2000). Accessory protein facilitated CFTR-CFTR interaction, a molecular mechanism to potentiate the chloride channel activity. *Cell* **103**(1), 169–179.

Williams, M. (1987). Purine receptors in mammalian tissues: Pharmacology and functional significance. *Annu. Rev. Pharmacol. Toxicol.* **27**, 315–345.

Wine, J. J. (1999). The genesis of cystic fibrosis lung disease. *J. Clin. Invest.* **103**(3), 309–312.

Wood, J. A., Wood, P. L., *et al.* (1993). Cytokine indices in Alzheimer's temporal cortex: No changes in mature IL-1 beta or IL-1RA but increases in the associated acute phase proteins IL-6, alpha 2-macroglobulin and C-reactive protein. *Brain Res.* **629**(2), 245–252.

Woodhouse, E. C., Amanatullah, D. F., *et al.* (1998). Adenosine receptor mediates motility in human melanoma cells. *Biochem. Biophys. Res. Commun.* **246**(3), 888–894.

Xaus, J., Mirabet, M., *et al.* (1999). IFN-gamma up-regulates the A2B adenosine receptor expression in macrophages: A mechanism of macrophage deactivation. *J. Immunol.* **162**(6), 3607–3614.

Yakel, J. L., Warren, R. A., *et al.* (1993). Functional expression of adenosine A2b receptor in Xenopus oocytes. *Mol. Pharmacol.* **43**(2), 277–280.

Zahniser, N. R., Simosky, J. K., *et al.* (2000). Functional uncoupling of adenosine A(2A) receptors and reduced response to caffeine in mice lacking dopamine D2 receptors. *J. Neurosci.* **20**(16), 5949–5957.

Zhao, Z. Q., Todd, J. C., *et al.* (1997). Adenosine inhibition of neutrophil damage during reperfusion does not involve K(ATP)-channel activation. *Am. J. Physiol.* **273**(4 pt. 2), H1677–1687.

Zhou, Q. Y., Li, C., *et al.* (1992). Molecular cloning and characterization of an adenosine receptor: The A3 adenosine receptor. *Proc. Natl. Acad. Sci. USA* **89**(16), 7432–7436.

# CHAPTER 6

# Relating the Structure of ATP-Gated Ion Channel Receptors to Their Function

**Terrance M. Egan, Keisuke Migita, and Mark M. Voigt**
Department of Pharmacological and Physiological Science, St. Louis University School of
Medicine, St. Louis, Missouri 63104

## I. INTRODUCTION

P2X receptors are ion channels that are opened and closed by extracellular adenosine triphosphate (ATP). All of the components necessary to ensure ion flux are contained in a single oligomeric assembly including the receptor for extracellular ATP and the ion channel pore. Functional receptors are formed from homo- or heteromeric combinations of individual family members that, at least in some cases, impart unique properties to the ATP response (Fig. 1). In this chapter, we review the

*Current Topics in Membranes, Volume 54*

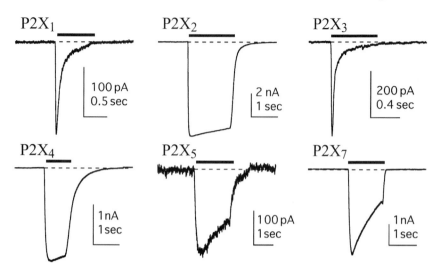

**FIGURE 1** ATP-gated current through homomeric receptors. Currents through six homomeric P2X receptors. In all cases, the solid line indicates the period of application of 30 $\mu M$ ATP to HEK-293 cells expressing one of the seven P2X receptor subunits. Each homomeric receptor has a distinct phenotype that is characterized by the size of the current, the rate of desensitization, and the sensitivity to an array of agonists (Khakh *et al.*, 2001). In our hands, cells expressing the P2X$_6$ receptor do not respond to ATP. The holding potential was $-40$ mV for experiments in all cases except that of the P2X$_5$ receptor. P2X$_5$ receptors display much smaller peak currents than do other family members and, therefore, to get a good picture of ATP-gated current through this channel, we increased the driving force for cation current across the membrane by holding the cell at $-80$ mV. The increase in baseline noise seen in this trace is a function of activation of an unknown ion channel at potentials equal to, or more negative than, about $-80$ mV in these cells. T. M. Egan, W. R. Haines, and M. M. Voigt (unpublished data).

current literature that deals with the relationship of structure to function for the P2X receptor family. By and large, the functional domains of these receptors have been identified using conventional mutagenesis to alter the phenotypic response of individual family members. Although P2X receptors bear little structural similarity to other ligand-gated ion channels, they do adhere to the general principles governing ion conduction through most transmembrane channels. Thus, like all ionotropic receptors, they assemble in a selective fashion, show a strong preference for some ions over others, and display equilibrium between the open and closed states of the channel that is modulated by selective agonists. A major assumption of the reports summarized in this chapter is that a mutation that alters the quality of one of these principles resides in a structural domain linked to a particular receptor–channel function. Although this assumption may hold true, no current structure–function study of P2X receptors can state without reservation that it conclusively defines a functional domain. The strengths and weaknesses of these studies are discussed below.

## II. SUBUNIT TOPOLOGY

The first clue that P2X receptors were different from other ionotropic receptors came from analysis of the primary sequences of the proteins encoded by the two original cloned cDNAs (Brake *et al.*, 1994; Valera *et al.*, 1994). Whereas the other classes of ionotropic receptors are made of proteins containing three or more putative transmembrane domains (Hollmann *et al.*, 1994; Methot *et al.*, 1994; Mukerji *et al.*, 1996), hydropathy plots of the P2X receptors predict that these proteins contain only two hydrophobic segments that are long enough to transverse the lipid membrane. Further, the deduced amino acid sequences lack canonical signal peptide sequences at their N-termini, leading to the suggestion that the N- and C-termini are positioned inside the cell and that the region between the two transmembrane domains are extracellular (Brake *et al.*, 1994) (Fig. 2). In support of that premise is the presence of multiple motifs for N-linked glycosylation in the putative extracellular domain. Experimental evidence from our laboratories (Torres *et al.*, 1998) and others (Newbolt *et al.*, 1998) verifies this hypothetical model of P2X receptors.

The two-transmembrane topology of the P2X receptor resembles that of another ATP-sensitive channel, the inwardly rectifying $K^+$ channel originally cloned by Ho *et al.* (1993). However, there is virtually no sequence homology between the two families. The ATP-sensitive $K^+$ channel is modulated by *intracellular* ATP and has a relatively small extracellular domain that lacks the multiple conserved cysteine residues found in the P2X receptor proteins. Further, the $K^+$ channel shares a significant sequence homology with non-ATP-sensitive $K^+$ channels, and this conservation is not found in P2X receptors (Ho *et al.*, 1993). P2X receptor topology is also quite different from other neurotransmitter-gated cation channels such as nicotinic cholinergic, serotonergic, and glutamatergic receptors (Barnard, 1992). In light of these facts, it would appear that the P2X receptors represent a novel structure for neurotransmitter-gated channels.

Although the P2X receptor clones might represent a new architecture for ionotropic receptors, they do share structural similarities with groups of other cloned ion channels (North, 1996). These proteins include four families of genes, one encoding a receptor for the peptide FMRFamide in

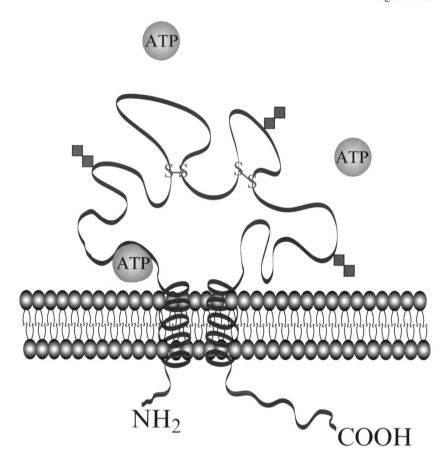

**FIGURE 2**  Subunit topology. The predicted topology of an individual subunit in the membrane is shown. Each subunit contains two transmembrane domains that link a large ectodomain. The N- and C-termini are intracellular.

snail (Lingueglia *et al.*, 1995), another encoding mechanosensitive channel subunits in *Caenorhabditis elegans* (Huang and Chalfie, 1994), the third encoding subunits of an amiloride-sensitive epithelial $Na^+$ channel (ENaC) (Canessa *et al.*, 1993), and the last the proton-gated channels (Waldmann *et al.*, 1997). In all cases, these proteins are thought to have two transmembrane domains and a large extracellular domain containing multiple conserved cysteine residues. However, comparison of the P2X primary sequences with those of the aforementioned channel proteins reveals no appreciable sequence identity.

## III. COMPOSITION AND STOICHIOMETRY OF RECEPTOR–CHANNEL COMPLEXES

Most ion channels are oligomeric assemblies. The stoichiometries of a number of ligand- and voltage-gated channels have been determined; these are usually found to encompass four or five subunits (Devillers-Thiery *et al.*, 1993; Yang *et al.*, 1995; Laube *et al.*, 1998). The stoichiometry of the PX receptor has not been firmly established. Nevertheless, initial studies that used immunoprecipitation of cross-linked $P2X_1$ receptors suggest that an assembly contains either three subunits or a multiple thereof (Nicke *et al.*, 1998). This hypothesis is supported by functional analysis of concatameric receptors (Stoop *et al.*, 1999), although the answer awaits the definitive resolution of the atomic structure of a P2X receptor, which, hopefully, will be forthcoming.

Individual receptor complexes are formed from homomeric or heteromeric assemblies of individual members of the family. Assembly occurs in a selective fashion with each subunit displaying a unique preference for some but not all members of the family. These patterns of assembly are discussed in greater detail below.

## IV. TRANSMEMBRANE DOMAINS

### A. The Wall of the Pore

P2X receptors exist to transduce a change in extracellular ATP concentration into a change in membrane potential. To do this, a part of the receptor must form a water-filled pore that allows cations to overcome the energy barrier to transmembrane ion flux provided by the lipid membrane. It is reasonable to assume that the transmembrane segments of the individual subunits form the conduction pathway. If, as discussed above, the oligomeric P2X receptor is composed of three subunits, then the protein has up to six transmembrane domains available to line the pore. Although it is possible that all six subunits contribute equally, it is also possible that the pore is formed from a subset of transmembrane domains or that the contributions of individual domains change as the channel goes through the gating cycle. Two different groups attempted to identify pore domains of $P2X_2$ receptor using the substituted cysteine accessibility method (SCAM) popularized by Karlin and Akabas (1998). Both groups performed scanning cysteine mutagenesis of the first (TM1) and second (TM2) transmembrane segments to make an array of mutant $P2X_2$ receptors that was probed using electrophysiology (Rassendren *et al.*, 1997; Egan *et al.*, 1998; Haines *et al.*,

2001b; Jiang *et al.*, 2001). ATP-gated current was measured before, during, and after modification of the engineered thiolates by water-soluble, sulfhydryl-reactive reagents, and a change in the size of the current was taken as an indication that the modified side chain of the substituted cysteine faced into the water-filled pore. Further, the groups assumed that a regular pattern of reactivity could be used to define the secondary structure of the pore lining domains. For example, an alternating pattern of reactivity in which every third or fourth residue reacts with reagent suggests that the pore is lined by transmembrane $\alpha$-helices (Karlin and Akabas, 1998). A key finding of both groups is that both the first (Haines *et al.*, 2001b; Jiang *et al.*, 2001) and second (Rassendren *et al.*, 1997; Egan *et al.*, 1998) transmembrane domains line the pore. However, some disagreement exists regarding the contribution of individual residues to the lining; these differences may reflect reagents and protocols unique to each laboratory. The disparity is greatest in data of TM1, where there is little overlap in the subsets of reactive residues identified in the different datasets (Haines *et al.*, 2001b; Jiang *et al.*, 2001). Many of the reactive residues of TM1 were positioned near the innermost and outermost aspects of the conduction pathway where the pore diameter is expected to be relatively large. These regions may be more closely associated with transduction of the ATP signal than with channel properties usually associated with narrower parts of the pore, for example, ion selectivity. Additional experiments are needed to clarify the role of this domain in channel function.

The picture is somewhat clearer for the second transmembrane domain. Here, both groups identified a number of pore-lining residues throughout TM2 that seem to lie on one or the other side of the channel gate (Rassendren *et al.*, 1997; Egan *et al.*, 1998). A regular pattern of reactivity that suggests an $\alpha$-helix is seen when MTSEA, a sulfhydryl-reactive methanethiosulfonate derivative, is applied in the absence of ATP (i.e., when the channel was closed) (Egan *et al.*, 1998) (Fig. 3). $\alpha$-Helical transmembrane domains are a common feature of ion channel proteins (Hille, 2001), and it would not be surprising if TM2 adopted this secondary structure. However, other experiments suggest that the helix is not static during gating. Egan *et al.* (1998) coapplied ATP and $Ag^+$ to label residues as the channel moved between the open and closed states. $Ag^+$ resembles $K^+$ in atomic radius and dehydration energy and therefore might be expected to interact with pore-lining residues in a manner approximating that of the physiologically relevant monovalent cations $Na^+$ and $K^+$ (Lu and Miller, 1995). It also covalently modifies hydrated thiolates. Coapplication of ATP and $Ag^+$ irreversibly modified current through 15 of 27 TM2 mutants (see Fig. 3). The most profound effects were seen in the stretch of amino acids spanning T336 to L347 where all but 2 of the 12 functional

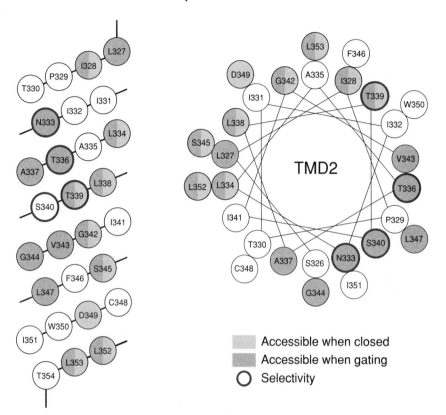

**FIGURE 3** Helical models of the second transmembrane domain of the $P2X_2$ receptor. Data come from our laboratories (Egan *et al.*, 1998; Migita *et al.*, 2001). The residues shaded green and yellow were identified using scanning cysteine mutagenesis. Green residues react to coapplication of $Ag^+$ and ATP with an increase or decrease in current amplitude. Yellow-shaded residues react to MTSEA in the absence of ATP. The finding that all of the residues modified in the close state (yellow residues) fall on one side of a helical model of TM2 suggests that this domain crosses the membrane as a helix. A less defined pattern is apparent when the channel is gating (green residues). The residues circled in red are involved in ion selectivity.

cysteine-substituted mutants had altered currents after application of $Ag^+$. This result seems to rule out the possibility that TM2 forms an immovable $\alpha$-helix or $\beta$-sheet during gating because both of these secondary structures would be expected to produce a more periodic pattern of reactivity (Karlin and Akabas, 1998). Many of the same residues (I328, N333, T336, L338, and D349) that react with $Ag^+$ are also modified by coapplication of ATP and MTSEA (Rassendren *et al.*, 1997). The reaction at T336 is particularly interesting because MTSEA blocked outward current faster than inward current, a result that is consistent with T336 lying in the ion-conducting pathway (Rassendren *et al.*, 1997). Further, a narrow constriction of the channel pore that impedes current flow in the absence of ATP (i.e., the channel gate) must lie on the extracellular side of D349 because this residue is accessible to MTSEA only when the channel is gating (Rassendren *et al.*, 1997; Egan *et al.*, 1998). The exact location of the channel gate is unknown. The gate is discussed in more detail below.

## B. The Selectivity Filter

Although a few native receptors show a measurable permeability to chloride (Thomas and Hume, 1990; Balachandran and Bennett, 1996), most P2X receptors are cation nonselective channels that are freely permeable to $Ca^{2+}$ (Bean, 1992). These receptors lack a common amino acid signature sequence analogous to the pore-loop domain of $K^+$-selective ion channels (MacKinnon, 1995) and there is no pattern of acidic amino acid repeats in any segment of any member of the P2X receptor family. This lack of a common structural motif has made identification of the selectivity filter difficult, although empirical data predict that such a binding site does exist. The $P2X_2$ receptor has a relative monovalent conductance sequence of $K^+ >$ $Rb^+ > Cs^+ > Na^+ > Li^+$ that corresponds to a Type IV Eisenman equilibrium ion-exchange sequence (Ding and Sachs, 1999a,b); this sequence is different from the relative mobility of the ions in water suggesting that cations interact with a anionic-binding site of low to moderate field strength located somewhere in the permeation pathway (Eisenman and Horn, 1983; Eisenman and Dani, 1987). This idea is supported by data that show that the inward flux of $Na^+$ through the $P2X_2$ receptor–channel saturates at high $[Na^+]_o$ as expected for a process involving binding and unbinding steps, and these data are well fit by a Michaelis–Menten model incorporating a single binding site located about 20% of the way through the transmembrane electric field (Ding and Sachs, 1999a,b). The binding site may be a fairly narrow part of the channel where the wall of the pore acts as a surrogate counterion to help the permeating ions shed water. The diameter of the narrowest region of the pore of homomeric $P2X_2$ receptors is estimated to be about 0.8 nm under near normal physiological conditions (e.g., extracellular $[Ca^{2+}]_o \geq 1$ m$M$) (Evans *et al.*, 1996). This size is large enough to permit small monovalent and divalent cations to pass through the channel but it is too small to permit appreciable conduction of larger cations such as *N*-methyl-D-glutamine$^+$ (NMDG$^+$) (Hille, 2001). However, the size of the channel pore increases dramatically during the course of a sustained application of ATP and low extracellular $Ca^{2+}$ to the point where large cations (including the 450-kDa cationic fluorescent dye YO-PRO-1) readily permeate the channel (Khakh *et al.*, 1999; Virginio *et al.*, 1999a). The molecular mechanism underlying this change in selectivity is unknown (Khakh and Lester, 1999). In some cases, pore dilation results in cell death (Virgilio *et al.*, 1998; Virginio *et al.*, 1999b), an effect that is absent in truncated receptors missing a part of the C-terminus (Surprenant *et al.*, 1996).

As mentioned above, the lack of a conserved amino sequence in the transmembrane segments makes it difficult to predict how P2X receptors attain cation selectivity. All family members contain two acidic amino acids (D315 and D349 of the $P2X_2$ receptor) positioned near the inner and outer extremes of the conduction pathway. Although it makes sense that the negative charge provided by these carboxylates could influence conduction by concentrating cations in the vicinity of the pore, neutralizing the charge does not affect relative calcium permeability (Migita *et al.*, 2001) or, at least in the case of a D349V mutant, single channel conductance (Nakazawa *et al.*, 1998). This is a somewhat surprising result in that the conservation of these charges suggests that the residues play important role(s) in P2X receptor–channel function. These roles may include regulation of agonist selectivity (Nakazawa *et al.*, 1998) and desensitization (Zhou *et al.*, 1998). All P2X receptors also contain a polar amino acid (N, Y, T, S) at a position equivalent to N333 of the P2X2 receptor. Single-channel studies show that mutating $P2X_2$–N333 to isoleucine reduces single-channel conductance (Nakazawa *et al.*, 1998) and relative $Ca^{2+}$ permeability (Migita *et al.*, 2001). However, mutating N333 to Q, Y, C did not change $Ca^{2+}$ permeability and the role that this residue makes in regulating ion conductance must still be determined. By contrast, site-directed mutagenesis of nonconserved polar residues dramatically alters the cation selectivity sequence of the $P2X_2$ receptor (Migita *et al.*, 2001). The effect is largest for the T339Y and S340Y mutations that reduced $P_{Ca}/P_{Cs}$ to less than 10% of control and shifted the permeability sequence of the alkali metal cations toward their relative mobility in water. All P2X receptors contain polar amino acids in their second transmembrane domains, although these residues are not positionally conserved. The variability in sequence may help to explain the range of relative $Ca^{2+}$ permeabilities reported for different members of the P2X receptor family (Khakh *et al.*, 2001). More experiments are needed to clarify the role of polar residues in regulating ion selectivity.

## C. The Gate

Agonist binding results in a conformational change that leads to ion flux through the aqueous pore. G342C is conserved in all members of the P2X receptor family and may play a critical role in channel function. Scanning cysteine mutagenesis suggests that it is accessible from either side of the membrane and therefore may form a part of the channel gate (Egan *et al.*, 1998). This hypothesis is supported by recent data using site-directed mutagenesis of the $P2X_4$ isoform. Khakh *et al.* (1999b; Khakh and Lester, 1999) showed that specific mutations at a position of the $P2X_4$ isoform corresponding to G342 of the $P2X_2$ receptor lead to changes in the ratio of the sizes of the currents through the normal-sized (carried by $Na^+$) and dilated (carried by $NMDG^+$) pores. They suggest that $P2X_4$–G347 imparts flexibility to a domain of the protein that may be the gate, and that mutations perturb the protein conformations that occur during gating. A more systematic investigation of the region at and around G342 is needed to fully understand the role this segment plays in channel gating.

## D. Assembly Domains

As discussed earlier, it is thought that P2X receptors must be oligomeric in nature to form functional ion-permeable channels. The fact that recombinant expression of single subunit cDNAs results in functional receptors demonstrates that these proteins are capable of forming homooligomers. Analysis of P2X heterooligomer formation has been carried out using coimmunoprecipitation assays of recombinantly expressed subunits (Torres *et al.*, 1999a), with the finding that most (five of the seven) subunits can form both homo- and heterooligomers. The outliers are the $P2X_6$ (which does not appear to form homooligomers) and $P2X_7$ (which does not appear to form heterooligomers). Of course, the assembly properties of any individual subunit may depend on the context of the cell type in which it is expressed. It is not known yet if all P2X receptor subunits participate in forming heteromeric assemblies *in vivo* or if some/most exist solely as homomeric channels. Evidence that two P2X subunit proteins that are coexpressed in a subpopulation of sensory ganglion neurons, $P2X_2$ and $P2X_3$, do coassemble and in doing so form a new phenotype that mimics a native receptor present on those neurons (Lewis *et al.*, 1995) supports the contention that heterooligomeric P2X receptors also form *in vivo*.

Biochemical and functional studies (Lewis *et al.*, 1995; Haines *et al.*, 1999; Torres *et al.*, 1999b) have demonstrated that the assembly of P2X subunits into oligomeric complexes appears to occur in a directed and specific fashion. That is, only certain combinations of subunits appear to be able to coassemble. This implies the existence of domains that confer selectivity for productive subunit interactions. Knowledge of this process may allow the development of techniques or protocols that can be used to disrupt assembly of particular heteromeric receptors (without altering levels of the particular homomers) and thus aid in clarifying the functional roles that those heteromers play in native cells and tissues. After all, particular combinations of subunits might yield unique context-specific functions based on cellular domain targeting or modification of intracellular or intrinsic regulatory processes that would not be detectable in a recombinant setting. Investigations into the assembly of $K^+$ channels (both inwardly rectifying and voltage-gated subtypes) (Xu *et al.*, 1995; Woodward *et al.*, 1997) and nicotinic acetylcholine receptors (nAChRs) (Green, 1999) lend support for the existence of assembly-directing domains in the extramembranous regions of subunit proteins. However, involvement of channel-forming protein transmembrane domains in subunit assembly has also been reported (Tucker *et al.*, 1996; Koster *et al.*, 1998). Previous work on potassium channels and nAChRs showed that the extracellular N-terminal regions of these channels carried critical motifs responsible for oligomeric assembly. As the P2X subunits have intracellular N-termini, it was not clear if they would assemble using the same rules. A first step into elucidating the mechanisms underlying P2X receptor assembly has been the use of deletional mutagenesis. The rationale behind these experiments was that if particular stretches of amino acids can be deleted without affecting function, they must not play a critical role in receptor assembly. We have shown that neither intracellular terminus of the $P2X_2$ subunit is critical for either homo- or heteromeric assembly (Torres *et al.*, 1999b). A construct in which the N-terminus was deleted to beyond the first transmembrane segment (TM1) and that included a signal peptide sequence from the nicotinic $\alpha_7$-subunit to maintain appropriate topology (as shown by glycosylation) also resulted in assembly-competent proteins. This suggested that it was either the extracellular domain or TM2 that is critical for assembly.

A second deletion mutant ($\Delta C304$) that terminated about 15 amino acids before TM2 did not assemble with itself or with full-length subunits, implying that TM2, perhaps together with an adjacent stretch of extracellular domain, plays a critical role in assembly. Although its glycosylation demonstrates that it has the correct topology, $\Delta C304$ could be assembly incompetent due to misfolding rather than lacking an assembly domain. Regardless, the deletion studies ruled out the N- and C-termini of the protein, and implicated the TM2 in both homo- and heteromeric assembly of P2X receptors. To test whether TM2 was indeed the sole determinant for these subunit-specific properties, we substituted the TM2 in $P2X_2$ with that of $P2X_6$ (construct 2.6). This chimera yielded a functional homomeric receptor, demonstrating that the lack of $P2X_6$ self-assembly was not due to a defective TM2. However, this chimera could not assemble with $P2X_6$. These findings strongly suggest that there must be another domain, besides TM2, that plays a role in assembly. Interestingly, the converse chimera (TM2 in $P2X_6$ replaced with that of $P2X_2$) does not assemble, also supporting involvement of an additional domain in assembly (data not shown). Heteromeric assembly of 2.6 with $P2X_2$, but not with $P2X_3$ or $P2X_6$ is in keeping with the wild-type $P2X_6$ phenotype and implies that TM2 does play a major role in directing selectivity of assembly. However, the function of an auxiliary domain(s) involved in assembly may also be important in the formation of heteromeric receptors as well. This is a possibility that needs to be addressed.

We propose that the first TM domain is an auxiliary domain involved in assembly. This is based on two sets of results. The first finding is that TM1 must lie in the membrane very near to TM2 as it contributes amino acids that are in, or close to, the pore-forming region of the protein (Haines *et al.*, 2001b), which we already know is partly formed by TM2 (Rassendren *et al.*, 1997; Egan *et al.*, 1998). Second, we identified five TM1 residues that when mutated to cysteine did not make functional receptors, and four of these did not make it to the cell surface or coassemble with $P2X_2$. HA (data not shown). This is in sharp contrast to results from the TM2 SCAM study, in which only two mutated residues resulted in nonfunctional receptors (S340C and D349C) and both could still coassemble with wild-type $P2X_2$. Another logical choice is the extracellular region immediately adjacent to TM2, as its tertiary structure may be influenced by that of TM2. Such a possibility is entirely consonant with the results from the $P2X_{1/3}$ chimeras presented earlier. Investigation of this domain should be a focus of future experiments.

## V. EXTRACELLULAR DOMAINS

ATP activates P2X receptor channels in a process that involves at least three separate but related steps: (1) agonist occupation of one or more binding domains, (2) transduction of the signal from the binding site to the channel gate, and (3) gating (i.e., opening and closing of the channel pore). Additional steps that modulate binding–gating–transduction are also possible, and these steps may involve functional domains that are different than those that underlie the base process itself. With the possible exception of the channel gate, all of the functional domains involved in activation of P2X receptors probably include a component of the extracellular loop. This is certainly true for sites that bind agonists (such as ATP and $\alpha,\beta$-MeATP) and antagonists [such as suramin and pyridoxal-phosphate-6-azophenyl-2′, 4′-disulfonate (PPADS)] because these reagents must be applied to the outside of a cell to be affective. It also seems likely that the loci underlying allosteric modulation of P2X receptors by $H^+$, $Ca^{2+}$, and $Zn^+$ are contained in an ectodomain as ATP-gated currents are sensitive to changes in extracellular concentrations of these ions.

## A. Agonist-Binding Site

The fact that P2X receptors are oligomers that respond to ATP in an allosteric fashion has led to the common assumption that each protomer contains a single ATP-binding site. Although this may be true, it is also possible that ATP binds at the interface of adjoining subunits; such a case is not without precedence as the acetylcholine-binding site of nicotinic receptors is interfacial (Karlin, 2002). Regardless of the exact location, two different lines of investigation support the idea that the receptor–channel complex contains more than one binding site. First, ATP concentration–response curves of homomeric receptors are best fit by the Hill equation when the slope is set to greater than 1 (North and Surprenant, 2000). Second, kinetic analysis of single-channel currents suggests that sequential binding of ATP shows positive cooperativity and that partially liganded channels do not open (Ding and Sachs, 1999a). Attempts to locate these binding sites are hampered by the lack of sequence homology between P2X receptors and other known ATP-binding proteins. However, all P2X receptors contain loci of conserved positive charge in the extracellular loop that may bind ATP by attracting the negative charge of the phosphate groups. This hypothesis was tested by measuring the effects of substituting neutral alanine for charged amino acids in the ectodomains of the P2X1 (Ennion et al., 2000) and $P2X_2$ receptors (Jiang et al., 2000). In both cases, the greatest changes occurred when the substitutions were made near the junctions of the extracellular loop and the transmembrane domains. Both receptors contain two positive charges near TM1 (K68 and K70 of P2X1; K69 and K71 of $P2X_2$) and three positive charges near TM2 (R292, R305, and K309 of P2X1; R290 R304, and K308 of $P2X_2$) that respond to alanine substitutions with altered receptor potency to ATP. For example, ATP is 1000-fold less effective in evoking membrane current through $P2X_1$–K309A receptors (Ennion et al., 2000) and is entirely ineffective at $P2X_2$–K308A receptors despite near-normal levels of receptor expression in the membrane (Jiang et al., 2000). The same loss of function is found in the human P2X7 receptor engineered to contain an alanine at a position analogous to $P2X_2$–K308 (Worthington et al., 2002).

One problem with these types of studies is that it is hard to directly equate a change in agonist potency with an effect on ATP binding; that is, it is equally likely that the mutation altered the ability of the channel to open in response to a normal amount of bound ATP (see Colquhoun, 1998). Jiang *et al.* (2000) addressed this problem by measuring the ability of a high concentration of ATP to protect cysteine-substituted residues within the putative ATP-binding site near TM1 from attack by water-soluble sulfhydryl reagents. Fifteen unique mutants were made, each containing a single cysteine substitution in the conserved domain spanning P2X$_2$–D57 to P2X$_2$–K71. Two of these (P2X$_2$–K69C and P2X$_2$–K71C) failed to respond to a 30 $\mu M$ ATP. Of the 13 remaining mutants that had a normal response, two (S65 and I67) showed a reduction in the size of the ATP-gated current after modification of the engineered cysteines by methanethiosulfonate reagents that added bulk to the —SH side chain. Of particular importance is the finding that the roughly 80% inhibition of the ATP-gated current through P2X$_2$–I67C by (5-sulfonatopentyl) methanethiosulfonate is prevented when this reagent is applied in the presence of a high concentration of ATP (100 $\mu M$). The ability of the agonist to protect I67C from modification suggests that this residue must lie in the binding site because the modifying reagent cannot gain access to the cysteine when the site is saturated with ATP.


## B. Antagonist-Binding Sites

Relatively little is known about where antagonists act to block the effects of ATP. Pyridoxal-5-phosphate (P5P) and its derivative PPADS cause a noncompetitive inhibition of some (P2X$_1$, P2X$_2$, P2X$_3$, and P2X$_5$) but not all (P2X$_4$) rat P2X receptors. In some cases (P2X$_1$, P2X$_2$, and P2X$_5$), the block develops slowly and is only partially reversible (Collo *et al.*, 1996). The lack of a complete recovery may reflect the formation of a Schiff base between the aldehyde group of the antagonist and a lysine in the extracellular domain of the receptor. Buell *et al.* (1996) tested this hypothesis by mutating the fixed positive charge of a lysine (K246 of the P2X$_2$ receptor) conserved in all three slowly reversing receptors to the negative charge of glutamate. They found that PPADS retained its ability to block the response of ATP but that the antagonism now occurred in a reversible fashion. The results suggest that PPADS binds to at least two sites in the extracellular domain. The first site may be involved in the initial binding of the antagonist and does not require interaction with the conserved lysine. This hypothesis is supported by the fact that rat P2X$_3$ (Lewis *et al.*, 1995) and human P2X$_4$ (Garcia-Guzman *et al.*, 1997) receptors are readily blocked by low micromolar concentrations of PPADS despite the fact that they lack the positionally conserved lysine. The second site centers on the conserved cysteine and imparts the long-lasting effect of aldehyde antagonists on the appropriate receptors.

## C. The Binding Site(s) for Allosteric Modulators

P2X receptors are differentially modulated by pH (Li *et al.*, 1996b; Stoop *et al.*, 1997; North and Surprenant, 2000), $Zn^{2+}$ (Cloues *et al.*, 1993; Li *et al.*, 1993), $Cd^{2+}$ (Acuna-Castillo *et al.*, 2000), $Cu^{2+}$ (Li *et al.*, 1996a), $Ag^+$ (Haines *et al.*, 2001a), $Hg^+$ (Acuna-Castillo *et al.*, 2000), $Ni^+$ (Ueno *et al.*, 2001), monovalent cations (Michel *et al.*, 1997), cibacron blue (Miller *et al.*, 1998), and ivermectin (Khakh *et al.*, 1999b). In most cases, the sites of action are unknown, although considerable attention has focused on the role of histidine in the effect of acidification. Histidine seems a likely candidate to constitute the proton-binding site based on the fact that its $pK_a$ is a good match to the apparent $pK_a$ of the pH titration curves for $P2X_2$ and $P2X_4$ receptors (Stoop *et al.*, 1997). Acidification inhibits ATP-gated current through rat (Stoop *et al.*, 1997) and human $P2X_4$ receptors (Clarke *et al.*, 2000). The human $P2X_4$ receptor contains four histidines, and mutating of one (H286) of these to alanine eliminates the pH sensitivity (Clarke *et al.*, 2000). A similar scenario occurs in rat $P2X_2$ receptors where mutation of only one (H319) of the nine histidines in the extracellular loop prevents the potentiation of ATP-gated current by acidification (Clyne *et al.*, 2002). This mutation did not eliminate the $Zn^{2+}$-induced potentiation of ATP-gated current. Rather, the effect of $Zn^{2+}$ was absent in two other mutants (H120A and H213A) suggesting that despite similar modes of action, $Zn^{2+}$ and pH work at two different sites (Clyne *et al.*, 2002).

## VI. INTRACELLULAR DOMAINS

The N- and C-terminus of any given P2X subunit reside intracellularly. All seven mammalian subunit proteins have relatively short N-termini of around 25 amino acids. The only obvious segment of conservation among these regions is a short domain (Y-D/E-T-X-K/R) that is an invariant nine amino acids upstream of the Gly residue that is usually taken as the beginning of TM1. Within this conserved region is a concensus motif for protein kinase C phosphorylation (T-X-K/R). Unpublished studies from our groups have found that $^{32}P$ is incorporated constitutively from radiolabeled $PP_i$ into the $P2X_2$ receptor but is not incorporated into a mutant $P2X_2$ receptor containing an alanine substitution at T18, lending support to this residue as a target for phosphorylation. Further, studies from Seguela's (Boue-Grabot *et al.*, 2000) and Evans' (Ennion and Evans, 2002) groups have shown that modification of the putative phosphate-accepting Thr residue into an Ala residue alters the desensitization kinetics of the mutant homooligomeric P2X receptor. Therefore, it has been proposed that some aspects of the receptor kinetics are under the influence of the phosphorylation state of this residue. Future studies are needed to determine what the mechanism(s) are by which phosphorylation of T18 can alter the kinetics of receptor function.

The C-terminus is the region of greatest diversity among the subunits, in that it ranges from around 30 amino acids in $P2X_6$ to about 215 residues in $P2X_7$. Our group (Torres *et al.*, 1999) and others (Werner *et al.*, 1996; Brandle *et al.*, 1997) have shown that this region of $P2X_2$ has an effect on the kinetics of receptor desensitization, as truncation or alternative splicing of the C-terminus results in a homooligomeric receptor that desensitized much more rapidly than the wild-type. Interestingly, there have been reports that a point mutation in the initial portion of the C-terminus of $P2X_7$ results in nonfunctional receptors—a finding that again may reflect alterations in channel kinetics (Gu *et al.*, 2001). This is the same region thought to be involved in formation of the superpore as discussed earlier.

## VII. FUTURE DIRECTIONS

At present, P2X receptors are considered to be the little brother of the ligand-gated ion channel superfamily. This is because we know much less about this branch of the family than we know about other members such as the nicotinic, glutamatergic, and gabaergic receptors. Like all members of the larger superfamily, P2X receptors require application of an agonist to achieve full activation of transmembrane ion current. This activation involves a conformational change in the receptor–channel protein that is triggered by agonist occupation of the ATP-binding site and results in the movement of the channel gate from a closed to an open position. Cations flow easily through the channel to produce a membrane depolarization and an increased $[Ca^{2+}]_i$ that have been found to underlie the pre- and postsynaptic actions of ATP. Much of what we know about how these proteins work comes from the comparison of the phenotypes of wild-type and mutated receptors. A primary goal of future experiments should be to extend these studies using more precise methods of measuring the static and dynamic properties of the protein structure. Elucidation of the crystal structure of other two-transmembrane spanning ion channels has led to advances in our understanding of how these channels work. Identification of a bacterial P2X receptor amenable to crystallization would be expected to lead to a similar revolution in the purine field. Further, it may be possible to accurately map the moving parts of the channel protein using electron spin resonance in studies that would help to determine how ATP evokes gating. Studies such as these would go a long way toward promoting the P2X receptor and allow it to stand side by side with other members of the ionotropic receptor superfamily.

### References

Acuna-Castillo, C., and Morales, B. (2000). Zinc and copper modulate differentially the P2X4 receptor. *J. Neurochem.* **74**(4), 1529–1537.

Balachandran, C., and Bennett, M. R. (1996). ATP-activated cationic and anionic conductances in cultured rat hippocampal neurons. *Neurosci. Lett.* **204**(1–2), 73–76.

Barnard, E. A. (1992). Receptor classes and the transmitter-gated ion channels. *Trends Biochem. Sci.* **17**, 368–374.

Bean, B. P. (1992). Pharmacology and electrophysiology of ATP-activated ion channels. *Trends Pharmacol. Sci.* **13**, 87–90.

Boue-Grabot, E., Archambault, V., *et al.* (2000). A protein kinase C site highly conserved in P2X subunits controls the desensitization kinetics of P2X(2) ATP-gated channels. *J. Biol. Chem.* **275**(14), 10190–10195.

Brake, A. J., Wagenbach, M. J., *et al.* (1994). New structural motif for ligand-gated ion channels defined by an ionotropic ATP receptor. *Nature* **371**, 519–523.

Brandle, U., Spielmanns, P., *et al.* (1997). Desensitization of the P2X(2) receptor controlled by alternative splicing. *FEBS Lett.* **404**(2–3), 294–298.

Buell, G., Lewis, C., *et al.* (1996). An antagonist-insensitive P2X receptor expressed in epithelia and brain. *EMBO J.* **15**(1), 55–62.

Canessa, C. M., Horisberger, J.-D., *et al.* (1993). Epithelial sodium channel related to proteins involved in neurodegeneration. *Nature* **361**, 467–470.

Clarke, C. E., Benham, C. D., *et al.* (2000). Mutation of histidine 286 of the human P2X4 purinoceptor removes extracellular pH sensitivity. *J. Physiol.* **523** (pt. 3), 697–703.

Cloues, R., Jones, S., *et al.* (1993). $Zn^{2+}$ potentiates ATP-activated currents in rat sympathetic neurons. *Pflugers Arch.* **424**(2), 152–158.

Clyne, J. D., LaPointe, L. D., *et al.* (2002). The role of histidine residues in modulation of the rat P2X(2) purinoceptor by zinc and pH. *J. Physiol.* **539**(pt. 2), 347–359.

Collo, G., North, R. A., *et al.* (1996). Cloning of P2X5 and P2X6 receptors and the distribution and properties of an extended family of ATP-gated ion channels. *J. Neurosci.* **16**(8), 2495–2507.

Colquhoun, D. (1998). Binding, gating, affinity and efficacy: The interpretation of structure-activity relationships for agonist and of the effects of mutating receptors. *Br. J. Pharmacol.* **125**, 923–947.

Devillers-Thiery, A., Galzi, J. L., *et al.* (1993). Functional architecture of the nicotinic acetylcholine receptor: A prototype of ligand-gated ion channels. *J. Membr. Biol.* **136**, 97–112.

Ding, S., Sachs, F., *et al.* (1999a). Ion permeation and block of P2X(2) purinoceptors: Single channel recordings. *J. Membr. Biol.* **172**(3), 215–223.

Ding, S., and Sachs, F. (1999b). Single channel properties of P2X2 purinoceptors. *J. Gen. Physiol.* **113**(5), 695–720.

Egan, T. M., Haines, W. R., *et al.* (1998). A domain contributing to the ion channel of ATP-gated P2X2 receptors identified by the substituted cysteine accessibility method. *J. Neurosci.* **18**(7), 2350–2359.

Eisenman, G., and Dani, J. A. (1987). An introduction to molecular architecture and permeability of ion channels. *Annu. Rev. Biophys. Biophys. Chem.* **16**, 205–226.

Eisenman, G., and Horn, R. (1983). Ionic selectivity revisited: The role of kinetic and equilibrium processes in ion permeation through channels. *J. Membr. Biol.* **76**(3), 197–225.

Ennion, S. J., and Evans, R. J. (2002). P2X(1) receptor subunit contribution to gating revealed by a dominant negative PKC mutant. *Biochem. Biophys. Res. Commun.* **291**(3), 611–616.

Ennion, S., Hagan, S., *et al.* (2000). The role of positively charged amino acids in ATP recognition by human P2X1 receptors. *J. Biol. Chem.* **275**, 35656.

Evans, R. J., Lewis, C., *et al.* (1996). Ionic permeability of, and divalent cation effects on, two ATP-gated cation channels (P2X receptors) expressed in mammalian cells. *J. Physiol.* **497**(pt. 2), 413–422.

Garcia-Guzman, M., Soto, F., *et al.* (1997). Characterization of recombinant human P2X4 receptor reveals pharmacological differences to the rat homologue. *Mol. Pharmacol.* **51**(1), 109–118.

Green, W. N. (1999). Ion channel assembly: Creating structures that function. *J. Gen. Physiol.* **113**, 163–169.

Gu, B. J., Zhang, W., *et al.* (2001). A glu-496 to ala polymorphism leads to loss of function of the human p2x7 receptor. *J. Biol. Chem.* **276**(14), 11135–11142.

Haines, W. R., Torres, G. E., *et al.* (1999). Properties of the novel ATP-gated ionotropic receptor composed of the P2X(1) and P2X(5) isoforms. *Mol. Pharmacol.* **56**(4), 720–727.

Haines, W. R., Migita, K., *et al.* (2001a). The first transmembrane domain of the P2X receptor subunit participates in the agonist-induced gating of the channel. *J. Biol. Chem.* **276**(35), 32793–32798.

Haines, W. R., Voigt, M. M., *et al.* (2001b). On the contribution of the first transmembrane domain to whole-cell current through an ATP-gated ionotropic P2X receptor. *J. Neurosci.* **21**(16), 5885–5892.

Hille, B. (2001). "Ionic Channels of Excitable Membranes." Sinauer Associates, Inc., Sunderland, MA.

Ho, K., Nichols, C. G., *et al.* (1993). Cloning and expression of an inwardly rectifying ATP-regulated potassium channel. *Nature* **362**, 31–38.

Hollmann, M., Maron, C., *et al.* (1994). N-glycosylation site tagging suggests a three transmembrane domain topology for the glutamate receptor GluR1. *Neuron* **13**(6), 1331–1343.

Huang, M., and Chalfie, M. (1994). Gene interactions affecting mechanosensory transduction in *Caenorhabditis elegans*. *Nature* **367**, 467–470.

Jiang, L. H., Rassendren, F., *et al.* (2000). Identification of amino acid residues contributing to the ATP-binding site of a purinergic P2X receptor. *J. Biol. Chem.* **275**(44), 34190–34196.

Jiang, L. H., Rassendren, F., *et al.* (2001). Amino acid residues involved in gating identified in the first membrane-spanning domain of the rat P2X2 receptor. *J. Biol. Chem.* **276**, 14902–14908.

Karlin, A. (2002). Emerging structure of the nicotinic acetylcholine receptors. *Nat. Rev. Neurosci.* **3**(2), 102–114.

Karlin, A., and Akabas, M. H. (1998). Substituted-cysteine accessibility method. *Methods Enzymol.* **293**, 123–145.

Khakh, B. S., and Lester, H. A. (1999). Dynamic selectivity filters in ion channels. *Neuron* **23**(4), 653–658.

Khakh, B. S., Bao, X. R., *et al.* (1999a). Neuronal P2X transmitter-gated cation channels change their ion selectivity in seconds. *Nat. Neurosci.* **2**(4), 322–330.

Khakh, B. S., Proctor, W. R., *et al.* (1999b). Allosteric control of gating and kinetics at P2X(4) receptor channels. *J. Neurosci.* **19**(17), 7289–7299.

Khakh, B. S., Burnstock, G., *et al.* (2001). International union of pharmacology. XXIV. Current status of the nomenclature and properties of P2X receptors and their subunits. *Pharmacol. Rev.* **53**(1), 107–118.

Koster, J. C., Bentle, K. A., *et al.* (1998). Assembly of ROMK1 (Kir 1.1a) inward rectifier $K_+$ channel subunits involves multiple interaction sites. *Biophys. J.* **74**, 1821–1829.

Laube, B., Kuhse, J., *et al.* (1998). Evidence for a tetrameric structure of recombinant NMDA receptors. *J. Neurosci.* **18**(8), 2954–2961.

Lewis, C., Neidhart, S., *et al.* (1995). Coexpression of $P2X_2$ and $P2X_3$ receptors subunits can account for ATP-gated currents in sensory neurons. *Nature* **377**, 432–435.

Li, C., Peoples, R. W., *et al.* (1993). $Zn^{2+}$ potentiates excitatory action of ATP on mammalian neurons. *Proc. Natl. Acad. Sci. USA* **90**, 8264–8267.

Li, C., Peoples, R. W., *et al.* (1996a). $Cu^{2+}$ potently enhances ATP-activated current in rat nodose ganglion neurons. *Neurosci. Lett.* **219**, 45–48.

Li, C., Peoples, R. W., *et al.* (1996b). Proton potentiation of ATP-gated ion channel responses to ATP and $Zn^{2+}$ in rat nodose ganglion neurons. *J. Neurophysiol.* **76**(5), 3048–3058.

Lingueglia, E., Champigny, G., *et al.* (1995). Cloning of the amiloride-sensitive FMRFamide peptide-gated sodium channel. *Nature* **378**, 730–733.

Lu, Q., and Miller, C. (1995). Silver as a probe of pore-forming residues in a potassium channel. *Science* **268**(5208), 304–307.

MacKinnon, R. (1995). Pore loops: An emerging theme in ion channel structure. *Neuron* **14**, 889–892.

Methot, N., McCarthy, M. P., *et al.* (1994). Secondary structure of the nicotinic acetylcholine receptor: Implications for structural models of a ligand-gated ion channel. *Biochemistry* **33**(24), 7709–7717.

Michel, A. D., Miller, K. J., *et al.* (1997). Radiolabeling of the rat P2X4 purinoceptor: Evidence for allosteric interactions of purinoceptor antagonists and monovalent cations with P2X purinoceptors. *Mol. Pharmacol.* **51**(3), 524–532.

Migita, K., Haines, W. R., *et al.* (2001). Polar residues of the second transmembrane domain influence cation permeability of the atp-gated p2x2 receptor. *J. Biol. Chem.* **276**(33), 30934–30941.

Miller, K. J., Michel, A. D., *et al.* (1998). Cibacron blue allosterically modulates the rat P2X4 receptor. *Neuropharmacology* **37**(12), 1579–1586.

Mukerji, J., Haghighi, A., *et al.* (1996). Immunological characterization and transmembrane topology of 5-hydroxytryptamine 3 receptors by functional epitope tagging. *J. Neurochem.* **66**(3), 1027–1032.

Nakazawa, K., Inoue, K., *et al.* (1998). An asparagine residue regulating conductance through P2X2 receptor/channels. *Eur. J. Pharmacol.* **347**(1), 141–144.

Newbolt, A., Stoop, R., *et al.* (1998). Membrane topology of an ATP-gated ion channel (P2X receptor). *J. Biol. Chem.* **273**(24), 15177–15182.

Nicke, A., Baumert, H. G., *et al.* (1998). P2X1 and P2X3 receptors form stable trimers: A novel structural motif of ligand-gated ion channels. *EMBO J.* **17**(11), 3016–3028.

North, R. A. (1996). Families of ion channels with two hydrophobic segments. *Curr. Opin. Cell Biol.* **8**(4), 474–483.

North, R. A., and Surprenant, A. (2000). Pharmacology of cloned P2X receptors. *Annu. Rev. Pharmacol. Toxicol.* **40**, 563–580.

Rassendren, F., Buell, G., *et al.* (1997). Identification of amino acid residues contributing to the pore of a P2X receptor. *EMBO J.* **16**(12), 3446–3454.

Stoop, R., Surprenant, A., *et al.* (1997). Different sensitivities to pH of ATP-induced currents at four cloned P2X receptors. *J. Neurophysiol.* **78**(4), 1837–1840.

Stoop, R., Thomas, S., *et al.* (1999). Contribution of individual subunits to the multimeric P2X(2) receptor: Estimates based on methanethiosulfonate block at T336C. *Mol. Pharmacol.* **56**(5), 973–981.

Surprenant, A., Rassendren, F., *et al.* (1996). The cytolytic $P_{2z}$ receptor for extracellular ATP identified as a $P_{2X}$ receptor (P2X$_7$). *Science* **272**, 735–737.

Thomas, S. A., and Hume, R. I. (1990). Permation of both cations and anions through a single class of ATP-activated ion channels in developing chick skeletal muscle. *J. Gen. Physiol.* **95**, 569–590.

Torres, G. E., Egan, T. M., *et al.* (1998). Topological analysis of the ATP-gated ionotropic P2X2 receptor subunit. *FEBS Lett.* **425**(1), 19–23.

Torres, G. E., Egan, T. M., *et al.* (1999a). Hetero-oligomeric assembly of P2X receptor subunits. Specificities exist with regard to possible partners. *J. Biol. Chem.* **274**(10), 6653–6659.

Torres, G. E., Egan, T. M., *et al.* (1999b). Identification of a domain involved in ATP-gated ionotropic receptor subunit assembly. *J. Biol. Chem.* **274**(32), 22359–22365.

Tucker, S. J., Bond, C. T., *et al.* (1996). Inhibitory interactions between two inward rectifier $K^+$ channel subunits mediated by the transmembrane domains. *J. Biol. Chem.* **271**, 5866–5870.

Ueno, T., Ueno, S., *et al.* (2001). Bidirectional modulation of P2X receptor-mediated response by divalent cations in rat dorsal motor nucleus of the vagus neurons. *J. Neurochem.* **78**(5), 1009–1018.

Valera, S., Hussy, N., Evans, R., Adami, N., North, R. A., Surprenant, A., and Buell, G. (1994). A new class of ligand-gated ion channel defined by $P_{2x}$ receptor for extracellular ATP. *Nature* **371**, 516–519.

Virgilio, F. D., Chiozzi, P., *et al.* (1998). Cytolytic P2X purinoceptors. *Cell Death Differ.* **5**(3), 191–199.

Virginio, C., MacKenzie, A., *et al.* (1999a). Kinetics of cell lysis, dye uptake and permeability changes in cells expressing the rat P2X7 receptor. *J. Physiol.* **519**, 335–346.

Virginio, C., MacKenzie, A., *et al.* (1999b). Pore dilation of neuronal P2X receptor channels. *Nat. Neurosci.* **2**(4), 315–321.

Waldmann, R., Bassilana, F., *et al.* (1997). Molecular cloning of a non-inactivating proton-gated $Na^+$ channel specific for sensory neurons. *J. Biol. Chem.* **272**(34), 20975–20978.

Werner, P., Seward, E. P., *et al.* (1996). Domains of P2X receptors involved in desensitization. *Proc. Nat. Acad. Sci. USA* **93**(26), 15485–15490.

Woodward, R., Stevens, E. B., *et al.* (1997). Molecular determinants for assembly of G-protein-activated inwardly rectifying $K^+$ channels. *J. Biol. Chem.* **272**, 10823–10830.

Worthington, R. A., Smart, M. L., *et al.* (2002). Point mutations confer loss of ATP-induced human P2X(7) receptor function. *FEBS Lett.* **512**(1–3), 43–46.

Xu, J., Yu, W., *et al.* (1995). Assembly of voltage-gated potassium channels. *J. Biol. Chem.* **270**, 24761–24768.

Yang, J., Jan, Y. N., *et al.* (1995). Determination of the subunit stoichiometry of an inwardly rectifying potassium channel. *Neuron* **15**, 1441–1447.

Zhou, Z., Monsma, L. R., *et al.* (1998). Identification of a site that modifies desensitization of P2X2 receptors. *Biochem. Biophys. Res. Commun.* **252**(3), 541–545.

# PART II

Physiology and Pathophysiology

# CHAPTER 7

# Epithelial Purinergic Receptors and Signaling in Health and Disease

**Gavin M. Braunstein and Erik M. Schwiebert**
Department of Physiology and Biophysics and Cell Biology and the Gregory Fleming James
Cystic Fibrosis Research Center, University of Alabama at Birmingham, Birmingham,
Alabama 35294

## I. INTRODUCTION

Purinergic signaling has been implicated in the regulation of many epithelial cell functions. In addition to triggering cell signaling, mainly through cytosolic calcium and phospholipase-coupled signal transduction and, possibly, via other cellular mechanisms, extracellular nucleotide signaling also modulates the potency of other autacoids and hormones that regulate epithelial cell function. Extracellular nucleotides and nucleosides also regulate transepithelial ion transport. In general, extracellular nucleotides and nucleosides, through P2Y, P2X, and P1 receptors, stimulate secretory $Cl^-$ and $H_2O$ transport, activate $K^+$ channels, inhibit absorptive $Na^+$ transport, modulate acid–base transport, and potentiate regulatory volume decrease following hypotonic cell swelling. We are careful to say "in general," because there are exceptions to this rule due, in large part, to where the epithelium resides and its function. For example, in highly secretory epithelium such as choroid plexus or ciliary processes of the ciliary body of the eye, purinergic signaling may also stimulate secretory $Na^+$ transport. Purinergic receptor-driven functions include triggering of calcium sparks and waves, potentiating ciliary beat frequency, and promoting mucus, glandular, and acinar secretion. Constitutive adenosine triphosphate (ATP) release and signaling have been implicated in the maintenance of cell signaling "setpoints" for cytosolic calcium, phosphoinositide turnover, and arachidonic acid metabolism. Purinergic signaling may even modulate gene expression in epithelial cells via specific transcription factors.

## II. THE EPITHELIUM AS AN IDEAL AUTOCRINE AND PARACRINE SIGNALING MICROENVIRONMENT

In epithelial cells, ATP is released from the cell in a polarized manner either at the apical or basolateral cell surfaces for the purpose of autocrine and paracrine regulation of the epithelial cell monolayer. In general, ATP release is directed apically. However, there may be physiological and pathophysiological exceptions to this rule as well (macula densa of the kidney as an example, see below). The mixture of P2Y, P2X, and P1 receptors in the apical and basolateral membrane domains may also differ. Nevertheless, a given epithelial cell commonly expresses all three purinergic receptor subfamilies and, often, multiple P2Y, P2X, and P1 receptor subtypes (see also below). That mixture of purinergic receptor subtypes governs what nature of signal is transduced from the released ATP and its metabolites, either in the luminal (apical) or interstitial (basolateral) microenvironments.

The two examples that our laboratory thinks about obsessively are the lung and airways epithelium and the nephron of the kidney. However, these systems and the microenvironments found in each can be extrapolated to epithelial cell models in other tissues discussed below. A third microenvironment in the intestine will also be hypothesized and presented. Sensory organ microenvironments will also be presented that provide unique and powerful media for autocrine and paracrine extracellular purinergic signaling.

## A. Lung and Airways

Figure 1 illustrates different anatomical representations of the lung and airways. There are two general sites of ATP release and signaling in the human lung and airways: the airway lumen and the submucosal gland secretions. Although it has not been investigated to date (and is not illustrated in Fig. 1), ATP secretion in the distal lung at the gas-exchange zone may occur with surfactant secretion. Airway surface epithelium in the smaller and larger conducting airways is capable of ATP and uridine triphosphate (UTP) release (Homolya *et al.*, 2000) in response to a myriad of stimuli (flow, touch, cyclic nucleotides, hypotonicity, and calcium agonists); moreover, secretions from submucosal glands that lie beneath the pseudostratified surface epithelium may also supply autocrine nucleotides and nucleosides. Our laboratory is currently engaged in dissecting the cellular mechanisms of ATP release under basal conditions and in response to the different stimuli above. These mechanisms are outlined in the chapter

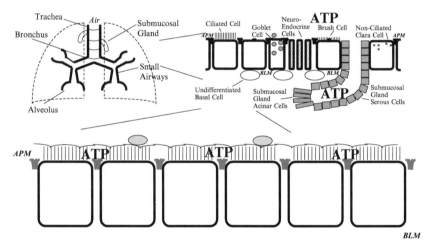

**FIGURE 1**   Lung and airways microenvironments for autocrine and paracrine purinergic signaling. Current hypotheses are presented for ATP secretion from submucosal glands and from the surface epithelium. Autocrine and paracrine nucleotide signaling is envisioned along the cells of the gland as well as the surface epithelium of the airways. The airway surface liquid microenvironment is an ideal setting for purinergic signaling.

on "Cellular Mechanisms of ATP Release" by Schwiebert, Zsembry, and Geibel. The mechanisms may be different in apical versus basolateral membranes of polarized human airway epithelial cell monolayers. Neuroendocrine cells, mast cells, and goblet cells along the airway may also release ATP along with other agonists from their granules. The analogy of the submucosal gland secreting purinergic agonists with mucins and other substances can be extended to the hepatic and pancreatic acini and the secretory glands/coils of the sweat gland and the salivary gland.

With regard to microenvironments, however, the airway surface liquid (ASL) bathing the cilia on ciliated airways epithelium in the large conducting airways is a microenvironment of interest in physiological paradigms and in pathophysiological contexts such as cystic fibrosis (CF) and primary ciliary dyskinesia (PCD) (Morse et al., 2001; Brown et al., 1991). Because purinergic signaling governs ciliary beat, epithelial cell signaling, and epithelial solute and water transport, extracellular nucleotide and nucleoside signaling in this microenvironment may be essential to regulating the composition of the ASL and the function of this critical extracellular space. Several laboratories including our own have found a loss of purinergic signaling in CF epithelial and heterologous cell models lacking cystic fibrosis transmembrane regulator (CFTR) (Braunstein et al., 2001); as such, loss of purinergic signaling may be critical to the loss of CFTR function, loss of ciliary beat, abnormal ASL composition, and overall loss of mucociliary clearance. These issues will be revisited below.

## B. Renal Nephron

Figure 2 provides a schematic of the nephron of the kidney. Once released in the lung, ATP is trapped in the airway lumen as a signaling molecule. However, unless the cilia beat to move that ATP around the lumen of the airway in the ASL, airway ATP signals in a static environment. Once filtered at the glomerulus or secreted into the lumen of the nephron (proximal tubule cells release micromolar quantities of ATP (Wilson et al., 1999; Schwiebert, 2002), the ATP is trapped in the tubular lumen as a signaling molecule, but it is propeled downstream by the tubular fluid rapidly and robustly where it can act on downstream tubules in a paracrine manner. Cells derived from multiple nephron segments express multiple P2Y, P2X, and P1 receptors, and Leipziger and colleagues and Satlin and co-workers have shown the presence of luminal P2 receptors in isolated and perfused cortical collecting tubule (CCT) or cortical collecting duct (CCD) preparations from mouse and rabbit (Deetjen et al., 2001; Satlin et al., 2001). Once ATP is excreted in the final urine and exits the nephron, it can

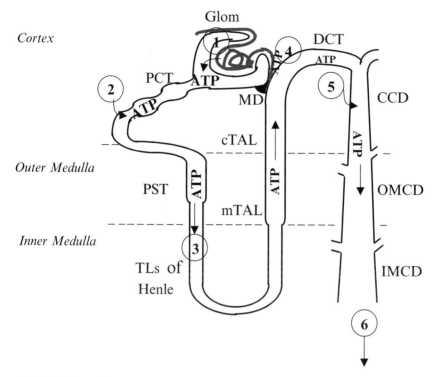

**FIGURE 2** Nephron of the kidney microenvironments for autocrine and paracrine purinergic signaling. The renal nephron is an ideal autocrine and paracrine signaling microenvironment. The circled numbers in the figure correspond to ways in which ATP may appear within renal tubules and/or the interstitium and how it might signal. (1) ATP is small enough to be filtered by the glomerulus if it enters through the afferent arteriolar circulation in sufficiently high concentration. (2) Because the proximal tubule is enriched with secretory mechanisms as well as ABC transporters (CFTR, mdr, MOAT, etc.), this may be a rich source of ATP, which would then be trapped with the lumen of the proximal tubule and downstream nephron segments since it is a multivalent anion. (3) ATP is then free to diffuse downstream within the tubular fluid, which is a convenient vehicle or medium to facilitate this signaling. (4) ATP may be released by the macula densa as an important signal for tubuloglomerular feedback. (5) Flow in the distal tubule (or any nephron segment for that matter) may stimulate ATP release. (6) Stretch of the tubules or the urinary bladder in particular (not pictured) may also trigger ATP release.

signal the uroepithelium or urothelium in the ureter and urinary bladder. Moreover, the urothelium releases ATP in response to increased distention of the ureter, the bladder, and associated structures (Knight *et al.*, 2002). Taken together, the lumen of the nephron and beyond provide an ideal microenvironment with a fluid medium that can transmit this autocrine and paracrine signaling molecule.

In addition to the lumen of the nephron, the interstitium between the cortical thick ascending limb (cTAL) and its plaque of macula densa cells and the adjoining glomerulus of the same nephron (altogether known as the "juxtaglomerular apparatus") is also a critical microenvironment for extracellular nucleotide and nucleoside signaling (Liu *et al.*, 2002). Purinergic signaling may, in fact, be essential for a process called "tubuloglomerular feedback" (see below).

## C. Intestinal Crypts

Hidden between and below the villi of the intestines are ideal autocrine and paracrine signaling microenvironments, the crypts of the intestine. Across the crypt epithelium chloride and fluid secretion are robust. This is driven, at least in part, by the expression of the CFTR. An elegant *in vivo* study by Valverde *et al.* (1995) comparing cell volume regulation in wild-type and CF knockout mice showed that CF $-/-$ intestinal crypts were impaired in their ability to volume regulate. We hypothesize that ATP may be secreted itself or along with chloride and water and signal in this specialized microenvironment (see Fig. 3).

## D. Fluid-Filled Microenvironment in the Eye and the Ear

ATP release and autocrine and paracrine purinergic signaling have been demonstrated in two major sensory organs, the eye and the ear. In particular, Mitchell, Civan, and colleagues have shown that retinal pigment epithelium that lie behind the neuronal layers of the retina in the inner chamber of the eye filled with vitreous humor release ATP via an ATP-permeable anion channel-mediated mechanism (Mitchell *et al.*, 1998). As such, ATP may act as an autocrine and paracrine agonist to modulate the retinal pigment epithelium, neuronal layers of the retina, and, perhaps, phototransduction itself. Mitchell (2001) has also shown more recently that highly secretory epithelial cells of the ciliary processes of the ciliary body, responsible for continuous secretion of aqueous humor in the outer chamber of the eye, also release ATP via multiple mechanisms. Autocrine and paracrine ATP signaling may play critical roles in the forward chamber of the eye as well.

The hair cells that line the cochlea of the inner ear respond to sound waves that enter the ear. ATP has been shown to be present in the endolymph, the fluid medium of the inner ear that bathes the hair cells (Housley, 2000). Moreover, P2X purinergic receptor channels are thought to be present on the

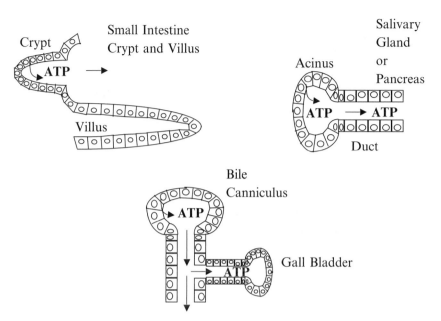

**FIGURE 3**   Gastrointestinal tract microenvironments for autocrine and paracrine purinergic signaling. The gastrointestinal tract is a secretory tissue with many secretory glands or tissues that may secrete ATP along with digestive enzymes and other hormones or peptides. The salivary glands upstream, as well as the pancreatic acinus, the biliary canniculus, and the intestinal crypt may all be rich sources of ATP. Novak and colleagues have already demonstrated ATP secretion by the pancreatic acinus in response to different stimuli (Sorensen and Novak, 2001).

cilia that lie atop each hair cell (Lee *et al.*, 2001); the cilia bend in response to each sound wave and the mechanical changes in the cilia are transduced from a mechanical signal to an electrical one. Purinergic signaling is emerging as a critical process in sound transduction by the cilia on the hair cell (Housley, 2000). Not only may ATP be released from the hair cells but ATP may bind to the P2X receptors on the cilia to augment ciliary beat. Figure 4 illustrates purinergic signaling in these essential sensory organs.

## III. INTEGRATED PURINERGIC SIGNALING IN MODEL EPITHELIUM

### A. Multiciliated Airways Epithelium

Nowhere have purinergic agonists, purinergic receptor expression, and purinergic-regulated functions been studied more ardently than in the lung and airways epithelium. As agonists were being screened to restore chloride

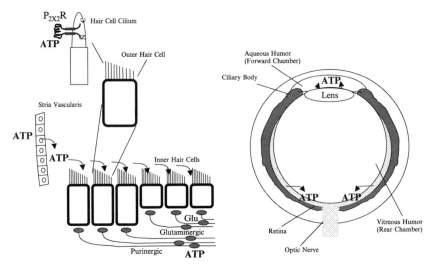

**FIGURE 4** Sensory organ microenvironments for autocrine and paracrine purinergic signaling. *Left:* Specialized epithelial cells secrete ATP into the endolymph filling the inner ear, which diffuses a short distance to the outer and inner hair cells whose cilia are enriched with P2X receptor channels. *Right:* Highly secretory epithelial cells of the ciliary processes of the ciliary body release ATP into the aqueous humor; absorptive and phagocytic retinal pigment epithelium cells also release ATP within the neuronal layers of the retina. P2 and P1 receptors are also found in these and other epithelial cell types within these sensory organs. Physiological consequences of this are discussed in the work of Mitchell, Civan, and colleagues (see references in text).

and fluid secretion in CF airways epithelium, nucleotide agonists emerged quickly as chloride and fluid secretagogues that stimulated chloride and fluid secretion independent of CFTR.

Leaders in this area of research are Ric Boucher, Kendall Harden, Eduardo Lazarowski and colleagues, who have demonstrated convincingly that P2Y G-protein-coupled receptors in general and the P2Y2 (P2U) receptor in particular play an important role in modulating airway epithelial function (Knowles *et al.*, 1991; Brown *et al.*, 1991). The reader is referred to the chapter by Lazarowski on P2Y receptors in this volume. Importantly, Knowles, Clarke, and Boucher showed that extracellular nucleotides stimulated chloride secretion in human patients with CF (Knowles *et al.*, 1991), triggering interest in targeting any and all purinergic receptors for CF therapy. Importantly, purinergic agonists in addition to ATP such as UTP, uridine diphosphate (UDP), and adenosine disphosphate (ADP) (as well as poorly hydrolyzable analogs of those listed) were also efficacious in CF as well as non-CF airway epithelial model systems (Lazarowski *et al.*, 1997;

Hwang *et al.*, 1996). This led to the identification of P2Y1, P2Y2, P2Y4, and P2Y6 receptors in human airway epithelial model systems (Communi *et al.*, 1999). In addition, adenosine receptors (the P1 receptors) also stimulated chloride secretion in airway epithelial cells (Lazarowski *et al.*, 1992); however, the major adenosine receptor, the A2 receptor, activates the cAMP/protein kinase A (PKA) signal transduction cascade and, eventually, CFTR. It is dysfunctional in CF. Several other laboratories have confirmed as well as extended these early findings in human epithelial cells or cells from other species (Cobb *et al.*, 2002; Huang *et al.*, 2001; Stutts *et al.*, 1995; Sitaraman *et al.*, 2002).

Our laboratory has added work on the expression and function of P2X purinergic receptor channels, the other principal P2 receptor subfamily (Taylor *et al.*, 1998, 1999). Biochemical characterization of P2X4 and P2X5 in human airway epithelial model systems with subtype-specific antibodies is currently in progress in our laboratory. A unifying theme in this chapter will be the fact that most, if not all, epithelial cell models express both P2Y G protein-coupled receptors and P2X receptor channels, sometimes in the same membrane domain of the polarized epithelium. Because each class of receptor increases cytosolic calcium, albeit via different mechanisms, both receptor subtypes are viable targets for pharmacotherapy of CF in the lung and airways (see below).

## 1. Cytosolic Calcium Regulation

Both P2Y G-protein-coupled receptors, via phospholipase-induced release of $Ca^{2+}$ from intracellular stores, and P2X receptor channel, via direct influx of $Ca^{2+}$ through the channel from extracellular stores, are capable of triggering calcium-dependent signal transduction cascades. P2Y-mediated effects on $Ca_i^{2+}$ have been studied extensively in human airway epithelial cell models by Paradiso and others (Van Scott *et al.*, 1995). Zsembery and Schwiebert, using an IB3-1 CF human airway epithelial cell model for therapeutic implications in CF airway, have shown that both P2Y and P2X subtypes increase cytosolic calcium (Zsembery *et al.*, 2002). Intriguingly, ATP stimulates both subtypes, although a fully transient increase in calcium is triggered by P2Y receptors, whereas P2X receptors cause a sustained increase in $Ca_i^{2+}$ [perhaps through the P2XR channels themselves and/or by opening store-operated or transient receptor polarization (TRP) calcium channels]. P2X receptor channel-mediated increases in cytosolic calcium are potentiated by removal of extracellular $Na^+$ by replacement with *N*-methyl-D-glutamine (NMDG) (Zsembery *et al.*, 2003). This suggests that $Na^+$ competes with $Ca^{2+}$ as each cation passes through the ATP-gated pore (each cation has a favorable gradient for entry). The P2X receptor-induced calcium influx is prevented by ethyleneglycoltetraacetic acid (EGTA) and augmented by raising extracellular $Ca^{2+}$ from 1 to 2 to 5 m$Mr$ (Zsembery et al., 2003).

Together, P2X and P2Y receptor activation leads to a pronounced, additive, and sustained increase in $Ca_i^{2+}$. This fact is critical for several reasons. First, either or both receptors could be targeted to trigger calcium-induced chloride secretion and potassium channel opening, which would be beneficial to the CF epithelium. Second, both receptors could be targeted to suppress absorptive $Na^+$ channels by calcium-dependent mechanisms. Such effects on all of the above would also correct abnormal cell volume regulation in the CF airways epithelium (see below). Local increases in calcium also augment ciliary beat frequency (see below), which would benefit attenuated mucociliary clearance in CF. Targeting either or both ATP receptor subfamilies for pharmacological therapy of the CF lung and airways is logical and feasible and takes full advantage of naturally expressed membrane proteins that are largely apical but are also basolateral.

### 2. Chloride and Fluid Secretion

Early work from Knowles, Stutts, and Boucher showed that purinergic agonists stimulated chloride secretion in CF as well as non-CF airway epithelial model systems (Knowles et al., 1991; Lazarowski et al., 1992). The most important demonstration of this was in in vivo nasal potential difference (PD) measurements in CF patients (Knowles et al., 1991). Sheldon Miller and colleagues subsequently showed that purinergic agonists stimulated fluid secretion in CF and non-CF airway epithelial cells (Jiang et al., 1993). Purinergic agonists emerged from a "kitchen sink" approach of adding a large panel of agonists to CF epithelial model systems to identify which agonist, if any, could stimulate chloride secretion independent of CFTR. Many studies confirmed this work in other airway epithelial model systems from human and other species. Guggino and colleagues showed that purinergic receptors stimulated multiple types of chloride channels, including CFTR and the outwardly rectifying chloride channels (ORCCs) from both the apical and basolateral membranes of primary rat tracheal epithelial cell monolayers (Hwang et al., 1996). It was concluded that the P2Y2 receptor was critical to the apical-mediated stimulation of chloride secretion; however, a rank order potency pharmacology most consistent with the P2Y3 receptor (a species homolog to P2Y1) stimulated chloride secretion from the basolateral side. As mentioned, UDP and ADP also stimulated chloride secretion in airway epithelia.

### 3. CFTR and ATP Release in Cell Volume Control

Epithelial cells release ATP in response to hypotonicity, dilution of the extracellular osmolality (Braunstein et al., 2001; Taylor et al., 1998;

Wilson *et al.*, 1999; Wang *et al.*, 1996; Roman *et al.*, 1999; Hazama *et al.*, 2000). Hypotonicity-induced ATP release is potentiated by the expression of CFTR in the epithelial cell and is dampened when CFTR is absent from the plasma membrane (Taylor *et al.*, 1998; Braunstein *et al.*, 2001). Because CFTR is a regulator of other ion channels, transporters, and vesicle trafficking and fusion, CFTR may augment ATP release by positively regulating any and all of those ATP release mechanisms (Braunstein *et al.*, 2001).

More recently, our laboratory has shown that CF human airway epithelial cells are sluggish in their recovery from a hypotonic insult, a process known as regulatory volume decrease (RVD) (Braunstein *et al.*, 2001). Transient or stable transfection of the wild-type CFTR cDNA into CF human airway epithelial cells corrects defective cell volume recovery in response to hypotonicity. Wild-type CFTR modulation of human airway epithelial cell volume regulation is blocked by the ATP scavengers, hexokinase or apyrase, and the global P2 receptor antagonist, suramin. Taken together, these data suggest that the presence of CFTR in the plasma membrane and CFTR-dependent ATP release govern epithelial cell RVD. This result begged the question: Would addition of selective purinergic agonists to P2Y and P2X receptor subtypes rescue defective volume regulation in CF human airway epithelial cells? This is indeed the case. Both P2X receptors and P2Y receptors stimulate chloride and, likely, potassium efflux from epithelial cell during RVD. This is one critical answer to the question: Why would a particular epithelial cell model express both P2Y and P2X receptors in the same cell? The answer is for protective redundancy against osmotic or other environmental insults (anoxia and ischemia, for example). Because epithelial cells are exposed to very different, dynamic microenvironments on their apical and basolateral sides, epithelial cells also express P2Y and P2X receptors in both membrane domains for this protective redundancy. A similar paradigm is evolving in hepatocytes and cholangiocytes in work by Fitz, Feranchak, and co-workers (see the chapter on "Purinergic Signaling in the Hepatobiliary System") (Roman *et al.*, 1999; Wang *et al.*, 1996).

## 4. Sodium Absorption

In addition to stimulating $Cl^-$ and $K^+$ efflux from the epithelial cell, ATP (and UTP) have also been observed to inhibit $Na^+$ influx pathways. Devor and Pilewski (1999) have shown that UTP, by binding to purinergic receptors and increasing intracellular calcium concentrations, leads to a long-term inhibition of $Na^+$ entry to the cell from the apical media. This was studied in cells with normal and mutant forms of CFTR. They conclude that nucleotides (ATP and/or UTP) could have a dual therapeutic role in the airway, stimulating $Cl^-$ efflux and inhibiting $Na^+$ entry. This conclusion has been borne out by a handful of studies in airway and other miscellaneous epithelial cell models, showing that stimulation of purinergic receptors stimulates $Cl^-$ secretion while, at the same time, attenuating $Na^+$ absorption. Inglis and co-workers (1999) studied these effects in distal bronchi and in rat fetal distal lung epithelial cells (Ramminger *et al.*, 1999). Iwase *et al.* (1997) showed similar results in rabbit tracheal epithelium. Although these were different preparations of distal airway epithelial cells, the conclusions were similar. ATP and UTP evoked a transient stimulation of $Cl^-$ secretion and a sustained inhibition of $Na^+$ absorption in the polarized monolayers of distal airway epithelia. In the nasal PD measurements of porcine distal bronchi, transepithelial PD was measured in cannulated and perfused preparations. UTP hyperpolarized the PD transiently; this stimulation was inhibited by serosal bumetanide. PD then declined to a sustained level lower than basal; this sustained inhibition was prevented by luminal amiloride. Because thapsigargin failed to inhibit UTP-stimulated $Cl^-$ secretion but did block the inhibition of $Na^+$ absorption, these results suggest that UTP stimulates $Cl^-$ secretion by a calcium-independent mechanism, whereas UTP inhibits $Na^+$ absorption via $Ca_i^{2+}$. Nevertheless, Inglis and colleagues argue that both effects of extracellular nucleotides should promote hydration of the airway.

### 5. Ciliary Beat and Mucociliary Clearance

The mucociliary system in the airway is necessary for optimal removal of mucus and inhaled pathogens and particles. Cilia on the surface of the airways epithelium beat at regular intervals and in a uniform direction toward the pharynx, and this facilitates the movement of materials along and up the "mucociliary escalator." A key stimulus for enhancing ciliary beat is increased intracellular $Ca^{2+}$ concentrations within the cilia themselves. If $Ca^{2+}$ influx is impaired, this could lead to jeopardized mucociliary clearance and disease. Of the hormones and neurotransmitters that are known to stimulate ciliary beat, the most potent is ATP. ATP acts primarily in this system by stimulating $Ca^{2+}$ influx from the external mileu, as well as stimulating internal stores to release $Ca^{2+}$. Silberberg, Priel, and co-workers have shown that extracellular sodium ions actually inhibit an ATP-gated calcium channel, perhaps through competition of both cations for the ATP-gated pore. $Na^+$, therefore, attenuates ATP-dependent ciliary beat. Their findings suggest a physiologically significant relationship between $Na^+$ concentrations and ciliary beat (Ma *et al.*, 1999). This has potentially significant therapeutic implications, suggesting that decreasing airway surface fluid $Na^+$, while increasing $Ca^{2+}$ in this fluid, may augment calcium influx into the cilium via P2X receptors and could increase mucociliary clearance in such diseases as cystic fibrosis, primary ciliary dyskinesia, or chronic bronchitis.

Figure 5 summarizes much of what is known regarding purinergic regulation of the airways epithelium. Although we focus a great deal on purinergic signaling in epithelial cell models and may be somewhat biased, purinergic signaling governs many functions that have been shown to be dysfunctional in CF. Addition of exogenous purinergic agonists to target P2Y and/or P2X receptors for CF therapy in the lung and airways is a viable paradigm that is being pursued by Boucher and co-workers (P2Y2 receptor) and by Schwiebert and colleagues in the United States and Priel, Silberberg, and co-workers in Israel (P2X receptors). The reader is referred to a discussion of purinergic signaling in the context of CF below.

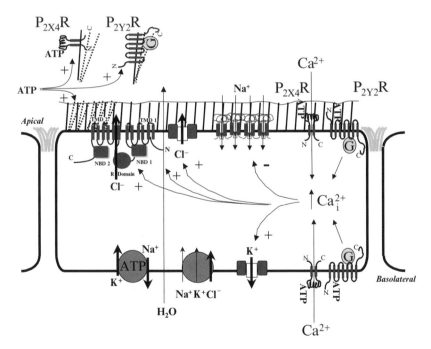

FIGURE 5 Purinergic regulation of the airways epithelium. Simplified view of a ciliated airway epithelium cell describing all of the physiological processes modulated by autocrine or paracrine purinergic signaling. See text for details.

## B. Epithelium along the Nephron of the Kidney

This topic has been reviewed in detail recently by Schwiebert (2001). A general overview is provided here with two specific paradigms that involve purinergic signaling. For more details, the reader should consult the chapters by Unwin and Leipziger and by Inscho in this volume.

Epithelial cell models from the kidney, cells from proximal tubule segments and the kidney cortex in particular, release ATP in near micromolar amounts under basal conditions and micromolar amounts under stimulated conditions (Wilson et al., 1999; Schwiebert et al., 2002). Taken together with ATP likely being filtered at the glomerulus, autocrine and paracrine signaling by ATP and its metabolites in the tubular lumen may occur constitutively along the entire nephron. Two paradigms are highlighted in this review, the macula densa and the collecting duct, which play critical roles in tubuloglomerular feedback and fine control of salt and water balance, respectively. All aspects of purinergic signaling along the renal epithelium and in the glomerulus and associated renal vasculature are discussed in companion chapters by Unwin and Leipziger and by Inscho in this volume.

### 1. Macula Densa

The process of tubuloglomerular (TG) feedback, where changes in the luminal content of salt and water are sensed by the dense plaque of macula densa (MD) cells in the cTAL and where that sensation is transduced to the glomerulus of the same nephron, is widely known but poorly understood. A long-held hypothesis is that a local mediator or paracrine agonist may be released by MD cells and regulate vascular tone of the glomerulus by causing contraction of glomerular mesangial cells and constriction of the afferent arteriole of that glomerulus. Past work has suggested that adenosine may be that mediator. Through an elegant, freshly dissected cTAL/MD/glomerular preparation, Bell and colleagues are examining more directly TG feedback signaling mechanisms. Briefly, they have found that changes in perfused NaCl concentrations are sensed in the lumen of the cTAL by the specialized MD cells, leading to ATP release from the basolateral surface of the MD plaque. ATP or its metabolites (adenosine, in particular) would then bind to glomerular mesangial cells in the interstitium and, ultimately, vascular smooth muscle cells in the afferent arteriole of the glomerulus (Liu et al., 2002). Via P2Y and/or P2X receptor channels, ATP would elevate cytosolic $Ca^{2+}$ in both cell types, leading to constriction of the afferent arteriole and a decrease in renal blood flow and glomerular filtration rate to that single glomerulus. In short, it is a classic paradigm for paracrine purinergic signaling

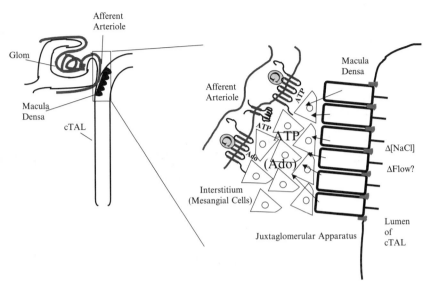

**FIGURE 6**   Purinergic regulation of renal tubuloglomerular feedback. Drawing adapted from the work of Bell and Peti-Peterdi describing the current view of how ATP, secreted from the macula densa in response to a reduction in NaCl in the cortical thick ascending limb (cTAL), or its metabolite, adenosine (Ado), signals the glomerular mesangium and the afferent arteriole.

involving specialized renal epithelial cells and their "cross-talk" with other renal glomerular cell types (see Fig. 6 for model).

## 2. Collecting Duct

Schwiebert, Stanton, and co-workers studied a mouse IMCD cell line (mIMCD-K2) to determine if nucleotides regulated NaCl transport and, if so, which purinergic receptor subtypes were involved. They observed that ATP, $\alpha\beta$-methylene ATP, and UTP stimulated $Cl^-$ secretion and inhibited $Na^+$ absorption. These P2 agonists as well as specific and degenerate reverse transcription polymerase chain reaction (RT-PCR) implicated four receptor subtypes: P2X3, P2X4, P2Y1, and P2Y2. Further analysis showed that apical nucleotide receptors were more effective in inhibiting $Na^+$ absorption and stimulating $Cl^-$ secretion (McCoy *et al.*, 1999). Leipziger and co-workers found similar results in M1-CCD cells and MDCK cells, and Deetjen and Leipziger and, later, Satlin and colleagues found an apical P2Y2 receptor in isolated, perfused CCDs that increases cytosolic $Ca^{2+}$ in these tubules (Gordjani *et al.*, 1997; Cuffe *et al.*, 2000; Deetjen *et al.*, 2000; Satlin *et al.*, 2001). Similar modulation of ion transport in this segment is likely. Purinergic receptors also modulate water transport in the collecting

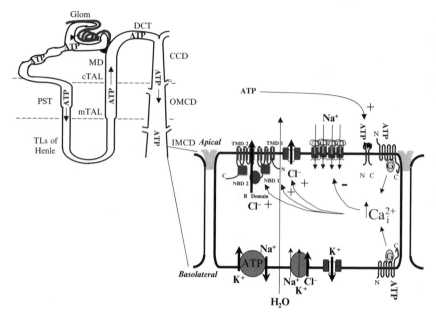

**FIGURE 7** Purinergic regulation of the "fine control" of salt and water balance. Drawing depicting purinergic regulation of ion and water transport in the inner medullary collecting duct (IMCD) cell. What is not pictured is the work of Kishore and Knepper, who showed that purinergic agonists inhibit vasopressin-induced water permeability in the IMCD (refer to review by Schwiebert and Kishore, 2001).

duct. Kishore and Knepper found that P2Y2 receptors were expressed along the nephron in general and in renal medulla in particular. ATP and UTP, via this receptor, inhibited vasopressin-induced water permeability across the renal collecting duct. This paradigm is described in detail in a recent review by Schwiebert and Kishore (2001). Figure 7 shows an inner medullary collecting duct cell model that integrates purinergic control of salt and water balance. Additional paradigms of purinergic signaling in the kidney are described by Unwin and Leipziger in their chapter entitled "Purinergic Signaling in the Kidney."

## C. Epithelium of Secretory Organs of the Gastrointestinal Tract

### 1. Hepatobiliary System

The idea that epithelial cells regulate their cell volume via autocrine ATP signaling is still considered a novel idea, despite convincing evidence. Roman and Fitz conducted earlier studies in a model liver cell line. They observed changes in $Cl^-$ permeability at the plasma membrane of rat hepatoma cells (HTC) in response to osmotic stress. This change in permeability was coupled with a reduction in cell volume. Both of these responses to osmotic stress were inhibited by apyrase, an ATP/ADP hydrolase (Wang *et al.*, 1996). Their experiments also showed that ATP was released from the cell, down its concentration gradient, as a result of cell swelling. Their findings concluded that cell swelling caused a downstream efflux of ATP through a conductive pathway thought to be a channel. This extracellular ATP was shown to be necessary for homeostatic recovery from this swelled state (Wang *et al.*, 1996).

Rather than a comprehensive summary and a model figure regarding purinergic signaling in the hepatobiliary epithelium, the reader is referred to the chapter on "Purinergic Signaling in the Hepatobiliary System" by Feranchak and Fitz. However, two paradigms are highlighted within this chapter that speak to purinergic modulation of hepatic epithelial cell function. Zsembery *et al.* (1998) studied the modulation of acid–base transport in cholangiocytes by purinergic agonists. Entering this study, it was known that cAMP agonists such as secretin promoted biliary $HCO_3^-$ secretion by activation of apical $Cl^-$ channels and $Cl^-/HCO_3^-$ (AE2) exchange. In contrast, ATP stimulated $Cl^-$ channels and $Na^+/H^+$ exchange (NHE). Using BCECF-loaded human Mz-CHA-1 and rat NRC cholangiocytes as epithelial cell model systems, Zsembery *et al.* showed that ATP, UTP, and ATP$\gamma$S all significantly increase NHE activity. Purinergic agonists alone did not affect AE2 activity; however, when added to cAMP agonists, they synergistically stimulated AE2 activity. Apical addition of the purinergic agonists ATP, UTP, and ATP$\gamma$S stimulated basolateral NHE activity, whereas, on the basolateral side, these purinergic agonists were without effect on NHE activity. In contrast, adenosine agonists (agonists that stimulated A1 and A2 receptors) stimulated NHE activity. These studies show that both P2 and P1 receptors regulate acid–base transport in liver cholangiocyte models.

Using liver epithelial cells, Frame and de Feijter (1997) showed that paracrine ATP release and signaling propagated $Ca^{2+}$ waves. The P2 receptor antagonist, suramin, completely blocked wave propagation in a cell line devoid of gap junctions. In another cell model that expressed gap junctions, a combination of suramin and a gap junction inhibitor completely blocked wave propagation. This study was followed by a more thorough study with regard to purinergic signaling by Wartenberg and colleagues (Sauer *et al.*, 2000; Schwiebert, 2000). $Ca^{2+}$ waves were inhibited by multiple P2 receptor antagonists, pretreatment and downregulation of P2Y and P2X receptors, depletion of intracellular ATP pools with 2-deoxyglucose, and the ATP scavenger apyrase. $Ca^{2+}$ waves and mechanically induced ATP release were also inhibited by a panel of anion transport and channel inhibitors. Taken together, these studies show that paracrine purinergic signaling can and may propagate $Ca^{2+}$ waves along the surface of a tissue epithelium and may be important for synchronized signaling along the epithelial barrier.

## 2. Pancreatic Acinus and Duct

Purinergic receptors have been found in the pancreas. Several functional subtypes of both the P2X and P2Y receptors are present on both the apical and basolateral membranes of pancreatic ductal cells (Luo et al., 1999). They are believed to communicate between the two membranes during ductal secretion, coordinating transport of osmoles. The acinar cells are solely secretory cells. The ductal cells can reabsorb and secrete salt and water, which determine the final composition of the pancreatic juice. It is known that CFTR drives the secretion of fluid and bicarbonate ($HCO_3^-$) in the pancreatic acini and ductal system, by its primary function as a cAMP-regulated $Cl^-$ channel. Therefore, in CF, there is impaired $Cl^-$, $HCO_3^-$ and fluid secretion in the acinus and impaired $Cl^-$ reabsorption in the duct. Muallem and co-workers recently studied CFTR-driven and $Cl^-$-dependent $HCO_3^-$ secretion in non-CF and CF pancreatic duct (Choi et al., 2001). This leads to an increased concentration of digestive enzymes, which can obstruct the ductal system.

ATP is a known agonist in the pancreatic duct system. It is stored in secretory granules by acinar cells upstream from the duct cells, and is released by these acinar cells into the luminal fluid. This fact has been demonstrated elegantly by Sorensen and Novak (2001), who imaged ATP release via bioluminescence detection with a confocal microscope. They showed that mechanical stimulation, hypotonic challenge, and the muscarinic calcium agonist, carbachol, stimulated ATP release in isolated pancreatic acini. The maximal ATP concentration measured after stimulation averaged 9 $\mu M$, which is similar to the maximal concentration of released ATP measured by similar or other detection assays in other cell or tissue preparation. The limiting factor is degradation of the released ATP that competes for the luciferase detection reagent. This is the first published study to measure ATP release with bioluminescence or any other detection assay in an in vivo-like tissue preparation.

Much work has also been done to define the actual subtypes of purinergic receptors present in the pancreatic ductal system. Muallem and co-workers described evidence for the expression of three P2Y subtypes and three P2X subtypes in microdissected intralobular pancreatic ducts. In the luminal membrane, P2Y2, P2Y4, and P2X7 were found. In contrast, the basolateral membrane expressed P2X1, P2X4, and P2X7 as well as P2Y1 (Luo et al., 1999). All were able to increase cytosolic calcium and, as such, activate

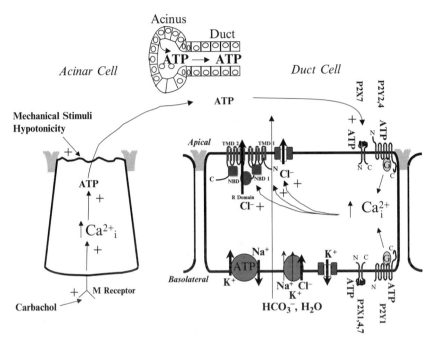

**FIGURE 8** Purinergic regulation of pancreatic secretion. Schematic view of the role of the pancreatic acinar cell in ATP secretion and the response of the pancreatic duct cell to that paracrine ATP signal.

Cl⁻ channels. It has recently been shown that upon interaction with purinergic receptors, ATP signaling increases cAMP and activates Cl⁻ channels (Nguyen *et al.*, 2001). This further supports the relationship between CFTR and ATP signaling in the pancreatic duct system; a summary of what is known about ATP release, ATP receptors, and ATP regulation of pancreatic epithelial cells is provided in Fig. 8.

### 3. Intestinal Epithelium

Much of the information regarding purinergic signaling in intestinal epithelium derives from human intestinal epithelial cell lines such as T84, Caco-2, HT-29, and 1407. These findings are synthesized into a model (Fig. 9). Nevertheless, uniform among these cell models is the expression of multiple types of P1 and P2 purinergic receptors on both apical and basolateral surfaces as well as multiple paradigms for stimulation of Cl⁻ channels in secretory Cl⁻ transport and in cell volume regulation.

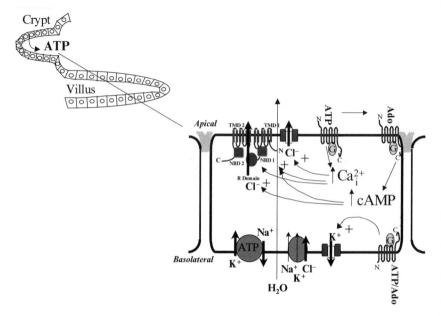

**FIGURE 9**  Purinergic regulation of intestinal secretion. Drawing of P2 and P1 receptor regulation of chloride and fluid secretion in intestinal epithelial cells mediated by CFTR chloride channels and other chloride channels modulated by cytosolic calcium, phospholipids, or other mediators.

Barrett and colleagues have published extensive work on nucleotide-mediated stimulation of chloride secretion in T84 cells (Barrett *et al.*, 1990). The latest study focused on the P2Y-selective agonist, UTP, and its effect on Cl⁻ secretion in T84 cells from the apical and basolateral side of polarized monolayers (Smitham and Barrett, 2001). UTP stimulated Cl⁻ secretion from either side of the epithelium that was largely a transient response. However, serosal UTP stimulation inhibited subsequent stimulatory effects of carbachol, thapsigargin, or even mucosal UTP; this may relate to the generation of inositol 3,4,5,6-tetrakisphosphate (IP$_4$). Other laboratories investigating purinergic regulation of ion transport in T84 monolayers have not been able to demonstrate nucleotide-mediated effects (Stutts *et al.*, 1995; Strohmeier *et al.*, 1995). Instead, they have focused on the nucleoside, adenosine, and its regulation of CFTR in an apical membrane-delimited manner. Madara and co-workers and Stutts and colleagues have published extensively in this area (Stutts *et al.*, 1995; Strohmeier *et al.*, 1995). More recent work by Stutts and Milgram show a close coupling between A2 adenosine receptors and CFTR in the apical membrane of T84 monolayers and Calu-3 airway epithelial monolayers (Huang *et al.*, 2001).

Purinergic receptor expression has also been addressed in these intestinal cell models as well as in duodenal villus. Burnstock and co-workers showed staining for P2X1 in the capillary plexus within the villus, whereas P2X5 and P2X7 were found in villus goblet cells (Groschel-Stewart *et al.*, 1999). P2X7 was also found in enterocytes at the villus tip, which are the most likely to be undergoing apoptosis. The work of Barrett and co-workers suggests apical P2Y2 expression in T84 cells. In Caco-2 cells, Collett and co-workers showed the presence of P2Y2, P2Y4, and P2Y6 mRNA as well as apical ATP, UTP, and UDP-mediated effects on $I_{SC}$ and $Ca_i^{2+}$ (McAlroy *et al.*, 2000). Okada and co-workers show the expression of the P2Y2 receptor in the I407 human intestinal cell line (see below).

Kunzelmann and Mall (2002) have shown that the colonic epithelium has secretory functions as well as its well-characterized absorptive functions. In the colon, there is a net absorption of water, NaCl, and short-chain fatty acids. This leads to very little salt and water content in the feces. Also, the colonic epithelium secretes KCl, bicarbonate, and mucus. CFTR, ENaC, and the Na, K-ATPase are believed to control the dual absorptive/secretory role of the colonic epithelium as well as the balance between absorption and secretion. Steroids and other modulating factors, such as serum- and glucocorticoid-regulated kinases (SGK), help regulate these processes. Mutations in CFTR lead to enhanced sodium absorption and impaired chloride secretion in the CF colon. This, in addition to defects in electrolyte transport, leads to the secretory diarrhea common to patients with CF.

Okada and colleagues have used the I407 human intestinal epithelial cell model to study the role of purinergic signaling, ion transport pathways, and CFTR in epithelial cell volume regulation (Okada *et al.*, 2001; Dezaki *et al.*, 2000). Early groundwork showed that hypotonic cell swelling increased $Ca_i^{2+}$ leading to activation of intermediate conductance and $Ca^{2+}$-dependent $IK^+$ channels. In contrast, they showed that a calcium-sensing receptor is activated by extracellular $Ca^{2+}$, leading to activation of outwardly rectifying $Cl^-$ channels. Together, efflux of $K^+$ and $Cl^-$ promotes RVD. One caviat is that it could be argued that $Cl^-$ would follow $K^+$ automatically to preserve macroscopic electroneutrality. Okada and co-workers subsequently showed that RVD was suppressed by the ATP scavenger, apyrase, and the P2 receptor antagonist, suramin. Extracellular ATP and UTP accelerated RVD and increased $Ca_i^{2+}$, effects that were suppressed by blocking extracellular purinergic signaling. Hypotonic cell swelling also increased $Ca_i^{2+}$, an effect that was partially inhibited by apyrase or suramin. They concluded that autocrine ATP release stimulated P2Y2 receptors that accelerated RVD, at least in part via $Ca_i^{2+}$. Finally, Okada and co-workers claim that I407 cells lack CFTR; however, ATP-regulated RVD is still intact. This is puzzling, as other laboratories have indeed found CFTR functionally or biologically in this cell line (E. M. Schwiebert and W. B. Guggino, unpublished observations). They do not show the data proving that this cell line lacks CFTR. Subsequent to this study, they performed similar work on C127 mouse mammary epithelial cells lacking or expressing CFTR and found that CFTR indeed augments ATP release and potentiates RVD (Hazama et al., 2000).

Despite the multitude of studies in intestinal epithelial cell models summarized in Fig. 9, very little is known about the sources of released ATP in the intestine and the roles of purinergic signaling in the intestine in vivo (or even in dissected intestinal tissue preparations). Valverde et al. (1995) showed that cell volume regulation, RVD in particular, was impaired in intestinal crypts from CF knockout mice versus their wild-type counterparts. It is interesting to speculate that a lack of autocrine and paracrine ATP signaling may underlie this defect.

## D. Sensory Epithelia

### 1. Retinal Pigment Epithelium and Cilary Epithelium

Sensory organs such as the eye and the ear present ideal environments for autocrine and paracrine signaling as described above. In 1998, Mitchell et al. (1998) performed seminal studies in ocular ciliary epithelial cells from the forward chamber of the eye on the ciliary processes of the ciliary body. First, they found that a fluorescence marker for ATP, quinacrine, labeled the ciliary epithelia but not innervating nerves. Pigmented and nonpigmented retinal epithelial cells were also labeled intensely, leading to a subsequent study (see below). Hypotonic challenge augmented ATP release that was inhibited by the $Cl^-$ channel blocking drug, 5-nitro-2-(3-phenylpropylamino)-benzoic acid (NPPB), but not by diphenylamine carboxylic acid or DPC and glibenclamide or the multidrug-resistant (mdr) inhibitors, verapamil and tamoxifen. This result suggested that ATP release occurs by a ATP-permeable anion channel that is separate from the ATP-binding cassette (ABC) transporters, CFTR and mdr-1. ATP release was also triggered by ionomycin by a mechanism insensitive to NPPB. Mitchell independently went on to test ATP release by a human retinal pigment epithelial cell line (Mitchell, 2001). The hypothesis was that autocrine ATP release and signaling also played a role in the retina in the posterior chamber of the eye. It was known that purinergic receptors stimulated fluid transport and inhibited phagocytosis by these cells, primarily involved in recycling of photopigment during phototransduction by retinal rod and cone photoreceptors. ATP release that had transient and sustained phases was stimulated by basic fibroblast growth factor, UTP, and by hypotonicity.

Interestingly, ionomycin was without effect in this retinal cell line. NPPB also inhibited ATP release in response to these stimuli. Substantial evidence was presented for degradation of ATP to its metabolites. In particular, Mitchell and Civan as well as other laboratories have shown effects of adenosine on these ocular epithelial cell lines (see below).

P2 and P1 adenosine receptor expression and function have been characterized extensively in epithelial cell models from the eye. An early study showed that both P2 and P1 receptors modulated adenylyl cyclase activity in both nonpigmented and pigmented ciliary epithelial cell models (Wax *et al.*, 1993). Parallel work by Blazynski characterized the presence of A2a and A2b adenosine receptors on retinal pigment epithelial cell membranes. This followed an older pharmacological study showing evidence for A2 receptors in human retinal pigment epithelial cells (Friedman *et al.*, 1989). Adenosine also stimulates $Cl^-$ channels in nonpigmented ciliary epithelial cells (Carre *et al.*, 1997). Although originally thought to be mediated by A1 or A2 receptors, A3 adenosine receptor-selective agonists stimulated $Cl^-$ channels via cytosolic $Ca^{2+}$-dependent processes in a subsequent study (Mitchell *et al.*, 1999). Of interest, adenosine stimulated the same $Cl^-$ currents as did hypotonic cell swelling in nonpigmented ciliary epithelial cells (Carre *et al.*, 1997). This result suggested that autocrine adenosine signaling may underlie ciliary epithelial cell volume regulation.

Epithelial cell models of the eye also express P2 nucleotide receptors. P2Y2 receptors have been characterized on corneal and lens epithelium (Merriman-Smith *et al.*, 1998; Fujihara *et al.*, 2001). Physiological, molecular biological, and biochemical evidence was presented for P2Y2 receptor expression in human retinal pigment epithelial cells (Sullivan *et al.*, 1997). In a thorough physiological analysis of purinergic-regulated ionic conductances in rat retinal pigment epithelial cells, Ryan *et al.* (1999) showed the presence of P2X receptor channels and P2Y G-protein-coupled receptors. Both patch-clamp recording and measurement of Fura-2 fluorescence were performed. ATP activated a fast inward current that was permeable to monovalent cations, where the inward current was carried mainly by $Na^+$. ATP also activated a slower, delayed outward current that was a calcium-activated $K^+$ conductance. Because these currents were modulated by ATP as well as UTP, ADP, and 2-MeS-ATP, these results suggest, at the very least, the presence of P2X, P2Y1, and P2Y2 receptors in retinal pigment epithelium.

Figure 10 illustrates the major findings with regard to autocrine and paracrine purinergic signaling within the eye. What remains unclear in these ocular epithelial cell models is the physiological roles for autocrine and paracrine purinergic signaling. They may play a prominent role in driving

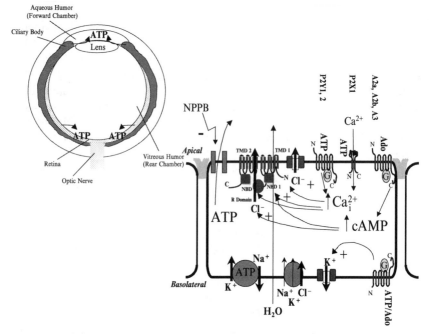

**FIGURE 10**  Purinergic signaling in the eye. Mitchell has shown evidence for an NPPB-sensitive ATP release pathway that is likely an ATP-permeable anion channel in ciliary body epithelial cells and in retinal pigment epithelial cells (Mitchell, 2001). Once released, that ATP is free to stimulate ATP receptors of these or other cells or can be metabolized to adenosine and activate adenosine receptors also expressed in ocular cells.

secretion of aqueous humor by the ciliary body, may play a role in modulating recycling of photopigment in the retina, and/or may mediate cell volume regulatory processes in all epithelia to respond to changes in the fluid-filled environment of the eye, a fluid-filled organ.

## 2. Hair Cell Epithelium of the Inner Ear

ATP is a critical autocrine and paracrine agonist in the endolymph and the scala media that fill the organs of the inner ear. P2 receptors have been found throughout the specialized epithelial cells and cell membranes that present in the cochlear and vestibular system. This topic has been reviewed recently by Housley (2000). Basal levels of ATP in the perilymph and endolymph are nanomolar, and they increase in response to noise stress, sound waves, and hypoxia. Ecto-ATPases and ectoapyrases may suppress the levels of extracellular ATP; however, the full range of stimuli for ATP release may not be appreciated and increases in ATP release (surrounding the stereocilia of the hair cells, for example) may not be readily detectable. Housley indicates that the sources of released ATP are not appreciated and present an important future direction for this work. However, a recent study by Munoz *et al.* (2001) showed that stores of ATP located in vesicles within marginal cells of the stria vascularis were released to increase ATP in the endolymph significantly following a sound stimulus. In contrast, hypoxia did not promote a statistically significant increase in endolymph ATP, although a slight increase was noted. It was also noted that substantial ATP was present under basal conditions, suggesting a role in the maintenance of inner ear and cochlear and vestibular function.

Despite continuing work on the sources of released ATP, it is known that extracellular ATP modulates transduction of sound waves by action of P2X receptors on the stereocilia and by P2Y receptors there and elsewhere. This is reviewed elegantly by Housley (2000). It is thought that P2X2 receptors function together with stretch-activated nonselective cation channels in the stereocilia to depolarize the outer hair cell membrane, suggesting that ATP may play a significant role in the inner ear signaling transduction pathway (Raybound and Housley, 1997; Kanjhan *et al.*, 1999; Housley *et al.*, 1999). However, it is possible that additional P2X receptor subtypes may contribute to the phenotype, in particular inner ear cell types. P2X receptors also play critical roles in Reissner's membrane and in intermediate cells of the stria vascularis, where they mediate $K^+$ shunt conductances (King *et al.*, 1998). It could be argued that the P2X receptor channels and other secretory $K^+$ channels contribute to the high $K^+$ concentrations in perilymph and endolymph. Indeed, mutation of the KvLQT channel in the stria vascularis marginal cell causes deafness in patients who also have long QT syndrome of the cardiac myocyte. The P2Y4 and P2Y2 receptors also play a critical role in regulating this KvLQT and other KCNE channels in strial marginal cells and vestibular dark cells (Marcus *et al.*, 1997; Sage and Marcus, 2002). Deeper in the inner ear, Marcus and co-workers found a role for P2X2 receptors in parasensory cation absorption by cochlear outer sulcus cells and vestibular transitional cells (Lee *et al.*, 2001). In fact, guinea pigs with vestibular imbalance disorder were corrected by infusion of ATP and other purinergic agonists into the scala tympani.

Purinergic signaling also affects other inner ear epithelial cell functions. Hensen's cells of the cochlea, specialized epithelial cells in the inner ear, respond to extracellular ATP with changes in ionic currents and concomitant increases in $Ca_1^{2+}$(Langostena *et al.*, 2001). Ionic currents that were activated included a rapid inward current, a more slowly rising inward current, and a slowly developing reduction in input conductance. ATP also appeared to inhibit cell–cell communication via gap junctions. The fast current activated by ATP were the P2X receptor channels located on these and other specialized sensory epithelial cells in the cochlea, whereas P2Y receptors mediated the slow responses in ionic currents and $Ca_i^{2+}$. The same group also performed similar work on Dieter's cells and showed that ATP promoted gap junctional communication between cells. Unfortunately, Fura-2 imaging of multiple cells within the tissue for Hensen's cells or Dieter's cells was not performed to monitor $Ca^{2+}$ wave propagation.

The inner ear and its specialized epithelial cells designed to sense and capture sound are a rich ground to study autocrine and paracrine nucleotide signaling. Seminal findings in these cells are shown in Fig. 11. Taken together, there are profound physiological roles for extracellular ATP signaling and both P2X and P2Y receptors in the specialized ciliated and nonciliated epithelial cells of the inner ear. Luciferase-based or other detection methods need to be applied to freshly dissected, endolymph-filled inner ear preparations to gauge the role of extracellular ATP in an *in vivo*-like preparation.

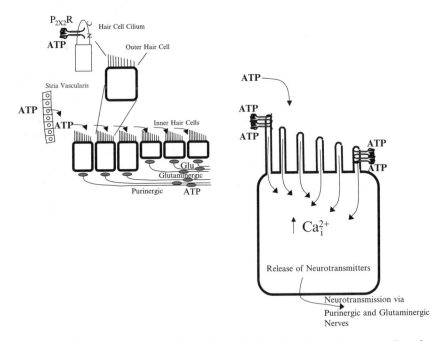

**FIGURE 11**  Purinergic regulation of hearing. Drawing of an elegant paradigm for paracrine purinergic signaling in the endolymph of the ear that is essential for sound transduction into an electrical signal that also involves purinergic nerves.

## IV. PARADIGMS FOR PURINERGIC SIGNALING IN DISEASED EPITHELIUM

### A. Cystic Fibrosis: A Loss of Extracellular Purinergic Signaling in the Lung Airways and in Other Tissues

The human airway is a prime example of a tissue that is modulated by purinergic signaling. Purinergic receptors have been localized to many cell types in the human airway. These receptors in the airway modulate many different epithelial cell functions, including regulation of salt, water, and acid–base homeostasis, cell signaling, ciliary beat, and cell volume regulation. Our laboratory as well as other laboratories have documented a role for CFTR and other ABC transporters in governing ATP release and autocrine and paracrine purinergic signaling. In cystic fibrosis (CF), purinergic signaling is attenuated. We believe that this defect, along with the primary defect in $Cl^-$ channel function and $Cl^-$ permeability (Welsh *et al.*, 1995), are major contributing factors to the abnormalities in the CF airway and, possibly, in other tissues affected in CF.

CF is a disease commonly associated with defects in secretory epithelia and is the most common, inherited, autosomal recessive disease among whites. CF affects other populations as well, but at a much lower frequency. The two most prevalent phenotypes of CF are pancreatic insufficiency and lung disease. CF is the result of one or more mutations in the CF gene. This gene encodes a 1480 amino acid protein, cystic fibrosis transmembrane conductance regulator (CFTR), which is located on the apical membrane of most epithelial cells in healthy individuals. CFTR is a member of a large superfamily of transporter proteins called ATP-binding cassette or ABC transporters. The CFTR protein is divided into five domains, including two membrane-spanning domains (MSDs), each containing six transmembrane segments. There is also a regulatory domain (R domain) in which phosphorylation by PKA and PKC occurs, and two nucleotide-binding domains (NBD) that flank the R domain that interact with, bind, and hydrolyze ATP weakly (Welsh *et al.*, 1995).

Currently, there are over 1000 mutations identified in the CFTR gene, the most common being a deletion of a phenylalanine residue at position 508 ($\Delta$F508). It occurs in the first NBD, causes a folding defect in this domain of the protein, and interferes with CFTR processing at the endoplasmic reticulum (ER). This mutation is the cause of 50–70% of CF cases in most white populations. In addition to the lungs and pancreas, other organs such as the liver, intestines, vas deferens, and sweat glands also show abnormalities as a result of CF mutations (Welsh and Smith, 1993; Collins, 1992). The distribution of these mutations is nonrandom: there are distinct "hotspots" in certain areas of the protein for these mutations. The region of CFTR with the highest density of mutations is NBD1, which is encoded by exon 10. These mutations can vary in their detrimental effects from severe to nonsymptomatic (Welsh et al., 1995).

The CFTR protein is widely believed to form an anion channel in the apical membrane, but there is speculation as to exactly what it transports. There is considerable evidence that CFTR is a $Cl^-$ channel activated by cyclic AMP (cAMP). Some believe that CFTR also has the capacity to conduct ATP (Reisin et al., 1994), but this is very controversial (Reddy et al., 1996). Past and current research is focusing on a separate ATP conductance channel that is closely associated with the CFTR protein (Braunstein et al., 2001; Sugita et al., 1998). Because CFTR also regulates numerous other processes in the airway epithelial cell, CFTR may regulate ATP release, mediated by ATP channels and ATP-filled vesicles, to accomplish these other regulatory functions.

It has been shown that CFTR is necessary for extracellular nucleotide signaling. In CF epithelial cells that lack functional CFTR at the apical membrane, nucleotide signaling is absent or insufficient. Therefore, the physiological consequences of this lost signal must be addressed. These issues are illustrated in Fig. 12. First, there is a loss of chloride conductance that may be necessary for the chloride secretion involved in regulatory volume decrease (RVD) and transepithelial transport (Schwiebert et al., 1995; Knowles et al., 1991). There could also be a loss of potassium and fluid secretion due to the loss of nucleotide signaling, two processes also critical in RVD. These consequences could lead to impaired intracellular cell volume regulation. An impairment in intracellular volume control, due to an inability of the epithelial cell to "sense" its external osmotic environment and/or to regulate its own volume, may also directly impact the abnormal apical surface liquid (ASL) microenvironment observed in CF. Smith and colleagues (1996) have observed a curious increase in ASL ionic strength in CF epithelia when compared to non-CF cells. This would also increase the tonicity of this microenvironment. Matsui and colleagues (1998) observed a reduced ASL depth and volume in CF epithelia, but no difference in ionic strength. Engelhardt and co-workers, using their elegant xenograft model of well-differentiated non-CF and CF epithelia, showed defects in ionic strength and volume in the CF ASL versus the normal counterpart (Zhang et al., 1999). These results and hypotheses relate back to an older and unappreciated study by Valverde and colleagues (1995), who observed defective cell volume regulation in the intestinal crypts of CF knockout mice when compared to wild-type mice. The possibility that these defects in ASL composition and cell volume regulation may be caused by defects in autocrine ATP release and signaling remains to be determined.

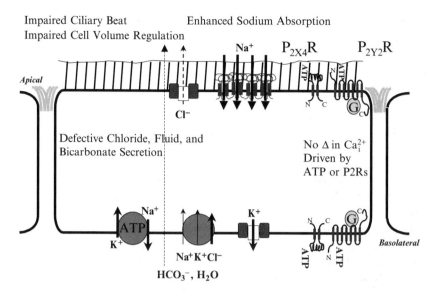

Impaired Ciliary Beat          Enhanced Sodium Absorption
Impaired Cell Volume Regulation

**FIGURE 12**   Loss of purinergic signaling in CF. In contrast to Fig. 5, many or all of the factors regulated by extracellular purinergic signaling in non-CF cells may be lost in CF cells due to a loss in apical-mediated ATP release. See text and pertinent references for details.

## B. Autosomal Dominant Polycystic Kidney Disease: A Change in Nephron Structure Creates a Detrimental Autocrine Signaling Environment

Purinergic receptors and signaling are also critical along the nephron of the kidney. The richest source of ATP release in the nephron is the proximal tubule. There are many reasons why this is the case, including an accelerated metabolic and transport rate in the proximal tubule and/or the expression of multiple ABC transporters such as CFTR, multidrug resistance (mdr), and multiple organic anion transporter (MOAT). Once released into the tubular lumen, ATP would be trapped as a charged anionic species and carried in the tubular fluid downstream to the other nephron segments, where it can bind to and interact with purinergic receptors. Autocrine and paracrine ATP signaling in the tubular fluid via purinergic receptors can affect any number of renal functions, due to the versatility of purinergic control over epithelial functions and signaling pathways (see recent reviews by Schwiebert and Kishore, 2001; Schwiebert, 2001).

Because purinergic signaling plays an autocrine and paracrine role in the regulation of renal tubular function, defects in purinergic signaling could contribute significantly to kidney pathophysiology. Extracellular purinergic signaling is known to modulate water and salt balance along the nephron.

Extracellular ATP in the apical environment or tubular lumen has been shown to inhibit water and salt reabsorption by the kidney. Therefore, the addition of purinergic agonists has been proposed to be of therapeutic benefit to decrease water and salt reabsorption, thus decreasing blood volume and pressure. A recent review examined this paradigm in some detail (Schwiebert and Kishore, 2001).

The relationship between autocrine and paracrine purinergic signaling and the renal disorder polycystic kidney disease (PKD) is not intuitively obvious. However, PKD is a progressive disorder chacracterized mainly by the slow and gradual remodeling of the tubules of the kidney and the ducts of the pancreas and liver. In autosomal dominant PKD (ADPKD), this remodeling results in the formation of a fluid-filled cyst, created from the closure of two ends of a renal tubule. In autosomal recessive PKD (ARPKD), the tubules and ducts dilate; however, they form cyst-like structures only in end-stage disease. Either way, an abnormal microenvironment encapsulated by a single monolayer of epithelial cells develops. As alluded to above and in a chapter on ATP release (see Chapter 2), such a microenvironment is ideal for autocrine and paracrine signaling.

Our laboratory has shown that ATP (and, likely, its metabolites) is present in micromolar quantities in a subset of ADPKD cyst fluid samples (Wilson *et al.*, 1999). In this study and in another very recent one (Schwiebert *et al.*, 2002), ATP release was found to be as or more robust in ADPKD epithelia than in normal controls. This could be due to increased metabolism and/or proliferation of ADPKD epithelia during their reversion to an undifferentiated phenotype. More recent work has shown that both P2Y and P2X receptors are present on the luminal membrane of human ADPKD monolayers that stimulate $Cl^-$ secretion by a cytosolic calcium-driven mechanism. As such, all of the elements are present for autocrine and paracrine ATP and ATP metabolite signaling to drive $Cl^-$ secretion into the cyst. Obviously, this would be detrimental to the progression of ADPKD. An additional problematic feature of "trapped" purinergic signaling within the cyst is the fact that purinergic agonists are mitogens or comitogens along with growth factors for renal epithelial cells and mesangial cells (Erlinge, 1998, Paller *et al.*, 1998; Wang *et al.*, 1992). In PKD, growth factors are also released instead into the cyst. There, they interact with growth factor receptors mislocalized to the luminal membrane, leading to a devastating positive feedback loop for growth and proliferation of the cysts (Murcia *et al.*, 1998; Wilson, 1997). Figure 13 illustrates a working hypothesis for the implied detrimental effects of extracellular purinergic signaling and PKD. The ADPKD results in distinct cysts or fluid-filled spheres lined by epithelial cells. Any normally apical ion or fluid transport is instead released into the encapsulated cyst. Therefore, any secretagogue could prove to be

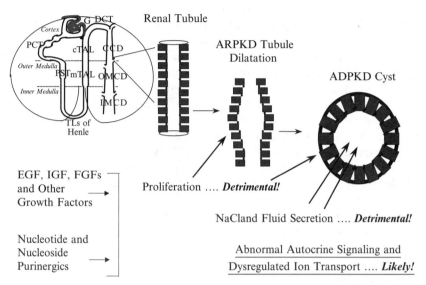

**FIGURE 13** Detrimental purinergic signaling in PKD. In autosomal recessive PKD (ARPKD), renal tubules are abnormally dilated. In autosomal dominant PKD (ADPKD), the dilated tubules "pinch off," causing fluid-filled cysts encapsulated by a single layer of renal epithelial cells. In the latter, NaCl and fluid secretion into the cyst is detrimental and may be augmented by nucleotide or nucleoside agonists and by growth factors released into the cyst lumen. In both disease versions of PKD, growth factors and purinergic factor also may act as mitogens to stimulate proliferation. As a result, cysts will grow faster and increase in volume, which is detrimental on both counts.

detrimental. Growth factors are normally released apically from the cell. Therefore, in some cell models, purinergic agonists were observed as mitogens or comitogens with growth factors. Therefore, ATP release into the cyst lumen could further prove to be detrimental.

## V. FUTURE DIRECTIONS

Our intent in this chapter was to show how purinergic signaling is shared ubiquitously by virtually all epithelia, regardless of the tissue or organ in which it functions. We introduced the concept of epithelial microenvironments to make the clear argument that the anatomy and composition of that microenvironment indicated, in large measure, how efficient and robust that purinergic signaling would be in that tissue. Although many epithelial cell functions were regulated in a similar manner either positively or negatively, we also tried to point out unique roles for purinergic signaling in specific tissues. Finally, we argued a primary or secondary role for epithelial purinergic signaling in diseases of epithelia, including CF and PKD.

Future directions are many. Much of the work that has been done on epithelial ATP release mechanisms, receptors, and purinergic-regulated functions has been done on *in vitro* preparations. In the early chapter on "Cellular Mechanisms of ATP Release," we declared that assays to detect released ATP must be adapted to *in vivo* or, at the very least, "*in vivo*-like" preparations such as freshly dissected tissue or even *in vivo* imaging where bioluminescence detection of released ATP is performed. Moreover, to assess the impact of purinergic signaling in *in vivo* preparations, imaging of cell height as a measure of cell volume with the aid of membrane dyes to label apical and basolateral surfaces and/or imaging key inorganic ions such as $Cl^-$, $Na^+$, and $Ca^{2+}$ or intracellular signaling ($Ca^{2+}$, pH) are necessary next steps. A logical progression of this work would be to compare normal and diseased mouse models to assess the impact of purinergic signaling (its loss, its detrimental presence, or its gain) on a pathophysiological epithelial paradigm. Purinergic receptor expression and purinergic regulation of epithelial function may be different in *in vitro* versus *in vivo* systems. A subtle difference in receptor expression may and would cause significant differences in signaling. In some respects, our own laboratory has published much of our work on *in vitro* systems, and, in this section, we are criticizing our own work. For us, it was critical to lay the foundations for the study of epithelial ATP release, ATP receptors, and ATP-regulated function on *in vitro* epithelial cell model systems from many tissues. We are still learning much from these systems and will continue to study them as we progress to new and important *in vivo* work.

One issue is quite clear. Numerous laboratories, working in their favorite epithelial model systems, will continue to learn important new information about epithelial purinergic signaling in health and disease. In particular, the epithelium as a source of released nucleotides and nucleosides, the epithelium as a robust location for purinergic receptor expression for those released purinergic agonists, and the epithelium as a tissue that is regulated by extracellular nucleotides and nucleosides will continue to be important foci for this research. Key questions include the following: (1) Why would the epithelium release ATP or adenosine across both its apical and basolateral surfaces? (2) Why does a given epithelial cell often express multiple P2Y, P2X, and P1 receptor subtypes and, often, in the same membrane domain? (3) Why does purinergic signaling govern so many different ion, solute, and water transport pathways in many different epithelial cell model systems? (4) Why is ATP release and signaling virtually ubiquitous in epithelial (and nonepithelial) cell systems? An integrated answer that we are pursuing for all of these questions is both functional and protective: to drive epithelial cell volume regulation as a protective mechanism for a cell that often faces very different microenvironments on each surface! Multiple, redundant receptors would also be needed for protection. Multiple ion, solute, and water channels and transporters would have to be regulated properly and at the same time in an integrated manner. We believe that integrated purinergic signaling fits.

## Acknowledgments

G.M.B. and E.M.S. are supported by R01 NIH Grants DK-54367-03 and HL-63934-03 to study ATP release mechanisms, ATP receptors (P2X receptor channels, in particular), and CFTR and the role of all three steps in autocrine control of human airway epithelial cell volume regulation. We are also grateful for past support from the Cystic Fibrosis Foundation for airway epithelial work and from the Polycystic Kidney Research Foundation and the American Heart Association (Southern Research Consortium) for renal epithelial work.

## References

Barrett, K. E., Cohn, J. A., et al. (1990). Immune-related intestinal chloride secretion. II. Effect of adenosine on T84 cell line. Am. J. Physiol. Cell Physiol. **258**(5), C902–C912.

Braunstein, G. M., Roman, R. M., et al. (2001). Cystic fibrosis transmembrane conductance regulator facilitates ATP release by stimulating a separate ATP release channel for autocrine control of cell volume regulation. J. Biol. Chem. **276**, 6621–6630.

Brown, H. A., Lazarowski, E. R., et al. (1991). Evidence that UTP and ATP regulate phospholipase C through common extracellular 5′-nucleotide receptor in human airway epithelial cells. Mol. Pharmacol. **40**, 648–655.

Carre, D. A., Mitchell, C. H., et al. (1997). Adenosine stimulates $Cl^-$ channels of non-pigmented ciliary epithelial cells. Am. J. Physiol. Cell Physiol. **273**, C1354–C1361.

Choi, J. Y., Muallem, D., et al. (2001). Aberrant CFTR-dependent $HCO_3^-$ transport in mutations associated with cystic fibrosis. Nature **410**, 94–97.

Cobb, B. R., Ruiz, F., et al. (2002). A2 adenosine receptors regulate CFTR through PKA and PLA2. Am. J. Physiol. Lung Cell Mol. Physiol. **282**(1), L12–L25.

Collins, F. S. (1992). Cystic fibrosis: Molecular biology and therapeutic implications. Science **256**, 774–779.

Communi, D., Paindavoine, P., et al. (1999). Expression of P2Y receptors in cell lines derived from the human lung. Br. J. Pharmacol. **127**, 562–568.

Cuffe, J. E., Bielfeld-Ackermann, A., et al. (2000). ATP stimulates chloride secretion and reduces sodium absorption in M-1 mouse cortical collecting duct cells. J. Physiol. **524**, 77–90.

Deetjen, P., Thomas, J., et al. (2000). The luminal P2Y receptor in the isolated perfused mouse cortical collecting duct. J. Am. Soc. Nephrol. **11**, 1798–1806.

Devor, D. C., and Pilewski, J. M. (1999). UTP inhibits $Na^+$ absorption in wild-type and DeltaF508 CFTR-expressing human bronchial epithelia. Am. J. Physiol. Cell Physiol. **276**, C827–C837.

Dezaki, K., Tsumura, T., et al. (2000). Receptor-mediated facilitation of cell volume regulation by swelling-induced ATP release in human epithelial cells. Jpn. J. Physiol. **50**, 235–241.

Erlinge, D. (1998). Extracellular ATP: A growth factor for vascular smooth muscle cells. Gen. Pharmacol. **31**, 1–8.

Frame, M. K., and de Feijter, A. W. (1997). Propagation of mechanically induced intercellular calcium waves via gap junctions and ATP receptors in rat liver epithelial cells. *Exp. Cell Res.* **230**, 197–207.

Friedman, Z., Hackett, S. F., *et al.* (1989). Human retinal pigment epithelial cells in culture possess A2-adenosine receptors. *Brain Res.* **492**, 29–35.

Fujihara, T., Murakami, T., *et al.* (2001). Improvement of corneal barrier function by the P2Y(2) agonist INS365 in a rat dry eye model. *Invest. Ophthalmol. Vis. Sci.* **42**, 96–100.

Gordjani, N., Nitschke, R., *et al.* (1997). Capacitative $Ca^{2+}$ entry (CCE) induced by luminal and basolateral ATP in polarized MDCK-C7 cells is restricted to the basolateral membrane. *Cell Calcium* **22**, 121–128.

Groschel-Stewart, U., Bardini, M., *et al.* (1999). P2X receptors in the rat duodenal villus. *Cell Tissue Res.* **297**, 111–117.

Hazama, A., Fan, H. T., *et al.* (2000). Swelling-activated, cystic fibrosis transmembrane conductance regulator-augmented ATP release and $Cl^-$ conductances in murine C127 cells. *J. Physiol.* **523**, 1–11.

Homolya, L., Steinberg, T. H., *et al.* (2000). Cell to cell communication in response to mechanical stress via bilateral release of ATP and UTP in polarized epithelia. *J. Cell Biol.* **150**, 1349–1360.

Housley, G. D. (2000). Physiological effects of extracellular nucleotides in the inner ear. *Clin. Exp. Pharmacol. Physiol.* **27**, 575–580.

Housley, G. D., Kanjhan, R., *et al.* (1999). Expression of the P2X(2) receptor subunit of the ATP-gated ion channel in the cochlea: Implications for sound transduction and auditory neurotransmission. *J. Neurosci.* **19**, 8377–8388.

Huang, P., Lazarowski, E. R., *et al.* (2001). Compartmentalized autocrine signaling to cystic fibrosis transmembrane conductance regulator at the apical membrane of airway epithelial cells. *Proc. Natl. Acad. Sci. USA* **98**, 14120–14125.

Hwang, T. H., Schwiebert, E. M., *et al.* (1996). Apical and basolateral ATP stimulates tracheal epithelial chloride secretion via multiple purinergic receptors. *Am. J. Physiol. Cell Physiol.* **270**, C1611–C1623.

Inglis, S. K., Collett, A., *et al.* (1999). Effect of luminal nucleotides on $Cl^-$ secretion and $Na^+$ absorption in distal bronchi. *Pflugers Arch.* **438**, 621–627.

Iwase, N., Sasaki, T., *et al.* (1997). ATP-induced $Cl^-$ secretion with suppressed $Na^+$ absorption in rabbit tracheal epithelium. *Respir. Physiol.* **107**, 173–180.

Jiang, C., Finkbeiner, W. E., *et al.* (1993). Altered fluid transport across airway epithelium in cystic fibrosis. *Science* **262**, 424–427.

Kanjhan, R., Housley, G. D., *et al.* (1999). Distribution of the P2X2 receptor subunit of the ATP-gated ion channel in the rat central nervous system. *J. Comp. Neurol.* **407**, 11–32.

King, M., Housley, G. D., *et al.* (1998). Expression of ATP-gated ion channels by Reissner's membrane epithelial cells. *Neuroreport* **9**, 2467–2474.

Kishore, B. K., Ginns, S. M., *et al.* (2000). Cellular localization of P2Y(2) purinoceptor in rat renal inner medulla and lung. *Am. J. Physiol. Renal Physiol.* **278**, F43–F51.

Knepper, M. A., Nielsen, S., *et al.* (1994). Mechanism of vasopressin action in the renal collecting duct. *Semin. Nephrol.* **14**, 302–321.

Knight, G. E., Bodin, P., *et al.* (2002). ATP is released from guinea pig ureter epithelium on distension. *Am. J. Physiol. Renal Physiol.* **282**(2), F281–F288.

Knowles, M. R., Clarke, L. L., *et al.* (1991). Activation by extracellular nucleotides of chloride secretion in airway epithelia of patients with cystic fibrosis. *N. Engl. J. Med.* **325**, 533–538.

Kunzelmann, K., and Mall, M. (2002). Electrolyte transport in the mammalian colon: Mechanisms and implications for disease. *Physiol. Rev.* **82**(1), 245–289.

Langostena, L., Ashmore, J. K., *et al.* (2001). Purinergic control of intercellular communication between Hensen's cells of the guinea-pig cochlea. *J. Physiol.* **531,** 693–706.

Lazarowski, E. R., Mason, S. J., *et al.* (1992). Adenosine receptors on human airway epithelia and their relationship to chloride secretion. *Br. J. Pharmacol.* **106,** 774–782.

Lazarowski, E. R., Paradiso, A. M., *et al.* (1997). UDP activates a mucosal-restricted receptor on human nasal epithelial cells that is distinct from the P2Y2 receptor. *Proc. Natl. Acad. Sci. USA* **94,** 2599–2603.

Lee, J. H., Chiba, T., *et al.* (2001). P2X2 receptor mediates stimulation of parasensory cation absorption by cochlear outer sulcus cells and vestibular transitional cells. *J. Neurosci.* **21,** 9168–9174.

Liu, R., Bell, P. D., *et al.* (2002). Purinergic receptor signaling at the basolateral membrane of macula densa cells. *J. Am. Soc. Nephrol.* **13**(5), 1145–1151.

Luo, X., Zheng, W., *et al.* (1999). Multiple functional P2X and P2Y receptors in the luminal and basolateral membranes of pancreatic duct cells. *Am. J. Physiol. Cell Physiol.* **277,** C205–C215.

Ma, W., Korngreen, A., *et al.* (1999). Extracellular sodium regulates airway ciliary motility by inhibiting a P2X receptor. *Nature* **400,** 894–897.

Marcus, D. C., Sunose, H., *et al.* (1997). P2U purinergic receptor inhibits apical IsK/KvLQT1 channel via protein kinase C in vestibular dark cells. *Am. J. Physiol. Cell Physiol.* **273,** C2022–C2029.

Matsui, H., Grubb, B. R., *et al.* (1998). Evidence for periciliary liquid layer depletion, not abnormal ion composition, in the pathogenesis of cystic fibrosis airway disease. *Cell* **95,** 1005–1015.

McAlroy, H. L., Ahmed, S., *et al.* (2000). Multiple P2Y receptor subtypes in the apical membranes of polarized epithelial cells. *Br. J. Physiol.* **131,** 1651–1658.

McCoy, D. E., Taylor, A. L. *et al.* (1999). Nucleotides regulate NaCl transport in mIMCD-K2 cells via P2X and P2Y purinergic receptors. *Am. J. Physiol. Renal Physiol.* **277,** F552–F559.

Merriman-Smith, R., Tunstall, M., *et al.* (1998). Expression profiles of P2-receptor isoforms P2Y1 and P2Y2 in the rat lens. *Invest. Ophthalmol. Vis. Sci.* **39,** 2791–2796.

Mitchell, C. (2001). Release of ATP by a human retinal pigment epithelial cell line: Potential for autocrine stimulation through subretinal space. *J. Physiol.* **534,** 193–202.

Mitchell, C. H., Carre, D. A., *et al.* (1998). A release mechanism for stored ATP in ocular ciliary epithelia. *Proc. Natl. Acad. Sci. USA* **95,** 7174–7178.

Mitchell, C. H., Peterson-Yantorno, K. *et al.* (1999). A3 adenosine receptors regulate Cl⁻ channels of nonpigmented ciliary epithelial cells. *Am. J. Physiol. Cell Physiol.* **276,** C659–C666.

Morse, D. M., Smullen, J. L., *et al.* (2001). Differential effects of UTP, ATP, and adenosine on ciliary activity of human nasal epithelial cells. *Am. J. Physiol. Cell Physiol.* **280,** C1485–C1497.

Munoz, D. J., Thorne, P. R., *et al.* (1999). P2X receptor-mediated changes in cochlear potentials arising from exogenous adenosine 5′ triphosphate in endolymph. *Hear. Res.* **138**(1–2), 56–64.

Murcia, N. S., Woychik, R. P., *et al.* (1998). The molecular biology of polycystic kidney disease. *Pediatr. Nephrol.* **12,** 721–726.

Nguyen, T. D., Meichle, S., *et al.* (2001). P2Y11, a purinergic receptor acting via cAMP, mediates secretion by pancreatic duct epithelial cells. *Am. J. Physiol. Gastrointest. Liver Physiol.* **280,** G795–G804.

Okada, Y., Maeno, E., *et al.* (2001). Receptor-mediated control of regulatory volume decrease (RVD) and apoptotic volume decrease (AVD). *J. Physiol.* **532,** 3–16.

Paller, M. S., Schnaith, E. J., *et al.* (1998). Purinergic receptors mediate cell proliferation and enhanced recovery from renal ischemia by adenosine triphosphate. *J. Lab. Clin. Med.* **131**, 174–183.

Ramminger, S. J., Collett, A., *et al.* (1999). P2Y2 receptor-mediated inhibition of ion transport in distal lung epithelial cells. *Br. J. Pharmacol.* **128**, 293–300.

Raybound, N. P., and Housley, G. D. (1997). Variation in expression of the outer hair cell P2X receptor conductance along the guinea-pig cochlea. *J. Physiol.* **498**, 717–727.

Reddy, M. M., Quinton, P. M., *et al.* (1996). Failure of the cystic fibrosis transmembrane conductance regulator to conduct ATP. *Science* **271**, 1876–1879.

Reisin, I. L., Prat, A. G., *et al.* (1994). The cystic fibrosis transmembrane conductance regulator is a dual ATP and chloride channel. *J. Biol. Chem.* **269**, 20584–20591.

Roman, R. M., Feranchak, A. P. *et al.* (1999). Endogenous ATP regulates $Cl^-$ secretion in cultured human and rat biliary epithelial cells. *Am. J. Physiol. Gastrointest. Liver Physiol.* **276**, G1222–G1230.

Ryan, J. S., Baldridge, W. H., *et al.* (1999). Purinergic regulation of cation conductances and intracellular $Ca^{2+}$ in cultured rat retinal pigment epithelial cells. *J. Physiol.* **520**, 745–759.

Sage, C. L., and Marcus, D. C. (2002). Immunolocalization of P2Y4 and P2Y2 purinergic receptors in strial marginal cells and vestibular dark cells. *J. Membr. Biol.* **185**, 103–115.

Satlin, L. M., Sheng, S., *et al.* (2001). Epithelial sodium channels are regulated by flow. *Am. J. Physiol. Renal Physiol.* **280**(6), F1010–F1018.

Sauer, H., Hescheler, J., *et al.* (2000). Mechanical strain-induced Ca(2+) waves are propagated via ATP release and purinergic receptor activation. *Am. J. Physiol. Cell Physiol.* **279**, C295–C300.

Schwiebert, E. M. (2000). Extracellular ATP-mediated propagation of Ca(2+) waves. Focus on "mechanical strain-induced Ca(2+) waves are propagated via ATP release and purinergic receptor activation". *Am. J. Physiol. Cell Physiol.* **279**, C281–C283.

Schwiebert, E. M. (2001). ATP release mechanisms, ATP receptors and purinergic signalling along the nephron. *Clin. Exp. Pharmacol. Physiol.* **28**, 340–350.

Schwiebert, E. M., and Kishore, B. K. (2001). Extracellular nucleotide signaling along the renal epithelium. *Am. J. Physiol. Renal Physiol.* **280**, F945–F563.

Schwiebert, E. M., Egan, M. E., *et al.* (1995). CFTR regulates outwardly rectifying chloride channels through an autocrine mechanism involving ATP. *Cell* **81**, 1063–1073.

Schwiebert, E. M., Wallace, D. P., *et al.* (2002). Autocrine extracellular purinergic signaling in epithelial cells derived from polycystic kidneys. *Am. J. Physiol. Renal Physiol.* **282**, F763–F775.

Sitaraman, S. V., Wang, L., *et al.* (2002). The adenosine A2b receptor is recruited to the plasma membrane and associates with E3KARP and ezrin upon agonist stimulation. *J. Biol. Chem.* **277**, 33188–33195.

Smith, J. J., Travis, S. M., *et al.* (1996). Cystic fibrosis airway epithelia fail to kill bacteria because of abnormal airway surface fluid. *Cell* **85**, 229–236.

Smitham, J. E., and Barrett, K. E. (2001). Differential effects of apical and basolateral uridine triphosphate on intestinal epithelial chloride secretion. *Am. J. Physiol. Cell Physiol.* **280**, C1431–C1439.

Sorensen, C. E., and Novak, I. (2001). Visualization of ATP release in pancreatic acini in response to cholinergic stimulus. Use of fluorescent probes and confocal microscopy. *J. Biol. Chem.* **276**, 32925–32932.

Strohmeier, G. R., Reppert, S. M., *et al.* (1995). The A2b adenosine receptor mediates cAMP responses to adenosine receptor agonists in human intestinal epithelia. *J. Biol. Chem.* **270**, 2387–2394.

Stutts, M. J., Lazarowski, E. R., *et al.* (1995). Activation of CFTR Cl⁻ conductance in polarized T84 cells by luminal extracellular ATP. *Am. J. Physiol. Cell Physiol.* **268**, C425–C433.

Sugita, M., Yue, Y., *et al.* (1998). CFTR Cl⁻ channel and CFTR-associated ATP channel: Distinct pores regulated by common gates. *EMBO J.* **17**, 898–908.

Sullivan, D. M., Erb, L., *et al.* (1997). Identification and characterization of P2Y2 nucleotide receptors in human retinal pigment epithelial cells. *J. Neurosci. Res.* **49**, 43–52.

Taylor, A. L., Kudlow, B. A., *et al.* (1998). Bioluminescence detection of ATP release mechanisms in epithelia. *Am. J. Physiol. Cell Physiol.* **275**(44), C1391–C1406.

Taylor, A. L., Schwiebert, L. M., *et al.* (1999). Epithelial P2X purinergic receptor channel expression and function. *J. Clin. Invest.* **104**, 875–884.

Valverde, M. A., O'Brien, J. A., *et al.* (1995). Impaired cell volume regulation in intestinal crypt epithelia of cystic fibrosis mice. *Proc. Natl. Acad. Sci. USA* **92**, 9038–9041.

Van Scott, M. R., Chinet, T. C., *et al.* (1995). Purinergic regulation of ion transport across nonciliated bronchial epithelial (Clara) cells. *Am. J. Physiol. Lung Cell Mol. Physiol.* **269**, L30–L37.

Wang, D. J., Huang, N. N., *et al.* (1992). Extracellular ATP and ADP stimulate proliferation and porcine aortic smooth muscle cells. *J. Cell Physiol.* **153**, 221–233.

Wang, Y., Roman, R. M., *et al.* (1996). Autosomal signaling through ATP release represents a novel mechanism for cell volume regulation. *Proc. Natl. Acad. Sci. USA* **93**, 12020–12025.

Wax, M., Sanghavi, D. M., *et al.* (1993). Purinergic receptors in ocular ciliary epithelial cells. *Exp. Eye Res.* **57**, 89–95.

Welsh, M. J., and Smith, A. E. (1993). Molecular mechanisms of CFTR chloride channel dysfunction in CF. *Cell* **73**, 1251–1254.

Welsh, M. J., Tsui, L. C., *et al.* (1995). Cystic fibrosis. In "The Metabolic and Molecular Bases of Inherited Disease," Vol. III, (7th Ed.) McGraw-Hill, New York.

Wilson, P. D. (1997). Epithelial cell polarity and disease. *Am. J. Physiol. Renal Physiol.* **272**, F434–442.

Wilson, P. D., Hovater, J. S., *et al.* (1999). ATP release mechanisms in primary cultures of epithelia derived from the cysts of polycystic kidneys. *J. Am. Soc. Nephrol.* **10**, 218–229.

Zhang, Y., and Engelhardt, J. F. (1999). Airway surface fluid volume and chloride content in cystic fibrosis and normal bronchial xenografts. *Am. J. Physiol. Cell Physiol.* **276**(2), C469–C476.

Zsembery, A., Spirli, C., *et al.* (1998). Purinergic regulation of acid/base transport in human and rat biliary epithelial cell lines. *Hepatology* **28**, 914–920.

Zsembery, A., Boyce, A. T., *et al.* (2002). Dissection of P2Y and P2X nucleotide receptor-triggered increases in cytosolic calcium in human cystic fibrosis airway epithelial cells. *J. Physiol.* (Submitted).

# CHAPTER 8

# Nucleotide Release and Purinergic Signaling in the Vasculature Driven by the Red Blood Cell

**Randy S. Sprague,**[*] **Mary L. Ellsworth,**[*] **and Hans H. Detrich**[†]
[*]St. Louis University School of Medicine, St. Louis, Missouri 63104
[†]Washington University School of Medicine, St. Louis, Missouri 63110

## I. INTRODUCTION

When exposed to any of a number of chemically dissimilar soluble mediators, the endothelium of blood vessels is capable of the synthesis and release of factors that relax the underlying vascular smooth muscle. Indeed, the use of agonists such as acetylcholine (Furchgott and Zawadzki, 1980), bradykinin (Bogle *et al.*, 1991), as well as adenosine triphosphate (ATP) and adenosine diphosphate (ADP) (Busse *et al.*, 1998) to stimulate endothelial cells to produce endothelium-derived relaxing factors (EDRFs) has been instrumental in establishing the role of the endothelium as a critical determinant of vascular caliber (Palmer *et al.*, 1987; Ignarro *et al.*, 1988). Although such studies establish that EDRFs are capable of producing relaxation of vascular smooth muscle, they do not necessarily address the mechanisms responsible for the stimulation of the synthesis and release of these relaxing factors under *in vivo* conditions or in intact organs.

In most intact vascular beds total blood flow and the distribution of that flow within an organ are dictated by tissue function and/or metabolic need. Therefore, if EDRFs are determinants of vascular caliber, and, thereby, the magnitude and distribution of flow within individual vascular beds, then it would be anticipated that local stimuli would be important determinants of their synthesis and release. One example of a local stimulus for EDRF synthesis would be the reduction in oxygen tension and/or increase in hydrogen ion concentration that occur in striated muscle when metabolic need exceeds delivery of oxygen and nutrients. Similarly, increases in shear stress applied to the endothelium of blood vessels as the result of increments in the velocity of blood flow have been shown to stimulate the local synthesis of EDRFs (Rubanyi *et al.*, 1986; Buga *et al.*, 1991).

## II. THE SOURCE OF SOLUBLE MEDIATORS THAT REGULATE EDRF SYNTHESIS

The hypothesis that soluble mediators participate in the local control of blood flow via the stimulation of EDRF synthesis demands that the source of these mediators be in close proximity to the endothelium and that the mediator be released in response to physiological stimuli. The possible sources of such mediators include the endothelium itself, vascular smooth muscle, nervous tissue, other organ-specific cell types, and circulating elements in the blood. Recent evidence suggests that one of the components of blood, the red blood cell, is an important source of a mediator that, when released locally, produces endothelium-dependent vasodilation (Bergfeld and Forrester, 1992; Sprague *et al.*, 1995, 1998a,b; McCullough *et al.*, 1997; Collins *et al.*, 1998). This mediator is ATP. ATP is of particular interest because it is synthesized in the red blood cell and is present in millimolar amounts (Miseta *et al.*, 1993; Bergfeld and Forrester, 1992; Sprague *et al.*, 1996, 1998a). Indeed, ATP is the major adenine nucleotide present in the red blood cell [ratios for intracellular concentrations of ATP, ADP, and adenosine monophosphate (AMP) are 100:10:1, respectively] (Miseta *et al.*, 1993).

When released from the red blood cell, ATP can interact with one or more purinergic receptors present in blood vessels (Houston *et al.*, 1987; Liu *et al.*, 1989; Communi *et al.*, 1991; Motte *et al.*, 1993). Members of the P2x purinergic receptor family are present primarily on vascular smooth muscle cells and their activation results in contraction of that cell (Gilman, 1987; Carthy and Iyenger, 1990; Meinkoth *et al.*, 1993). In contrast, the P2y receptor subtypes are found primarily on the endothelium (Houston *et al.*, 1987; Liu *et al.*, 1989; Communi *et al.*, 1991; Motte *et al.*, 1993). The binding of ATP to the endothelial P2y receptor results in the synthesis of nitric oxide (NO) (McCullough *et al.*, 1997; Collins *et al.*, 1998) and/or vasodilator arachidonic acid metabolites (Offermanns and Simon, 1996; Clapham and Neer, 1997). Thus, ATP applied directly to the vascular smooth muscle of an intact vessel, e.g., that released from nerve terminals, would be expected to produce vasoconstriction via activation of P2x receptors. In contrast, ATP applied to the luminal side of a vessel, e.g., that released within the circulation from red blood cells, would be expected to produce endothelium-dependent vasodilation through interaction with the P2y receptors present on the endothelial cell and the subsequent release of EDRFs.

## III. EVIDENCE FOR THE RELEASE OF ATP FROM RED BLOOD CELLS IN RESPONSE TO PHYSIOLOGICAL STIMULI

In 1992, Bergfeld and Forrester reported that ATP was released from isolated human red blood cells when they were exposed to reduced oxygen tension in the presence of increased carbon dioxide tension and acidosis. Although this study demonstrated that ATP is released from red blood cells in response to physiological stimuli, the experimental design did not permit assessment of the relative contribution of individual gas tensions or pH to that release. These issues were clarified in the studies of Ellsworth *et al.* (1995) in which it was demonstrated that ATP was released from hamster red blood cells in response to *either* reduced oxygen tension *or* reduced pH in the presence of a physiological carbon dioxide tension. Importantly, it was subsequently determined that in addition to hamster red blood cells, the erythrocytes of rats, rabbits, and humans release ATP in response to reductions in oxygen tension (Ellsworth *et al.*, 2000). These observations are consistent with the hypothesis that red blood cell-derived ATP could serve as an important determinant of blood flow distribution in tissues in which oxygen delivery is well regulated.

In contrast to other vascular beds, in the lung, red blood cells are rarely exposed to marked reductions in oxygen tension or pH. Indeed, under normal conditions, the oxygen tension and pH of mixed venous blood entering the lung are greater than that required for release of significant amounts of ATP (Ellsworth, 2000). In addition, the pulmonary circulation differs from other circulations in the body in another aspect, i.e., unlike other organs, the lung must, at all times, accept the entire cardiac output.

However, despite large variations in total blood flow, vascular resistance in the lung is 10-fold less than that of the systemic circulation. Therefore, if red blood cell-derived ATP is a determinant of vascular resistance in the lung, some stimulus for the release of that ATP other than reductions in oxygen tension and pH must be operative.

Several lines of evidence suggest that mechanical forces applied to the red blood cell are capable of stimulating the release of ATP. In 1931, Zipf reported that defibrinated blood released a "depressor substance" that when administered to anesthetized rabbits, resulted in a decrease in systemic blood pressure. The substance was suggested to be an adenine nucleotide. Some 20 years later it was recognized that when blood was agitated by passage through a perfusion pump, a substance was released that produced vasodilation in the hind limb of the cat (Folkow, 1952). The red blood cell was proposed to be the source of this substance and it was suggested that the vasodilator released was ATP. These observations provide support for the hypothesis that red blood cells contain a potent vasodilatory substance, likely ATP, that can be released in the intact circulation. Moreover, these studies suggested that amounts of ATP released from red blood cells increased *pari passu* with the strength of the stimulus, i.e., with increased agitation (Zipf, 1931; Folkow, 1952). One interpretation of these findings is that mechanical deformation of red blood cells, as would occur when these cells traverse the microcirculation, results in the release of ATP. To more directly address this hypothesis, experiments were performed in which red blood cells of humans, rabbits, and dogs were deformed by passage through filters with calibrated pores of decreasing size, i.e., increasing mechanical deformation (Sprague *et al.*, 1996, 1998b). These species were chosen because in both rabbits and humans, NO *is* a determinant of pulmonary vascular resistance, whereas, in the dog, NO *does not* subserve this role in the lung (Nishiwaki *et al.*, 1992; Leeman *et al.*, 1994; Sprague *et al.*, 1996). It was determined that mechanical deformation of red blood cells of humans and rabbits results in the release of ATP and that amounts of ATP released increased in a stimulus-dependent fashion, i.e., as the degree of deformation applied to the red blood cell increased (Sprague *et al.*, 1996). Importantly, amounts of ATP released in response to passage of rabbit red blood cells through the intact pulmonary circulation were of the same magnitude as that released in response to mechanical deformation produced by filtration (Sprague *et al.*, 1996, 1998a). In contrast to human and rabbit red blood cells, dog erythrocytes neither released ATP in response to mechanical deformation nor stimulated NO synthesis in the isolated perfused rabbit lung (Sprague *et al.*, 1996). These results demonstrate that in addition to reductions in oxygen tension and/or pH, increases in mechanical deformation of human and rabbit erythrocytes, as would occur with increments in the velocity of flow through a vessel, is a stimulus for ATP release from these cells. Moreover, this ATP could then serve as a stimulus for endogenous NO synthesis. Thus, under this hypothesis, ATP released from red blood cells in response to physiological stimuli would act on endothelial cells to stimulate endogenous EDRF synthesis enabling the erythrocyte to participate in the local regulation of vascular caliber.

## IV. ROLE OF RED BLOOD CELL-DERIVED ATP IN SPECIFIC VASCULAR BEDS

### A. Striated Muscle

To make a case that red blood cell-released ATP is a component of blood flow regulation, intraluminal application of ATP must produce vasodilation of peripheral resistance vessels because it is these vessels that are responsible for the regulation of tissue perfusion. Not only must the ATP induce a local vasodilation, but the change in vascular caliber must also be conducted upstream for perfusion to be markedly enhanced (Kurjiaka and Segal, 1995; Segal and Duling, 1987). By definition, conducted vasomotor responses are alterations in vascular caliber that transcend the point of initiation, enabling the coordination of microvascular changes. Although the mechanism for such communication is complex, a component of the cellular response likely involves electronic spread through gap junctions. (Segal and Duling, 1987, 1989; Christ $et\ al.$, 1996).

In 1997, McCullough $et\ al.$ described a series of studies carried out in the hamster cheek pouch retractor muscle evaluating the effectiveness of intraluminal ATP as an initiator of conducted vasomotor responses. Using $in\ vivo$ video microscopy it was determined that the intraluminal micropressure application of 40 and 400 pl (10 psi for durations of 50 or 500 ms, respectively) of ATP at concentrations ranging from $10^{-8}$ to $10^{-4}\ M$ produced local vasodilation as well as vasodilation at sites 45–1750 $\mu$m upstream from the site of application. The maximum increase in diameter resulted from applications of $10^{-6}\ M$ ATP. Neither a buffer control solution nor $10^{-6}\ M$ adenosine had any influence on upstream diameter, although adenosine did induce a local response. Because flow is related to vessel radius by the fourth power, as described by Poiseuille's law, the observed 8% increase in vascular diameter produced by intraluminal ATP administration would be associated with an approximate 17% increase in flow. In addition, because the response was conducted, the increase in perfusion would, according to Segal and Duling (1987), likely be significantly greater than this, a finding previously confirmed by Kurjiaka and Segal (1995). McCullough $et\ al.$ (1997) also reported, based on literature values for flow

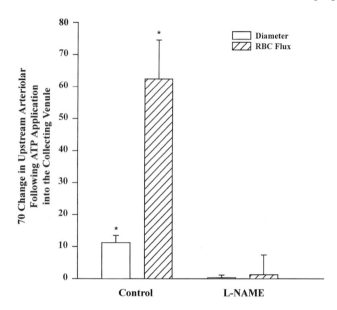

**FIGURE 1**   Percent change in upstream arteriolar diameter (open boxes) and red blood cell flux (hatched boxes) following application of 400 pl $10^{-6}$ $M$ ATP into the collecting venule. Changes are presented for control vessels and for vessels following the administration of 0.4 mg/ 100 g body weight L-NAME (Sigma). Values are mean $\pm 1$ SD. *Significant change $p < 0.05$.

in the retractor muscle (Kuo and Pittman, 1988, 1990; Swain and Pittman, 1989) and values for ATP release from hamster red blood cells in response to physiological stimuli (Ellsworth *et al.*, 1995), that the local concentration of ATP in a vessel of similar size would be $10^{-6}$ $M$, the ATP concentration at which they observed the maximum response in their micropressure experiments. Because oxygen tension on the arteriolar side of the skeletal muscle circulation would not be expected to be at the levels previously demonstrated to be associated with ATP release, a similar set of studies was conducted in which $10^{-6}$ $M$ ATP was applied into the collecting venule. Vessel diameter changes in the associated feed arteriole were monitored and blood flow in these vessels determined using fluorescently labeled red blood cells (Sarelius and Duling, 1982). The results of these studies demonstrated that ATP applied to the collecting venule produced a conducted vasodilator response in the feed arteriole of magnitude similar to that observed with intraarterial application. Following ATP administration, red blood cell flow increased markedly consistent with predictions based on Poiseuille's law (Fig. 1). The finding that ATP applied to a venule produced arteriolar dilation implies that the stimulus for arteriolar vasodilation had to traverse the capillary network. Thus, these studies provided evidence in support of the hypothesis that endothelial cells participate in the transmission of the signal for the conducted vasodilator response to ATP.

It is clear that physiological levels of ATP can induce a conducted vasodilation, although the mechanisms by which this response occurs are unclear. The EDRF, NO, has been reported to be involved in the responsiveness of the vasculature to ATP. Thus, a logical extension of the previous studies was to ascertain if NO is involved in the response of the hamster striated muscle microvasculature to intraluminal ATP. To evaluate the role of NO, animals were treated with $N^{\omega}$-nitro-L-arginine methyl ester [L-NAME; 0.4 mg/100 g body weight, intravenously (iv); Sigma], an inhibitor of NO synthesis followed by an assessment of the responsiveness of the microvasculature to ATP. It was determined that in the presence of L-NAME, the conducted vasodilation induced by both intra-arteriolar (McCullough *et al.*, 1997) and intra-venular (Collins *et al.*, 1998) $10^{-6}$ $M$ ATP was eliminated (Fig. 1). In animals administered a 100-fold excess of L-arginine (50 mg/100 g body weight, iv; Sigma) prior to the administration of L-NAME the response to $10^{-6}$ $M$ ATP administration was not different from that observed in the absence of L-NAME. Both studies concluded that NO was involved in the response although neither was able to conclude the mechanism by which NO exerted its effect. These results do not preclude a role for other EDRFs in the response, especially as at higher concentrations of ATP ($10^{-4}$ $M$), the administration of L-NAME was unable to completely inhibit the conducted vasomotor response.

To evaluate the ability of intraluminal ATP to alter tissue oxygen levels, a 400 pl pulse of $10^{-6}$ $M$ ATP was injected into the lumen of a venule via a micropressure system. In addition to measurements of capillary red blood cell supply rate, tissue $P_{O_2}$ was measured using a microelectrode positioned outside of a capillary within the network that feeds the collecting venule. We found that intraluminal ATP induced an 8% increase in upstream arteriolar diameter and a 33% increase in capillary red blood cell supply rate coincident with a 2–3 mm Hg increase in tissue $P_{O_2}$ (Fig. 2). When the ATP was replaced with buffer no change in red blood cell supply rate or in tissue $P_{O_2}$ was observed (Ellsworth, 2000). These results further support the idea that ATP released from red blood cells can serve as a regulator of blood flow distribution enhancing oxygen supply to the tissue.

## B. Cerebral Circulation

In the cerebral circulation ATP is a natural agonist that displays prominent vasoactivity. Sources for ATP in the brain include corelease from

250                                                                      Sprague *et al.*

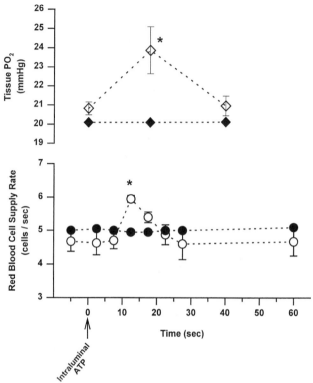

**FIGURE 2** Capillary red blood cell supply rate and tissue Po$_2$ following intraluminal application of 400 pl $10^{-6}$ *M* ATP (open symbols) and buffer control (closed symbols) into collecting venule at time 0. *Significant change from baseline.

perivascular sympathetic nerves to modulate pressor responses (Muramatsu and Kigoshi, 1987; Muramatsu *et al.*, 1981). Such neuronally liberated ATP could overflow into the vessel lumen to elicit vasomotor responses (Bevan *et al.*, 1987). Release from cerebral nerves (Kuroda and McIlwain, 1974) may act on cerebral vessels both directly via diffusion or indirectly through astrocyte stimulation (Centemeri *et al.*, 1997; Harder *et al.*, 1998). ATP is also released from red blood cells during reduced oxygen tension and/or low pH (Ellsworth *et al.*, 1995; Bergfeld and Forrester, 1992) resulting in vasodilation and as such may be a mechanism for sensing metabolic tissue need (Bergfeld and Forrester, 1992). Thus, intraluminally released ATP may be involved in maintaining blood flow in the brain by stimulating endogenous NO synthesis resulting in vasodilation in response to metabolic need.

The hypothesis that ATP released from red blood cells can produce vasodilation of cerebral microvessels was tested by measurement of the change in diameter of isolated and cannulated rat penetrating arterioles in response to reductions in oxygen tension. The isolated vessels were perfused with either physiological salt solution (PSS) alone, PSS containing dextran to increase viscosity to that of blood, or PSS containing red blood cells. Vessels were perfused during normoxia ($Po_2 \sim 21\%$) and reduced oxygen tension ($Po_2 \sim 10\%$). The ATP content of the microvessel effluent was determined at both oxygen tensions in the presence of the various perfusates (Dietrich et al., 2000). Only vessels perfused with red blood cells dilated in response to reduced oxygen tension. Importantly, only in the presence of red blood cells did reductions in oxygen tension result in an increase in ATP efflux from the isolated vessels. These results provide support for the hypothesis that ATP released from red blood cells in response to metabolic need may be a mechanism for the regulation of cerebral microvascular blood flow.

In cerebral microvessels, purinergic receptor distribution has only recently been characterized. Thus, it was reported that ATP elicits both vasodilator and vasoconstrictor responses following intra- or extra-luminal administration in vivo and in vitro (Dietrich et al., 1996; Janigro et al., 1996, 1997; Kajita et al., 1996; Mayhan, 1992; Rosenblum et al., 1994; You et al., 1999). Importantly, ATP locally applied to cerebral penetrating arterioles not only caused vasomotor responses at the site of stimulation, but, in addition, vasoactivity was conducted along these vessels (Dietrich et al., 1996) to upstream sites. The latter finding provides additional support for the hypothesis that ATP is a participant in the regulation of cerebral blood flow.

In addition to cerebral penetrating arterioles, purinergic receptor distribution in pial and cortical arterioles has also been investigated (Lewis et al., 2000; You et al., 1999; Horiuchi et al., 2001). In isolated nonpressurized pial arterioles, Lewis et al. (2000) found that ATP produced transient and sustained vasoconstriction, and suggested that these responses were mediated by $P2X_1$-, $P2Y_2$-, and $P2Y_6$-like receptors, respectively. In addition, Horiuchi et al. (2001) reported that in pressurized rat penetrating arterioles, transient constriction to ATP was mediated by smooth muscle $P2X_1$ receptors whereas dilation was mediated by endothelial $P2Y_2$ receptors. Similarly, You et al. (1999) concluded that in larger size penetrating arterioles dilation to intraluminal ATP results from $P2Y_2$ stimulation.

Purinergic endothelial stimulation may cause the release of a number of mediators responsible for the observed vessel responses. Most prominent are EDRFs, which comprise a group of structurally different substances that includes NO (Kelm et al., 1988), vasodilatory prostaglandins such as

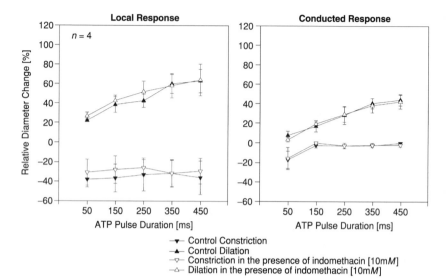

**FIGURE 3** Relative diameter responses of rat penetrating cerebral arterioles to pulses of micro applied ATP. ATP was pressure ejected with pulse durations ranging from 50 ms (weak stimulus) to 450 ms (strong stimulus) and the diameter responses observed at the site of stimulation (local) and distant (at least 500 μm) from the site of stimulation (conducted). ATP caused a transient vessel constriction followed by a dilation. Only the dilation was markedly conducted. Indomethacin (10 mM, cyclooxygenase inhibitor) had no effect on local or conducted vasomotor responses thus excluding prostaglandins as possible mediators for the observed vessel responses.

prostacyclin (PGI$_2$) (De Nucci *et al.*, 1988), and endothelium-derived hyperpolarization factor (EDHF) (Chen and Suzuki, 1991). The latter has not yet been completely characterized, but it may be a metabolite of the cytochrome P-450 oxygenase system (Campbell *et al.*, 1996). We tested the possible contribution of either mediator on ATP-induced local and conducted vasomotor responses by inhibiting key enzymes in the production of either EDRF. Inhibition of prostaglandin production with indomethacin had no effect on the vessel responses (Fig. 3). Inhibition of NO synthesis with N$^W$-monomethyl-L-arginine (L-NMMA) decreased dilation at short ATP pulses but not at strong stimulation (Fig. 4). Inhibition of cytochrome P-450 with clotrimazole had little effect on local responses at short ATP pulses but reduced dilation to long pulses. However, conducted responses were attenuated for all pulse durations (Fig. 5). These initial results indicate that prostaglandins are not involved in ATP-induced vasomotor responses. NO appears to be released at short stimulation pulses and contributes both to local and conducted vasomotor responses. In contrast, EDHF appears to

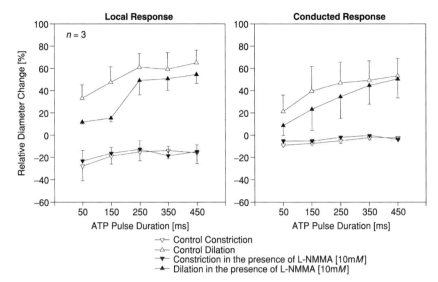

**FIGURE 4** Local and conducted vasomotor responses to ATP (see Fig. 3) before and after inhibition of nitric oxide synthesis by L-NMMA (10 m$M$). Local and conducted vessel responses were attenuated at short pulse durations but not at long pulses indicating that nitric oxide may be released at low ATP concentrations, but another mediator may be released at higher pulses, possibly EDHF (Fig. 5).

be released locally at higher ATP pulses and is involved in the conducted vasomotor response. Thus ATP may initially release NO at low concentrations, which is supplanted by EDHF release at greater ATP stimuli. Our observation is consistent with the hypothesis that for weak stimulation there is a calcium threshold of up to 200 nmol/liter at which NO is produced in response to endothelial stimulation. Stronger stimulations cause greater increases in calcium activity and the production of arachidonic acid derivatives (Parsaee *et al.*, 1992).

In rat cerebral arterioles, intraluminal ATP could contribute to the control of local blood flow by mediating both local and conducted vasodilation in response to metabolic need (Parsaee *et al.*, 1992). ATP may cause the release of both NO at low concentration and at higher concentration EDHF in addition to NO to cause the observed vessel responses via P2Y$_2$ stimulation. Because ATP can diffuse from the vessel lumen to arteriolar smooth muscle cells (McCullough *et al.*, 1997) transient smooth muscle constriction, caused by P2X receptor stimulation, may counteract an overdilation due to large amounts of intraluminal ATP and, thereby, prevent shear-induced damage of the intima of these microvessels

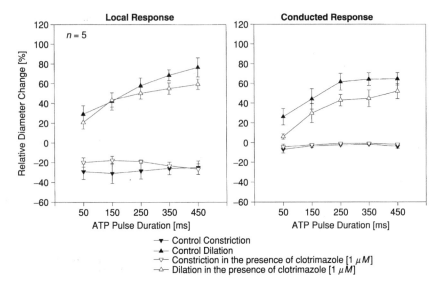

**FIGURE 5**   Local and conducted vasomotor responses to ATP (see Fig. 3) before and after inhibition of cytochrome P-450 by clotrimazole (1 $\mu M$). Inhibition of EDHF release affected local responses only at high pulse durations. However, conducted responses were attenuated at all pulses, indicating that EDHF is involved in the observed conducted vessel dilation to ATP.

## C. Pathological Aspects of Cerebrovascular Regulation by ATP

If ATP released from blood cells is of physiological importance, then it should be possible to demonstrate that such regulation is disturbed in pathological conditions such as subarachnoid hemorrhage. During sub-arachnoid hemorrhage, blood flows from a ruptured vessel into the subarachnoid space. Due to red blood cell lysis, massive amounts of oxyhemoglobin are released. In a model of subarachnoid hemorrhage produced by the addition of oxyhemoglobin to isolated rat penetrating arterioles, Kajita *et al.* (1996) determined that conducted vasomotor responses are greatly attenuated, demonstrating that a mechanism important for physiological regulation of the cerebral circulation was impaired.

In addition to the release of oxyhemoglobin, red blood cell lysis also results in the release of large amounts of ATP that interact with purinergic receptors. In the cerebral circulation, composition of arterial smooth muscular P2 receptors varies with vessel size. Although both cerebral macro- and microvessels contain endothelial $P2Y_2$ receptors (You *et al.*, 1999), P2 receptors that mediate constriction of smooth muscle cells vary with vessel size. Cerebral arterioles possess P2X receptors that quickly desensitize

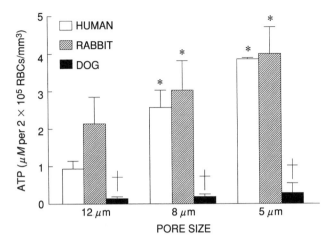

**FIGURE 6**  ATP concentration in filter effluent per $2 \times 10^5$ red blood cells (RBCs)/mm$^3$ in response to passage of human ($n = 5$), rabbit ($n = 12$), and dog ($n = 5$) RBCs through filters with average pore sizes of 12, 8, or 5 $\mu$m. Values are mean $\pm$ SE. $^*p < 0.05$ compared to respective 12 $\mu$m value; $^\dagger p < 0.05$ compared to human and rabbit values.

leading to a transient constriction (Horiuchi *et al.*, 2001). Larger cerebral arteries, however, do not constrict in response to P2X stimulation (You *et al.*, 1999). Indeed, in these vessels, ATP produces a sustained vessel constriction via the P2Y$_2$ receptors (Macdonald *et al.*, 1998; You *et al.*, 1999). With these considerations in mind, it would be anticipated that massive ATP release from brain tissue, such as following subarachnoid hemorrhage, trauma, or platelet activation, may have only a short-term effect on arteriolar vessel constriction (due to P2X$_1$ desensitization). However, in large cerebral arteries, such large amounts of ATP could produce vasospasm due to smooth muscle P2Y$_2$ stimulation. Thus, in the cerebral circulation, differences in purinergic receptor distribution between macro- and microvessels may have important consequences in pathological conditions in which ATP is released.

## D. Pulmonary Circulation

When traversing microvascular beds such as in the pulmonary circulation, red blood cells are subjected to mechanical deformation, a stimulus reported to result in ATP release from rabbit and human red blood cells, but not dog erythrocytes (Fig. 6) (Sprague *et al.*, 1996,

1998a,b). To demonstrate that the ability to release ATP in response to mechanical deformation was associated with the stimulation of endogenous NO synthesis in the pulmonary circulation, experiments were conducted in which rabbit, human, or dog red blood cells were added to the perfusate of isolated rabbit lungs. In lungs perfused with a physiological salt solution in the absence of red blood cells, the addition of the NO synthase inhibitor L-NAME had no effect on pressure–flow relationships, i.e., increases in flow rate alone did not stimulate endogenous NO synthesis. However, when rabbit (Sprague *et al.*, 1996) or human (Sprague *et al.*, 1998a,b) red blood cells were present in the perfusate of isolated lungs, the administration of L-NAME was associated with a shift in the pressure–flow relationship consistent with a decrease in vascular caliber (Fig. 7A and B). In contrast, when dog red blood cells were added to the perfusate, L-NAME was without effect on pressure–flow relationships (Fig. 7C) (Sprague *et al.*, 1996). Although the erythrocytes of humans, rabbits, and dogs all contain millimolar amounts of ATP, as noted above, of the three cell types, only human and rabbit red blood cells release ATP in response to physiological stimuli and induce endogenous NO synthesis in the isolated rabbit lung. Moreover, in additional studies, it was determined that a minimum number of red blood cells that release ATP (hematocrit 20%) was required for stimulation of NO synthesis in this model (Sprague *et al.*, 1995, 1996).

   In support of the hypothesis that the mechanism by which human and rabbit red blood cells stimulate endogenous NO synthesis in the pulmonary circulation is via the release of ATP, studies were performed in which the

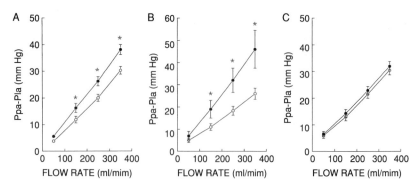

**FIGURE 7**   Pressure–flow relationships before ($\circ$) and after ($\bullet$) administration of $N^\omega$-nitro-L-arginine methyl ester (L-NAME, 100 $\mu M$) in isolated rabbit lungs perfused with physiological salt solution containing red blood cells of either rabbits ($n = 6$, hematocrit $20 \pm 1$, A), humans ($n = 4$, hematocrit $17 \pm 2$, B), or dogs ($n = 5$, hematocrit $32 \pm 2$, C). Values are mean $\pm$ SE. *$p < 0.05$ compared to respective value before L-NAME.

concentration of ATP in the effluent of isolated lungs was determined (Sprague *et al.*, 1998a). In the absence of red blood cells in the perfusate, minimal amounts of ATP were detected in the lung effluent at flow rates ranging from 50 to 300 ml/min. In contrast, when washed rabbit red blood cells were present in the perfusate (hematocrit 20 ± 1%) ATP was present in the effluent of the lung and, importantly, the concentration increased *pari passu* with increments in flow rate (Fig. 8).

If ATP is an important regulator of pulmonary NO synthesis and, thereby, pulmonary vascular resistance, *in vivo*, a hemodynamic response to ATP should be demonstrable in the intact circulation of that organ. Hassessian and Burnstock (1995) reported that in isolated rat lungs perfused with a physiological salt solution, the addition of ATP to the perfusate resulted in a decrease in perfusion pressure, presumably via the activation of P2Y receptors present on the endothelium. Following the administration of an inhibitor of endogenous NO synthesis, the same concentration of ATP evoked an increase in pressure, presumably via activation of P2X receptors present on the vascular smooth muscle now unopposed by the action of endothelium-derived NO. Recently, we have confirmed that the addition of ATP into the perfusate of isolated rabbit lungs perfused with a physiological salt solution results in reductions in pulmonary vascular resistance. In addition, we determined that reduction in total pulmonary

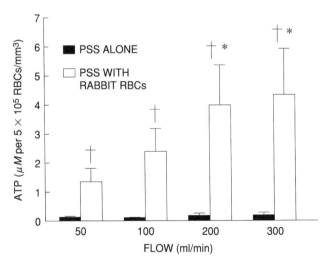

**FIGURE 8**  ATP concentration in the effluent of isolated rabbit lungs perfused at various flow rates with physiological salt solution alone (PSS, $n = 6$) or PSS to which washed rabbit RBCs were added (hematocrit 20 ± 1, $n = 8$). Data are mean ± SE. *$p < 0.05$ compared to respective 50 ml/min value; †$p < 0.05$ compared to PSS alone.

vascular resistance associated with ATP infusion is primarily the result of a decrease in upstream (arterial) resistance. Importantly, in isolated rabbit lungs perfused with rabbit blood, the administration of an inhibitor of NO synthesis (L-NAME) resulted in increases in vascular resistance confined to the upstream segment (Sprague *et al.*, 1994). Thus, both exogenously administered ATP and ATP released from rabbit red blood cells produced similar effects on the intrapulmonary distribution of vascular resistance.

The finding that red blood cells that release ATP in response to physiological stimuli are required for flow-induced NO synthesis in the lung, coupled with the findings that mechanical deformation of red blood cells results in ATP release, which, in turn, can stimulate NO synthesis in the lung, suggests a novel mechanism for the control of pulmonary vascular caliber. In this construct, as the red blood cell is increasingly deformed by increments in the velocity of blood flow through a vessel and/or by reductions in vascular caliber, it releases ATP, which stimulates endothelial synthesis of NO resulting in relaxation of vascular smooth muscle and, thereby, an increase in vascular caliber. This vasodilation results in a decrease in pulmonary vascular resistance as well as a decrease in the stimulus for red blood cell deformation and ATP release. Thus, under this hypothesis, red blood cell-derived ATP contributes to the low resistance to blood flow present in the healthy lung.

## V. PROPOSED SIGNAL-TRANSDUCTION PATHWAY FOR ATP RELEASE FROM RED BLOOD CELLS

If erythrocyte-derived ATP is an important regulator of endogenous EDRF synthesis, it would be anticipated that a signal-transduction pathway exists for the regulation of ATP release from red blood cells in response to physiological stimuli. ATP is a highly charged molecule and, as such, cannot pass easily through cell membranes. Recently a family of membrane-associated proteins grouped together as the ATP binding cassette has been described (Abraham *et al.*, 1993; Reisin *et al.*, 1994; Schwiebert *et al.*, 1995; Al-Awqati, 1995). A member of the family, cystic fibrosis transmembrane conductance regulator (CFTR) (Abraham *et al.*, 1993; Reisin *et al.*, 1994; Schwiebert *et al.*, 1995), was suggested to be a transporter of ATP from the interior of cells to the external milieu. Although not all properties of the CFTR channel have been fully characterized, the activity of this channel has been shown to be inhibited by sulfonylureas such as tolbutamide and glibenclamide (Sheppard *et al.*, 1993; Schultz *et al.*, 1996) as well as by some nonsteroidal antiinflammatory agents such as niflumic acid and ibuprofen

(Cuthbert *et al.*, 1994; Devor and Schultz, 1998). To determine if the activity of CFTR was required for ATP release from red blood cells, experiments were conducted in which rabbit red blood cells were incubated with either glibenclamide or niflumic acid. In the presence of these inhibitors of the activity of CFTR, the release of ATP in response to mechanical deformation was prevented (Sprague *et al.*, 1998b).

The finding that human red blood cells, like those of rabbits, release ATP in response to mechanical deformation (Sprague *et al.*, 1996) presented a unique opportunity for the study of the contribution of CFTR to deformation-induced ATP release from human red blood cells. Cystic fibrosis (CF) is a genetic disorder characterized clinically by a constellation of symptoms related, in large part, to chronic pulmonary disease and pulmonary hypertension (Davis *et al.*, 1996). In spite of the fact that CF patients may have different patterns of organ involvement, severity of disease, and genetic mutation, they share a common pathophysiological problem; the activity of CFTR is markedly diminished or lost (Cheng *et al.*, 1990; Purchelle *et al.*, 1992; Davis *et al.*, 1996). Importantly, in contrast to healthy humans and humans with chronic lung disease not related to CF, red blood cells of CF patients did not release ATP in response to mechanical deformation (Sprague *et al.*, 1998b).

Although these studies implicate CFTR as a component of the signal-transduction pathway for deformation-induced ATP release from rabbit and human erythrocytes, they do not permit determination of its exact role in the process. Thus, CFTR could function as a conduit for ATP, however, it is also possible that CFTR activity is required for activation of another channel that functions as an ATP conduit. The role of CFTR as a conduit for ATP release from several cell types is under active investigation at this time (Reisin *et al.*, 1994; Schwiebert *et al.*, 1995; Hanrahan *et al.*, 1996; Prat *et al.*, 1996; Reddy *et al.*, 1996). Notwithstanding, the finding that CFTR activity is required for deformation-induced ATP release from red blood cells permits the formulation of hypotheses regarding the nature of the signal-transduction pathway responsible for ATP release from red blood cells of rabbits and humans.

The activity of CFTR is regulated by a cAMP-dependent protein kinase (protein kinase A, PKA) (Schwiebert *et al.*, 1995; Hanrahan *et al.*, 1996). Following its activation by cAMP, PKA acts to phosphorylate several sites on CFTR resulting in activation of that channel (Tabcharani *et al.*, 1990). It was reported that both adenylyl cyclase (Sheppard *et al.*, 1969; Rodan *et al.*, 1976; Nakagawa *et al.*, 1984) and PKA (Rubin *et al.*, 1972; Dreyfuss *et al.*, 1978) are present in red blood cells of humans. Incubation of rabbit and human red blood cells with an active cAMP analogue (Sp cAMP) or with agents that stimulate the activity of adenylyl cyclase (forskolin) and prevent cAMP degradation (isobutyl-methyl xanthine) was reported to stimulate ATP release (Sprague *et al.*, 2001). In addition, deformation-induced ATP release from rabbit red blood cells was shown to be inhibited by an inactive cAMP analogue and inhibitor of PKA, Rp cAMP (Sprague *et al.*, 2001). Taken together, the results of these studies provide strong support for the hypothesis that adenylyl cyclase and PKA are components of a signal-transduction pathway that controls the release of ATP from red blood cells.

If increases in cAMP are required for the activation of CFTR, then the regulation of adenylyl cyclase activity would be a critical control point in the pathway that relates physiological stimuli to ATP release from red blood cells. Heterotrimeric G-proteins are major regulators of the activity of adenylyl cyclase and, thereby, the intracellular concentration of cAMP (Gilman, 1987; Offermanns and Simon, 1996; Clapham, 1997). When G-proteins are activated by binding with GTP, they dissociate into $\alpha$ and $\beta/\gamma$ subunits (Gilman, 1987). The $\alpha$ subunit of the G-protein termed Gs can then bind to adenylyl cyclase resulting in its activation. Heterotrimeric G-proteins of the Gs subclass have been reported to be present in the membrane of human and rabbit red blood cells (Wand *et al.*, 1994, Olearczyk *et al.*, 2001). To determine if the heterotrimeric G-protein, Gs, is a component of a signal-transduction pathway for ATP release from erythrocytes, experiments were performed in which rabbit red blood cells were incubated with one of two chemically dissimilar receptor-mediated activators of this G-protein. Incubation of rabbit erythrocytes with the stable prostacyclin analogue, iloprost, resulted in the release of ATP. In addition, ATP release from these cells was stimulated by incubation with epinephrine. Importantly, the effect of epinephrine was prevented by pretreatment of the red blood cells with propranolol, an $\alpha$-adrenergic receptor antagonist. The results of these studies are consistent with the hypothesis that stimulation of the activity of heterotrimeric G-proteins of the Gs subclass could participate in a signal-transduction pathway for ATP release from red blood cells.

In addition, to the $\alpha$ subunit of Gs, it was reported recently that the $\beta/\gamma$ subunit of another G-protein subclass, namely, Gi/o, is capable of activating some isoforms of adenylyl cyclase (Clapham and Neer, 1997). Importantly, G-proteins of the Gi subclass have been shown to be present in the membrane of human red blood cells (Kaslow *et al.*, 1980; Hanski and Gilman, 1982; Codina *et al.*, 1984; Carthy and Iyengar, 1990). The hypothesis that heterotrimeric G-proteins are involved in deformation-induced release of ATP from red blood cells is supported by reports that activation of G-proteins, and in particular Gi/o, occurs when cells are exposed to shear stress (Ohno *et al.*, 1993; Kuchan *et al.*, 1994; Gudi *et al.*,

1996). Moreover, it was reported that a pertussis toxin-sensitive G-protein, presumably of the Gi/o subclass, interacts with CFTR in bovine tracheal membranes (Ismailov *et al.*, 1996). To determine if activation of Gi would stimulate ATP release from red blood cells, experiments were performed in which rabbit and human erythrocytes were incubated with the activator of Gi/o, mastoparin. In the presence of mastoparin, red blood cells from both species released ATP. Pretreatment of rabbit red blood cells with pertussis toxin, an inhibitor of the activity of Gi/o, prevented mastoparin-induced increases in ATP release. Moreover, deformation-induced ATP release from rabbit red blood cells was attenuated by pretreatment with pertussis toxin. Taken together, the results of these studies suggest that in addition to Gs, the heterotrimeric G-protein, Gi, is a component of a pathway that results in the release of ATP from erythrocytes in response to physiological stimuli such as mechanical deformation.

As noted above, the ability of CFTR to function as an ATP conduit remains highly controversial. However, Bergfeld and Forrester (1992) suggested that another channel present in the red blood cell membrane, the nucleoside transporter, could facilitate the movement of ATP out of erythrocytes. Although often thought of as a means for transport of nucleosides into the erythrocyte, the nucleoside transporter was reported to facilitate the movement of uridine both into and out of red blood cells, i.e., the channel is capable of bidirectional transport (Tse *et al.*, 1985). Although the role of this channel as a conduit for ATP egress from the red blood cell has not been well defined, it was reported that incubation of human red blood cells with a selective inhibitor of the nucleoside transporter, nitrobenzyl-6-thioinosine (NBTI) (Jarvis *et al.*, 1981, 1982; Plagemann and Wohlhueter, 1984; Van Belle *et al.*, 1991), inhibited ATP release in response to reduced oxygen tension in the presence of hypercapnia (Bergfeld and Forrester, 1992). The relationship between CFTR and the nucleoside transporter is unknown. However, the hypothesis that CFTR could control the activity of other channels present in the red blood cell membrane, such as the nucleoside transporter, is not unique, i.e., it was suggested that CFTR controls chloride flux out of cells by regulating a distinct outward rectifying chloride channel (Schwiebert *et al.*, 1995; Hanrahann *et al.*, 1996). The role of the nucleoside transporter as well as members of the ABC transporter family as ATP conduits in red blood cells remains under active investigation at this time.

The results of the studies presented above suggest that the red blood cell contains several components that, when linked together, could constitute a signal-transduction mechanism for deformation-induced release of ATP from that cell (Fig. 9).

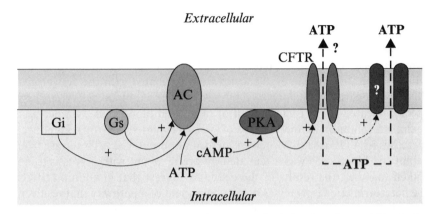

**FIGURE 9**  A proposed signal-transduction pathway for ATP release from red blood cells. +, stimulated; Gs, stimulatory heterotrimeric G-protein; Gi, inhibitory heterotrimeric G-protein; AC, adenylyl cyclase; PKA, protein kinase A; CFTR, cystic fibrosis transmembrane conductance regulator.

## VI. CONCLUSIONS

The finding that red blood cells synthesize ATP and release this vasoactive mediator in response to physiological stimuli suggests that these cells possess the capacity to contribute to the regulation of vascular resistance as well as the distribution of blood flow within vascular beds. Red blood cell-derived ATP could contribute to the control of vascular resistance in striated muscle, the brain, and the lung under both physiological and pathophysiological conditions. In striated muscle and the cerebral circulations, the ability of the red blood cell to release ATP in response to reduced oxygen tension and/or pH or mechanical deformation could aid in the distribution of blood flow to segments of the tissue in which oxygen and nutrient delivery is insufficient to meet metabolic need. Although in the pulmonary circulation erythrocytes are not exposed to the reductions in oxygen tension and pH encountered in exercising muscle, these cells do undergo mechanical deformation. Thus, ATP released from red blood cells as they traverse the extensive microcirculation of the lung could contribute to the low levels of vascular resistance in that organ.

If red blood cell-derived ATP is an important determinant of vascular resistance under physiological conditions, it would be anticipated that derangements in ATP release from erythrocytes would have pathophysiological consequences. Excessive release of ATP associated with lysis of red blood cells, as would occur in subarachnoid hemorrhage, would be anticipated to produce intense vasospasm leading to increased organ injury.

In contrast, the failure of red blood cells to release ATP in response to physiological stimuli would be anticipated to result in the loss of local control of the distribution of blood flow in striated muscle and the cerebral circulation and total pulmonary vascular resistance in the lung. In support of the latter hypothesis, it was reported that red blood cells of humans with primary pulmonary hypertension, a disorder of unknown etiology, fail to release ATP in response to mechanical deformation (Sprague *et al.*, 2001).

In summary, in addition to the traditional view of red blood cells as vehicles for the transport of oxygen from the lung to the peripheral circulation, recent evidence suggests that by virtue of their ability of release ATP in a regulated fashion, these cells could serve as mobile sensors of the adequacy of the distribution of blood flow and vascular resistance under physiological conditions. Moreover, derangements of this mechanism for the local control of vascular caliber could contribute to the pathophysiology of such diverse processes as subarachnoid hemorrhage and pulmonary hypertension.

## References

Abraham, E. H. (1993). The multidrug resistance (mdr1) gene product functions as an ATP channel. *Proc. Natl. Acad. Sci. USA* **90**, 312–316.

Al-Awqati, Q. (1995). Regulation of ion channels by ABC transporters that secrete ATP. *Science* **269**, 805–806.

Bergfeld, G. R., and Forrester, T. (1992). Release of ATP from human erythrocytes in response to a brief period of hypoxia and hypercapnia. *Cardiovasc. Res.* **26**, 40–47.

Bevan, J. A., Lather, I., and Rowan, R. (1987). Some implications of the high intrasynaptic norepinephrine concentrations in resistance arteries. *Blood Vessels* **24**, 137–140.

Bogle, R. G., Coade, S. B., Moncada, S., Pearson, J. D., and Mann, G. E. (1991). Bradykinin and ATP stimulate L-arginine uptake and nitric oxide release in vascular endothelial cells. *Biochem. Biophys. Res. Commun.* **80**, 926–932.

Buga, G. M., Gold, M. E., Fukuto, J. M., and Ignarro, L. J. (1991). Shear stress-induced release of nitric oxide from endothelial cells grown on beads. *Hypertension* **17**, 187–193.

Busse, R., Ogilvie, A., and Pohl, U. (1998). Vasomotor activity of diadenosine triphosphate and diadenosine tetraphosphate in isolated arteries. *Am. J. Physiol.* **254**, H828–H832.

Campbell, W. B., Gebremedhin, D., Pratt, P. F., and Harder, D. R. (1996). Identification of epoxyeicosatrienoic acids as endothelium-derived hyperpolarizing factors. *Circ. Res.* **78**, 415–423.

Carthy, D., and Iyengar, R. (1990). A 43 kDa form of the GTP-binding protein Gi3 in human erythrocytes. *FEBS Lett.* **262**, 101–103.

Centemeri, C., Bolego, C., Abbracchio, M. P., Cattabeni, F., Puglisi, L., Burnstock, G., and Nicosia, S. (1997). Characterization of the $Ca^{2+}$ responses evoked by ATP and other nucleotides in mammalian brain astrocytes. *Br. J. Pharmacol.* **121**, 1700–1706.

Chen, G., and Suzuki, H. (1991). Endothelium-dependent hyperpolarization elicited by adenine compounds in rabbit carotid artery. *Am. J. Physiol.* **260**, H1037–H1042.

Cheng, S. H., Gregory, R. J., Marshall, J., Paul, S., Souza, D. W., White, G. A., O'Riordan, C. R., and Smith, A. E. (1990). Defective intracellular transport and processing of CFTR is the molecular basis of most cystic fibrosis. *Cell* **63**, 827–834.

Christ, G. J., Spray, D. C., El-Sabban, M., Moore, L. K., and Brink, P. R. (1996). Gap junctions in vascular tissues—Evaluating the role of intercellular communication in the modulation of vasomotor tone. *Circ. Res.* **79**, 631–646.

Clapham, D. E., and Neer, E. J. (1997). G protein $\beta\gamma$ subunits. *Annu. Rev. Pharmacol. Toxicol.* **37**, 167–203.

Codina, J., Hildebrant, J. D., Sekura, R. D., Birnbaumer, M., Bryan, J., Manclark, C. R., Iyengar, R., and Birnbaumer, L. (1984). Ns and Ni, the stimulatory and inhibitory regulatory components of adenylyl cyclases. *J. Biol. Chem.* **259**, 5871–5886.

Collins, D. M., McCullough, W. T., and Ellsworth, M. L. (1998). Conducted vascular responses: Communication across the capillary bed. *Microvasc. Res.* **56**, 43–53.

Communi, D., Raspe, E., Pirotton, S., and Boeynaems, J. M. (1991). Coexpression of $P_{2y}$ and $P_{2u}$ receptors on aortic endothelial cells: Comparison of cell localization and signaling pathways. *Circ. Res.* **76**, 191–198.

Cuthbert, A. W., Evans, M. J., Colledge, W. H., MacVinish, L. J., and Ratcliff, R. (1994). Kinin-stimulated chloride secretion in mouse colon requires the participation of CFTR chloride channels. *Braz. J. Med. Biol. Res.* **27**, 1905–1910.

Davis, P. B., Drumm, M., and Konstan, M. W. (1996). Cystic fibrosis. *Am. J. Respir. Crit. Care Med.* **154**, 1229–1256.

De Nucci, G., Gryglewski, R. J., Warner, T. D., and Vane, J. R. (1988). Receptor-mediated release of endothelium-derived relaxing factor and prostacyclin from bovine aortic endothelial cells is coupled. *Proc. Nat. Acad. Sci. USA* **85**, 2334–2338.

Devor, D. C., and Schultz, B. D. (1998). Ibuprofen inhibits cystic fibrosis transmembrane conductance regulator-mediated Cl secretion. *J. Clin. Invest.* **102**, 679–687.

Dietrich, H. H., Kajita, Y., and Dacey, R. G., Jr. (1996). Local and conducted vasomotor responses in isolated rat cerebral arterioles. *Am. J. Physiol. Heart Circ. Physiol.* **271**, H1109–H1116.

Dietrich, H. H., Ellsworth, M. L., Sprague, R. S., and Dacey, R. G. (2000). Red blood cell regulation of microvascular tone through adenosine triphosphate. *Am. J. Physiol.* **278**, H1294–H1298.

Dreyfuss, G., Schwartz, K. J., and Blout, E. R. (1978). Compartmentalization of cyclic AMP-dependent protein kinases in human erythrocytes. *Proc. Natl. Acad. Sci. USA* **75**, 5926–5930.

Ellsworth, M. L. (2000). The red blood cell as an oxygen sensor: What is the evidence? *Acta Physiol. Scand.* **168**, 551–559.

Ellsworth, M. L., Forrester, T., Ellis, C. G., and Dietrich, H. H. (1995). The erythrocyte as a regulator of vascular tone. *Am. J. Physiol. Heart Circ. Physiol.* **269**, H2155–H2161.

Folkow, B. (1952). A critical study of some methods used in investigations on the blood circulation. *Acta Physiol. Scand.* **27**, 118–129.

Furchgott, R. F., and Zawadzki, J. V. (1980). The obligatory role of endothelial cells in the relaxation of arterial smooth muscle by acetylcholine. *Nature* **288**, 373–376.

Gilman, A. (1987). G proteins: Transducers of receptor-generated signals. *Annu. Rev. Biochem.* **56**, 615–649.

Gudi, S.R.P., Clark, C. B., and Frangos, J. A. (1996). Fluid flow rapidly activates G proteins in human endothelial cells. *Circ. Res.* **79**, 834–839.

Hanrahan, J. W., Mathews, C. J., Grygorczyk, R., Tabcharani, J. A., Grzelczak, Z., Chang, X.-B., and Riordan, J. R. (1996). Regulation of the CFTR chloride channel from humans and sharks. *J. Exp. Zool.* **275**, 283–291.

Hanski, E., and Gilman, A. G. (1982). The guanine nucleotide-binding regulatory component of adenylate cyclase in human erythrocytes. *J. Cyclic. Nucleotide Res.* **8**, 323–336.

Harder, D. R., Alkayed, N. J., Lange, A. R., Gebremedhin, D., and Roman, R. J. (1998). Functional hyperemia in the brain—Hypothesis for astrocyte-derived vasodilator metabolites. *Stroke* **29**, 229–234.

Hassessian, H., and Burnstock, G. (1995). Interacting roles of nitric oxide and ATP in the pulmonary circulation of the rat. *Br. J. Pharmacol.* **114**, 846–850.

Horiuchi, T., Dietrich, H. H., Tsugane, S., and Dacey, R. G., Jr. (2001). Analysis of purine- and pyrimidine-induced vascular responses in the isolated rat cerebral arteriole. *Am. J. Physiol. Heart Circ. Physiol.* **280**, H767–H776.

Houston, D. A., Burnstock, G., and Vanhoutte, P. M. (1987). Different $P_2$-purinergic receptor subtypes of endothelium and smooth muscle in canine blood vessels. *J. Pharmacol. Exp. Ther.* **241**, 501–506.

Ignarro, L. J., Buga, G., and Chaudhuri, G. (1988). EDRF generation and release from perfused bovine pulmonary artery and vein. *Eur. J. Pharmacol.* **149**, 79–88.

Ismailov, I. I., Jovov, B., Fuller, C. M., Berdiev, B. K., Keeton, D. A., and Benos, D. J. (1996). G-protein regulation of outward rectified epithelial chloride channels incorporated into planar bilayer membranes. *J. Biol. Chem.* **271**, 4776–4780.

Janigro, D., Nguyen, T. S., Gordon, E. L., and Winn, H. R. (1996). Physiological properties of ATP-activated cation channels in rat brain microvascular endothelial cells. *Am. J. Physiol. Heart Circ. Physiol.* **270**, H1423–H1434.

Janigro, D., Nguyen, T. S., Meno, J., West, G. A., and Winn, H. R. (1997). Endothelium-dependent regulation of cerebrovascular tone by extracellular and intracellular ATP. *Am. J. Physiol. Heart Circ. Physiol.* **273**, H878–H885.

Jarvis, S. M., and Young, J. D. (1981). Extraction and partial purification of the nucleoside-transport system from human erythrocytes based on the assay of nitrobenzylthioinosine-binding activity. *Biochem. J.* **194**, 331–339.

Jarvis, S. M., Hammond, J. R., Paterson, A. R. P., and Clanachan, A. S. (1982). Species differences in nucleoside transport. *Biochem. J.* **208**, 83–88.

Kajita, Y., Dietrich, H. H., and Dacey, R. G., Jr. (1996). Effects of oxyhemoglobin on local and propagated vasodilatory responses induced by adenosine, adenosine diphosphate, and adenosine triphosphate in rat cerebral arterioles. *J. Neurosurg.* **85**, 908–916.

Kaslow, H. R., Johnson, G. L., Brothers, V. M., and Bourne, H. R. (1980). A regulatory component of adenylate cyclase from human erythrocyte membranes. *J. Biol. Chem.* **255**, 3736–3741.

Kelm, M., Feelisch, M., Spahr, R., Piper, H. M., Noack, E., and Schrader, J. (1988). Quantitative and kinetic characterization of nitric oxide and EDRF released from cultured endothelial cells. *Biochem. Biophys. Res. Commun.* **154**, 236–244.

Kuchan, M. J., Jo, H., and Frangos, J. A. (1994). Role of G proteins in shear stress-mediated nitric oxide production by endothelial cells. *Am. J. Physiol.* **267**, C753–C758.

Kuo, L., and Pittman, R. N. (1988). Effect of hemodilution on oxygen transport in arteriolar networks of hamster striated muscle. *Am. J. Physiol.* **254**, H331–H339.

Kuo, L., and Pittman, R. N. (1990). Influence of hemoconcentration on arteriolar oxygen transport in hamster striated muscle. *Am. J. Physiol.* **259**, H1694–H1702.

Kuroda, Y., and McIlwain, H. (1974). Uptake and release of (14C)adenine derivatives at beds of mammalian cortical synaptosomes in a superfusion system. *J. Neurochem.* **22**, 691–699.

Kurjiaka, D. T., and Segal, S. S. (1995). Conducted vasodilation elevates flow in arteriole networks of hamster striated muscle. *Am. J. Physiol.* **269**, H1723–H1728.

Leeman, M., Zegers de Beyl, V., Delcroix, M., and Naeije, R. (1994). Effects of endogenous nitric oxide on pulmonary vascular tone in intact dogs. *Am. J. Physiol.* **266**, H2343–H2347.

Lewis, C. J., Ennion, S. J., and Evans, R. J. (2000). P2 purinoceptor-mediated control of rat cerebral (pial) microvasculature; contribution of P2x and P2y receptors. *J. Physiol.* **527**, 315–324.

Liu, S. F., McCormack, D. G., Evans, T. W., and Barnes, P. J. (1989). Characterization and distribution of P$_2$-purinoceptor subtypes in rat pulmonary vessels. *J. Pharmacol. Exp. Ther.* **251**, 1204–1210.

Macdonald, R. L., Zhang, J., Weir, B., Marton, L. S., and Wollman, R. (1998). Adenosine triphosphate causes vasospasm of the rat femoral artery. *Neurosurgery* **42**, 825–832.

Mayhan, W. G. (1992). Endothelium-dependent responses of cerebral arterioles to adenosine 5′-diphosphate. *J. Vasc. Res.* **29**, 353–358.

McCullough, W. T., Collins, D. M., and Ellsworth, M. L. (1997). Arteriolar responses to extracellular ATP in striated muscle. *Am. J. Physiol. Heart Circ. Physiol.* **272**, H1886–H1891.

Meinkoth, J. L., Alberts, A. S., Went, W., Fantozzi, D., Taylor, S. S., Hagiwara, M., Montminy, M., and Feramisco, J. R. (1993). Signal transduction through the cAMP-dependent protein kinase. *Mol. Cell. Biochem.* **127/128**, 179–186.

Miseta, A., Bogner, P., Berenyi, M., Kellermayer, M., Galambos, C., Wheatley, D., and Cameron, I. (1993). Relationship between cellular ATP, potassium, sodium and magnesium concentrations in mammalian and avian erythrocytes. *Biochim. Biophys. Acta* **1175**, 133–139.

Motte, S., Pirotton, S., and Boeynaems, J. M. (1993). Heterogeneity of ATP receptors in aortic endothelial cells: Involvement of P$_{2y}$ and P$_{2u}$ receptors in inositol phosphate response. *Circ. Res.* **72**, 504–510.

Muramatsu, I., and Kigoshi, S. (1987). Purinergic and non-purinergic innervation in the cerebral arteries of the dog. *Br. J. Pharmacol.* **92**, 901–908.

Muramatsu, I., Fujiwara, M., Miura, A., and Sakakibara, Y. (1981). Possible involvement of adenine nucleotides in sympathetic neuroeffector mechanisms of dog basilar artery. *J. Pharmacol. Exp. Ther.* **216**, 401–409.

Nakagawa, M., Willner, J., Cerri, C., and Reydel, P. (1984). The effect of membrane preparation and cellular maturation on human erythrocyte adenylate cyclase. *Biochim. Biophys. Acta* **770**, 122–126.

Nishiwaki, K., Nyhan, D. P., Rock, P., Desai, P. M., Peterson, W. P., Pribble, C. G., and Murray, P. A. (1992). N$^G$-nitro-L-arginine and pulmonary vascular pressure-flow relationship in conscious dogs. *Am. J. Physiol.* **262**, H1331–H1337.

Offermanns, S., and Simon, M. I. (1996). Organization of transmembrane signalling by heterotrimeric G proteins. *Cancer Surv.* **27**, 177–198.

Ohno, M., Gibbons, G. H., Dzau, V. J., and Cooke, J. P. (1993). Shear stress elevates endothelial cGMP: Role of potassium channel and G protein coupling. *Circulation* **88**, 193–197.

Olearczyk, J. J., Stephenson, A. H., Lonigro, A. J., and Sprague, R. S. (2001). Receptor-mediated activation of the heterotrimeric G-protein Gs results in ATP release from erythrocytes. *Med. Sci. Monit.* **7**, 669–674.

Palmer, R., Ferrige, M. J., and Moncada, S. (1987). Nitric oxide release accounts for the biological activity of endothelium-derived relaxing factor. *Nature* **327**, 524–526.

Parsaee, H., McEwan, J. R., Joseph, S., and MacDermot, J. (1992). Differential sensitivities of the prostacyclin and nitric oxide biosynthetic pathways to cytosolic calcium in bovine aortic endothelial cells. *Br. J. Pharmacol.* **107**, 1013–1019.

Plagemann, P. G. W., and Wohlhueter, R. M. (1984). Kinetics of nucleoside transport in human erythrocytes: Alterations during blood preservation. *Biochim. Biophys. Acta* **778**, 176–184.

Prat, A. G., Reisen, I. L., Ausiello, D. A., and Cantiello, H. F. (1996). Cellular ATP release by the cystic fibrosis transmembrane conductance regulator. *Am. J. Physiol.* **270**, C538–C545.

Purchelle, E., Gaillard, D., Ploton, D., Hinnrasky, J., Fuchy, C., Boutterin, M.-C., Jacquot, J., Dreye, D., Pavirani, A., and Dalemans, W. (1992). Differential localization of the cystic fibrosis transmembrane conductance regulator in normal and cystic fibrosis airway epithelium. *Am. J. Respir. Cell Mol. Biol.* **7**, 485–491.

Reddy, M. M., Quinton, P. M., Haws, C., Wine, J. J., Grygorczyk, R., Tabcharani, J. A., Hanrahan, J. W., Gunderson, K. L., and Kopito, R. R. (1996). Failure of the cystic fibrosis transmembrane conductance regulator to conduct ATP. *Science* **271**, 1876–1879.

Reisin, I. L., Prat, A. G., Abraham, E. H., Amara, J. F., Gregory, R. J., Ausiello, D. A., and Cantiello, H. F. (1994). The cystic fibrosis transmembrane conductance regulator is a dual ATP and chloride channel. *J. Biol. Chem.* **269**, 20584–20591.

Rodan, S. B., Rodan, G. A., and Sha'afi, R. I. (1976). Demonstration of adenylate cyclase activity in human red blood cell ghosts. *Biochim. Biophys. Acta* **428**, 509–515.

Rosenblum, W. I., Nelson, G. H., and Murata, S. (1994). Endothelium dependent dilation by purines of mouse brain arterioles *in vivo*. *Endothelium* **1**, 287–294.

Rubanyi, G. M., Romero, J. C., and Vanhoutte, P. M. (1986). Flow-induced release of endothelium-derived relaxing factor. *Am. J. Physiol.* **250**, H1145–H1149.

Rubin, C. S., Erlichman, J., and Rosen, O. M. (1972). Cyclic adenosine 3′,5′-monophosphate-dependent protein kinase of human erythrocyte membranes. *J. Biol. Chem.* **247**, 6135–6139.

Sarelius, I. H., and Duling, B. R. (1982). Direct measurement of microvessel hematocrit, red cell flux, velocity, and transit time. *Am. J. Physiol.* **243**, H1018–H1026.

Schultz, B. D., DeRoos, A. D. G., Venglarik, C. J., Singh, A. K., Frizzell, R. A., and Bridges, R. J. (1996). Glibenclamide blockade of CFTR chloride channels. *Am. J. Physiol.* **271**, L192–L200.

Schwiebert, E. M., Egan, M. E., Hwang, T.-H., Fulmer, S. B., Allen, S. S., Cutting, G. R., and Guggino, W. (1995). CFTR regulated outward rectifying chloride channels through an autocrine mechanism involving ATP. *Cell* **81**, 1063–1073.

Segal, S. S., and Duling, B. R. (1987). Propagation of vasodilation in resistance vessels of the hamster: Development and review of a working hypothesis. *Circ. Res.* **61**(Suppl. II), II-20–II-25.

Segal, S. S., and Duling, B. R. (1989). Conduction of vasomotor responses in arterioles: A role for cell-to-cell coupling? *Am. J. Physiol.* **25**, H838–H845.

Sheppard, H., and Burghardt, C. (1969). Adenyl cyclase in non-nucleated erythrocytes of several mammalian species. *Biochem. Pharmacol.* **18**, 2576–2578.

Sheppard, D. N., and Welsh, M. J. (1993). Inhibition of the cystic fibrosis transmembrane conductance regulator by ATP-sensitive $K^+$ channel regulators. *Ann. N.Y. Acad. Sci.* **707**, 275–284.

Sprague, R. S., Stephenson, A. H., Dimmitt, R. A., Weintraub, N. L., McMurdo, L., and Lonigro, A. J. (1995). Effect of L-NAME on pressure-flow relationships in isolated rabbit lungs. *Am. J. Physiol.* **269**, H1941–H1948.

Sprague, R. S., Ellsworth, M. L., Stephenson, A. H., and Lonigro, A. J. (1996). ATP: The red blood cell link to NO and local control of the pulmonary circulation. *Am. J. Physiol.* **271**, H2717–H2722.

Sprague, R. S., Ellsworth, M. L., Stephenson, A. H., and Lonigro, A. J. (1998). Increases in flow rate stimulate adenosine triphosphate release from red blood cells in isolated rabbit lungs. *Exp. Clin. Cardiol.* **3**, 73–77.

Sprague, R. S., Ellsworth, M. L., Stephenson, A. H., and Lonigro, A. J. (2001). Impaired release of ATP from red blood cells of humans with primary pulmonary hypertension. *Exp. Biol. Med.* **226**, 434–439.

Sprague, R. S., Ellsworth, M. L., Stephenson, A. H., and Lonigro, A. J. (2001). Participation of cAMP in a Signal-transduction pathway relating erythrocyte deformation of ATP release. *Am. J. Physiol.* **281**, C1158–C1164.

Sprague, R. S., Stephenson, A. H., Dimmitt, R. A., Weintraub, N. L., Branch, C. A., McMurdo, L., and Lonigro, A. J. (1994). Inhibition of nitric oxide synthesis results in a selective increase in arterial resistance in rabbit lungs. *Pol. J. Pharmacol.* **46**, 579–585.

Sprague, R. S., Ellsworth, M. L., Stephenson, A. H., Kleinhenz, M. E., and Lonigro, A. J. (1998). Deformation-induced ATP release from red blood cells requires cystic fibrosis transmembrane conductance regulator activity. *Am. J. Physiol.* **275**, H1726–H1732.

Swain, D. P., and Pittman, R. N. (1989). Oxygen exchange in the microcirculation of hamster retractor muscle. *Am. J. Physiol.* **256**, H247–H255.

Tabcharani, J. A., Chang, X.-B., Riordan, J. R., and Hanrahan, J. W. (1990). Phosphorylation-regulated Cl channel in CHO cells stably expressing the cystic fibrosis gene. *J. Biol. Chem.* **352**, 157–164.

Tse, C.-M., Wu, J.-S. R., and Young, J. D. (1985). Evidence for the asymmetrical binding of *p*-chloromercuriphenyl sulphonate to the human erythrocyte nucleoside transporter. *Biochim. Biophys. Acta* **818**, 316–324.

Van Belle, H. J., Wynants, J., and Ver Donck, K. (1991). Comparison of the existing nucleoside transport inhibitors: *In vitro* and *in vivo* data. *In* "Purine and Pyrimidine Metabolism in Man VII" ( Harkness, ed.), Part A, pp. 419–422. Plenum Press, New York.

Wand, G. S., Waltman, C., Martin, C. S., McCaul, M. E., Levine, M. A., and Wolfgang, D. (1994). Differential expression of guanosine triphosphate binding proteins in men at high and low risk for the future development of alcholism. *J. Clin. Invest.* **94**, 1004–1011.

You, J. P., Johnson, T. D., Marrelli, S. P., and Bryan, R. M., Jr. (1999). Functional heterogeneity of endothelial P2 purinoceptors in the cerebrovascular tree of the rat. *Am. J. Physiol. Heart Circ. Physiol.* **277**, H893–H900.

Zipf, K. (1931). Die chemische natur der "depressorischen substance" des blutes. *Arch. Pathol. Pharmakol.* **160**, 579–598.

# CHAPTER 9

# A Purine Signal for Functional Hyperemia in Skeletal and Cardiac Muscle

**Thomas Forrester**

Department of Pharmacological and Physiological Sciences, St. Louis University Medical Center, St. Louis, Missouri 63104

*Current Topics in Membranes, Volume 54*

## I. INTRODUCTION

> *Except perhaps in stark terror there is no other vasodilatation to*
> *compare with that in the skeletal muscles in exercise.*[1]
>                     (H. Barcroft, Smolenice, Czechoslovakia, 1968)

The nature of local blood flow control in cardiac and skeletal muscle has been the subject of much investigation (and lively debate) ever since the proposition of Latchenberger and Deahna (1876) and Gaskell (1878) that locally released "metabolites" could be responsible for the functional hyperemia seen in exercising skeletal muscles. It may be timely at this juncture in the light of new knowledge to restate the phenomenon in more modern terms.

Contracting muscle poses two problems that can be stated in terms of the classic concept of Claude Bernard's "milieu interieur." First, the activated muscle immediately starts to alter its own local environment by virtue of the alteration of membrane permeability, resulting in efflux and influx of substances across the cell membrane. The second problem is, of course, one of energy supply for the sudden increase in metabolic activity. The two problems seem to be solved adequately, judging by animal survival, with a dramatic and sudden increase in blood flow through the active musculature. An increase in delivery of oxygen is ensured, as is the efficient washout of noxious substances from the muscle, thus restoring the local environment of the muscle. When this is achieved a new steady-state blood flow is set for minimum change in the local environment. What has been just described is a type of negative feedback system, with the obvious difference that when the local environment is restored the muscle work does not cease automatically.

Figure 1 indicates the sequence of events leading to functional hyperemia. The *signal* component is divided into biophysical, chemical, and purely physical, each giving rise to release of purines, followed by the response, via $P_1$ and $P_2$ receptors (see review by Ralevic and Burnstock, 1998). An example of a biophysical signal would be membrane depolarization leading to purine release and smooth muscle relaxation (Section IV). An example of a chemical signal would be a reduction in local oxygen tension, resulting in vascular smooth muscle relaxation, either directly or via release of purines. Physical signals would include *shear forces* impinging on vascular

---

[1]A neural vasodilatory pathway to skeletal muscle originates in the cerebral cortex, is blocked by atropine and is abolished by sympathectomy. It is thought to be activated in extreme mental stress and responsible for vasodepressor syncope.

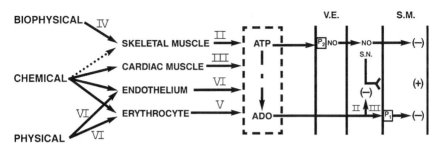

FIGURE 1    Purine signaling in functional hyperemia of skeletal and cardiac muscle. Roman numerals, relevant section in text; dotted arrow, no direct evidence; interrupted arrow, ectoATPase activity; V.E., vascular endothelium; S.M., vascular smooth muscle; NO, nitric oxide; S.N., sympathetic nerve supply; ATP, adenosine triphosphate; ADO, adenosine; $P_1$, $P_2$, purine receptors; +, smooth muscle contraction; −, smooth muscle relaxation or inhibition of S.N. by adenosine (sympatholytic effect); dashed lines, extracellular purine pool.

endothelial cells and erythrocytes. Relaxation of smooth muscle in response to *heat* production from active muscle would constitute another example of a physical signal. Heat production from active muscle can be regarded as a type of "metabolite" that must be cleared rapidly from the local environment before enzyme systems are irreversibly compromised.

## A. Aspects of Signal

### 1. Response Time

Examination of the response time of the hyperemic phenomenon can provide some information concerning the nature of the signal. How sensitive is the system? Figure 2 shows an example of a precise measurement of the rapidity of onset of increase in blood flow in human forearm musculature in response to a very mild exercise (Corcondilas *et al.*, 1964). It shows strain gauge plethysmography records from the forearms of a human subject. The right forearm flow was 3.9 ml min$^{-1}$ 100 g$^{-1}$ and the left was 3.0 ml min$^{-1}$ 100 g$^{-1}$. The subject then squeezed a stiff metallic ring for the short duration of 0.3 s in the right hand. The force developed ranged from 1 to 60 pounds. The contraction was more "isometric" than isotonic because of the stiffness of the ring. The increase in blood flow was easily recorded, even for such a brief contraction. An increase in flow could be detected within a second after completion of the contraction. The arterial blood pressure remained constant throughout the duration of the experiment. It was also shown that the increase in flow was proportional to the strength of contraction and

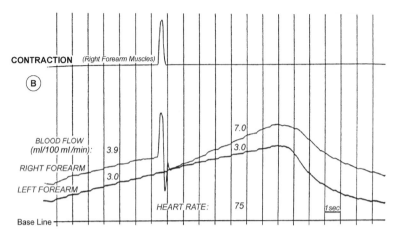

**FIGURE 2** Plethysmography recording of blood flow through the forearms of a human volunteer. An increase in blood flow in the right forearm can be detected within 2 s after commencement of exercise. Left forearm acted as control. Reproduced from Corcondilas *et al.* (1964), with permission.

that sympathectomy had no effect on the response. No change in oxygen saturation occurred in the blood obtained from the deep veins and the fact that the whole exercise took place well within the time before oxygen storage in the muscles would be called upon (0.3 s versus 1 min). The general conclusion was that the dilation began almost the instant that contraction commenced. A good example of this is given in Fig. 6.

## 2. Intensity of Impulse Flow

It cannot be assumed that the strength of contraction is directly responsible for the degree of blood flow. This was demonstrated clearly by the work of Khayutin (1968) using cat gastrocnemius muscle. Figure 3 shows the relationship between the force of contraction and peak blood flow in isometric and isotonic contractions. It is obvious that there is little proportional relationship between the strength of contraction and the degree of vasodilation. At the rate of 4–8 impulses $s^{-1}$ the contraction strength is only 20% of maximal, whereas the increase in the peak flow has already reached 60–80% of the maximum. At the rate of 16 $s^{-1}$ maximal dilation has been reached, whereas the strength of contraction is only half maximum. It did not seem to matter whether the contraction was isotonic or isometric. Khayutin (1968) concluded that neither the tension nor the degree of shortening nor the external physical work were significant factors determining the intensity of vasodilation.

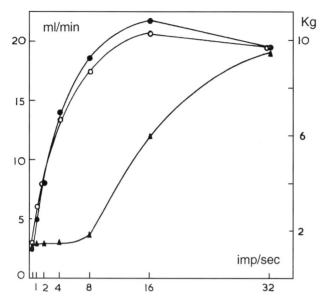

**FIGURE 3** Relationship between peak blood flow (o, isotonic contraction; •, isometric contraction) and force of contraction (▲) in cat gastrocnemius muscle. Peak blood flow was the rate of blood outflow at the moment of maximal vasodilation after muscle contraction. Ordinate: left side, blood flow (ml min⁻¹); right side, force of contraction in kilograms. Abscissa: number of impulses per second. From Khayutin (1968), with permission.

Khayutin (1968) also showed that the dilation does not just extend to the blood vessels of active muscle units, it affects the vessels of the whole muscle in proportion to the frequency of motor impulses and the number of motor units contracting. The product of these two quantities is termed the *intensity of impulse flow*. Figure 4 shows the linear relationship between peak blood flow and the logarithm of intensity of impulse flow at three different contraction levels. Evidently the intensity of impulse flow is the only factor that bears a constant relationship to the degree of vasodilation. This correlated with earlier work of Fales *et al.* (1960) who demonstrated that the oxygen consumption of the dog gastrocnemius-plantaris muscle was dependent primarily upon the external work being performed by the muscle. They related the extra oxygen consumption to the number of nerve impulses delivered to the muscle.

What can these findings tell us about the nature of the signal? The very rapid response would require the earliest possible signal. The apparent dissociation between the contractile activity of the muscle and the degree of blood flow would suggest that the signal is an event occurring before the contractile process goes into operation. There could be no earlier signal in

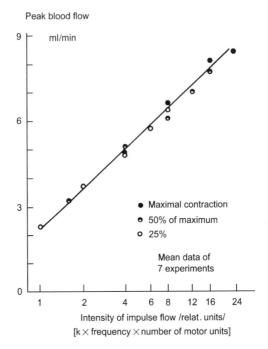

**FIGURE 4**   Relationship between peak blood flow and logarithm of intensity of impulse flow at three levels of contraction strength in cat gastrocnemius. From Khayutin (1968), with permission.

the natural sequence of events than membrane *depolarization*. This biophysical signal, followed by the excitation–contraction coupling mechanism, would certainly constitute a very reliable indicator of impending increase demand of oxygen.

## B. Aspects of Mechanism

In the past a common approach used to investigate the composition of "local metabolite" has been to try to mimic the dilatory response with likely candidates by infusing them into the arterial line supplying the muscle. The potency of adenosine triphosphate (ATP) as a vasodilator in both cardiac and skeletal muscle has been well documented (Drury and Szent-Gyorgyi, 1929; Duff, *et al.*, 1954; Wolf and Berne, 1956). These studies were based on infusion experiments, thus bringing into play the influence of vascular endothelial cells and the participation of nitric oxide (NO) in the relaxation of smooth muscle (Furchgott and Zawadski, 1980; Palmer *et al.*, 1987; Ignarro *et al.*, 1987). De Mey and Vanhoutte (1981) showed that ATP can induce an endothelium-dependent vasodilation of vascular smooth muscle, even after blockade of prostacyclin synthesis (Gordon and Martin, 1983). Kelm *et al.* (1988) demonstrated that ATP could directly release NO from cultured aortic endothelial cells.

## II. FUNCTIONAL HYPEREMIA IN SKELETAL MUSCLE

### A. ATP Release from Skeletal Muscle

In 1965 it was shown that an active frog sartorius liberated a substance that had a pronounced stimulatory effect on an *in situ* perfused frog heart preparation (Boyd and Forrester, 1965). This was a surprising result in a project that originally set out to use the frog heart as a bioassay system for acetylcholine released from the motor nerve terminals in the frog sartorius muscle (Forrester, 1967). It was expected that the solution bathing the active muscle would have an inhibitory (acetylcholine-like) effect upon the frog heart. Figure 5 shows the effect of the perfused stimulated muscle solution on the arterial blood pressure of the heart, as compared to two solutions of ATP. The inset shows the effect of the same solution on extract of firefly. When a stimulated muscle solution was incubated with the enzyme apyrase the stimulatory effect on the heart and the light signal from firefly were abolished. Apyrase converts ATP to adenosine monophosphate (AMP), which does not stimulate the frog heart. These and other tests were used to establish that the stimulatory substance released from frog skeletal muscle was ATP (Forrester, 1966). The same identification procedures were applied to the perfusate from stimulated hindlimb musculature of the frog. Table I shows the relation of ATP output to the frequency of motor nerve

**TABLE I**

Relationship of ATP Output to Frequency of Stimulation of Perfused Frog Hindlimb[a]

| Hz | pmol volley$^{-1}$ 100 g$^{-1}$ | pmole 100 g$^{-1}$ min$^{-1}$ |
|----|----|----|
| 1 | 4.0 | 240 |
| 2 | 1.5 | 180 |
| 5 | 1.4 | 420 |
| 10 | 3.0 | 1800 |

[a]Data from Forrester and Hassan (1973).

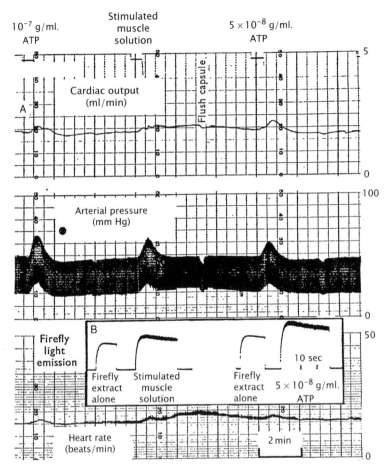

**FIGURE 5** (A) Response of frog heart perfused *in situ* with ATP solutions and with Ringer's solution in which an active frog sartorius was bathed. (B) The emission of light from firefly tail extract when exposed to ATP and the same stimulated muscle solution as in (A). Reproduced from Forrester (1966), with permission of The Physiological Society.

stimulation (Forrester and Hassan, 1973). These amounts compare to an output of 50 nmol min$^{-1}$ 100 g$^{-1}$ from an isolated sartorius muscle stimulated at 2 Hz (Boyd and Forrester, 1968). Presumably the great reduction in amounts measured in the perfusates was due to *in situ* ectoATPase activity (see below) that was not encountered in the bathing solution.

Using these identification procedures ATP was identified in fresh human plasma (Forrester, 1969) and in the venous effluent from human forearms before, during, and after sustained voluntary muscle contractions (Forrester and Lind, 1969; Forrester, 1972).

## B. EctoATPase Activity

Jorgensen (1956) found that exogenous ATP was rapidly degraded in whole blood. W. A. Engelhardt (1957) reported the occurrence of an ATPase that seemed to be localized on the surface of nucleated erythrocytes and he proposed the term *ectoenzyme*. Earlier Acs *et al.* (1954) had noted that over 90% of the ATPase activity of Ehrlich ascites cells was closely associated with the surface of the cell. Soon ectoATPase activity was found in a variety of tissues: intact cells of rat hemidiaphragm (Marsh and Haugaard, 1957), *Amoeba proteus* (Sells *et al.*, 1961), and intact glia cells (Cummins and Hyden, 1962). At about the same time as ATP was identified as being released from active frog skeletal muscle, an enzyme system on the outer surface of intact frog skeletal muscle was identified (Dunkley *et al.*, 1966). The properties of the enzymes involved (ATPase, adenylate kinase AMP deaminase) have been extensively studied (Manery *et al.*, 1968) and it also became evident that 5′-nucleotidase has its active site facing *outward* from the membrane in skeletal muscle (Woo and Manery, 1975). This significant finding was soon verified in many other tissues (DePierre and Karnovsky, 1974; Gurd and Evans, 1974; Trams and Lauter, 1974). Extracellular enzyme activity of this type was also observed in cardiac tissue (Williamson and DiPietro, 1965; Frick and Lowenstein, 1976). Thus the concept gradually emerged that there existed in the body a system that could utilize ATP *outside* the cellular compartment

The ubiquitous occurrence of ectoATPase activity in the body made it difficult in many instances to dissect out individual effects due to adenosine, AMP, adenosine diphosphate (ADP), and ATP. This problem has been largely solved by the identification and classification of ATP and adenosine receptors in various tissues (see review by Ralevic and Burnstock, 1998).

## C. Role for Adenosine

### 1. As a Vasodilator Agent

Scott *et al.* (1965) were the first to demonstrate release of adenosine from active skeletal muscle; however, the adenosine could be detected only after prolonged heavy work or exercise in the presence of ischemia (Bockman *et al.*, 1975, 1976; Haddy and Scott, 1975; Phair and Sparks, 1979). Honig and Frierson (1980) concluded that adenosine does not influence the rate of

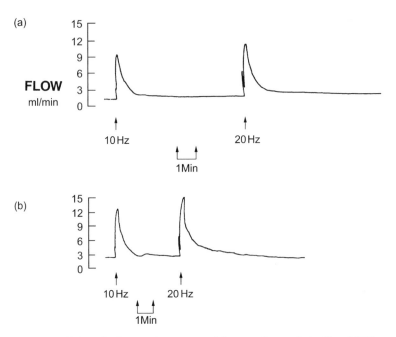

**FIGURE 6**    (a) Exercise hyperemia response of dog gracilis muscle to 10 and 20 Hz nerve stimulation for 5 s before $P_1$ receptor saturation; (b) after $P_1$ receptor saturation. Note the fast initial component of blood flow, unaffected by receptor saturation. Reproduced from Hester *et al.* (1982), with permission.

development of exercise vasodilation, but may prolong vasodilation after a heavy work load (at a constant flow rate). Tachyphylaxis to adenosine was demonstrated in renal tissue (Hashimoto and Kokubun, 1971). Interference between mechanism and response was used to explore a role for adenosine. Hester *et al.* (1982) saturated the adenosine ($P_1$) receptor population in dog gracilis muscle by infusion of adenosine at concentrations of more than 1000 times the normal resting adenosine levels found in the blood. They reasoned that any major blood flow responses in exercise could hardly be ascribed to endogenous adenosine when there was so much exogenous adenosine present. Infused adenosine did cause dilation but the flow returned to control levels after 1–3 h of continuous adenosine infusion. Figure 6 shows the responses of a dog gracilis muscle to stimulation for a period of 5 s at 10 and 20 Hz before and during adenosine infusion. In both cases a flow rate of at least four times the resting flow is achieved *within a few seconds* of stimulation onset. Compilation of their results is given in Fig. 7. Clearly densensitization of the adenosine receptor did not affect functional

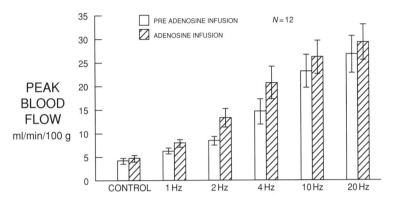

**FIGURE 7**    Peak blood flow in dog gracilis muscle in response to stimulation frequencies of 1, 2, 4, 10, and 20 Hz before (open bars) and after (closed bars) saturating infusions of adenosine. Values are mean + SEM. Reproduced from Hester *et al.* (1982), with permission.

hyperemia. One obvious feature is that at all frequencies of stimulation peak flow in the presence of adenosine infusion was greater than the control. This may indicate another role for adenosine if indeed ATP is a causal agent. The presence of adenosine may have the effect of inhibiting ectoATPase activity, thus potentiating the effect of released ATP.

### 2. As a Sympatholytic Agent

Suppression of the sympathetic vasoconstrictor effect on skeletal muscle blood vessels during muscle exercise has been known for a long time (Rein, 1930; Remensnyder *et al.*, 1962; Kjellmer, 1965) and was termed the "sympatholysin" effect. Verhaeghe *et al.* (1977) reported that adenosine, as with ADP and ATP, depresses the output of norepinephrine in doses that appear to be too small to have a direct effect on the vascular smooth muscle (Fig. 8). This effect was antagonized by the $P_1$ blocker aminophylline. This finding was confirmed by Su (1978). Using the degradation-resistant analogue of ATP (APPCP) De Mey *et al.* (1979) showed neatly that the inhibitory effect of ATP on noradrenergic transmission is most likely a result of its rapid degradation to adenosine, which acts on prejunctional $P_1$ receptors.

### D. Surgical Exclusion of a Role for Sympathetic Nerves

The participation of the sympathetic division of the autonomic nervous system in the generation of exercise hyperemia was ruled out in early studies

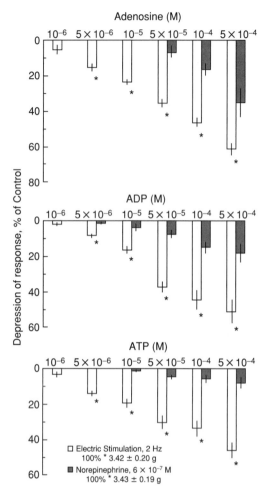

**FIGURE 8**   Depression by adenosine, ADP, and ATP of the response of saphenous veins to
sympathetic nerve stimulation (open bars) and norepinephrine (shaded bars). Data expressed as
mean + SE. Differences between electrical stimulation and norepinephrine responses were
significant ($p < 0.01$). Reproduced from Verhaeghe *et al.* (1977), with permission.

by Grant (1938). The experiment was performed on a patient 6 months after
right cervical ganglionectomy for relief of symptoms due to arterial
obstruction in the right hand. The circulation to both forearms was normal.
The subject gripped a bar tightly with both hands for half a second once per
second up to a period of 4 min, when his forearms were nearly tired out.
Table II shows the forearm blood flows before and immediately after

TABLE II

Effect of Right-Sided Sympathectomy on Human Forearm Muscle Blood Flow
(ml 100 g$^{-1}$ min$^{-1}$) before and Immediately after Rhythmic Contraction[a]

| Duration of exercise | Forearm | Before exercise | Immediately after |
| --- | --- | --- | --- |
| 1 s | L | 1.3 | 11.4 |
| | R | 2.1 | 15.6 |
| 5 s | L | 1.5 | 13.2 |
| | R | 1.8 | 18.9 |
| 20 s | L | 1.6 | 13.8 |
| | R | 2.1 | 20.7 |
| 40 s | L | 1.2 | 18.6 |
| | R | 2.1 | 20.4 |
| 4 min | L | 1.2 | 24.0 |
| | R | 1.8 | 22.2 |

[a]Data from Grant (1938).

exercise. With a duration of exercise of 1 s (one contraction of 0.5 s duration) the blood flow in the sympathectomized right arm increased from 2.1 to 15.6 ml 100 g$^{-1}$ min$^{-1}$ compared to the left arm control increase from 1.3 to 11.4 ml 100 g$^{-1}$ min$^{-1}$. After 4 min (240 successive contractions) the increase in blood flow through the sympathectomized arm was little different from control. It is interesting that in a patient with cervical sympathectomy for Raynaud's disease, reactive (or occlusion) hyperemia remained intact (Patterson and Shepherd, 1954). Later, Donald et al. (1970) showed that there was no difference in the blood flow between sympathetically innervated and denervated dog hind limbs. The duration of the exercise was short. This effect wore off after 20–30 min of exercise (Rowlands and Donald, 1968)

## III. FUNCTIONAL HYPEREMIA IN CARDIAC MUSCLE

### A. Adenosine Release from Myocardium

Purine signaling in the coronary circulation has long been the subject of much debate since Jacob and Berne (1960) showed rapid influx of the myocardial cell membrane by the purine adenosine. They surmised that efflux of adenosine from the myocardial cell could occur with similar ease. Berne (1963) demonstrated release of inosine and hypoxanthine in response to myocardial hypoxia. The assumption was made that these were the degradation products of adenosine that had been released from the myocardial cell during a period of hypoxia. The enzymatic activity for such extracellular degradation had already been defined (Conway and Cooke, 1939; Balis, Marrian, and Brown, 1951). Finally, Richman and Wyborny (1964) were able to reveal the extent of release of adenosine from hypoxic myocardium by inhibiting the action of adenosine deaminase with 8-azaguanine. Katori and Berne (1966) provided significant data indicating that increased myocardial oxygen demand resulted in an increase of released adenosine. Later, increased sensitivity for adenosine detection allowed the use of 8-azaguanine to be discarded. Rubio and Berne (1969) showed that quantitatively there seemed to be enough adenosine for appropriate coronary dilatation in response to hypoxia. Wiedmeier and Spell (1977) showed no release of adenosine when coronary dilation was evoked *without* an increase in myocardial oxygen demand, emphasizing the significance of hypoxia as a trigger for adenosine release.

### 1. Aminophylline Experiments

Up to that point the evidence for adenosine being a principal vasodilator substance in the coronary circulation was attractive. However, experiments using the drug aminophylline, which inhibits the coronary vasodilation action of adenosine, raised some formidable problems. If hypoxia-induced release of adenosine was responsible for relaxation of the vascular smooth muscle, would administration of aminophylline *inhibit* hypoxia-induced vasodilation? It did not (Afonso et al., 1972). This result was confirmed by Wadsworth (1972). Giles and Wilcken (1977) clearly showed that the action of ATP on the coronary arteries is unaffected by aminophylline.

The experiments with aminophylline were criticized on the basis that a true physiological situation did not pertain. It was suggested that the metabolism of the cell might be affected in some way by the drug and that normal conditions were not present for the experiment. Also, it was suggested that there was a greater increase in the amount of adenosine available to the coronary vessels in the presence of aminophylline, thus effectively overcoming the blocking action of aminophylline (Olsson, 1981). These criticisms were addressed by Jones et al. (1982). Figure 9 shows a recording of the left coronary blood flow in a dog heart. Infusion of adenosine superimposed upon the dilation produced by norepinephrine resulted in additional hyperemia. Aminophylline was then administered in a $P_1$ blocking dose. This was followed by another infusion of norepinephrine. After this a dose of adenosine did not cause an additional hyperemia, presumably because the $P_1$ receptor population was blocked by the aminophylline. It was plain that the functional hyperemic response did not involve the receptor population available for blocking by exogenously

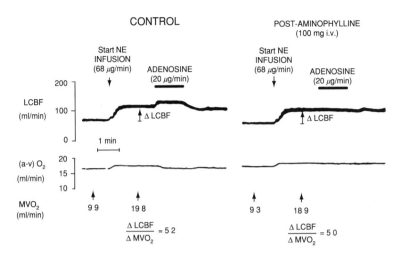

**FIGURE 9**    Record of left coronary blood flow (LCBF) in dog heart. Functional hyperemia during norepinephrine infusion was not altered by aminophylline. Coronary response to adenosine during functional hyperemic response was completely blocked by aminophylline. Reproduced from Jones *et al.* (1982), with permission.

applied aminophylline. Therefore, under conditions of a physiological hyperemia (the response to norepinephrine) it would appear that $P_1$ receptor blockage did not affect the vasodilator response. Jones *et al.* (1982) also showed that aminophylline did not influence the levels of adenosine in the myocardium during the hyperemia, effectively ruling out the possibility that there was a greater increase in the adenosine available to the coronary vessels in the presence of aminophylline.

Currently theophylline is becoming popular in the treatment of angina, peripheral vascular disease, acute renal failure, and erythrocytosis (Collis *et al.*, 1994).

## 2. Role for Adenosine

A role for adenosine seems evident in postocclusion hyperemia of cardiac muscle. Figure 10 shows the effect of aminophylline on postocclusion hyperemia of myocardium in two separate studies (Giles and Wilcken, 1977; Wadsworth, 1972). In both cases there is a clear diminution of flow in the postocclusion phase. The adenosine contribution has presumably been eliminated by the $P_1$ blockade. This situation is analogous to that seen in skeletal muscle when adenosine could be detected only after prolonged heavy work (see above, Honig and Frierson, 1980). A sequential appearance of ATP and adenosine in the coronary perfusate from hypoxic hearts was

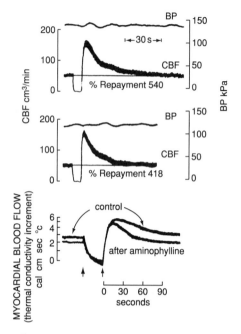

**FIGURE 10** Upper figure shows postocclusion hyperemia after 8 s of coronary artery occlusion (greyhound). Middle figure shows response after infusion of aminophylline (200 μg min$^{-1}$). The hyperemic period is shortened by the $P_1$ blockade. From Giles and Wilcken (1977), with permission. Lower figure shows postocclusion hyperemia in left anterior descending artery in cat heart before and after aminophylline. Reproduced from Wadsworth (1972), with permission.

clearly demonstrated by Vial *et al.* (1987); the adenosine source may be directly from cardiac cells or indirectly from ectoATPase activity.

Similar to skeletal muscle, a *sympatholytic* role for adenosine seems likely for the coronary vasculature. In conditions of greatly reduced metabolic vasodilation (ventricular fibrillation) of the coronary arteries, stimulation of the coronary sympathetic nerve supply showed pronounced initial *vasoconstriction* (Berne *et al.*, 1965). In conditions of great stress, this effect of adenosine on the coronary blood vessels may be highly significant.

## B. Release of ATP from Heart

### 1. Langendorff Preparations

Paddle and Burnstock (1974), using the Langendorff perfusion technique with guinea pig heart, detected ATP in the perfusate when the hearts were

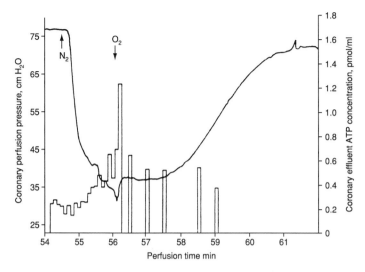

**FIGURE 11**  Effect of a 90-s period of hypoxia on the effluent concentration of ATP (bars) and on the coronary perfusion pressure in a Langendorff-perfused guinea pig heart. ATP concentrations are expressed as mean values in 6-s collection periods. Reproduced from Paddle and Burnstock (1974), with permission.

exposed to an hypoxic gas mixture for a period of 90 s. This time of exposure was found to produce maximum vasodilation of the coronary circulation. Figure 11 gives the result of one such experiment. Perfusion pressure falls rapidly after introduction of the hypoxia, indicating sudden vasodilation of the coronary circulation in this constant flow arrangement. This result was soon verified by Stowe *et al.* (1974). The source of the ATP in those perfusion experiments could have been endothelial cells, smooth muscle, or cardiac muscle cells. Vascular endothelial and smooth muscle cells in culture have been shown to release ATP selectively in response to potentially damaging stimuli (Pearson and Gordon, 1979).

## 2. Working Heart Preparations

Hearts perfused in the Langendorff mode perform a low and variable amount of work that is impossible to measure. Working heart preparations give rise to roughly twice the ATP concentrations in the coronary effluent (see Forrester, 1990, for review). A precise relationship between work load and rate of release of ATP (Fig. 12) was established by Doyle and Forrester (1985). The release of more ATP in response to an increase in work load may indicate that an increase in work load may have increasing hypoxia as the intermediate stimulus.

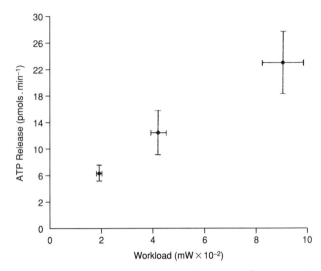

**FIGURE 12**   Relationship between work load (milliwatts $\times 10^{-2}$) and rate of ATP release (pmols min$^{-1}$) into the perfusate from isolated frog heart. Samples were taken 30 s after commencement of increased work load. Horizontal and vertical bars at each point represent one SEM. From Doyle and Forrester (1985).

## C. Release of ATP from Isolated Cardiac Myocytes

The isolated adult myocardial cell releases ATP in response to a brief period of hypoxia (Forrester and Williams, 1977). ATP was not detected in cell suspensions obtained from hearts that had been left asystolic for 10 min before the cell harvesting procedure. This was taken to indicate that dead cells in suspension did not contribute to the light signal evoked from firefly extract in response to hypoxia.

A constant concern was the possibility that the enzyme isolation procedure in some way damaged the cell membranes. Many tests, including assessment of oxygen consumption, insensitivity to calcium ion, activity of cellular lactate dehydrogenase (LDH) and normal behavior of the insulin receptor (Williams and Gold, 1975; Williams and Forrester, 1983) all suggested that the myocytes in suspension were undamaged. Figure 13 gives the time course of ATP release in both hypoxic and oxygenated cells. Within 1 min of exposure there is almost a 5-fold increase in ATP release over control. It was calculated that over a period of 1 min the release of ATP from cells exposed to hypoxia amounted to 0.5% of the total intracellular ATP. In the oxygenated state the proportion released was 0.05% (Forrester and Williams, 1977)

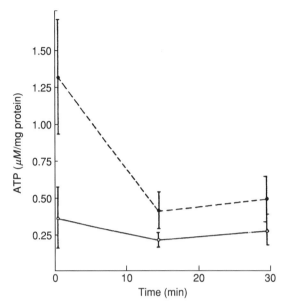

**FIGURE 13** ATP amounts in fluid suspending isolated adult cardiocytes from rat heart. Cells suspensions were exposed to hypoxia (dashed line) and tested for ATP (firefly) at 1, 15, and 30 min. Oxygenated controls (solid line). Each point is the mean of six measurements; vertical bars, one SD. Temperature, $37°C$. From Forrester and Williams (1977), with permission of The Physiological Society.

It was necessary to exclude the possibility that the isolated cells simply extruded ATP as a function of time, perhaps an indication of steady cell deterioration. Figure 14 shows the response of firefly extract in a cell suspension exposed alternately to hypoxic and oxygenated buffer solutions. Each time the hypoxic cells were introduced into the oxygenated environment, the levels of ATP release approximately equaled the amounts of ATP released by the cells in a normoxic environment.

An assessment of ATP degradation in the suspension fluid gave an indication of the great dynamism of the ATP release process. When cells were present a signal equivalent to $0.39 \mu M$ ATP was produced but following removal of cells by a 10-s centrifugation interval the cell-free fluid sample gave a signal of only $0.11 \mu M$ equivalent. In fact this meant that $0.022 \mu M$ per second was lost after removal of the cells. The rate of ATP hydrolysis was measured by adding ATP to the cell-free suspension fluid. Between $0.009$ and $0.04 \mu M$ per second was hydrolyzed (Table III). Therefore the disappearance of $0.274 \mu M$ ATP after the 10-s centrifugation seemed to be accounted for by this "ectoATPase" activity. A preliminary

**TABLE III**

Rate of ATP Hydrolysis

| Initial ATP concentration ($\mu M$) | Rate of hydrolysis ($\mu M$/h) | | |
|---|---|---|---|
| | $0°C$ (in $O_2$:$CO_2$) | $37°C$ (in $O_2$:$CO_2$) | $37°C$ (in $N_2$) |
| 0.5 | 4 | 33 | 77 |
| 1.0 | 10 | 147 | 88 |
| 2.0 | 18 | 598 | 185 |
| 5.0 | — | — | 540 |
| 10.0 | 111 | 690 | 1050 |
| 15.0 | 147 | 900 | 1230 |
| 25.0 | 132 | 2220 | 2520 |
| 50.0 | 180 | 2160 | 1257 |
| | $K_m$ 10.5–11 $\mu M$ | $K_m$ 12–13 $\mu M$ | $K_m$ 12.5 $\mu M$ |
| | V/2 1.5 $\mu M$/min | V/2 18.3 $\mu M$/min | V/2 20.8 $\mu M$/min |

characterization of this ATPase activity was performed and the results are shown in Table III. The $K_m$ value was 13 $\mu M$ and the $Q_{10}$ was found to be between 25 and 37°C. This activity was unaffected by hypoxic conditions and ouabain (Forrester and Williams, 1977).

The possibility that the increased amounts of ATP detected from cells in the hypoxic environment were due to an *inhibition* of ATPase activity had to be ruled out, therefore ATPase activity was determined in suspension fluid samples derived from cells incubated in hypoxic buffer for 1 h. The results indicated that the enzyme is no less active in the hypoxic condition than in normoxia (Table III). At that time Williams and Forrester (1983) proposed that the ATP was released from a membrane surface protein–ATP-$Ca^{2+}$ complex and speculated that the signal for release might be alteration in adjacent membrane charge (membrane depolarization?).

## IV. ATP RELEASE ASSOCIATED WITH MEMBRANE DEPOLARIZATION

In an early report Abood et al. (1962) showed *in vitro* that ATP is liberated from both nerve and muscle cells in association with membrane depolarization.

### A. Release of ATP from Electric Organ

An elegant example of ATP release associated with membrane depolarization was provided by Israel et al. (1976). They recorded the postsynaptic response to presynaptic nerve stimulation in the electric organ of *Torpedo*

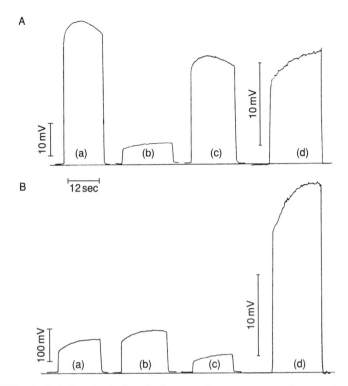

**FIGURE 14**   Emission of light from firefly extract in response to a cell suspension exposed alternately to hypoxic and oxygenated buffer solution. (A) (a) Signal after exposure to hypoxia; (b) response after cells were returned to oxygenated medium; (c) response on returning cells to hypoxic buffer; (d) response when finally returned to oxygenated buffer. Note alteration in pen recorder amplification. (B) (a)–(d) Paired oxygenated controls matching solutions tested in (A). From Forrester and Williams (1977), with permission of The Physiological Society.

*marmorata.* The surface of the organ was continuously superfused with extract of firefly tail. A discrete light signal was emitted from the surface of the organ in response to each nerve stimulation (Fig. 15). These light signals were diminished by application of curare and potentiated with eserine. The question of whether the ATP was released postsynaptically as a result of membrane depolarization or as a result of activation of the acetylcholine receptor was answered by first blocking the receptor with curare and then superfusing a high concentration of potassium to depolarize the post-synaptic membrane. A light signal was produced under those circumstances, indicating that the ATP release was associated with membrane depolarization.

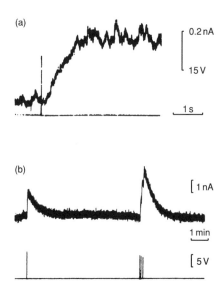

**FIGURE 15** Discrete release of ATP from electric organ of *Torpedo* in response to (a) a single nerve impulse and (b) single and triple nerve impulses. Note the slower sweep speed in (b). Upper traces, light emission; lower traces, electrical responses. Reproduced with permission from Israel *et al.* (1976), and the Royal Society, London.

## B. Release of ATP from Synaptosomes

A similar technique was used by T. D. White (1977) to demonstrate ATP release from depolarized isolated synaptosomes prepared from rat brain. The synaptosomes were incubated in a Krebs–Henseleit solution containing extract of firefly. Depolarization of the synaptosomal membrane was achieved in two different ways: first, by elevation of extracellular potassium and second, by adding the drug veratridine, which selectively opens sodium channels in the membrane. In both cases a light signal was emitted from the fluid surrounding the suspended synaptosomes (Fig. 16). Neither AMP nor ADP elicited a light signal from the luciferin suspension.

### 1. Potassium Depolarization

As the extracellular $K^+$ concentration was increased, the amount of ATP released was increased. This gradual process of depolarization was thought to avoid the sudden permeability changes of the membrane to both sodium and potassium (Hodgkin and Huxley, 1952a,b,c,d). Prior exposure of the synaptosomes to tetrodotoxin (TTX), which selectively blocks sodium channels, had no effect on the light signal emitted in response to high

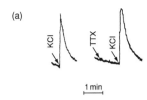

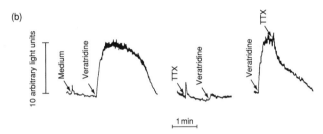

**FIGURE 16**  (a) Light signals evoked from synaptosomes suspended in firefly solution in response to elevation of $K^+$ by 47 m$M$. Prior application of TTX (final concentration $4 \times 10^{-7}$ $M$) did not affect the response. (b) Light signals evoked from a firefly suspension of synaptosomes in response to the depolarizing agent veratridine (final concentration $5 \times 10^{-5}$ $M$). Prior exposure to TTX blocks the light signal, suggesting that the veratridine-evoked release of ATP was due to membrane depolarization *subsequent* to opening of sodium channels. From White (1977), reproduced with permission and reprinted with permission from *Nature*.

extracellular K, indicating that the ATP release was the result of membrane depolarization *independent* of the opening of the sodium channels (Fig. 16a). The addition of a calcium chelator [ethylene glycoltetraacidic acid (EGTA)] attenuated the light signal evoked by the extracellular K suggesting a partial dependence on calcium ion.

### 2. Veratridine Depolarization

ATP release evoked by veratridine was blocked by the *prior* administration of TTX. When TTX was added at the peak of the veratridine response the light signal fell rapidly (Fig. 16b). These findings strongly suggested that the veratridine-evoked release of ATP was initiated by the opening of the sodium channels, leading to membrane depolarization and ATP release. The slower developed signal in the case of veratridine compared to the K-induced signal presumably indicates the time it takes for veratridine to open the sodium channels and the time it takes for membrane depolarization to follow this event. This release, in contrast to that seen with high K, was augmented in the presence of EDTA. It is known that calcium exerts an antagonism on the action of veratridine upon the sodium channel (Ulbricht, 1969; Catterall, 1975). White (1978) suggested that the apparent lack of a calcium requirement differentiated this process from the veratridine-induced release of a number of putative transmitters from synaptosomes (Blaustein et al., 1972).

## C. Lack of Evidence Supporting Corelease

Potter and White (1980) went on to show that the magnitude of ATP release depended upon the region of the brain from which the synaptosomes were prepared. Clearly a variable factor was the spectrum of nerve fiber types. They further eliminated the possibility that the ATP was coreleased with noradrenaline or dopamine (Potter and White, 1982). Depolarization of synaptosomes prepared from the cerebral cortex released ATP that could not be associated with acetylcholine release (White et al., 1980). Zhang, et al. (1988) showed that cultured neostriatal neurons (that had undergone synaptogenesis) released ATP in response to veratridine. Of this release 80% was blocked by TTX. High $K^+$ depolarization also triggered ATP release from these neurons. It was concluded that the source of ATP was not vesicular but cytoplasmic, Sawynok et al. (1993) demonstrated that the bulk of ATP release from dorsal spinal cord synaptosomes did not originate from descending noradrenergic nerve terminals.

## V. A PURINE SIGNAL FROM ERYTHROCYTES

Bergfeld and Forrester (1992) found that the human erythrocyte releases ATP in response to a brief exposure (50 s) to an hypoxic/hypercapnic environment that might be encountered in vigorously exercising skeletal and cardiac muscle. It was proposed that erythrocytes circulating into such an environment could release amounts of ATP sufficient to influence (via endothelial NO release) local blood flow. This release was blocked in two different ways: first by application of the specific band 3 anion channel blocking agents *niflumic acid* (a translocation inhibitor), DIDS (a transport site inhibitor), and *dipyridamole* (a channel blocker); second by application of the nucleoside transport blocker *nitrobenzylthioinosine* (NBTI). This provided strong evidence that the ATP efflux was closely associated with the function of band 3 protein as an anion channel.

The extreme potency of NBTI in blocking ATP efflux may indicate that the primary pathway for ATP release is via the nucleoside transporter.

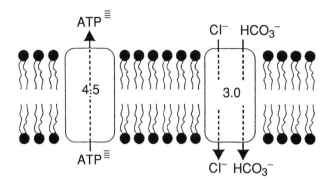

**FIGURE 17**   Model proposed for ATP release through erythrocyte membrane in response to hypoxia. Numbers indicate bands classified by electrophoresis. Hypoxia operates via band 4.5 protein. Maintenance of electrical balance is by passage of anions chloride and bicarbonate through band 3 protein. Electrical equilibrium could also be supported by efflux of $Mg^{2+}$ along with ATP. From Bergfeld and Forrester (1992), with permission.

A comprehensive analysis of the binding characteristics of nitrobenzylthioinosine can be found in the review by Gati and Paterson (1989). The NBTI-sensitive nucleoside transporter has been cloned (Griffiths *et al.*, 1997) and it has been concluded that it is a member of a previously unrecognized group of integral membrane proteins (see review by Cass *et al.*, 1998).

Blockade of ATP efflux was also achieved by the substitution of permeant extracellular anions with an impermeant methanesulfonate anion, indicating that an exchange mechanism pertained, preserving electrical balance across the membrane.

Based upon the three modes of blockade (band 3 anion blockade, nucleoside transporter blockade, and impermeant anion substitution) a model for the release of ATP was proposed (Fig. 17). Can the erythrocyte band 3 protein be equated with the cystic fibrosis transmembrane conductance regulator (CFTR) protein? These proteins have been cloned and sequenced and are quite dissimilar (Kopito and Lodish, 1985; Riordan *et al.*, 1989).

Does this red blood cell (RBC) purine signal have any role to play in functional hyperemia? It is unlikely that such a mechanism is responsible for the extremely rapid initial response so clearly demonstrated by Hester *et al.* (1982) (Fig. 6) and shown earlier by Corcondilas *et al.* (1964) (Fig. 2). A further problem is that the *degree* of hypoxia/hypercapnia that is needed to evoke ATP release from erythrocytes could never have been achieved in the short time of skeletal muscle exercise (0.3 s) in these experiments, bearing in mind the oxygen storage capacity of skeletal muscle. The situation with cardiac muscle, without significant oxygen storage capacity, may be quite different with regard to the erythrocyte ATP release. Perhaps this purine signal from erythrocytes could have a significant *maintainance* part to play, once appropriate vasodilation has been initiated by ATP release from depolarizing membranes.

## VI. PHYSICAL FORCES CONTRIBUTING TO THE PURINE SIGNAL

### A. Deformation of Erythrocytes

Human erythrocytes release ATP in amounts proportional to the degree of deformation, as applied by passing the cells through filters of different pore size (Sprague *et al.*, 1998). With the use of CFTR inhibitors (glibenclamide, niflumic acid) they concluded that membrane CFTR activity was required for such a release. Certainly mechanical deformation occurs *in vivo*, enabling the cell to pass through openings with smaller diameter (e.g., some capillaries) without an increase in flow resistance (Schmid-Schonbein, 1976; Lipowsky *et al.*, 1980; Cokelet *et al.*, 1980). It is possible that when flow converts from laminar to turbulent some degree of erythrocyte deformation may occur. Such turbulence may occur in exercise hyperemia. It is unlikely that this ATP release could make any contribution to the first, fast phase of exercise hyperemia, but may play a part in further augmentation of dilation.

### B. Effect of Shear Stress on Endothelial Cells

An increase in shear stress brought about by a sudden increase in flow liberates vasoactive substances from vascular endothelial cells (Pohl *et al.*, 1986; Rubanyi *et al.*, 1986; Christie *et al.*, 1989; Ralevic *et al.*, 1992). Release of ATP into the pulmonary vascular bed within seconds after an increase of flow rate was detected by Hassessian *et al.* (1993). Probably turbulent blood flow can release ATP from endothelial cells.

Long and Stone (1985) and Lewis and Smith (1991) showed that hemodynamic shear stress in endothelial cells resulted in an increase of intracellular calcium concentration, which is now known to be required for NO release. Mo *et al.* (1991) further demonstrated that extracellular adenine nucleotides were needed for this action of shear stress. The CFTR protein is expressed in human endothelial cells from umbilical vein and lung microvasculature (Tousson *et al.*, 1998).

A clear indication that shear stress forces could provide a purine signal was demonstrated by Grygoczyk and Hanrahan (1997). They used the firefly technique to directly detect significant amounts of ATP released from epithelial and nonepithelial cell lines *not* expressing the CFTR membrane protein. Minimum mechanical disturbance, such as changing the bath solution, enhanced ATP release in a very short period (seconds). They suggested that "physical perturbation" may be a physiological stimulus for ATP release. This was a much-delayed verification of an old observation that samples of whole blood, when shaken vigorously developed vasodilatory properties (Zipf, 1931). More recently, van Wamel *et al.* (2000) have shown that controlled stretching of many cell types, including endothelial cells, promotes secretion of factors that can alter proteins of gene expression.

## VII. DISCUSSION

### A. Why Should Muscle Cells Release ATP?

In view of the energetic importance of intracellular ATP this question has been continuously posed over the past 40 years. It is usually followed by the query: Why would cells benefit from such a loss of ATP? In the case of functional hyperemia the *raison d'être* seems clear considering the potency of ATP as a vasodilator agent. Muscle cells facing an impending negative balance between metabolic demand and oxygen supply need a mechanism for immediate increase of oxygen supply, i.e., rapid dilation of local blood vessels. It is proposed here that this urgent need is met by the extracellular appearance of ATP released from depolarizing muscle cells. A third query naturally arises from this observation: Can the muscle cell afford to lose ATP in the face of an incipient hypoxic situation? Quantitatively the amounts of ATP required for vasodilation are around 0.001% of the amounts of ATP intracellularly. In view of the very large and rapid turnover of ATP in healthy muscle cells, particularly skeletal muscle cells, this loss would appear to be a wise investment to ensure an adequate oxygen supply in the immediate future.

### B. How Do Cells Release ATP?

The ATP released can only come from the intracellular compartment (implying a membrane channel or ATP transporter) or from off the membrane (implying unknown responses of membrane to hypoxia) or from outside the membrane (implying unknown involvement of ATP-generating systems adjacent to the outside of the membrane).

## 1. Is There an ATP Channel?

Passage of ATP across the inner mitochondrial membrane has been known for a long time. An adenine nucleotide carrier ("translocase") acts as an antiporter with ADP. ATP is transported out of the matrix about 30 times more rapidly than ADP. This carrier constitutes about 14% of the inner mitochondrial membrane protein. It is specifically inhibited by atractyloside, a plant glycoside. Dixon (1986) found that ATP efflux from amphibian erythrocytes was inhibited by atractyloside. A complicating feature was that incubation of the cells with atractyloside reduced intracellular ATP levels, thus reducing the concentration gradient for passage of ATP through a channel. This disadvantage precludes exploring the direct effect of atractyloside on functional hyperemia. Dixon (1986) also tested whether high potassium or veratridine evoked ATP release from frog erythrocytes, in view of the synaptosomal results of White (see Section IV). They did not.

The model proposed by Bergfeld and Forrester (1992) for ATP release from human erythrocytes (Section V) favors the involvement of nucleoside channel in extracellular ATP appearance, since the specific blocker NBTI was so effective. It could be speculated that efflux of adenosine through this channel could give rise to extracellular ATP by ectoenzyme phosphorylation.

ATP efflux has been associated with the multidrug resistance efflux pump P-glycoprotein (Pgp) expressed in the membrane of cultured mammalian cells (Abraham et al., 1993). CFTR has been shown to act as a dual ATP and chloride channel in the membrane of mammalian cells (Reisin et al., 1994). However, Sugita et al. (1998) concluded that the permeation pathways for $Cl^-$ and ATP are distinct and that an ATP conduction pathway is not necessarily associated with the expression of CFTR. Both of these membrane proteins are members of the ATP-binding cassette (ABC) superfamily of proteins. Schwiebert et al. (1995) proposed that one role of CFTR could be to provide extracellular ATP that stimulates outwardly rectifying chloride channels via purinergic receptors in airway epithelial cells. That the CFTR protein can act as a conduit for ATP has been challenged (Reddy et al., 1996; Li et al., 1996; Grygorczyk et al., 1996). For recent developments in this field the review by Schwiebert (1999) is recommended.

The CFTR protein is present in mammalian hearts (Nagel et al., 1992; Ehara and Matsuura, 1993; Gadsby et al., 1995), but to date has not been identified in the plasma membranes of skeletal muscle cells or synaptosomes. It would be interesting to find the effect of acute cellular hypoxia, which is a powerful stimulus for ATP release from myocardial cells (Forrester and Williams, 1977), on the CFTR protein properties.

## 2. Can Mechanical Stresses open ATP Channels in the Membrane?

A firm association between mechanical distortion forces and membrane depolarization has long been known in sensory physiological studies (see review by Sackin, 1995). There is much evidence suggesting that membrane channels can be affected by external mechanical events. Stretch-activated channels have been observed in frog erythrocytes (Hamill, 1983). Brehm *et al.* (1984), making patch-clamp recordings of current through acetylcholine-activated channels in myotonal muscle from *Xenopus laevis*, found that the event frequency could be altered by adjusting the negative pressure applied to the patch. Guharay and Sachs (1984) showed that the frequency of channel opening greatly increased with increased suction force of the patch-clamp pipette; the channels became more sensitive to stretch following exposure to cytochalasin B, which disrupts the cytoskeleton (Weber and Osborn, 1981). The same phenomenon was demonstrated by Lansman *et al.* (1987) who found in aortic endothelial cells that the opening frequency of a calcium channel increased with stretching the membrane by increasing the suction through the electrode. Indeed Schwiebert (1999) made the point that the degree of suction pressure in patch-clamp experiments may be critical for the appearance of ATP. This variable may be responsible for disparate results concerning the presence or absence of ATP channels. Flow-evoked release of ATP from endothelial cells was shown to be blocked by glibenclamide (Hassessian *et al.*, 1993). The finding that ATP release by deformation of erythrocytes is blocked by glibenclamide may indicate a CFTR channel that can be affected by external mechanical forces (Sprague *et al.*, 1998).

## 3. Other Possible Local Sources of Extracellular ATP

ATP can be produced extracellularly by normal and neoplastic cells in culture (Agren *et al.*, 1971). The neoplastic cell types synthesized three to six times more ATP than the corresponding normal cell types. It was concluded that the ATP was formed at the outer surface of the cell membranes. An ATP-generating system on the outside of single amoebas of *Dictyostelium discoideum* cells was characterized by Parish and Weibel (1980).

Although highly unlikely, it must not be overlooked that enzyme systems are bidirectional, the direction of the cascade being determined by the substrate profile. Thus the well-characterized ectoATPase systems in skeletal and cardiac muscle could theoretically produce small amounts of ATP extracellularly.

## C. Release of ATP in Association with Membrane Depolarization

Although the evidence (Section IV) for such an *association* appears sound, our knowledge of underlying mechanisms is scanty, apart from the exclusion of the sodium channel. It is unlikely that mechanical effects, hypoxia, or the CFTR protein are involved in the synaptosomal preparation.

## VIII. SUMMARY AND CONCLUSIONS

Evidence for participation of the purine compounds ATP and adenosine in functional hyperemia of skeletal and cardiac muscle is presented.

In skeletal muscle functional hyperemia exhibits an initial increase in blood flow that can be detected within a second of the onset of exercise. This increase occurs under normoxic conditions. It is proposed that the release of ATP from depolarizing muscle membranes is responsible for this fast phase of flow increase. Further dilation is proportional to the frequency of motor impulses and the number of motor units contracting ["intensity of impulse flow"—Khayutin (1968)], which in turn would indicate the degree of total amount of membrane depolarization occurring. Subsequent increase in blood flow would be sustained by the release of both ATP and adenosine in response to the increasing hypoxic stimulus following the increasing oxygen demand.

In cardiac muscle a powerful trigger for a purine signal seems to be hypoxia, which can set off release of ATP within a few seconds of onset (Fig. 13). Accompanying release of adenosine, along with extracellular conversion of ATP to adenosine, would supplement the initial vasodilation due to ATP.

In both tissues the sympatholytic action of adenosine (suppression of sympathetic nerve vasoconstriction) would operate, being particularly significant in cardiac muscle.

EctoATPase activity in both tissues may play a significant dual role: (1) to keep the action of these powerful vasodilators a local one and (2) to boost local adenosine concentrations, thus augmenting the sympatholytic effect.

Release of ATP from mechanical distortion of vascular endothelium and erythrocytes has been established. It is possible that ATP channels may be opened directly by mechanical forces or indirectly via the CFTR membrane protein. The underlying cytoskeleton may also come into play. It is a daunting thought to consider how much investigation may be required to unravel this transduction mechanism. One example of the complexity of a mechanical transduction process is described by Ruwhof and van der Laarse (2000) regarding the hypertrophic response of cardiac muscle to mechanical stretching. Causal mechanisms linking ATP release to hypoxia and membrane depolarization may prove equally complex.

## References

Abood, L. G., Koketsu, K., and Miyamoto, S. (1962). Outflux of various phosphates during membrane depolarization of excitable tissues. *Am. J. Physiol.* **202,** 469–474.

Abraham, E. H., Prat, A. G., Gerweck, L., Seneveratne, T., Arceci, R. J., Kramer, R., Guidotti, G., and Cantiello, H. F. (1993). The multidrug resistance (mdr1) gene product functions as an ATP channel. *Proc. Natl. Acad. Sci. USA* **90,** 312–316.

Afonso, S., Ansfield, T. S., Berndt, T. B., and Rowe, G. C. (1972). Coronary vasodilator response to hypoxia before and after aminophylline. *J. Physiol.* **221,** 589–600.

Agren, G., Ponten, J., Ronquist, G., and Westermark, B. (1971). Demonstration of an ATPase at the cell surface of intact normal and neoplastic human cells in culture. *J. Cell Biol.* **78,** 171–176.

Balis, M. E., Marrian, D. H., and Brown, G. B. (1951). On the utilization of guanine by the rat. *J. Am. Chem. Soc.* **73,** 3319–3320.

Barcroft, H. (1968). Blood flow and metabolism in skeletal muscle. *In* "Circulation in Skeletal Muscle" (O. Hudlicka, ed.) pp. 121–135. Pergamon Press, Oxford.

Bergfeld, G. R., and Forrester, T. (1992). Release of ATP from human erythrocytes in response to a brief period of hypoxia and hypercapnia. *Cardiovasc. Res.* **26,** 40–47.

Berne, R. M. (1963). Cardiac nucleotides in hypoxia: Possible role in regulation of coronary blood flow. *Am. J. Physiol.* **204,** 317–322.

Berne, R. M., DeGeest, H., and Levy, M. N. (1965). Influence of the cardiac nerves on coronary resistance. *Am. J. Physiol.* **208,** 763–769.

Blaustein, M. P., Johnson, E. M., and Needleman, P. (1972). Calcium-dependent norepinephrine release from presynaptic nerve endings *in vitro. Proc. Natl. Acad. Sci. USA* **69,** 2237–2240.

Bockman, E. L., Berne, R. M., and Rubio, R. (1975). Release of adenosine and lack of release of ATP from contracting skeletal muscle. *Pfluegers Arch.* **355,** 229–241.

Bockman, E. L., Berne, R. M., and Rubio, R. (1976). Adenosine and active hyperemia in dog skeletal muscle. *Am. J. Physiol.* **230,** 1531–1537.

Boyd, I. A., and Forrester, T. (1965). The release of acetylcholine and other metabolic products from skeletal muscle. *J. Physiol.* **176,** 25–26P.

Boyd, I. A., and Forrester, T. (1968). The release of adenosine triphosphate from frog skeletal muscle *in vitro. J. Physiol.* **199,** 115–135.

Brehm, P., Kullberg, R., and Moody-Corbett, F. (1984). Properties of non-junctional acetylcholine receptor channels on innervated muscle of *Xenopus laevis. J. Physiol.* **350,** 631–648.

Cass, C. E., Young, J. D., and Baldwin, S. A. (1998). Recent advances in the molecular biology of nucleoside transporters of mammalian cells. *Biochem. Cell Biol.* **76,** 761–770.

Catterall, W. A. (1975). Activation of the action potential Na ionophore of cultured neuroblastoma cells by veratridine and batrachotoxin. *J. Biol. Chem.* **250,** 4053–4059.

Christie, M. T., Griffith, T. M., and Lewis, M. J. (1989). Comparison of basal and agonist-stimulated release of endothelium-derived relaxing factor from different arteries. *Br. J. Pharmacol.* **98,** 397–406.

Cokelet, G. R., Meiselman, H. J., and Brooks, D. E. (1980). *"Erythrocyte Mechanics and Blood Flow."* A.R. Liss, New York.

Collis, M. G., Bowman, C. J., and Yates, M. S. (1994). Drugs acting on purinergic receptors. *In* "Cardiovascular Pharmacology and Therapeutics" (Singh, B. N., Dzau, V. J., Vanhoutte, P. M., and Woosky, R. L., eds.) Churchill Livingstone, New York.

Conway, E. J., and Cooke, R. (1939). The deaminases of adenosine and adenylic acid in blood and tissues. *Biochem. J.* **33**, 479–492.

Corcondilas, A., Koroxenidas, G. T., and Shepherd, J. T. (1964). Effect of a brief contraction of forearm muscles on forearm blood flow. *J. Appl. Physiol.* **19**, 142–146.

De Mey, J., and Vanhoutte, P. M. (1981). Role of the intima in cholinergic and purinergic relaxation of isolated canine femoral arteries. *J. Physiol.* **316**, 347–355.

De Mey, J., Burnstock, G., and Vanhoutte, P. M. (1979). Modulation of the evoked release of noradrenaline in canine saphenous vein via presynaptic receptors for adenosine but not ATP. *Eur. J. Pharmacol.* **55**, 401–405.

DePierre, J. W., and Karnovsky, M. L. (1974). Ecto-enzyme of granulocytes: 5' nucleotidase. *Science* **183**, 1096–1098.

Dixon, J. P. (1986). Extracellular adenosine triphosphate associated with amphibian erythrocytes: Inhibition of ATP release by anion channel blockers. Ph.D. Thesis. University of Glasgow, Glasgow, Scotland.

Donald, D. E., Rowlands, D. J., and Ferguson, D. A. (1970). Similarity of blood flow in the normal and the sympathectomized dog hind limb during graded exercise. *Circ. Res.* **23**, 185–199.

Doyle, T. B., and Forrester, T. (1985). Appearance of adenosine triphosphate in the perfusate from working frog heart. *Pfluegers Arch.* **405**, 80–82.

Drury, A. N., and Szent-Gyorgyi, A. (1929). The physiological activity of adenine compounds with especial reference to their action upon the mammalian heart. *J. Physiol.* **68**, 213–237.

Duff, F., Patterson, G. C., and Shepherd, J. T. (1954). A quantitative study of the response to adenosine triphosphate of the blood vessels of the human hand and forearm. *J. Physiol.* **125**, 581–589.

Dunkley, C. R., Manery, J. F., and Dryden, E. E. (1966). The conversion of ATP to IMP by muscle surface enzymes. *J. Cell. Physiol.* **68**, 241–248.

Ehara, T., and Matsuura, H. (1993). Single channel study of the cyclic AMP-regulated chloride current in guinea pig ventricular myocytes. *J. Gen. Physiol.* **464**, 307–320.

Fales, J. T., Heisey, S. R., and Zierler, K. L. (1960). Dependency of oxygen consumption of skeletal muscle on number of stimuli during work in the dog. *Am. J. Physiol.* **198**, 1333–1342.

Forrester, T. (1966). Release of adenosine triphosphate from active skeletal muscle. *J. Physiol.* **186**, 107–109P.

Forrester, T. (1967). The identification and assay of acetylcholine and adenosine triphosphate released from active skeletal muscle. Ph.D. Thesis. University of Glasgow, Glasgow, Scotland.

Forrester, T. (1969). The identification of adenosine triphosphate in fresh human plasma. *J. Physiol.* **232**, 53–54P.

Forrester, T. (1972). An estimate of adenosine triphosphate release into the venous effluent from exercising human forearm muscle. *J. Physiol.* **244**, 611–628.

Forrester, T. (1990). Release of ATP from heart. Presentation of a release model using human erythrocyte. *Ann. N.Y. Acad. Sci.* **603**, 335–352.

Forrester, T., and Hassan, M. O. (1973). Appearance of adenosine triphosphate in the perfusate from active frog skeletal muscle. *J. Physiol.* **232**, 86–88P.

Forrester, T., and Lind, A. R. (1969). Identification of adenosine triphosphate in human plasma and the concentration in the venous effluent of forearm muscles before, during and after sustained contractions. *J. Physiol.* **204**, 347–364.

Forrester, T., and Williams, C. A. (1977). Release of adenosine triphosphate from isolated adult heart cells in response to hypoxia. *J. Physiol.* **268**, 371–390.

Frick, J. P., and Lowenstein, J. M. (1976). Studies of 5′ nucleotidase in the perfused rat heart. *J. Biol. Chem.* **251**, 6372–6378.

Furchgott, R. F., and Zawadzki, J. V. (1980). The obligatory role of endothelial cells in the relaxation of arterial smooth muscle by acetylcholine. *Nature* **288**, 373–376.

Gadsby, D. C., Nagel, G., and Hwang, T.-C. (1995). The CFTR chloride channel of mammalian heart. *Annu. Rev. Physiol.* **57**, 387–416.

Gaskell, W. H. (1878). Further researches on the vasomotor nerves of ordinary muscles. *J. Physiol.* **1**, 262–302. Marcel Dekker Inc., New York and Basel.

Gati, W. P., and Paterson, A. R. P. (1989). Nucleoside transport. In "Red Blood Cell Membranes, Structure, Function, Clinical Implications" (P. Agre and J. C. Parker, eds.), pp. 635–661. Marcel Dekker Inc., New York and Basel.

Giles, R. W., and Wilcken, D. E. L. (1977). Reactive hyperemia in the dog heart: Interrelations between adenosine, ATP and aminophylline and the effect of indomethacin. *Cardiovasc. Res.* **11**, 113–121.

Gordon, J. L., and Martin, W. (1983). Endothelium-dependent relaxation of the pig aorta: Relationship to stimulation of Rb efflux from isolated endothelial cells. *Br. J. Pharmacol.* **79**, 531–542.

Grant, R. T. (1938). Observations on the blood circulation in voluntary muscle in man. *Clin. Sci.* **3**, 157–173.

Griffiths, M., Beaumont, N., Yao, S. Y. M., Sundaram, M., Boumah, C. E., Davies, A., Kwong, F. Y. P., Coe, I., Cass, C. E., Young, J. D., and Baldwin, S. A. (1997). Cloning of a human nucleoside transporter implicated in the cellular uptake of adenosine and chemotherapeutic drugs. *Nat. Med.* **3**, 89–93.

Grygoczyk, R., and Hanrahan, J. W. (1997). CFTR-independent ATP release from epithelial cells triggered by mechanical stimuli. *Am. J. Physiol.* **272**, C1058–C1066.

Grygorczyk, R., Tabcharani, J. A., and Hanrahan, J. W. (1996). CFTR channels expressed in CHO cells do not have detectable ATP conductance. *J. Membr. Biol.* **151**, 139–148.

Guharay, F., and Sachs, F. (1984). Stretch-activated single ion channel currents in tissue-cultured embryonic chick skeletal muscle. *J. Physiol.* **352**, 685–701.

Gurd, J. W., and Evans, W. H. (1974). Distribution of liver plasma membrane 5′ nucleotidase as indicated by its reaction with anti-plasma membrane serum. *Arch. Biophys.* **164**, 305–311.

Haddy, F. J., and Scott, J. B. (1975). Metabolic factors in peripheral circulatory regulation. *Fed. Proc.* **34**, 2006–2011.

Hamill, O. P. (1983). Potassium and chloride channels in red blood cells. In "Single Channel Recording" (B. Sakmann and E. Neher, eds.), p. 451. Plenum Press, New York.

Hashimoto, K., and Kokubun, H. (1971). Adenosine-catecholamine interaction in the renal vascular response. *Proc. Soc. Exp. Biol. Med.* **135**, 1125–1128.

Hassessian, H., Bodin, P., and Burnstock, G. (1993). Blockade by glibenclamide of the flow-evoked endothelial release of ATP that contributes to vasodilatation in the pulmonary vascular bed of the rat. *Br. J. Pharmacol.* **109**, 466–472.

Hester, R. L., Guyton, A. C., and Barber, B. J. (1982). Reactive and exercise hyperemia during high levels of adenosine infusion. *Am. J. Physiol.* **243**, H181–H186.

Hodgkin, A. L., and Huxley, A. F. (1952a). Currents carried by sodium and potassium ions through the membrane of the giant axon of *Loligo*. *J. Physiol.* **116**, 449–472.

Hodgkin, A. L., and Huxley, A. F. (1952b). The components of membrane conductance in the giant axon of *Loligo*. *J. Physiol.* **116**, 473–496.

Hodgkin, A. L., and Huxley, A. F. (1952c). The dual effect of membrane potential on sodium conductance in the giant axon of *Loligo*. *J. Physiol.* **116**, 497–506.

Hodgkin, A. L., and Huxley, A. F. (1952d). A quantitative description of membrane current and its application to conduction and excitation in nerve. *J. Physiol.* **117**, 500–544.

Honig, C. R., and Frierson, J. L. (1980). Role of adenosine in exercise vasodilation in dog gracilis muscle. *Am. J. Physiol.* **238**, H703–H715.

Ignarro, L. J., Byrns, R. E., Buga, G. M., and Wood, K. S. (1987). Endothelium-derived relaxing factor from pulmonary artery and vein possesses pharmacologic and chemical properties identical to those of nitric oxide radical. *Circ. Res.* **61**, 866–879.

Israel, M., Lesbats, B., Meunier, F. M., and Stinnakre, J. (1976). Postsynaptic release of adenosine triphosphate induced by single impulse transmitter action. *Proc. R. Soc. London Ser. B* **193**, 461–468.

Jacob, M. I., and Berne, R. M. (1960). Metabolism of purine derivatives by the isolated cat heart. *Am. J. Physiol.* **198**, 322–326.

Jones, C. E., Hurst, T. W., and Randall, J. R. (1982). Effect of aminophylline on coronary functional hyperemia and myocardial adenosine. *Am. J. Physiol.* **243**, H480–H487.

Jorgensen, S. (1956). Breakdown of adenine and hypoxanthine nucleotides and nucleosides in human plasma. *Acta Pharmacol. Toxicol.* **12**, 294–302.

Katori, M., and Berne, R. M. (1966). Release of adenosine from anoxic hearts: Relationship of coronary flow. *Circ. Res.* **19**, 420–425.

Kelm, M., Feelisch, M., Spahr, R., Piper, H.-M., Noack, E., and Schrader, J. (1988). Quantitative and kinetic characterization of nitric oxide and EDRF released from cultured endothelial cells. *Biochim. Biophys. Acta. Res. Commun.* **154**, 236–244.

Khayutin, V. M. (1968). Determinants of working hyperemia in skeletal muscle. *In* "Circulation in Skeletal Muscle" (O. Hudlicka, ed.), pp. 145–157. Pergamon Press, Oxford.

Kjellmer, I. (1965). On the competition between metabolic vasodilation and neurogenic vasoconstriction in skeletal muscle. *Acta Physiol. Scand.* **63**, 450–459.

Kopito, R. R., and Lodish, H. F. (1985). Primary structure and transmembrane orientation of the murine anion exchange protein. *Nature* **316**, 234–238.

Lansman, J. B., Hallam, T. J., and Rink, T. J. (1987). Single stretch-activated ion channels in vascular endothelial cells as mechanotranducers. *Nature* **325**, 811–813.

Latchenberger, J., and Deahna, A. (1876). Beitrage zur Lehre von der reflectorischen Erregung der Gefassmuskeln. *Pfluegers Arch.* **12**, 157–204.

Lewis, M. T., and Smith, J. A. (1991). Factors regulating the release of endothelium-derived relaxing factor. *In* "Endothelial Regulation of Vascular Tone" (U. S. Ryan and G. M. Rubanyi, eds.), pp. 139–154. Marcel Dekker, New York.

Li, C., Ramjeesingh, M., and Bear, C. E. (1996). Purified cystic fibrosis transmembrane conductance regulator (CFTR) does not function as an ATP channel. *J. Biol. Chem.* **271**, 11623–11626.

Lipowsky, H. H., Usami, S., and Chicu, S. (1980). *In vivo* measurements of "apparent viscosity" and microvessel hematocrit in the mesentery of the cat. *Microvasc. Res.* **19**, 297–319.

Long, C. J., and Stone, T. W. (1985). The release of endothelium-derived relaxant factor is calcium dependent. *Blood Vessels* **22**, 205–208.

Manery, J. F., Riordan, J. R., and Dryden, E. E. (1968). Characteristics of nucleotide-converting enzymes at muscle surfaces with special reference to ion sensitivity. *Can. J. Physiol. Pharmacol.* **46**, 537–547.

Merrill, G. F., Haddy, F. J., and Dabney, J. M. (1978). Adenosine, theophylline and perfusate pH in the isolated perfused guinea pig heart. *Circ. Res.* **42**, 225–229.

Mo, M., Eskin, S. G., and Schilling, W. P. (1991). Flow-induced changes in calcium signalling of vascular endothelial cells: Effect of shear stress and ATP. *Am. J. Physiol.* **260**, H1698–H1707.

Nagel, G., Hwang, T.-C., Nastiuk, K. L., Nairn, A. C., and Gadsby, D. C. (1992). The protein kinase A-regulated cardiac chloride channel resembles the cystic fibrosis transmembrane conductance regulator. *Nature* **360**, 81–84.

Olsson, R. A. (1981). Local factors regulating cardiac and skeletal muscle blood flow. *Annu. Rev. Physiol.* **43**, 385–395.

Paddle, B. M., and Burnstock, G. (1974). Release of ATP from perfused heart during coronary vasodilation. *Blood Vessels* **11**, 110–119.

Palmer, R. M. J., Ferrige, A. G., and Moncada, S. (1987). Nitric oxide release accounts for the biological activity of endothelium-derived relaxing factor. *Nature* **327**, 524–526.

Parish, R. W., and Weibel, M. (1980). Extracellular ATP, ecto-ATPase and calcium influx in *Dictyostelium discoideum* cells. *FEBS Lett.* **118**, 263–266.

Patterson, G. C., and Shepherd, J. T. (1954). The blood flow in the human forearm following venous congestion. *J. Physiol.* **125**, 501–507.

Pearson, J. D., and Gordon, J. L. (1979). Vascular endothelial and smooth muscle cells in culture selectively release adenine nucleotides. *Nature* **281**, 384–386.

Phair, R. D., and Sparks, H. V. (1979). Adenosine content of skeletal muscle during active hyperemia and ischemia contraction. *Am. J. Physiol.* **237**, H1–H9.

Pohl, U., Busse, R., Kuon, E., and Basseny, E. (1986). Pulsatile perfusion stimulates the release of endothelial autocoids. *J. Appl. Cardiol.* **1**, 215–235.

Potter, P. E., and White, T. D. (1980). Release of adenosine 5'-triphosphate from synaptosomes from different regions of the brain. *Neuroscience* **5**, 1351–1356.

Potter, P. E., and White, T. D. (1982). Lack of effect of 6-hydroxydopamine pretreatment on depolarization-induced release of ATP from rat brain synaptosomes. *Eur. J. Pharmacol.* **80**, 143–147.

Ralevic, V., and Burnstock, G. (1998). Receptors for purines and pyrimidines. *Pharmacol. Rev* **50**, 413–492.

Ralevic, V., Lincoln, J., and Burnstock, G. (1992). Release of vasoactive substances from endothelial cells. *In* "Endothelial Regulation of Vascular Tone" (U. S. Ryan and G. M. Rubanyi, eds.), pp. 297–328. Marcell Dekker, New York.

Reddy, M. M., Quinton, P. M., Haws, C., Wine, J. J., Grygorczyk, R., Tabcharani, J. A., Hanrahan, J. W., Gunderson, K. L., and Kopito, R. R. (1996). Failure of the cystic fibrosis transmembrane conductance regulator to conduct ATP. *Science* **271**, 1876–1879.

Rein, H. (1930). Die Interferenz der vasomotorischen Regulationen. *Klin. Wochenschr.* **9**, 1485–1489.

Reisin, I. L., Prat, A. G., Abraham, E. H., Amara, J. F., Gregory, R. J., Ausiello, D. A., and Contiello, H. F. (1994). The cyctic fibrosis transmembrane conductance regulator is a dual ATP and chloride channel. *J. Biol. Chem.* **269**, 20584–20591.

Remensnyder, J. P., Mitchell, J. H., and Sarnoff, S. J. (1962). Functional sympatholysis during muscular activity. *Circ. Res.* **11**, 370–380.

Richman, H. G., and Wyborny, L. (1964). Adenine nucleotide degradation in the rabbit heart. *Am. J. Physiol.* **207**, 1139–1145.

Riordan, J. R., Rommens, J. M., Kerem, B.-S., Alon, N., Rozmahel, R., Grzelczak, Z., Zielenski, J., Lok, S., Plavsic, N., Chou, J.-L., Drumm, M. L., Iannuzzi, M. C., Collins, F. S., and Tsui, L.-C. (1989). Identification of the cystic fibrosis gene: Cloning and characterization of complimentary DNA. *Science* **245**, 1066–1073.

Rowlands, D. J., and Donald, D. E. (1968). Sympathetic vasoconstrictive responses during exercise or drug-induced vasodilation. A time-dependent response. *Circ. Res.* **23**, 45–60.

Rubanyi, G. M., Romero, J. C., and Vanhoutte, P. M. (1986). Flow-induced release of endothelium-derived relaxing factor. *Am. J. Physiol.* **250**, H1145–H1149.

Rubio, R., and Berne, R. M. (1969). Release of adenosine by the normal myocardium in dogs; its relationship to the regulation of coronary resistance. *Circ. Res.* **25,** 407–415.

Ruwhof, C., and van der Laarse, A. (2000). Mechanical stress-induced cardiac hypertrophy: Mechanisms and signal transduction pathways. *Cardiovasc. Res.* **47,** 23–37.

Sackin, H. (1995). Mechanosensitive channels. *Annu. Rev. Physiol.* **57,** 333–353.

Sawynok, J., Downie, J. W., Reid, A. R., Cahill, C. M., and White, T. D. (1993). ATP release from dorsal spinal cord synaptosomes: Characterization and neuronal origin. *Brain Res.* **610,** 32–38.

Schmid-Schonbein, G. W. (1976). Microrheology of erythrocytes, blood viscosity and the distortion of blood flow in the microcirculation. *Int. Rev. Physiol.* **9,** 1–62.

Schwiebert, E. M. (1999). ABC transporter-facilitated ATP conductive transport. *Am. J. Physiol.* **276,** C1–C8.

Schwiebert, E. M., Egan, M. E., Hwang, T. H., Fulmer, S. B., Allen, S. A., Cutting, G. R., and Guggino, W. B. (1995). CFTR regulates outwardly rectifying chloride channels through an autocrine mechanism involving ATP. *Cell* **81,** 1063–1073.

Scott, J. B., Daugherty, R. M., Dabney, J. M., and Haddy, F. J. (1965). Role of chemical factors in regulation of flow through kidney, hindlimb and heart. *Am. J. Physiol.* **298,** 813–824.

Sprague, R. S., Ellsworth, M. L., Stephenson, A. H., Kleinhenz, M. E., and Lonigro, A. J. (1998). Deformation-induced ATP release from red blood cells requires CFTR activity. *Am. J. Physiol.* **275,** H1726–H1732.

Stowe, D. F., Sullivan, T. E., Dabney, J. M., Scott, J. B., and Haddy, F. J. (1974). Role of ATP in coronary flow regulation in isolated perfused guinea pig heart. *Physiologist* **17,** 339.

Su, C. (1978). Purinergic inhibition of adrenergic transmission in rabbit blood vessels. *J. Pharmacol. Exp. Ther.* **204,** 351–361.

Sugita, M., Yue, Y., and Foskett, J. K. (1998). CFTR Cl⁻ channel and CFTR-associated ATP channel: Distinct pores regulated by common gates. *EMBO J.* **17,** 898–908.

Tousson, A., Van Tine, B. A., Naren, A. P., Shaw, G. M., and Schwiebert, L. M. (1998). Characterization of CFTR expression and chloride channel activity in human endothelia. *Am. J. Physiol.* **275,** C1555–C1564.

Trams, E. G., and Lauter, C. J. (1974). On the sidedness of plasma membrane enzymes. *Biochim. Biophys. Acta.* **345,** 180–197.

Ulbricht, W. (1969). The effect of veratridine on exitable membranes of nerve and muscle. *Ergebn. Physiol.* **61,** 18–71.

van Wamel, A. J., Ruwhof, C., Valk-Kokshoorn, L. J., Schrier, P. I., and van der Laarse, A. (2000). Rapid effects of stretched myocardial and vascular cells on gene expression of neonatal rat cardiomyocytes with emphasis on autocrine and paracrine mechanisms. *Arch. Biochem. Biophys.* **381,** 67–73.

Verhaeghe, R. H., Vanhoutte, P. M., and Shepherd, J. T. (1977). Inhibition of sympathetic neurotransmission in canine blood vessels by adenosine and adenine nucleotides. *Circ. Res.* **40,** 208–215.

Vial, C., Owen, R., Opie, L. H., and Posel, D. (1987). Significance of release of adenosine triphosphate and adenosine induced by hypoxia and adrenaline in perfused rat heart. *J. Mol. Cell. Cardiol.* **19,** 187–197.

Wadsworth, R. M. (1972). The effects of aminophylline on the increased myocardial blood flow produced by systemic hypoxia or by coronary occlusion. *Eur. J. Pharmacol.* **20,** 130–132.

Weber, K., and Osborn, M. (1981). Microtubule and intermediate filament networks in cells viewed by immunofluorescence microscopy. *In* "Cytoskeletal Elements and Plasma

Membrane Organization" (G. Poste and G. L. Nicolson, eds.), pp. 1–55. Elsevier/North-Holland Biomedical Press, Amsterdam.

White, T. D. (1977). Direct detection of depolarization-induced release of ATP from a synaptosomal preparation. *Nature* **267**, 67–68.

White, T. D. (1978). Release of ATP from a synaptosomal preparation by elevated extracellular K and by veratridine. *J. Neurochem.* **30**, 329–336.

White, T. D., Potter, P., and Wonnacott, S. (1980). Depolarization-induced release of ATP from cortical synaptosomes is not associated with acetylcholine release. *J. Neurochem.* **34**, 1109–1112.

Wiedmeier, V. T., and Spell, L. H. (1977). Effects of catecholamines, histamine and nitroglycerine on flow, oxygen utilization and adenosine production in the perfused guinea pig heart. *Circ. Res.* **41**, 503–508.

Williams, C. A., and Gold, A. H. (1975). The effect of insulin on glycogen metabolism of rat heart cells in suspension culture. *Fed. Proc.* **34**, 262.

Williams, C. A., and Forrester, T. (1983). Possible source of adenosine triphosphate released from rat myocytes in response to hypoxia and acidosis. *Cardiovasc. Res.* **17**, 301–312.

Williamson, J. R., and DiPietro, D. L. (1965). Evidence for extracellular enzymatic activity of the isolated perfused rat heart. *Biochem. J.* **95**, 226–232.

Wolfe, M. M., and Berne, R. M. (1956). Coronary vasodilator properties of purine and pyrimidine derivatives. *Circ. Res.* **4**, 343–348.

Woo, Y.-T., and Manery, J. F. (1975). 5'-Nucleotidase: An ecto-enzyme of frog skeletal muscle. *Biochim. Biophys. Acta.* **397**, 144–152.

Zipf, K. (1931). Die chemische Natur der 'depressorischen Substanz' des blut. *Arch. Exp. Pharmakol.* **160**, 579–598.

# CHAPTER 10

# Purinergic Receptors in the Nervous System

**Geoffrey Burnstock**

Autonomic Neuroscience Institute, Royal Free and University College London Medical School, London NW3 2PF, UK

## I. INTRODUCTION

Pamela Holton provided the first hint of a transmitter role for adenosine triphosphate (ATP) in the nervous system by demonstrating release of ATP during antidromic stimulation of sensory nerves (Holton, 1959). Then, in my laboratory in Melbourne in 1970, we proposed that nonadrenergic, noncholinergic (NANC) nerves supplying smooth muscle of the gut and bladder utilized ATP as a neurotransmitter (Burnstock et al., 1970, 1972). The experimental evidence included mimicry of the NANC nerve-mediated response by ATP; measurement of release of ATP during stimulation of NANC nerves with luciferin-luciferase luminometry; histochemical labeling of subpopulations of neurons in the gut and the bladder with quinacrine, a fluorescent dye known to selectively label high levels of ATP bound to peptides; and the demonstration that the slowly degradable analogue of ATP, $\alpha\beta$-methylene ATP, which produces selective desensitization of the ATP receptor, blocks the responses to NANC nerve stimulation. The term "purinergic" and the evidence for purinergic transmission in a wide variety of systems were presented in an early pharmacological review (Burnstock, 1972).

Implicit in the concept of purinergic neurotransmission is the existence of postjunctional purinergic receptors. A basis for distinguishing two types of purinoceptor, identified as P1 and P2 [for adenosine and ATP/adenosine diphosphate (ADP), respectively], was proposed (Burnstock, 1978), but it was not until 1985 that a basis for distinguishing two types of P2 receptor (P2X and P2Y) was suggested, largely on the basis of pharmacological criteria. Further P2 receptor subtypes followed including a P2T receptor selective for ADP on platelets and a P2Z receptor on macrophages (Gordon, 1986), and a P2U receptor that could recognize pyrimidines such as uridine triphosphate (UTP) as well as ATP (O'Connor et al., 1991). Abbracchio and Burnstock (1994), on the basis of studies of transduction mechanisms (Dubyak, 1991) and the cloning of P2Y (Lustig et al., 1993; Webb et al., 1993) and later P2X purinoceptors (Brake et al., 1994; Valera et al., 1994), proposed that purinoceptors should be considered to belong to two major families: a P2X family of ligand-gated ion channel receptors and a P2Y family of G-protein-coupled purinoceptors. This nomenclature has been widely adopted and currently seven P2X subtypes and about eight P2Y receptor subtypes are recognized (Burnstock, 2001) (Table I). The current consensus is that three P2X subtypes combine either as homomultimers or heteromultimers to form ion pores and there is growing recognition that heterodimers might form between P2Y receptor subtypes (see Chapter 1 of this volume (Burnstock, ATP and its metabolites as potent extracellular agonists)). In addition hetero-oligomerization of adenosine A1 receptors with $P2Y_1$ receptors in rat brain has been proposed (Yoshioka et al., 2002).

**TABLE I**

Comparison of Fast Ionotropic and Slow Metabotropic Receptors for Acetylcholine (ACh),
γ-Aminobutyric acid (GABA), Glutamate, and 5-Hydroxytryptamine (5-HT) with Those for
Purines and Pyrimidines[a]

| | Receptors | |
|---|---|---|
| Messenger | Fast ionotropic | Slow metabotropic |
| ACh | Nicotinic | Muscarinic |
| | Muscle type | $M_1$–$M_5$ |
| | Neuronal type | |
| GABA | GABA A | GABA B |
| Glutamate | AMPA[b] | mGlu$_1$–mGlu$_7$ |
| | Kainate | |
| | NMDA[b] | |
| 5-HT | 5-HT$_3$ | 5-HT$_{1A-F}$ |
| | | 5-HT$_{2A-C}$ |
| | | 5-HT$_4$ |
| | | 5-HT$_{5A-B}$ |
| | | 5-HT$_6$ |
| | | 5-HT$_7$ |
| ATP | P2X$_{1-7}$ | P2Y$_1$, P2Y$_2$, |
| | | P2Y$_4$, P2Y$_6$, |
| | | P2Y$_{11}$, P2Y$_{12}$, |
| | | P2Y$_{13}$, P2Y$_{14}$ |

[a]Modified from Abbracchio and Burnstock (1998). [b]AMPA, 2-(aminomethyl)phenylacetic acid; NMDA, N-methyl-D-aspartate. Reproduced with permission from The Japanese Pharmacological Society.

Most studies of fast signaling in the nervous system have been concerned with the role of ATP acting postjunctionally as a transmitter orcontransmitter (see Burnstock, 1976; 1990a,b), whereas adenosine, after ectoenzymatic breakdown of released ATP, acts largely as a prejunctional modulator of transmitter release (see Dunwiddie, 1985; Ribeiro, 1995). In addition, there are many examples of the potent long-term (trophic) effects of ATP, UTP, and related compounds on neurons and glial cells (see Neary et al., 1996) and on peripheral nerve, smooth muscle, and epithelial cell proliferation, growth, and differentiation (Fig. 1) (see Abbracchio and Burnstock, 1998).

In this chapter, I will focus on the localization and roles of P2 receptor subtypes in the central nervous system (CNS), as comprehensive reviews of purinergic signaling in the peripheral nervous system have been published recently (Burnstock, 1996, 1999a, 2000, 2001b,c; Ralevic and Burnstock, 1998; Williams and Burnstock, 1997; Kennedy, 2001; Dunn et al., 2001; King and North, 2000), although reviews on limited aspects of purinergic

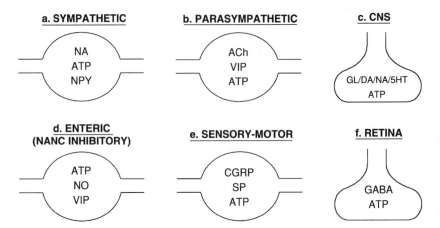

**FIGURE 1** Schematic showing the chemical coding of cotransmitters in autonomic, sensory motor, and retinal nerves and in neurons in the CNS. ATP has been shown recently to be a cotransmitter with noradrenaline, dopamine, or 5-HT as well as with glutamate. Modified from Burnstock, 1999b.

signaling in the CNS are available (Burnstock, 1977, 1996; Phillis and Wu, 1981; Inoue *et al.*, 1996; Dunwiddie *et al.*, 1996; Gibb and Halliday, 1996; Abbracchio, 1997; Robertson, 1998); the recent reviews by Nörenberg and Illes (2000) and by Masino and Dunwiddie (2001) are particularly useful. The recent volume of *Progress in Brain Research* edited by Illes and Zimmermann also contains valuable articles on both the peripheral and central nervous systems (Illes and Zimmermann, 1999).

## II. PERIPHERAL NEUROEFFECTOR TRANSMISSION

### A. Sympathetic Nerves

The first hint about sympathetic purinergic cotransmission was in a paper published by Burnstock and Holman, 1962, in which they recorded excitatory junction potentials (EJPs) in smooth muscle cells of the vas deferens in response to stimulation of the hypogastric nerves. Although these junction potentials were blocked by guanethidine, which prevents the release of sympathetic neurotransmitters, we were surprised at the time that adrenoceptor antagonists were ineffective (Burnstock and Holman, 1962). It was over 20 years before selective desensitization of the ATP receptor by $\alpha\beta$-methylene ATP was shown to block the EJPs, when it became clear that we were looking at the responses to ATP as a cotransmitter in sympathetic nerves (Sneddon and Burnstock, 1984). Release of ATP was shown to be abolished by tetrodotoxin and guanethidine, and after destruction of sympathetic nerves by 6-hydroxydopamine, but not by reserpine, which blocked the second slow noradrenergic phase of the response, but not the initial fast phase (Kirkpatrick and Burnstock, 1987). Spritzing ATP onto single smooth muscle cells of the vas deferens mimicked the EJP, whereas spritzed noradrenaline (NA) did not (Sneddon and Westfall, 1984).

Sympathetic purinergic cotransmission has also been clearly demonstrated in a variety of blood vessels (Burnstock, 1988, 1990a). The proportion of NA to ATP is extremely variable in the sympathetic nerves supplying the different blood vessels. The purinergic component is relatively minor in rabbit ear and rat tail arteries, is more pronounced in the rabbit saphenous artery, and has been claimed to be the sole transmitter in sympathetic nerves supplying arterioles in the mesentery and the submucosal plexus of the intestine, whereas NA release from these nerves acts as a modulator of ATP release (Ramme *et al.*, 1987; Evans and Surprenant, 1992). ATP-evoked noradrenaline release has been detected from both rat (Boehm, 1999) and guinea pig (Sperlágh *et al.*, 2000) sympathetic nerve terminals.

## B. Parasympathetic Nerves

Parasympathetic nerves supplying the urinary bladder utilize acetylcholine (ACh) and ATP as cotransmitters, in variable proportions in different species (Burnstock *et al.*, 1978; Burnstock, 2001c) and by analogy with sympathetic nerves, ATP again acts through P2X ionotropic receptors, whereas the slow component of the response is mediated by a metabotropic receptor, in this case muscarinic (Hoyle and Burnstock, 1985). There is some evidence to suggest parasympathetic, purinergic cotransmission to resistance vessels in the heart and airways (Inoue and Kannan, 1988; Saffrey *et al.*, 1992).

## C. Sensory Nerves

Since the seminal studies of Lewis (1927) it has been well established that transmitters released following the passage of antidromic impulses down sensory nerve collaterals during "axon reflex" activity produce vasodilatation of skin vessels. We know now that axon reflex activity is widespread in autonomic effector systems and forms an important physiological component of autonomic control (Maggi and Meli, 1988; Burnstock, 1993b; Rubino and Burnstock, 1996). Calcitonin gene-related peptide (CGRP) and substance P are well established to coexist in sensory motor nerves and, in some subpopulations, ATP is also a cotransmitter (Holton, 1959; Sweeney *et al.*, 1989; Burnstock, 1993a).

$P2X_3$ receptors were cloned in 1995 and shown to be predominantly localized in small nociceptive sensory neurons (Chen *et al.*, 1995; Lewis *et al.*, 1995; see also Section II.A). $P2X_3$ receptors have been localized on sensory nerve terminals in skin, tongue (Bo *et al.*, 1999; Rong *et al.*, 2000), knee (Dowd *et al.*, 1998), and tooth pulp (Alavi *et al.*, 2001), and in the subepithelial nerve plexus of the urinary bladder (Cockayne *et al.*, 2000; Vlaskovska *et al.*, 2001; Rong *et al.*, 2001).

When ATP was applied to a blister base or injected intradermally, it caused pain in humans (Bleehen and Keele, 1977; Coutts *et al.*, 1981). The pain-producing effects of ATP were greatly potentiated by acute capsaicin treatment and ultraviolet (UV) irradiation (Hamilton *et al.*, 2000). In animal models, subplantar injection of ATP and $2',3'-0$-(4-benzoylbenzoyl)-ATP (BzATP) produced nocifensive behavior (hindpaw lifting and licking) in the rat and mouse (Bland-Ward and Humphrey, 1997; Hamilton *et al.*, 1999; Cockayne *et al.*, 2000; Jarvis *et al.*, 2001).

In addition to nociceptors, ATP has been shown to excite a variety of other primary afferent neurons. ATP released by oxygen sensing chemo-receptors in carotid body activates P2X receptors present on nerve endings of rat sinus nerve, and the hypoxic signaling in the carotid body is mediated by the corelease of ATP and ACh (Zhang *et al.*, 2000). Intraarterial injection of ATP and $\alpha\beta$-methylene ATP ($\alpha\beta$-meATP) excited mesenteric afferent nerves in the rat (Kirkup *et al.*, 1999). Functional P2X receptors have also been demonstrated to be present on canine pulmonary vagal C fibers (Pelleg and Hurt, 1996) and vagal afferent nerves participating in the homeostatic mechanism for cardiovascular and respiratory regulation in the rat (McQueen *et al.*, 1998).

Studies of transgenic mice lacking the $P2X_3$ subunit provided direct evidence for the physiological roles of homo- and or heteromeric P2X receptors containing the $P2X_3$ subunit (Cockayne *et al.*, 2000; Souslova *et al.*, 2000; Vlaskovska *et al.*, 2001). A new hypothesis for purinergic mech-anosensory transduction in visceral organs involved in initiation of pain has been proposed (Burnstock, 2001a) in which it is suggested that distension of tubes such as the ureter, salivary ducts, and gut, and sacs such as urinary and gallbladders, leads to the release of ATP from the lining epithelial cells, which diffuses to the subepithelial sensory nerve plexus to stimulate $P2X_3$ and/or $P2X_{2/3}$ receptors, which mediate messages to the sensory ganglia and to pain centers in the central nervous system (CNS). It has been clearly shown that ATP is released from the epithelial cells in the distended bladder

(Ferguson *et al.*, 1997; Vlaskovska *et al.*, 2001) and ureter (Knight *et al.*, 1999) and P2X$_3$ receptors have also been identified in subepithelial nerves in the ureter (Lee *et al.*, 2000) and in the bladder (Cockayne *et al.*, 2000). Recording in a P2X$_3$ knockout mouse, we have shown that the micturition reflex is impaired and that responses of sensory fibers to P2X$_3$ agonists are gone, suggesting that P2X$_3$ receptors on sensorynerves in the bladder have a physiological as well as a nociceptive role (Cockayne *et al.*, 2000).

Purinoceptors also have a strong presence on special sensory nerve terminals and associated cells in the ear (see Housley, 1997; Chen *et al.*, 2000) and eye (see Pintor, 1999; Pannicke *et al.*, 2000). For an excellent recent review of purinergic transmission in visual, cochlear, and vestibular systems see Housley (2001).

## D. Intramural Nerves

Intrinsic neurons exist in most of the major organs of the body. Many of these are part of the parasympathetic nervous system, but certainly in the gut and perhaps also in the heart, some of these intrinsic neurons are derived from neural crest tissue and differ from those that form the sympathetic and parasympathetic systems and appear to represent an independent control system. In the heart, subpopulations of intrinsic nerves in the atrial and intraatrial septum have been shown to contain ATP as well as nitric oxide (NO), neuropeptide Y, ACh, and 5-hydroxytryptamine. Many of these nerves project to the coronary microvasculature and produce potent vasomotor actions (Burnstock, 1990a,b; Saffrey *et al.*, 1992).

A subpopulation of intramural enteric nerves provides NANC inhibitory innervation of gastrointestinal smooth muscle. Three major cotransmitters are released from these nerves: (1) ATP produces fast inhibitory junction potentials (IJPs); (2) NO also produces IJPs, but with a slower time course; and (3) vasoactive intestinal peptide (VIP) produces slow tonic relaxations (see Belai and Burnstock, 1994). The proportions of these three transmitters vary considerably in different regions of the gut and in different species; forexample, in some sphincters the NANC inhibitory nerves primarily utilize VIP, in others they utilize NO, and in nonsphincteric regions of the intestine, ATP is more prominent (see Burnstock, 2001b).

## E. Motor Nerves to Skeletal Muscle

It has long been known that ATP is stored in and together with ACh released from motor nerve terminals. In the adult skeletal neuromuscular junction, ACh appears to be the sole neurotransmitter acting through nicotinic receptors, whereas the released ATP acts both postjunctionally to potentiate the action of ACh and prejunctionally after breakdown toadenosine via $P1(A_1)$ receptors to modulate the release of ACh (Silinsky, 1984; Hamilton and Smith, 1991; Lu and Smith, 1998). However, in the developing myotube, ATP as well as ACh act through ion channel receptors (Kolb and Wakelam, 1983; Häggblad and Heilbronn, 1988). ATP responsiveness also reappears after denervation (Wells *et al.*, 1995). Recent studies from our laboratory have shown sequential expression of $P2X_5$, $P2X_6$, and $P2X_2$ receptor subtypes in developing rat skeletal muscle (Ryten *et al.*, 2001).

## III. SENSORY AND AUTONOMIC GANGLIA

An effect of ATP on autonomic ganglia was first reported in 1948 when Feldberg and Hebb demonstrated that intraarterial injection of ATP excited neurons in the cat superior cervical ganglion (SCG) (Feldberg and Hebb, 1948). Later work from deGroat's laboratory showed that in the cat vesical parasympathetic ganglia and rat SCG, purines inhibited synaptic transmission through $P_1$ receptors, but high doses of ATP depolarized and excited the postganglionic neurons (Theobald and de Groat, 1977, 1989). The earliest intracellular recordings ofthe action of ATP on neurons were obtained in frog sympathetic ganglia (Siggins *et al.*, 1977; Akasu *et al.*, 1983). ATP produced a depolarization through a reduction in $K^+$ conductance, which was probably mediated through P2Y receptors. ATP was shown to excite mammalian dorsal root ganglia (DRG) neurons and some neurons from the dorsal horn of the spinal cord (Jahr and Jessell, 1983; Krishtal *et al.*, 1983). These responses were associated with an increase in membrane conductance, which we now know was due to the activation of P2X receptors (see Dunn *et al.*, 2001).

### A. Sensory Ganglia

Sensory neurons of the DRG share with neurons of the sympathetic, parasympathetic, and enteric ganglia, along with adrenomedullary chromaffin cells, a common embryological origin in the neural crest. In contrast, cranial sensory neurons are derived from the placodes. Although these sensory and autonomic neurons exhibit some common properties, they also show very diverse phenotypes commensurate with their diverse physiological roles (see Dunn *et al.*, 2001). The nociceptive roles of $P2X_3$ receptors on sensory nerves have also been reviewed recently (see Burnstock and Wood, 1996; Burnstock, 2000, 2001; Salter and Sollevi, 2001).

There have been many reports characterizing the native P2X receptors in sensory neurons, including those from dorsal root, trigeminal, nodose, and petrosal ganglia. DRG and trigeminal ganglia contain primary somatosensory neurons, receiving nociceptive, mechanical, and proprioceptive inputs. Nodose and petrosal ganglia, on the other hand, contain cell bodies of afferents to visceral organs (Lindsay, 1996).

P2X receptors on the cell bodies of sensory neurons have been studied extensively using voltage-clamp recordings from dissociated neurons of the DRG (Bouvier *et al.*, 1991; Burgard *et al.*, 1999; Grubb and Evans, 1999; Li *et al.*, 1999; Rae *et al.*, 1998; Robertson *et al.*, 1996; Ueno *et al.*, 1999), trigeminal and nodose ganglia (Khakh *et al.*, 1995; Khalil *et al.*, 1994; Thomas *et al.*, 1998), and petrosal ganglia (see Dunn *et al.*, 2001; Khakh *et al.*, 1997). Rapid application of ATP to acutely dissociated or cultured sensory neurons evokes action potentials and under voltage clamp, a fast-activating inward current. The activation of P2X receptors results in depolarization and an increase in intracellular $Ca^{2+}$ concentration (Bean *et al.*, 1990; Bouvier *et al.*, 1991). Negative cross-talk between anionic $GABA_A$ and cationic P2X ionotropic receptors of rat DRG neurons has been demonstrated (Sokolova *et al.*, 2001).

The $P2X_3$ subunit that was first cloned using a cDNA library from neonatal rat DRG neurons shows a selectively high level of expression in a subset of sensory neurons, including those in DRG and trigeminal and nodose ganglia (Chen *et al.*, 1995; Lewis *et al.*, 1995; Collo *et al.*, 1996). In DRG and trigeminal ganglia, although mRNA transcripts of $P2X_{1-6}$ have been detected, the level of $P2X_3$ transcript is the highest. Sensory neurons from nodose ganglia express, in addition to $P2X_3$, significant levels of $P2X_1$, $P2X_2$, and $P2X_4$ RNAs, and some of these RNAs are present in the same cell. 5YBR Green Fluorescence has been used recently to quantitate P2X receptor in RNA in DRG (Ueno *et al.*, 2002). The expression pattern of $P2X_3$ receptors in sensory ganglia has also been studied by immunohistochemistry at both the light microscope (Vulchanova *et al.*, 1997, 1998; Bradbury *et al.*, 1998; Xiang *et al.*, 1998a; Novakovic *et al.*, 1999; Barden and Bennett, 2000) and electron microscope (Llewellyn-Smith and Burnstock, 1998) levels. In DRG, intensive $P2X_3$ immunoreactivity is found predominantly in a subset of small- and medium-diameter neurons, although absent from most large neurons. The $P2X_3$ subunit is predominantly located in the nonpeptidergic subpopulation of nociceptors that binds the isolectin B4 (IB4), and is greatly reduced by neonatal capsaicin treatment (Vulchanova *et al.*, 1998). The $P2X_3$ subunit is present in an approximately equal number of neurons projecting to skin and viscera but in very few of those innervating skeletal muscle (Bradbury *et al.*, 1998). P2X3 receptors are stongly represented in sensory ganglia during rat embryonic neurogenesis (Cheung and Burnstock, 2002).

$P2X_2$ receptor immunoreactivity is observed in many small and large DRG neurons, although the level is lower than that of $P2X_3$ (Vulchanova et al., 1997; Labrakakis et al., 2000; Petruska et al., 2000a). Some neurons seem to contain both $P2X_2$ and $P2X_3$ immunoreactivity. Although $P2X_3$ immunoreactivity is the predominant type detected, variable levels of immunoreactivity of $P2X_{1,2}$ and $P2X_{4-6}$ have also been detected in DRG neurons (Xiang et al., 1998a; Barden and Bennett, 2000; Petruska et al., 2000a,b). These receptors take the form of clusters 0.2–0.5 $\mu$m in diameter, and rarely appear to colocalize (Barden and Bennett, 2000).

In trigeminal ganglia, $P2X_3$ immunoreactivity is found in the cell bodies of both small and large neurons (Cook et al., 1997; Llewellyn-Smith and Burnstock, 1998; Xiang et al., 1998a). Lower levels of immunoreactivity to $P2X_{1,2}$ and $P2X_{4-6}$ appear to be present in these neurons. Concurrent release of ATP and substance P from neurons in guinea pig trigeminal ganglia in vivo has been reported (Matsuka et al., 2001). In nodose ganglia, $P2X_2$ and $P2X_3$ immunoreactivities are both present (Xiang et al., 1998a) and are colocalized in the same neurons (Vulchanova etal., 1997). This is consistent with the colocalization of their mRNAs (Lewis et al., 1995). Both transient and sustained responses to P2 agonists occur in DRG neurons (see Dunn et al., 2001). The transient response in DRG neurons is activated by ATP, $\alpha\beta$-meATP and 2-methylthio ATP (2-meSATP) (Robertson et al., 1996). The pharmacological evidence to date generally supports the hypothesis that this rapidly desensitizing transient response is mediated by homomeric $P2X_3$ receptors.

ATP, $\alpha\beta$-meATP, and 2-meSATP also evoke sustained currents in rat nodose neurons. These responses are inhibited by suramin, pyridoxal-phosphate-6-azophenyl-2′,4′-disulfonic acid (PPADS), Cibacron blue, trinitrophenyl (TNP)-ATP, and $Ca^{2+}$ (Khakh et al., 1995; Thomas et al., 1998; Virginio et al., 1998), but not by diinosine pentaphosphate ($Ip_5I$) at concentrations up to 100 $\mu M$ (Dunn et al., 2000). Therefore, the $\alpha\beta$-meATP-sensitive persistent responses in nodose neurons resemble the recombinant $P2X_{2/3}$ receptors (Lewis et al., 1995). Neurons of the mouse nodose ganglion give persistent responses to both ATP and $\alpha\beta$-meATP similar to those seen in the rat (Cockayne et al., 2000; Souslova et al., 2000). In $P2X_3$-deficient mice, no nodose neurons respond to $\alpha\beta$-meATP at concentrations up to 100 $\mu M$, whereas the response to ATP is significantly reduced. The residual persistent responses to ATP have all the characteristics of recombinant $P2X_2$ homomers. Thus, the pharmacological evidence is consistent with the notion that both heteromeric $P2X_{2/3}$ and homomeric $P2X_2$ receptors are present in significant amounts in nodose neurons, although the proportions may vary from cell to cell. Recently, a novel potent and selective nonnucleotide antagonist of $P2X_3$ and $P2X_{2/3}$

receptors, A-317491, has been shown to reduce chronic inflammatory and neuropathic pain in the rat (Jarvis et al., 2002).

There are also species differences. Transient responses are the predominant type evoked by P2X agonists from DRG neurons of rat and mouse, with persistent and biphasic types seen less frequently (Burgard et al., 1999; Grubb and Evans, 1999; Cockayne et al., 2000). In contrast, only sustained inward currents have been reported on DRG neurons from bullfrog (Bean, 1990; Li et al., 1997). On the other hand, nearly all of nodose neurons from rat (Li et al., 1993; Khakh et al., 1995), mouse (Cockayne et al., 2000), and guinea pig (Zhong et al., 2001) responded to ATP and $\alpha\beta$-meATP with persistent responses.

It is possible that distinct P2X receptors may be differentially distributed at cell soma and nerve terminals of the same neuron (for review see Khakh and Henderson, 2000). The physiological significance of the heterogeneity in P2X receptor expression in sensory neurons is not yet clear.

Recent studies have shown that ATP and UTP produce slow and sustained excitation of sensory neurons in DRG via P2Y2 receptors (Molliver et al., 2002). $P2Y_1$ receptors have been demonstrated immunohistochemically in rat and human nodose ganglia (Fong et al., 2002). Colocalisation of $P2Y_1$ and $P2X_3$ immunoreactivity has been described in a subpopulation of DRG neurons (Borvendeg et al., 2003).

## B. Sympathetic Ganglia

A number of studies have demonstrated the presence of P2X receptors in sympathetic ganglia by immunohistochemistry. Immunoreactivity for $P2X_1$, $P2X_2$, $P2X_3$, $P2X_4$, and $P2X_6$ receptors was detected in SCG and coeliac ganglia of the rat (Xiang et al., 1998a). In a study of cultured SCG neurons, $P2X_2$ was the most highly expressed receptor; lower, though detectable levels of all the other subunits, except $P2X_4$, were present (Li et al., 2000). However, the extent to which the expression of P2X receptors may be influenced by tissue culture conditions is at present unclear. In a study of the guinea pig, SCG, $P2X_2$, and $P2X_3$ immunoreactivity was detected (Zhong et al., 2000a).

P2X1-green fluorescent protein has been used to study the time course of P2X1 receptor clustering (about $1 \mu m$ diameter) in plasma membranes of sympathetic neurons and internalisation of receptors following prolonged exposure to ATP (Li et al., 2000). In keeping with the histochemical evidence, mRNA for most P2X subunits have been detected in sympathetic neurons. $P2X_5$ and $P2X_6$ receptors were first isolated by polymerase chain reaction (PCR) from coeliac and SCG mRNAs, respectively (Collo et al., 1996). Fragments of $P2X_3$ and $P2X_4$ receptors have also been cloned from a rat SCG cDNA library (Lewis *et al.*, 1995; Buell *et al.*, 1996). Three splice variants of the rat $P2X_2$ receptor have been cloned, and all three were detected in SCG neurons by *in situ* hybridization (Simon *et al.*, 1997a). Other *in situ* hybridization studies have detected $P2X_1$, $P2X_4$, and $P2X_6$ mRNA in rat SCG neurons (Buell *et al.*, 1996; Collo *et al.*, 1996).

Purinergic synaptic transmission was not demonstrated between sympathetic neurons until the early 1990s (Evans *et al.*, 1992; Silinsky and Gerzanich, 1993). A number of subsequent studies have characterized the receptors present on sympathetic neurons, and it is now clear that there is a species difference between rat and guinea pig. In the guinea pig, $\alpha\beta$-meATP is an effective agonist on SCG (Reekie and Burnstock, 1994; Zhong *et al.*, 2000a) and coeliac ganglion neurons (Khakh *et al.*, 1995). In contrast, $\alpha\beta$-meATP evoked only a small slowly desensitizing response in a proportion of neurons from rat SCG (Cloues *et al.*, 1993; Khakh *et al.*, 1995; Rogers *et al.*, 1997). However in a study of rat and mouse coeliac ganglion neurons, no responses to 100 $\mu M$ $\alpha\beta$-meATP were detected (Zhong *et al.*, 2000b).

Most of the properties described for P2X receptors in rat sympathetic neurons (kinetics, agonist and antagonist profile, effect of $Zn^{2+}$ and pH) are consistent with those of the recombinant $P2X_2$ receptor. The presence of a small slowly desensitizing $\alpha\beta$-meATP response in rat SCG neurons can most easily be explained by the coexistence of some heteromeric $P2X_{2/3}$ receptors.

## C. Parasympathetic Ganglia

Because of their smaller size and more diffuse location, parasympathetic ganglia are much harder to study. Consequently there is less immunohistochemical or molecular biological information on the presence of P2X receptors in these neurons.

Prior to the cloning of P2X receptors, ATP was found to produce excitation in the vesical parasympathetic ganglion of the cat (Theobald and de Groat, 1989). Responses to ATP have been recorded from dissociated neurons from the chick ciliary ganglia (Abe *et al.*, 1995), rabbit vesical parasympathetic ganglion (Nishimura and Tokimasa, 1996), intramural ganglia from the guinea pig urinary bladder (Burnstock *et al.*, 1987), and guinea pig and rat cardiac neurons in culture (Allen and Burnstock, 1990; Fieber and Adams, 1991). In general, the results are very similar to those obtained in sympathetic neurons. Thus, application of ATP evokes a rapid depolarization or inward current through the activation of P2X receptors. Although 2-meSATP and ATP are approximately equipotent, $\alpha\beta$-meATP evoked only small responses when applied at high concentrations to rat neurons.

The neurons providing motor innervation to the bladder and other pelvic organs originate in the pelvic plexus. In the rat and mouse, this plexus consists of a pair of major pelvic ganglia and a number of small accessory ganglia (Purinton *et al.*, 1973). In the guinea pig there are additional intramural ganglia within the wall of the bladder (Crowe *et al.*, 1986). The pelvic ganglia receives sympathetic and parasympathetic inputs from preganglionic axons within the hypogastric and pelvic nerves, respectively (Keast, 1995), and is therefore unique in being neither sympathetic nor parasympathetic.

Using polyclonal antibodies specific for P2X receptor subunits, $P2X_2$ immunoreactivity has been detected in pelvic ganglion neurons of the rat (Zhong *et al.*, 1998). Although many neurons showed low levels of staining, a small percentage showed strong and specific staining. In keeping with these observations, $P2X_2$ but not $P2X_1$ immunoreactivity was detected in axons and nerve terminals in the vas deferens (Vulchanova *et al.*, 1996). In the guinea pig pelvic ganglion, in addition to staining for the $P2X_2$ subunit, $P2X_3$ immunoreactivity has also been detected (Zhong *et al.*, 2000a). Studies using *in situ* hybridization have detected high levels of $P2X_2$ mRNA in rat pelvic ganglion neurons (Zhong *et al.*, 1998). Although some $P2X_4$ message was also detectable, no staining was observed using probes directed against $P2X_1$ and $P2X_3$ mRNA.

In summary, autonomic neurons express $P2X_2$ and $P2X_3$ subunits, which can coassemble to form three functionally identifiable receptors. However, the level of $P2X_3$ expression varies greatly between species (rat vs. guinea pig) and even from cell to cell. Thus, rat autonomic neurons express homomeric $P2X_2$ receptors almost exclusively, whereas guinea pig neurons express a mixture of homomeric $P2X_2$, heteromeric $P2X_{2/3}$, and in some cases homomeric $P2X_3$ receptors. Adrenomedullary chromaffin cells appear to be an exception to this general pattern as those of the rat lack P2X receptors, whereas those in the guinea pig appear to express only the $P2X_2$ subtype.

## D. Enteric Ganglia

Katayama and Morita (1989) were the first to study the effects of ATP on single myenteric neurons from guinea pig small intestine, using the intracellular electrophysiological recording technique. Myenteric neurons are classified into two groups electrophysiologically and ATP elicited hyperpolarization in 80% of AH (type II) neurons and depolarization in 90% of S (type I) neurons in a dose-dependent manner. Quinidine reversibly depressed both the ATP-induced responses.

Four laboratories led by Jackie Wood, Carlos Barajas-López, James Galligan, and Michael Mulholland have extended these studies of purinergic signaling in guinea pig myenteric neurons (see Surprenant, 1994; Furness *et al.*, 1998; Burnstock, 2001b; Galligan, 2002). Elegant whole-cell and outside-out patch-clamp recordings were used to characterize the physiological and pharmacological properties of P2X receptors on myenteric neurons of the guinea pig ileum (Barajas-López *et al.*, 1996; Zhou and Galligan, 1996; LePard *et al.*, 1997). ATP and analogues evoked rapid inward currents in over 90% of the neurons studied.

P2X receptors and nicotinic cholinergic receptors are linked in a mutually inhibitory manner in guinea pig myenteric neurons (Zhou and Galligan, 1998). Mulholland's group carried out studies of purinergic signaling in dispersed primary cultures of guinea pig myenteric plexus. Extracellular ATP was shown to mediate $Ca^{2+}$ signaling via a phospholipase C (PLC)-dependent mechanism (Kimball *et al.*, 1996). Enteric neurons differed from one another in their ability to respond to combinations of ATP with ACh, ATP with SP, ATP with ACh, ATP with ACh and SP, ATP with bombesin, or ATP with ACh and bombesin (Kimball and Mulholland, 1995). These authors have shown that extracellular ATP also acts on cultured guinea pig enteric *glia* via an inositol triphosphate ($IP_3$) mechanism (probably involving a P2Y receptor) to mobilize intracellular $Ca^{2+}$ (Sarosi *et al.*, 1998). Evidence has been presented that two distinct types of P2 receptors are linked to a rise in $[Ca^{2+}]_i$ in guinea pig intestinal myenteric neurons of both AH and S neuronal phenotypes and are not restricted to calbindin immunoreactive sensory neurons (Christofi *et al.*, 1996, 1997). Enhanced excitability of myenteric AH neurons has been demonstrated recently in the inflamed guinea pig distal colon (Linden *et al.*, 2003).

ATP regulates synaptic transmission by pre- as well as postsynaptic mechanism in guinea pig myenteric neurons, i.e., ATP augments nicotinic fast depolarization produced by ACh, but inhibits muscarinic and SP-mediated depolarizations in both AH and S neurons (Kamiji *et al.*, 1994).

Exogenous and endogenous ATP, released during increase in intraluminal pressure, inhibits intestinal peristalsis in guinea pig via different apamine-sensitive purine receptor mechanisms. Exogenous ATP depresses peristalsis mostly via suramin- and PPADS-insensitive P2 receptors, whereas endogenous purines act via P2 receptors sensitive to both suramin and PPADS (Heinemann *et al.*, 1999). A preliminary report claims that purinergic transmission is involved in a descending excitatory reflex in guinea pig small intestine (Clark *et al.*, 1996). The authors present evidence that excitatory motoneurons have P2X receptors that are excited by anally directed purinergic interneurons. The possible role(s) of ATP in slow synaptic transmission in enteric ganglia remains to be confirmed. Evidence has been presented recently that ATP plays a major role in excitatory neuroneuronal transmission in both ascending and descending reflex pathways to the longitudinal and circular muscles of the guinea pig ileum triggered by mucosal stimulation (Spencer *et al.*, 2000). It has been proposed that ATP is a sensory mediator via P2X$_3$ receptors on intrinsic sensory neurons in the enteric plexuses (Burnstock, 2001a; Bertrand and Borstein, 2002).

Synaptosomal preparations from the guinea pig ileum myenteric plexus were first described by Dowe *et al.* (1980) and Briggs and Cooper (1981). ATP and adenosine were equipotent in their ability to inhibit the nicotinically induced release of [$^3$H]ACh; the inhibition by both ATP and adenosine was reversed by theophylline, indicating that a P1 receptor was involved (Reese and Cooper, 1982). However, high concentrations of ATP caused a marked increase in the release of [$^3$H]ACh, presumably mediated by a prejunctional P2 receptor.

Intracellular recordings from *submucosal* neurons in guinea pig small intestine showed that ATP induced fast transient depolarization of most AH-type neurons and fast transient depolarization followed by slower onset, longer lasting depolarization of S-type neurons (Barajas-López *et al.*, 1994). When whole-cell patch recordings were employed, superfusion of ATP and analogues evoked rapidly desensitizing inward current and ATP-induced single channel currents were also recorded. In a whole-cell patch-clamp study of ATP-induced membrane currents in guinea pig small intestinal submucous neurons by another group (Glushakov *et al.*, 1996), the currents activated by ATP were not blocked by suramin and were often enhanced by Reactive blue 2. This could indicate the involvement of P2X$_4$ or P2X$_6$ receptors (see Burnstock, 1997). The functional interaction between nicotinic and P2X receptors has been investigated in freshly dissociated guinea pig submucosal neurons in primary culture: whole-cell currents induced by ATP were blocked by PPADS and showed some interdependence on ACh-induced nicotinic currents blocked by hexamethonium (Glushakov *et al.*, 1996; Barajas-López *et al.*, 1998; Zhou and Galligan, 1998). Inhibitory interactions between 5HT$_3$ and P2X channels in submucosal neurons have been described in a recent publication (Barajas-Lopez, 2002). Evidence has also been presented for two subtypes of P2X receptor in neurons of guinea pig ileal submucosal plexus (Glushakov *et al.*, 1998).

Recent immunohistochemical studies have demonstrated the P2X$_2$ receptor (Castelucci *et al.*, 2002), P2X$_3$ receptor (Van Nassauw *et al.*, 2002; Poole *et al.*, 2002) and P2X$_7$ receptors (Hu et al., 2001) on subpopulations of guinea pig myenteric and submucous ganglionic neurons. P2X$_2$ and P2X$_5$ receptors have also been immunolocalised on interstitial cells of Cajal (Burnstock and Lavin, 2002).

## IV. CENTRAL NERVOUS SYSTEM

### A. Introduction

The actions of adenosine in the CNS have been recognized for many years beginning with the work of Feldberg and Sherwood (1954), which included ataxia and a tendency to sleep following intracerebroventricular injections (see reviews by Reddington et al., 1983; Williams, 1984; Snyder, 1985; Dunwiddie, 1985; Marangos and Boulenger, 1985; Phillis et al., 1986; Ribeiro, 1995; Fredholm, 1995). However, consideration of the role(s) of ATP in the CNS received less attention until recently (see Burnstock, 1996, 1977; Phillis and Wu, 1981; Inoue et al., 1996; Dunwiddie et al., 1996; Gibb and Halliday, 1996; Abbracchio, 1997; Robertson, 1998; Illes and Zimmermann, 1999; Nörenberg and Illes, 2000). In particular, purinergic synaptic transmission has been clearly identified in the brain (Khakh, 2001). It was first observed in the medial habenula (Edwards et al., 1992) and has now been described in a number of other brain areas, including locus coeruleus (Nieber et al., 1997), hippocampus (Pankratov et al., 1999; Mori et al., 2001), and somatic-sensory cortex (Pankratov et al., 2002). Although adenosine, following ectoenzymatic breakdown of ATP, is the predominant, presynaptic modulator of transmitter release in the CNS (see Dunwiddie, 1985; Ribeiro, 1995), ATP itself can also act presynaptically (Cunha and Ribeiro, 2000a).

ATP is present in high concentrations within the brain, varying from approximately $2\,\mathrm{m}M$ in the cortex to $4\,\mathrm{m}M$ in the putamen and hippocampus (Kogure and Alonso, 1978). Release of ATP from synaptosomal preparations of discrete areas of the rat and guinea pig brain, including cortex, hypothalamus, medulla, and habenula, has been measured (Barberis and McIlwain, 1976; White, 1978; Potter and White, 1980; White et al., 1980, Sperlágh et al., 1988a,b). In cortical synaptosomes, a proportion of the ATP appears to be coreleased with acetylcholine, and a smaller proportion with noradrenaline, but most is not released from either cholinergic or adrenergic varicosities (Potter and White, 1980; White et al., 1980). In preparations of affinity-purified cholinergic nerve terminals from the rat cuneate nucleus, ATP and acetylcholine are coreleased (Richardson and Brown, 1987). There is also evidence for corelease of ATP with catecholamines from neurons in the locus coerulus (Poelchen et al., 2001), for corelease of ATP with GABA in dorsal horn and lateral hypothalamic neurons (Jo and Schlichter, 1999; Jo and Role, 2002a) and evidence for corelease of ATP with glutamate in the hippocampus (Mori et al., 2001) as well as widespread and pronounced modulatory effects of ATP on glutamatergic mechanisms (Illes et al., 2001). Colocalisation of functional nicotinic and ionotropic nucleotide receptors have also been identified in isolated cholinergic synaptic terminals in midbrain (Diaz-Hernandez *et al.*, 2002).

Nerve terminals *in vivo* or *in vitro* take up adenosine, which is then mostly converted into nucleotides, predominantly ATP and cyclic AMP (Pull and McIlwain, 1972a,b; Kuroda and McIlwain, 1974). The passage of [$^3$H]adenosine, which is taken up by neurons and converted into nucleotides, has been traced in the brain. In the manner of a neurotransmitter, the radiolabeled substances are transported rapidly along the axons, are released from terminals, and are taken up by postsynaptic neurons (Schubert and Kreutzberg, 1974; Schubert *et al.*, 1976).

ATP was shown to cause excitation of neurons in the cuneate nucleus, when applied focally by iontophoresis from a microelectrode (Galindo *et al.*, 1967). Since then ATP has also been shown to cause excitation of neurons in many other regions of the brain (see below).

*In situ* hybridization of P2 receptor subtype mRNA and immunohisto-chemistry of receptor subtype proteins have been carried out in recent years to show wide, but heterogeneous, distribution in the CNS of both P2X receptors (Tanaka *et al.*, 1996; Vulchanova *et al.*, 1997; Lê *et al.*, 1998; Xiang *et al.*, 1998b; Llewellyn-Smith and Burnstock, 1998; Loesch and Burnstock, 1998; Kanjhan *et al.*, 1999; Atkinson *et al.*, 2000a; Yao *et al.*, 2000; Rubio and Soto, 2001) and P2Y receptors (Simon *et al.*, 1997b; Morán-Jiménez and Matute, 2000; Moore *et al.*, 2000a; Laitinen *et al.*, 2001; Lee *et al.*, 2001). As yet, central purinergic pathways involved in particular behavioral or homeostatic mechanisms have not been identified, with a few exceptions (see below).

In addition to P2 receptors and the release of ATP from neurons in the brain, there is abundant evidence for P2 receptors on, and release of ATP from, glial cells, suggesting that both short- and long-term (trophic) neuronal–glial cell interactions are taking place (see Neary *et al.*, 1996; Abbracchio and Burnstock, 1998; Fields and Stevens, 2000; John *et al.*, 2001). Adenosine plays an important neuroprotective role (Wardas, 2002) in cerebral ischaemia (see Rudolphi *et al.*, 1992) following nucleotide release from damaged tissue (Braun *et al.*, 1998).

## B. Spinal Cord

There was early identification of dense areas of acid phosphatase and 5'-nucleotidase activity in the substantia gelatinosa of the spinal cords of rats and mice and the possible implication for purinergic transmission was raised (see Suran, 1974). Although P1 (A$_1$ and A$_2$) receptors in neurons in the dorsal and ventral spinal cord mediating modulation of neuronal activity by adenosine was shown in the early 1980s (see Geiger et al., 1984; Choca et al., 1987; Santicioli et al., 1993; Li and Perl, 1994; Sawynok, 1998), the presence and roles of P2 purinoceptors were not recognized until much later (see Salter et al., 1993; Abbracchio, 1997). ATP-evoked increases in intracellular calcium were demonstrated in neurons and glia of the dorsal spinal cord (Salter and Hicks, 1994). It was later shown that the ATP-evoked release of $Ca^{2+}$ from astrocytes was via the phospholipase $C\beta/IP_3$ pathway (Salter and Hicks, 1995), suggesting mediation via a P2Y receptor. It was proposed that ATP released in synaptic regions acts as a synaptic modulator by augmenting the actions of excitatory amino acids (Li and Perl, 1995). ATP was shown to activate the $K^+$ channel responsible for outwardly rectifying currents via a P2Y purinoceptor linked to a pertussis toxin-insensitive G-protein in cultured rat spinal neurons (Ikeuchi and Nishizaki, 1996). In a recent paper, ATP was shown to inhibit slow depolarisation via P2Y receptors in substantia gelatinosa neurons (Yoshida et al., 2002). Properties of P2Y receptors in Xenopus spinal neurons related to motor pattern generation have also been reported (Brown and Dale, 2002).

mRNA for $P2X_2$, $P2X_4$, and $P2X_6$ receptors have been identified within spinal motor nuclei (Collo et al., 1996). $P2X_3$ immunoreactivity is apparent on the axon terminals of DRG neurons that extend across the entire mediolateral extent of inner lamina II of the dorsal horn (Vulchanova et al., 1997; Bradbury et al., 1998; Nakasuka and Gu, 2001). The immunolabeled nerve profiles in lamina II for $P2X_3$ receptors are located largely on terminals with ultrastructural characteristics of sensory afferent terminals (Llewellyn-Smith and Burnstock, 1998). In contrast, although $P2X_2$ immunoreactivity is most prominent in lamina II, it is also seen in deeper layers, and only rarely overlaps with $P2X_3$ immunoreactivity (Vulchanova et al., 1997). At central terminals of primary afferent neurons, ATP has been shown to act either presynaptically facilitating glutamate release (Gu and MacDermott, 1997; Khakh and Henderson, 1998; Li et al., 1998; Nakasuka and Gu, 2001) or postsynaptically (Fyffe and Perl, 1984; Salter and Henry, 1985; Sawynok and Sweeney, 1989; Li andPerl, 1995; Bardoni et al., 1997). ATP facilitates spontaneous glycinergic inhibitory postsynaptic currents (IPSCs) in neurons from rat substantia gelatinosa, mechanically dissociated from the dorsal horn (Rhee et al., 2000), and P2X receptors are also expressed on glycinergic presynaptic nerve terminals (Jang et al., 2001).

ATP has been shown to be released from dorsal spinal cord synaptosomes (Sawynok et al., 1993). Morphine and capsaicin release purines from capsaicin-sensitive primary afferent nerve terminals in the spinal cord (Sweeney et al., 1989).

There have been a number of papers concerned with purinoceptor involvement in nociceptive pathways in the spinal cord (see Hamilton andMcMahon, 2000; Burnstock, 2000). For example, intrathecally administeredP2-purinoceptor antagonists, suramin and PPADS, produced antinociceptive effects in rats (Driessen *et al.*, 1994). Substance P released by sensory nerve stimulation potentiates purinergic inhibition of nociceptive dorsal horn neurons induced by peripheral vibration (De Koninck *et al.*, 1994). ATP-activated P2X receptors in lamina II of the rat spinal cord have been claimed to play a role in transmitting or modulating nociceptive information (Bardoni *et al.*, 1997). Presynaptic $P2X_2$ receptors modulate excitatory synaptic transmission between primary afferent fibers and spinal cord dorsal horn neurons of the lumbar spinal cord of rats (Li *et al.*, 1998). The vanilloid receptor, VR1, is colocalized with the $P2X_3$ receptor and withbinding sites for the lectin $IB_4$ within a large proportion of dorsal root ganglion neurons and their terminals in the dorsal horn of the spinal cord (Guo *et al.*, 1999). $\alpha\beta$-meATP-induced thermal hyperalgesia may be mediated by spinal $P2X_3$ receptors, perhaps by evoking glutamate release (Tsuda *et al.*, 1999b). Evidence has been presented that spinal endogenous ATP may play a role in capsaicin-induced neurogenic pain via $P2X_3$ or $P2X_{2/3}$ receptors and formalin-induced inflammatory pain via different P2Xand/or P2Y receptors (Tsuda *et al.*, 1999a). A paper from another group also suggested that spinal P2X receptors may play a role in the modulation of spinal nociceptive transmission following the development ofinflammation (Stanfa *et al.*, 2000).

Within the spinal cord, P2X receptors are present in a subpopulation of dorsal horn neurons (Bardoni *et al.*, 1997; Jo and Schlichter, 1999; Hugel and Schlichter, 2000; Rhee *et al.*, 2000; Laube, 2002), and ATP is coreleased with GABA (Jo and Schlichter, 1999). In addition to acting as a fast excitatory synaptic transmitter, ATP facilitates excitatory transmission by increasing glutamate release and enhances inhibitory neurotransmission mediated by both GABAand glycine (Hugel and Schlichter, 2000; Rhee *et al.*, 2000).

## C. Brainstem

### 1. Rostral Ventrolateral Medulla (RVLM)

Iontophoretic application of ATP excited the spinal cord-projection neurons in the RVLM of the medulla oblongata causing powerful vasopressor actions, a response that was mimicked and then blocked by$\alpha\beta$-meATP as well as by suramin (Sun *et al.*, 1992). A further study suggested that two P2X receptor subtypes might be present in RVLM neurons, one sensitive to both ATP and $\alpha\beta$-meATP and the other sensitive to ATP, but not to $\alpha\beta$-meATP (Ralevic et al., 1996). Activation of P2X receptors in the ventrolateral medulla (VLM) was shown to be capable of producing marked excitation of both sympathoexcitatory and sympathoinhibitory neurons (Horiuchi et al., 1999). Evidence has been presented to suggest that $CO_2$-evoked changes in respiration are mediated, at least in part, by P2X purinoceptors in the retrofacial area of the VLM (Thomas et al., 1999). In a follow-up paper, $CO_2$-P2X-mediated actions were observed only in inspiratory neurons that have purinoceptors with pH sensitivity (characteristic of the $P2X_2$ receptor subtype) that could account for the actions of $CO_2$ in modifying ventilatory activity (Thomas and Spyer, 2000). Not surprisingly, adenosine was shown to be a neuromodulator in the RLVM (Thomas and Spyer, 1999).

## 2. Trigeminal Mesencephalic Nucleus (MNV)

Although the MNV is located in the CNS, it contains cell bodies of primary afferent neurons that relay proprioceptive information exclusively (Liem et al., 1991). An early paper reported excitation of single sensory neurons in the rat caudal trigeminal nucleus by iontophoretically applied ATP (Salt and Hill, 1983).

The MNV is known to contain mRNA for $P2X_2$, $P2X_4$, $P2X_5$, and $P2X_6$ subtypes (Collo et al., 1996; Kanjhan et al., 1999). With in situ hybridization studies, higher levels of mRNA for $P2X_5$ were found in this nucleus than in any other brain area (Buell et al., 1996). ATP-gated ion channels (P2X receptors) were described in rat trigeminal MNV proprioceptive neurons from whole-cell and outside-out patch-clamp recording (Khakh et al., 1997; Patel et al., 2001), possibly mediated by $P2X_5$ receptor homomultimers and $P2X_{2/5}$ heteromultimers (see Patel et al., 2001). It has been suggested that in the MNV there is ATP receptor-mediated enhancement of fast excitatory glutamate release onto trigeminal mesencephalic motor nucleus neurons (Khakh and Henderson, 1998).

## 3. Area Postrema (AP)

ATP was shown to have transient excitatory effects on neurons within the AP that are associated with the vomiting reflex (Borison et al., 1975). $P2X_2$ immunoreactive nerve fibers and cell bodies have been demonstrated in the AP as well as in the dorsal vagal nucleus.

## 4. Locus Coeruleus (LC)

There were early reports of modulation of neurons in the LC by adenosine (Taylor and Stone, 1980; Shefner and Chiu, 1986; Regenold and Illes, 1990). The first report of the action (depolarization) of ATP on P2 receptors in LC was by Harms *et al.* (1992) and later papers examined the ionic mechanism and receptor characterization of these responses (Shen and North, 1993; Illes *et al.*, 1994; Masaki *et al.*, 2001). $\alpha\beta$-meATP and $\alpha\beta$-mADP were shown toincrease the firing rate of rat LC neurons (Trezise *et al.*, 1996). Pharmacological studies of P2 purinoceptor subtypes in rat LC neurons suggested that both P2X and P2Y might be present (Frohlich *et al.*, 1996) and a study by another group supported this suggestion (Sansum *et al.*, 1998). Intracellular recordings from slices of rat LC led the authors to suggest that ATP may be released either as the sole transmitter from purinergic neurons terminating in the LC or as a cotransmitter with noradrenaline from recurrent axon collaterals or dendrites of the LC neurons themselves (Nieber *et al.*, 1997), the latter proposal being supported experimentally in a later paper (Poelchen *et al.*, 2001). Microinjection of ATP or $\alpha\beta$-meATP into locus coeruleus and periaqueductal grey matter) led to changes in bladder function and arterial blood pressure (Rocha *et al.*, 2001).

## 5. Nucleus Tractus Solitarius (NTS)

The NTS (particularly neurons in the caudal NTS) is a central relay station for viscerosensory information to respiratory, cardiovascular, and digestive neuronal networks. Extracellular purines have been claimed to be the primary mediators signaling emergency changes in the internal environment in the CNS (Braun *et al.*, 1998). Neuromodulatory actions of adenosine have been described in the NTS (Tseng *et al.*, 1988; Barraco *et al.*, 1991; Mosqueda-Garcia *et al.*, 1991; Tao and Abdel-Rahman, 1993; Thomas *et al.*, 2000). Patch-clamp studies of neurons dissociated from rat NTS revealed P2 receptor-mediated responses (Ueno *et al.*, 1992) and microinjection of P2 receptor agonists into the subpostremal NTS in anesthetized rats produced reduction of arterial blood pressure (Ergene *et al.*, 1994; Kitchen *et al.*, 2001) probably via a $P2X_1$ or $P2X_3$ subtype, as $\alpha\beta$-meATP was particularly potent. It was suggested that the actions of ATP and adenosine in the NTS may be functionally linked to selectively coordinate the regulation of regional vasomotor tone (Barraco *et al.*, 1996). Purines applied in the NTS have been shown to affect baroreceptor and cardiorespiratory functions (Phillis *et al.*, 1997; Carey and Jordan, 1998; Scislo *et al.*, 2001; Paton *et al.*, 2002). Activation of P2X and P1 ($A_{2A}$) receptors in the NTS also elicit differential control of lumbar and renal sympathetic nerve activity (Scislo *et al.*, 1997; Scislo and O'Leary, 1998). In a later paper from this group, they concluded that the fast responses to stimulation of NTS P2X receptors were mediated via glutamatergic ionotropic mechanisms, whereas the slow responses to stimulation of NTS P2X and $A_{2A}$ receptors were not (Scislo and O'Leary, 2000).

The immunohistochemical distribution of P2X receptor subtypes in the NTS of the rat has been described (Yao *et al.*, 2001). Colocalization of P2X$_2$ and P2X$_3$ immunoreactivity has been described in the NTS (Vulchanova *et al.*, 1997). At the electron microscope level, P2X$_3$ receptor-positive boutons have been shown to synapse on dendrites and cell bodies and have complex synaptic relationships with other axon terminals and dendrites (Llewellyn-Smith and Burnstock, 1998). P2X$_2$ receptors have been localized presynaptically in vagal afferent fibers in rat NTS (Atkinson *et al.*, 2000b). A recent whole-cell patch-clamp study of neurons in the caudal NTS led to the conclusion that ATP activates (1) presynaptic P1(A$_1$) receptors after breakdown to adenosine, reducing evoked release of glutamate from the primary afferent nerve terminals and (2) presynaptic P2X receptors on the axon terminals of intrinsic excitatory caudal NTS neurons, facilitating spontaneous release of glutamate (Kato and Shigetomi, 2001).

**6. Motor and Sensory Nuclei**

The mRNA for P2X$_4$ and P2X$_6$ receptors as well as three P2X$_2$ receptor subunit isoforms has been identified in the hypoglossal nucleus and this was taken to indicate modulation of inspiratory hypoglossal activity and perhaps a general role in modulatory motor outflow in the CNS (Collo *et al.*, 1997; Funk *et al.*, 1997). Adenosine suppresses excitatory glutaminergic inputs to rat hypoglossal motoneurons (Bellingham and Berger, 1994).

Evidence for multiple P2X and P2Y subtypes in the rat medial vestibular nucleus has been presented (Chessell *et al.*, 1997). A P2Y receptor subtype activated by ADP was identified in medulla neurons isolated from neonatal rat brain (Ikeuchi *et al.*, 1995), perhaps, with hindsight, P2Y$_1$ or P2Y$_{12}$.

Adenosine-5'-tetraphosphate (Ap$_4$) was shown to be active on rat midbrain synaptosomal preparations, probably acting via P2X$_1$ or P2X$_3$ receptor subtypes or possibly by another unidentified P2X subtype (Pintor *et al.*, 2000).

The actions of ATP and ACh were examined with patch-clamp recording on dissociated preganglionic neurons in the dorsal motor nucleus of the vagus (DMV); the results suggested that these neurons functionally colocalize nicotinic and P2X receptors (Nabekura *et al.*, 1995). Over 90% of the preganglionic neurons in this nucleus respond to ATP and RT-PCR showed mRNA in the DMV encoding P2X$_2$ and P2X$_4$ receptors (Ueno *et al.*, 2000). In a later full paper from the group, they suggest that the functional receptors expressed in DMV neurons are characterised mainly by P2X$_2$ and P2X$_{2/6}$ subtypes (Ueno *et al.*, 2001).

P1(A$_1$) adenosine receptor agonists presynaptically inhibit both GABAergic and glutamatergic synaptic transmission in periaqueductal gray neurons (Bagley *et al.*, 1999).

## D. Diencephalon

### 1. Thalamus

Adenosine promotes burst activity in guinea pig geniculocortical neurons (Pape, 1992). Adenosine can downregulate inhibitory postsynaptic responses in thalamus and exert antioscillatory effects (Ulrich and Huguenard, 1995).

P2X receptors have been localized in thalamus using $\alpha,\beta$-[$^3$H]meATP binding (Bo and Burnstock, 1994) and P2Y receptors have also been described in thalamic neurons (Mironov, 1994). Nociceptive activity was elicited by electrical stimulation of afferent C fibers in the sural nerve and recorded from single neurons in the rat ventrobasal complex of the thalamus; the P2 receptor antagonist reactive red, administered intrathecally, produced significant reduction of the evoked activity in thalamic neurons (Driessen et al., 1998). A lower density of P1(A$_1$) receptors in the nucleus reticularis thalami in rats with genetic absence epilepsy has been reported (Ekonomou et al., 1998).

### 2. Hypothalamus

Presynaptic P1 receptors mediate inhibition of GABA release in suprachiasmatic and arcuate nucleus neurons (Chen and van den Pol, 1997). Adenosine-induced presynaptic inhibition of IPSCs and EPSCs in rat hypothalamus supraoptic nucleus neurons via A1 receptors indicates inhibition of release of both GABA and glutamate (Oliet and Poulain, 1999). Release of $^3$H-labeled nucleosides from [$^3$H]adenine-labeled hypothalamic synaptosomes was first described in 1979 (Fredholm and Vernet, 1979). ATP releases luteinizing hormone-releasing hormone (LHRH) from isolated hypothalamic granules (Burrows and Barnea, 1982). Adenosine deaminase-containing neurons in the posterior hypothalamus innervate mesencephalic primary sensory neurons, perhaps indicating purinergic control of jaw movements (Nagy et al., 1986). A hypothalamic role has been suggested for extracellular ATP to facilitate copper uptake and copper stimulation of the release of LHRH from medium eminence, via an interaction with purinergic receptors (Barnea et al., 1991).

The hypothalamic suprachiasmatic nucleus (SCN) is regarded as the site of the endogenous biological clock controlling mammalian circadian rhythms. Long-term communication between glial cells, reflected by waves of fluorescence indicating $Ca^{2+}$ movements, probably via gap junctions, canbe induced by ATP [as well as by glutamate and 5-hydroxytryptamine (5-HT)] (van den Pol et al., 1992).

ATP injected into the paraventricular nucleus stimulates release of arginine-vasopressin (AVP) resulting in antidiuretic action through renal

AVP ($V_2$) receptors (Mori *et al.*, 1992). In a later study this group showed that ATP (but not ADP, AMP, or adenosine) injected into the supraoptic nucleus (SON) also decreased urine outflow (Mori *et al.*, 1994). ATP and $\alpha\beta$-meATP excite neurosecretory vasopressin cells in the SON, an effect blocked by suramin (Day *et al.*, 1993).

The magnocellular neurons of the supraoptic and paraventricular nuclei receive a dense plexus of fiber originating from the $A_1$ noradrenergic neurons in the ventrolateral medulla. Although $A_1$ noradrenaline (NA)-containing neurons of the caudal medulla provide a direct excitatory input to supraoptic vasopressin cells, they do not use NA as their primary transmitter (Day *et al.*, 1990); later, ATP was shown to be acting as a cotransmitter with NA in these neurons (Buller *et al.*, 1996). Corelease of ATP and NA from superfused rat hypothalamic slices was also demonstrated by Sperlágh *et al.* (1998b), although they concluded that they were probably not released from the same nerve endings. Further support for corelease of ATP with NA at synapses in the hypothalamus comes from evidence that purinergic and adrenergic agonists synergize when stimulating vasopressin and oxytocin release (Kapoor and Sladek, 2000). Candidates forcoreleased transmitters in $A_1$ neurons also include substance P and neuropeptide Y (NYP) as well as ATP, and it has been proposed that substance P and NYP differentially potentiate ATP- and NA-stimulated vasopressin and oxytocin release (Kapoor and Sladek, 2001). Purinergic and GABAergic cotransmission has also been claimed in the lateral hypothalamus of the chick embryo (Jo and Role, 2000) with cholinergic modulation of cotransmitter roles (Jo and Roles, 2002b). Evidence has been presented that ATP may be released from magnocellular neurons (Troadec *et al.*, 1998).

A study of the effects of ATP in increasing $[Ca^{2+}]_i$ in cultured rat hypothalamic neurons was taken to support the action of ATP as an excitatory neurotransmitter (Chen *et al.*, 1994). Excitatory effects of ATP on P2X receptors in acutely dissociated ventromedial hypothalamic neurons have also been described (Sorimachi *et al.*, 2001). A role for adenosine $A_1$ receptors in mediating cardiovascular changes evoked during stimulation of the hypothalamic defense area (HDA) has been postulated (Dawid-Milner *et al.*, 1994). Application of ATP and UTP (but not adenosine) produced TTX-insensitive depolarizations accompanied by increases in input conductance in supraoptic magnocellular neurosecretory cells; both P2X and P2Y receptors have been suggested to be involved (Hiruma and Bourque, 1995).

Ultrastructural localization of both $P2X_2$ and $P2X_6$ immunoreactivity at both pre- and postsynaptic sites in the rat hypothalamo-neurohypophysial system has been described (Loesch *et al.*, 1999; Loesch and Burnstock,

2001). Purinergic regulation of stimulus-secretion coupling in the neurohypophysis has been discussed recently (Troadec and Thirion, 2002). From a study of the expression of P2X receptor subtypes in the supraoptic nucleus using RT-PCR, *in situ* hybridization, $Ca^{2+}$ imaging, and whole-cell patch-clamp techniques, it was concluded that mRNAs for P2X$_3$ and P2X$_4$ were predominant, but that P2X$_7$ mRNA was also present (Shibuya *et al.*, 1999).

It has been suggested that ATP, cosecreted with vasopressin and oxytocin, may play a key role in the regulation of stimulus-secretion coupling in theneurohypophysis by acting through P2X$_2$ receptors increasing AVP release,and after breakdown to adenosine acting via P2(A$_1$) receptors (inhibiting N-type $Ca^{2+}$ channels) to decrease neuropeptide release (Lemos and Wang, 2000). Adenosine was also shown to modulate activity in SON neurons by inhibiting N-type $Ca^{2+}$ channels via A$_1$ receptors (Noguchi and Yamashita, 2000). Evidence for the involvement of purinergic signalling in hypothalamus and brain stem nuclei in body temperature regulation has been presented recently (Gourine *et al.*, 2002).

### 3. Habenula

The first clear demonstration of ATP receptor-mediated fast synaptic currents in the CNS was described in the rat medial habenula (Edwards *et al.*, 1992). These synaptic currents were mimicked by ATP and reversibly blocked by suramin and by $\alpha\beta$-meATP desensitization. The evidence was extended by demonstration of ATP release from an isolated rat habenula preparation during electrical field stimulation (Sperlágh *et al.*, 1995; Vizi *et al.*, 1997). This group later showed that the projections from the triangular septal and septofimbrial nucleus to the habenula are the major source of ATP in the rat habenula and utilize ATP as a fast transmitter, probably with glutamate as a cotransmitter (Sperlágh *et al.*, 1998a). However, glutamate and ATP were claimed to be released from separate neurons in this system by another group (Robertson and Edwards, 1998), who further concluded that the P2X receptor-mediated synaptic currents are the onlycalcium-permeable fast-transmitter gated curents in these neurons (Robertson *et al.*, 1999). In a recent study from this group, long-term potentiation of glutamatergic synaptic transmission induced by activation of presynaptic P2Y receptors in the rat medial habenula nucleus has been claimed (Price *et al.*, 2003).

### E. Cerebellum

P1(A$_1$) receptors were identified on cerebellar granule cells (Wojcik and Neff, 1983; Sanz *et al.*, 1996) and adenosine was shown to selectively block parallel fiber-mediated synaptic potentials in the rat cerebellar cortex (Kocsis *et al.*, 1984) and inhibit Purkinje cell firing (Bickford *et al.*, 1985) and glutamate release from cultured cerebellar neurons (Dolphin and Prestwich, 1985).

A high density of P2X binding was found in the cerebellar cortex (Balcar *et al.*, 1995), and with the aid of RT-PCR technology, mRNAs for P2X$_{1-4}$ and P2X$_6$ subunits were identified in the cerebellum during the first postnatal week with coexpression of two units in Purkinje cells demonstrated with patch-clamping (Garcia-Lecea *et al.*, 2001). A lack of correlation between glutamate-induced release of ATP and neuronal death in cultured cerebellar neurons was reported (Marcaida *et al.*, 1995). Cerebellar Purkinje neurons were shown experimentally to have both P2Y receptors (Kirischuk *et al.*, 1996) and P2X (probably P2X$_2$) receptors (Mateo *et al.*, 1998; Garcia-Lecea *et al.*, 1999). Patch-clamp studies of dissociated cerebellar neurons revealed a P2Y purinoceptor-operated potassium channels (Ikeuchi and Nishizaki, 1996).

ATP increased the release of aspartate from cultured cerebellar granule neurons and also potentiated its release by glutamate; it was concluded that this was consistent with a cotransmitter role of ATP in the cerebellum (Merlo and Volonté, 1996). This group also showed that a P2 purinoceptor antagonist prevented glutamate-evoked cytotoxicity in these cultured neurons (Volonté and Merlo, 1996; Amadio *et al.*, 2002).

P2X$_2$ receptor immunoreactivity in the cerebellum was demonstrated and claimed to be consistent with a role for extracellular ATP acting as a fast transmitter in motor learning and coordination of movement (Kanjhan *et al.*, 1996). Ecto-5′-nucleotidase has been localized on cell membranes in cultures of cerebellar granule cells (Maienschein and Zimmermann, 1996) and also ectophosphorylated protein (Merlo and Volonté, 1996) and ecto-ATPase (Zinchuk *et al.*, 1999), consistent with purinergic signaling. Single-channel properties of P2X receptors in rat cerebellar slices suggested that they may be P2X$_{4/6}$ heteromultimer receptors (Halliday and Gibb, 1997). Both ADP and adenosine prevent apoptosis of cultured rat cerebellar granule cells via P2X and A$_1$ receptors, respectively (Vitolo *et al.*, 1998). In a recent study by Florenzano *et al.* (2002), it is claimed that cerebellar lesion upregulates P2X1 and P2X2 receptors in precerebellar nuclei.

## F. Midbrain/Basal Ganglia

### 1. Striatum

*In vivo* release of adenosine from cat basal ganglia was taken as early support for the existence of purinergic nerves in the brain (Barberis *et al.*,

1984). Autoradiographic labeling of $P1(A_2)$ receptors showed them to be exclusively restricted to the human caudate nucleus, putamen, nucleus accumbens, and globus pallidus as well as the olfactory tubercle and these receptors were dramatically decreased in the basal ganglia of patients with Huntington's chorea (Martinez-Mir *et al.*, 1991). Adenosine $A_{2A}$ receptor modulation of electrically evoked GABA release from slices of rat globus pallidus was described (Mayfield *et al.*, 1993). $A_{2A}$ receptors were also shown to be prominent in dopamine-innervated areas of the basal ganglia (Svenningsson *et al.*, 1997) and adenosine–dopamine receptor–receptor interactions were proposed to be an integrative mechanism for motor stimulation actions in basal ganglia (Ferre *et al.*, 1997). Dopaminergic principal neurons in the ventral tegmental area do not possess somatic P2 purinoceptors, in contrast to peripheral and central noradrenergic neurons (Poelchen *et al.*, 1998).

The extracellular actions of adenosine via P1 receptors on the striatum were the first instance in which active purines were recognised; they werelargely involved in presynaptic modulation of release of dopamine, ACh, GABA, and glutamate (Harms *et al.*, 1979; see Fuxe *et al.*, 1998; Svenningsson *et al.*, 1999). The P1 receptor subtype involved is predominantly $A_1$, but $A_2$ receptors were shown to mediate stimulation of dopamine synthesis (Chowdhury and Fillenz, 1991).

ATP release was demonstrated from affinity-purified rat cholinergic nerveterminals from rat caudate nucleus and adenosine resulting from ectoenzymatic breakdown of ATP, which acted on prejunctional $A_1$ receptors to inhibit ACh release (Richardson and Brown, 1987). The ATP-induced increase in cytosolic $Ca^{2+}$ concentration (Sorimachi *et al.*, 2000) and dopamine release (Krügel *et al.*, 1999; 2001a) has been demonstrated in the rat nucleus accumbens. It was shown that ATP was released from cultured mouse embryonic neostriatal neurons (Zhang *et al.*, 1988) and that adenosine is produced from extracellular ATP at the striatal cholinergic synapse (James and Richardson, 1993). ATP-evoked potassium current in rat striatal neurons was shown to be mediated by a P2 purinergic receptor (Ikeuchi and Nishizaki, 1995). ATP increases the extracellular dopamine level in the rat striatum through stimulation of P2Y purinoceptors (Zhang *et al.*, 1995), although it has been claimed to inhibit dopamine release in the neostriatum (Trendelenburg and Bultmann, 2000). Dopamine facilitates activation of P2X receptors by ATP (Inoue *et al.*, 1992b). Activation of ATP receptors increases cytosolic $Ca^{2+}$ concentration in nucleus accumbens neurons, involved in locomotion and motivation (Sorimachi *et al.*, 2000). Intraaccumbens injection of 2-meSATP leads to release of dopamine (Kittner *et al.*, 2000).

A recent study demonstrated ATP-induced *in vivo* neurotoxicity in the rat striatum via P2 receptors and it was suggested that ATP may be an important mediator of neuropathological events following brain injury (Ryu *et al.*, 2002).

## 2. Ventral Tegmentum

Extracellular ATP increases cytosolic $Ca^{2+}$ concentraion on ventral tegmental neurons of rat brain (Sorimachi *et al.*, 2002). Stimulation of P2Y1 receptors enhance dopaminergic mechanisms *in vivo* (Krügel *et al.*, 2001b). GABAergic synaptic terminals from rat midbrain exhibit functional P2X and dinucleotide receptors, able to induce GABA secretion (Gomez-Villafuertes *et al.*, 2001) and GABA receptors modulate the activity of P2X receptors (Gomez-Villafuertes *et al.*, 2003). Using synaptasomal preparations of rat midbrain, it was shown that aminergic nerve terminals possess modulatory presynaptic nucleotide and dinucleotide receptors (Giraldez *et al.*, 2001).

## G. Hippocampus

There are many papers describing the actions of adenosine via P1 receptors in the hippocampus, including prejunctional modulation and both inhibition and enhancement of transmitter release, influencing long-term potentiation in CA1 neurons and synaptic plasticity (see, for example, Fujii *et al.*, 1999; Costenla *et al.*, 1999; Cunha and Ribeiro, 2000b). Most workers recognize that adenosine is produced following rapid ectoenzymatic breakdown of released ATP (see Dunwiddie *et al.*, 1997).

The first evidence that adenine nucleotides as well as adenosine might be involved in synaptic transmission in the hippocampus was presented in 1981 (Lee *et al.*, 1981). Nicotinamide adenosine dinucleotide (NAD) was shown to depress synaptic transmission in the hippocampus via receptor binding sites that were distinct from those for adenosine (Richards *et al.*, 1983). Iontophoretic application of ATP depressed glutamate-evoked activity of hippocampal neurons, but experiments to resolve whether the action was via adenosine after breakdown of ATP were not carried out(Di Cori and Henry, 1984). P2 receptor-mediated inhibition of noradrenaline release in the rat hippocampus has been claimed (Koch *et al.*, 1997).

Field stimulation of hippocampal slices released [$^3$H]- or [$^{14}$C]adenosine mainly as adenine nucleotides (Jonzon and Fredholm, 1985). ATP release was also shown from rat hippocampal mossy fiber synaptosomes (Terrian *et al.*, 1989). 5'-Nucleotidase was shown to be widespread in the hippocampus of several mammalian species (Lee *et al.*, 1986).

Stimulation of Schaffer collaterals of rat and mouse hippocampal slices resulted in release of ATP and a stable increase in the size of the population spike known as long-term potentiation (LTP) (Wieraszko and Seyfried, 1989a; Wieraszko and Ehrlich, 1994). The amount of ATP released was significantly greater in seizure-prone mice (Wieraszko and Seyfried, 1989b). Extracellular ATP evoked glutamate release from cultured hippocampal neurons (Inoue *et al.*, 1992a), although it was later claimed that ATP inhibits release of excitatory glutamate, but stimulates release of inhibitory GABA (Inoue, 1998). Evidence for the presence of a metabotropic P2Y receptor in hippocampus neurons has been presented (Mironov, 1994; Ikeuchi *et al.*, 1996). ATP inhibited $K^+$ channels in cultured rat hippocampal neurons through P2Y receptors that showed equipotency to UTP and ATP (Nakazawa *et al.*, 1994), suggesting that a $P2Y_2$ or $P2Y_4$ subtype was involved. It was proposed that glutamate releases ATP from hippocampal neurons to act as a neurotransmitter (Inoue *et al.*, 1995). Evidence using P2 purinoceptor antagonists was presented to suggest that both presynaptic and postsynaptic P2 purinoceptors in the hippocampus modulate the release and action of endogenous glutamate (Motin and Bennett, 1995). ATP inhibits synaptic release of glutamate by acting on P2Y receptors on pyramidal neurons in hippocampal slices (Mendoza-Fernandez *et al.*, 2000). The latest work from this group suggests that there is cooperativity between extracellular ATP and activiation of NMDA receptors in LTP induction in hippocampal CA1 neurons (Fujii *et al.*, 2002). Presynaptic inhibitory P2 receptors in the hippocampus that inhibit calcium oscillations produced by release of glutamate have also been examined (Koizumi and Inoue, 1997). ATP released from presynaptic terminals during burst stimulation (or applied extracellularly) is involved in the induction of LTP in CA1 neurons of guinea pig hippocampus through phosphorylation of extracellular domains of synaptic membrane proteins as the substrate for ectoprotein kinase (Fujii *et al.*, 1995, 2000). The latest work from this group suggests that there is cooperativity between extracellular ATP and activation of NMDA receptors in LTP induction in hippocampal CA1 neurons (Fujii *et al.*, 2002). Examination of the effect of ATP on voltage-clamped, dissociated rat hippocampal neurons showed that over 30% possessed P2X receptors (Balachandran and Bennett, 1996). ATP can produce facilitation of transmission in rat hippocampal slice neurons, which may require the simultaneous activation of P1 and P2 receptors (Nikbakht and Stone, 2000). Diadenosine pentaphosphate enhanced the activity of N-type $Ca^{2+}$ channels in rat CA3 hippocampal neurons (Panchenko *et al.*, 1996).

In an analysis of the role of ATP as a transmitter in the hippocampus andits role in synaptic plasticity, Wieraszko (1996) concluded that the purinergic system is particularly involved in the long-term maintenance, rather than the initial induction, of LTP. In a recent paper, it has been suggested that by controlling the background calcium and thus the activity of NMDA receptors at low firing frequencies, P2X receptors act as a dynamic low-frequency filter so that weak stimuli do not induce LTP (Pankratov *et al.*, 2002b). Inhibitory avoidance training led to decreased ATP diphosphohydrolase activity in hippocampal synaptosomes, suggesting involvement of this enzyme in the formation of inhibitory avoidance memory (Bonan *et al.*, 2000).

A purinergic component of the excitatory postsynaptic current (EPSC) was identified mediated by P2X receptors in CA1 neurons of the rat hippocampus; EPSCs, which were blocked by PPADS, were elicited by stimulating the Schaffer collateral at a frequency below 0.2 Hz (Pankratov *et al.*, 1998). Evidence for fast synaptic transmission mediated by P2X receptors in CA3 pyramidal cells of rat hippocampal slice cultures has been reported recently (Mori *et al.*, 2001). Epileptiform activity in the CA3 region of rat hippocampal slices was modulated by adenine nucleotides, probably acting via an excitatory P2X receptor (Ross *et al.*, 1998).

Binding studies have shown mRNA for several P2 receptor subtypes in the hippocampus (Kidd *et al.*, 1995), including binding to $\alpha\beta$-[$^3$H]meATP (Bo and Burnstock, 1994; Balcar *et al.*, 1995). Immunolocalization of the P2X$_4$ receptor was widespread in the hippocampus; immunopositive cells were prominent in the pyramidal cell layer (in interneurons as well as pyramidal cells), scattered through CA1, CA2, and CA3 subfields as well aswithin the granule cell layer and hilus of the gentate gyrus (Lê *et al.*, 1998).

Arguments have been presented that ATP may have a role in the protection of the function of the hippocampus from overstimulation by glutamate (Inoue, 1998; Inoue and Koizumi, 2001). ATP produces an initial rise and later reduction in serotonin release from perfused rat hippocampus, mediated by P2X and P1(A$_1$) receptors, respectively (Okada *et al.*, 1999). Intrahippocampal infusion of suramin, acting either by blocking purinergic neurotransmission or as an inhibitor of ATP degradation, modulated inhibitory avoidance learning in rats (Bonan *et al.*, 1999). Evidence waspresented to support the view that ATP and glutamate acting through *N*-methyl-D-aspertate (NMDA) receptors are required to induce LTP in CA1 neurons of guinea pig hippocampal slices (Fujii *et al.*, 1999).

There is some exciting recent data indicating a trophic role for ATP in the hippocampus (Skladchikova *et al.*, 1999). It was shown that ATP and its slowly hydrolyzable analogues strongly inhibited neurite outgrowth and also inhibited aggregation of hippocampal neurons; it was suggested that the results indicate that extracellular ATP may be involved in synaptic plasticity through modulation of neural cell adhesion molecule (NCAM)-mediated adhesion and neurite outgrowth. Utilization of green fluorescent protein (GFP)-tagged P2X$_2$ receptors on embryonic hippocampal neurons has led to the claim that ATP application can lead to changes in dendritic morphology and receptor distribution (Khakh et al., 2001). Changes in [Ca$^{2+}$] during ATP-induced synaptic plasticity in guinea pig hippocampal CA1 neurons have been claimed (Yamazaki et al., 2002).

We have recently carried out a study of the single-channel properties of P2X receptors in outside-out patches from rat hippocampal granule cells that suggests the presence of P2X$_{2/4/6}$ heteromultimers (Wong et al., 2000). This has been supported by expression studies (Soto and Rubio, 2001). A recent paper implicates P2X7 receptors in the regulation of neurotransmitter release in the rat hippocampus (Sperlágh et al., 2002). Activation of presynaptic P2X$_7$-like receptors depress mossy fibre-CA3 synaptic transmission through P38 mitogen-activated protein kinase (Armstrong et al., 2002; Armstrong and MacVicar, 2000).

## H. Cortex

The seminal study by Sattin and Rall (1970), showing that adenosine acted via adenylate cyclase to produce cyclic AMP, was carried out on cerebral cortex slices. There were early studies of purinergic transmission in the cortex; electrically evoked release of nucleotides and nucleosides from both brain slices and synaptosomes prepared from cerebral cortex raised the possibility that they may participate in intercellular transmission (Pull and McIlwain, 1972a; Kuroda and McIlwain, 1973, 1974). About the same time, microiontophoretic application of adenine nucleotides was shown to depress the firing of corticospinal and other unidentified cerebral cortical neurons, although ATP had an additional excitant action on some neurons (Phillis et al., 1974, 1979a; Stone and Taylor, 1978). In another study, adenosine and adenine nucleotides were shown to have an inhibitory action on the N wave (a postsynaptic potential) amplitude in neurons of guinea pig olfactory cortex slices, but not on postsynaptic potentials in superior colliculus (Okada and Kuroda, 1975). This was confirmed by Scholfield (1978). Schubert and Kreutzberg (1974) showed that after injections of tritiated adenosine into the visual cortex of rabbits, radioactive material subsequently appeared in the thalamocortical relay cells of the lateral geniculate nucleus, consistent with synaptic transmission. This was supported by similar experiments in the somatosensory cortex (Wise and Jones, 1976). Antidepressant drugs such as imipramine and amitriptyline potentiated the suppression of neuronal firing in rat cerebral cortex by adenosine (Stone and Taylor, 1978; Phillis, 1984). Competitive interactions between adenosine and benzodiazepines in cerebral cortical neurons wererecognized early (see Phillis and Wu, 1981). It was suggested thatadenosine was involved in the initial phase of seizure-induced functional hyperemia in the cortex (Schrader et al., 1980). Evidence has beenpresented to suggest thatmorphine releases ATP, and that after breakdown to adenosine, depresses neurotransmission in the cortex (Phillis et al., 1979b, 1980; Stone, 1981).

Adenosine has been shown to inhibit release of transmitters from slices of rat cerebral cortex, including noradrenaline (Taylor and Stone, 1980),acetylcholine (Giovannelli et al., 1988), GABA (Hollins and Stone, 1980), andamino acids (Simpson et al., 1992; von Kügelgen et al., 1992), probably involving both $A_1$ and $A_3$ receptors (Brand et al., 2001). A decrease of adenosine $A_1$ receptor gene expression in the cortex of aged rats has been claimed (Cheng et al., 2000). In contrast, P2X receptors can act presynaptically in olfactory bulbs to enhance the release of glutamate (Bobanovic et al., 2002). P2X receptors have been identified in rat cortical synaptosomes (Lundy et al., 2002).

Although two earlier papers mentioned some direct excitatory effect of ATP in the cortex (Phillis and Wu, 1981; Stone and Taylor, 1978), serious attention to these actions rather than those of adenosine did not appear until the 1990s. The maximum increase in nucleotide hydrolysis coincided with maximum brain growth, which was taken to indicate a role for ATPand ADP hydrolyzing ectoenzyme activity in neurotransmission (Müller et al., 1990). ATP diphosphohydrolase was shown to be one of the ectoenzymes involved in breakdown of ATP released from synaptosomes (Battastini et al., 1991). ATP-induced inward current in neurons freshly dissociated from the tuberomammary nucleus has been described (Furukawa et al., 1994). Recently, P2X receptors have been shown to mediate synaptic transmission in rat pyramidal neurons of layer II/III of somatosensory cortex (Pankratov et al., 2002a).

Although the presence of presynaptic P1($A_1$) receptors in the cortex is well established, evidence was also presented for P2Y purinoceptor-mediated inhibition of noradrenaline release in the rat cortex (von Kügelgen et al., 1994). Release of glutamate from depolarized nerve terminals has also been claimed to be inhibited by both P1 and P2Y receptors (Bennett and Boarder, 2000). Release of ATP from cortical synaptosomes is decreased by vesamicol, which also inhibits acetylcholine, but not glutamate release (Salgado et al., 1996).

ATP receptor-mediated $Ca^{2+}$ concentration changes in pyramidal neocortical neurons in rat brain slices from 2-week-old rats were mediated by both P2X and P2Y receptors (Lalo and Kostyuk, 1998; Lalo et al., 1998). Responses of cultured rat cerebral cortical neurons to UTP (Nishizaki and Mori, 1998) indicated the presence of $P2Y_2$ and/or $P2Y_4$ receptors. P2Y receptors were identified on synaptosomal membranes from rat cortex (Schafer and Reiser, 1999). A P2Y purinoceptor from the adult bovine corpus callosum was cloned and characterized and the mRNA was shown to be present in the frontal cortex (Deng et al., 1998). $P2Y_1$ receptors were shown to be present in cerebral cortical cell cultures from embryonic rats (Bennett et al., 1999). Evidence has been presented that $AP_5A$ and ATP activate dinucleotide and P2X receptors, respectively, on human cerebro-cortical synaptic terminals (Pintor et al., 1999). P2Y purinergic and muscarinic receptor activation evoke a sustained increase in intracellular calcium in rat neocortical neurons and glial cells and it was proposed that a common calcium entry pathway was involved (Prothero et al., 2000). Antagonists of P2Y receptors modulate glutamate, kainate, and 2-(aminomethyl)phenylacetic acid (AMPA)-induced currents in cultured cortical neurons, and this was taken to suggest a potential use of these compounds as neuroprotective agents (Zona et al., 2000). Interaction between P2Y and NMDA receptors in layer V pyramidal neurons of rat prefrontal cortex have been demonstrated recently (Wirkner et al., 2002; Luthardt et al., 2003).

## I. Physiology and Pathophysiology of Purinoceptors in Brain

The depressant actions of adenosine, and the relation of adenosine to the central actions of methylxanthines such as caffeine (Daly et al., 1981), and the role of adenosine as an endogenous anticonvulsant were recognized early (Dragunow, 1988). Subsequently adenosine receptor involvement in anxiety, sleep, depression, neuroprotection, mood regulation, circadian rhythms, cerebral ischemia, epilepsy, analgesia, and movement disorders has been explored (see Williams, 1987; Barraco, 1991; Chagoya, 1995; Sebastiao and Ribeiro, 1996; Guieu et al., 1996; Müller, 1997). More recently, there has been interest in the potential role of adenosine antagonist therapy for Parkinson's disease (see Mally and Stone, 1996; Richardson et al., 1997; Grondin et al., 1999), and in adenosine receptors as targets for therapeutic intervention in neurodegenerative diseases (see Abbracchio and Cattabeni, 1999). The therapeutic potential of adenosine kinase inhibitors is also attracting current attention (see Kowaluk and Jarvis, 2000).

Interest in the physiological and pathophysiological roles of ATP in the brain is only just beginning, although there was some early exploration of the effect of ATP on the area postrema, the chemoreceptor trigger zone for vomiting (Borison, 1974), and of ATP involvement in depression (Naylor et al., 1976; Whalley et al., 1980; Rybakowski et al., 1981; Agren et al., 1983). Physiological roles for ATP and adenosine in both lateral inhibition and central facilitation in the CNS provoked the hypothesis that purines are involved in memory (Kuroda, 1989; Chen *et al.*, 1996; Bonan *et al.*, 2000). Inoue *et al.* (1996) reviewed evidence for the involvement of ATP in schizophrenia (see also Lara and Souza, 2000) and in central regulation of the autonomic nervous system, as well as in memory and learning. Evidence for the modulation of epileptiform activity in the CA3 region of rat hippocampus by adenine nucleotides has been presented (Ross *et al.*, 1998). Chronic stress affects hydrolysis of adenine nucleotides in the frontal cortex and hypothalamus of rats (Torres *et al.*, 2002).

Escalating recent studies of the functional roles of P2 receptors in the brain include the neuroprotective effects of P2 receptor antagonists (Volonté *et al.*, 1999), immunohistochemical localization of $P2Y_1$ purinoceptors in Alzheimer's disease (Moore *et al.*, 2000b), and central nervous (brainstem) control of cardiovascular, respiratory, urogenital, and gut function (Phillis *et al.*, 1997; Horiuchi *et al.*, 1999; Thomas *et al.*, 1999; Thomas and Spyer, 2000; King *et al.*, 2000; Rocha *et al.*, 2001).

## V. FUTURE DEVELOPMENTS

The field of nucleotides and their receptors in the nervous system is still in its infancy and I predict a rapid expansion of interest in this field in the coming years. Some of the directions these studies are likely to take are the development of transgenic mice with absent or enhanced P2 receptor subtypes, behavioral studies of the effects of purines and pyrimidines and related compounds applied to the brain, trophic roles of nucleotides in cell development and death in various pathophysiological conditions, resolution of ATP membrane transport mechanisms, and the development of selective agonists and antagonists that work *in vivo* for therapeutic application in a variety of diseases.

### References

Abbracchio, M. P. (1997). ATP in brain function. *In* "Purinergic Approaches in Experimental Therapeutics" (K. A. Jacobson and M. F. Jarvis, eds.), pp. 383–404. Wiley-Liss, New York.

Abbracchio, M. P., and Burnstock, G. (1994). Purinoceptors: Are there families of $P_{2x}$ and $P_{2y}$ purinoceptors? *Pharmacol. Ther.* **64**, 445–475.

Abbracchio, M. P., and Burnstock, G. (1998). Purinergic signalling: Pathophysiological roles. *Jpn. J. Pharmacol.* **78**, 113–145.

Abbracchio, M. P., and Cattabeni, F. (1999). Brain adenosine receptors as targets for therapeutic intervention in neurodegenerative diseases. *Ann. N.Y. Acad. Sci.* **890**, 79–92.

Abe, Y., Sorimachi, M., Itoyama, Y., Furukawa, K., and Akaike, N. (1995). ATP responses in the embryo chick ciliary ganglion cells. *Neuroscience* **64**, 547–551.

Agren, H., Niklasson, F., and Hallgren, R. (1983). Brain purinergic activity linked withdepressive symptomatology: Hypoxanthine and xanthine in CSF of patients with major depressive disorders. *Psychiatry Res.* **9**, 179–189.

Akasu, T., Hirai, K., and Koketsu, K. (1983). Modulatory actions of ATP on membrane potentials of bullfrog sympathetic ganglion cells. *Brain Res.* **258**, 313–317.

Alavi, A. M., Dubyak, G. R., and Burnstock, G. (2001). Immunohistochemical evidence for ATP receptors in human dental pulp. *J. Dent. Res.* **80**, 476–483.

Allen, T. G. J., and Burnstock, G. (1990). The actions of adenosine 5′-triphosphate on guinea-pig intracardiac neurones in culture. *Br. J. Pharmacol.* **100**, 269–276.

Amadio, S., D'Ambrosi, N., Cavaliere, F., Murra, B., Sancesario, G., Bernardi, G., Burnstock, G., and Volonte, C. (2002). P2 receptor modulation and cytotoxic function in cultured CNS neurones. *Neuropharmacology* **42**, 489–501.

Armstrong, J. N., and Mac Vicar, B. A. (2000). P2X–receptor activation depresses synaptic transmission at hippocampal mossy fiber-CA3 synapses. *Soc. Neurosci. Abstr.* **332**, 15.

Armstrong, J. N., Brust, T. B., Lewis, R. G., and MacVicar, B. A. (2002). Activation of presynaptic P2X7-like receptors depresses mossy fiber-CA3 synaptic transmission through p38 mitogen-activated protein kinase. *J. Neurosci.* **22**, 5938–5945.

Atkinson, L., Batten, T. F., and Deuchars, J. (2000a). P2X(2) receptor immunoreactivity in the dorsal vagal complex and area postrema of the rat. *Neuroscience* **99**, 683–696.

Atkinson, L., Batten, T. F. C., and Deuchars, J. (2000b). The $P2X_2$ receptor is localised presynaptically in vagal afferent fibers in the nucleus tractus solitarii of the rat. *J. Physiol.* **523**, 259P.

Bagley, E. E., Vaughan, C. W., and Christie, M. J. (1999). Inhibition by adenosine receptor agonists of synaptic transmission in rat periaqueductal grey neurons. *J. Physiol.* **516**, (pt. 1), 219–225.

Balachandran, C., and Bennett, M. R. (1996). ATP-activated cationic and anionic conductances in cultured rat hippocampal neurons. *Neurosci. Lett.* **204**, 73–76.

Balcar, V. J., Li, Y., Killinger, S., and Bennett, M. R. (1995). Autoradiography of P2x ATP receptors in the rat brain. *Br. J. Pharmacol.* **115**, 302–306.

Barajas-López, C., Espinosa-Luna, R., and Gerzanich, V. (1994). ATP closes a potassium and opens a cationic conductance through different receptors in neurons of guinea pig submucous plexus. *J. Pharmacol. Exp. Ther.* **268**, 1397–1402.

Barajas-López, C., Huizinga, J. D., Collins, S. M., Gerzanich, V., Espinosa-Luna, R., and Peres, A. L. (1996). $P_{2x}$-purinoceptors of myenteric neurones from the guinea-pig ileum and their unusual pharmacological properties. *Br. J. Pharmacol.* **119**, 1541–1548.

Barajas-López, C., Espinosa-Luna, R., and Zhu, Y. (1998). Functional interactions between nicotinic and P2X channels in short-term cultures of guinea-pig submucosal neurons. *J. Physiol.* **513**, 671–683.

Barajas-López, C., Montano, L. M., and Espinosa-Luna, R. (2002). Inhibitory interactions between 5-HT3 and P2X channels in submucosal neurons. *Am. J. Physiol Gastrointest. Liver Physiol.* **283**, G1238–G1248.

Barberis, C., and McIlwain, H. (1976). 5′-Adenine mononucleotides in synaptosomal preparations from guinea pig neocortex: Their change on incubation, superfusion and stimulation. *J. Neurochem.* **26**, 1015–1021.

Barberis, C., Guibert, B., Daudet, F., Charriere, B., and Leviel, V. (1984). In vivo release of adenosine from cat basal ganglia—studies with a push pull cannula. *Neurochem. Int.* **6**, 545–551.

Barden, J. A., and Bennett, M. R. (2000). Distribution of P2X purinoceptor clusters on individual rat dorsal root ganglion cells. *Neurosci. Lett.* **287**, 183–186.

Bardoni, R., Goldstein, P. A., Lee, C. J., Gu, J. G., and MacDermott, A. B. (1997). ATP $P_{2x}$ receptors mediate fast synaptic transmission in the dorsal horn of the rat spinal cord. *J. Neurosci.* **17**, 5297–5304.

Barnea, A., Cho, G., and Katz, B. M. (1991). A putative role for extracellular ATP: Facilitation of [67]copper uptake and of copper stimulation of the release of luteinizing hormone-releasing hormone from median eminence explants. *Brain Res.* **541**, 93–97.

Barraco, R. A. (1991). Behavioral actions of adenosine and related substances. *In* "Adenosine and Adenine Nucleotides as Regulators of Cellular Function" (J. W. Phillis, ed.), pp. 339–366. CRC Press, Boca Raton, FL.

Barraco, R. A., el Ridi, M. R., Ergene, E., and Phillis, J. W. (1991). Adenosine receptor subtypes in the brainstem mediate distinct cardiovascular response patterns. *Brain Res. Bull.* **26**, 59–84.

Barraco, R. A., O'Leary, D. S., Ergene, E., and Scislo, T. J. (1996). Activation of purinergic receptor subtypes in the nucleus tractus solitarius elicits specific regional vascular response patterns. *J. Auton. Nerv. Syst.* **59**, 113–124.

Battastini, A. M., da Rocha, J. B., Barcellos, C. K., Dias, R. D., and Sarkis, J. J. (1991). Characterization of an ATP diphosphohydrolase (EC 3.6.1.5) in synaptosomes from cerebral cortex of adult rats. *Neurochem. Res.* **16**, 1303–1310.

Bean, B. P. (1990). ATP-activated channels in rat and bullfrog sensory neurons: Concentration dependence and kinetics. *J. Neurosci.* **10**, 1–10.

Bean, B. P., Williams, C. A., and Ceelen, P. W. (1990). ATP-activated channels in rat and bullfrog sensory neurons: Current-voltage relation and single-channel behavior. *J. Neurosci.* **10**, 11–19.

Belai, A., and Burnstock, G. (1994). Evidence for coexistence of ATP and nitric oxide in non-adrenergic, non-cholinergic (NANC) inhibitory neurones in the rat ileum, colon and anococcygeus muscle. *Cell Tissue Res.* **278**, 197–200.

Bellingham, M. C., and Berger, A. J. (1994). Adenosine suppresses excitatory glutamatergic inputs to rat hypoglossal motoneurons *in vitro*. *Neurosci. Lett.* **177**, 143–146.

Bennett, G. C., and Boarder, M. R. (2000). The effect of nucleotides and adenosine on stimulus-evoked glutamate release from rat brain cortical slices. *Br. J. Pharmacol.* **131**, 617–623.

Bennett, G. C., Ford, A. P. D. W., and Boarder, M. R. (1999). Characterisation of P2 receptors on cerebral cortical cell cultures from embryonic rats. *Br. J. Pharmacol.* **126**, 12P.

Bertrand, P. P., and Bornstein, J. C. (2002). ATP as a putative sensory mediator: activation of intrinsic sensory neurons of the myenteric plexus via P2X receptors. *J. Neurosci.* **22**, 4767–4775.

Bickford, P. C., Fredholm, B. B., Dunwiddie, T. V., and Freedman, R. (1985). Inhibition of Purkinje cell firing by systemic administration of phenylisopropyl adenosine: Effect ofcentral noradrenaline depletion by DSP4. *Life Sci.* **37**, 289–297.

Bland-Ward, P. A., and Humphrey, P. P. A. (1997). Acute nociception mediated by hindpaw P2X receptor activation in the rat. *Br. J. Pharmacol.* **122**, 365–371.

Bleehen, T., and Keele, C. A. (1977). Observations on the algogenic actions of adenosine compounds on human blister base preparation. *Pain* **3**, 367–377.

Bo, X., and Burnstock, G. (1994). Distribution of [$^3$H]$\alpha$,$\beta$-methylene ATP binding sites in rat brain and spinal cord. *Neuroreport* **5**, 1601–1604.

Bo, X., Alavi, A., Xiang, Z., Oglesby, I., Ford, A., and Burnstock, G. (1999). Localization of ATP-gated $P2X_2$ and $P2X_3$ receptor immunoreactive nerves in rat taste buds. *Neuroreport* **10**, 1107–1111.

Bobanovic, L. K., Royle, S. J., and Murrell-Lagnado, R. D. (2002). P2X receptor trafficking in neurons is subunit specific. *J. Neurosci.* **22**, 4814–4824.

Boehm, S. (1999). ATP stimulates sympathetic transmitter release via presynaptic P2X purinoceptors. *J. Neurosci.* **19**, 737–746.

Bonan, C. D., Roesler, R., Quevedo, J., Battastini, A. M., Izquierdo, I., and Sarkis, J. J. (1999). Effects of suramin on hippocampal apyrase activity and inhibitory avoidance learning of rats. *Pharmacol. Biochem. Behav.* **63**, 153–158.

Bonan, C. D., Roesler, R., Pereira, G. S., Battastini, A. M., Izquierdo, I., and Sarkis, J. J. (2000). Learning-specific decrease in synaptosomal ATP diphosphohydrolase activity from hippocampus and entorhinal cortex of adult rats. *Brain Res.* **854**, 253–256.

Borison, H. L. (1974). Area postrema: Chemoreceptor trigger zone for vomiting—is that all? *Life Sci.* **14**, 1807–1817.

Borison, H. L., Hawken, M. J., Hubbard, J. I., and Sirett, N. E. (1975). Unit activity from cat area postrema influenced by drugs. *Brain Res.* **92**, 153–156.

Borvendeg, S. J., Gerevich, Z., Gillen, C., and Illes, P. (2003). P2Y receptor-mediated inhibition of voltage-dependent Ca2+ channels in rat dorsal root ganglion neurons. *Synapse* **47**, 159–161.

Bouvier, M. M., Evans, M. L., and Benham, C. D. (1991). Calcium influx induced by stimulation of ATP receptors on neurons cultured from rat dorsal root ganglia. *Eur. J.Neurosci.* **3**, 285–291.

Bradbury, E. J., Burnstock, G., and McMahon, S. B. (1998). The expression of P2X₃ purinoceptors in sensory neurons: Effects of axotomy and glial-derived neurotrophic factor. *Mol. Cell. Neurosci.* **12**, 256–268.

Brake, A. J., Wagenbach, M. J., and Julius, D. (1994). New structural motif for ligand-gated ion channels defined by an ionotropic ATP receptor. *Nature* **371**, 519–523.

Brand, A., Vissiennon, Z., Eschke, D., and Nieber, K. (2001). Adenosine A₁ and A₃ receptors mediate inhibition of synaptic transmission in rat cortical neurons. *Neuropharmacology* **40**, 85–95.

Braun, N., Zhu, Y., Krieglstein, J., Culmsee, C., and Zimmermann, H. (1998). Upregulation of the enzyme chain hydrolyzing extracellular ATP after transient forebrain ischemia in the rat. *J. Neurosci.* **18**, 4891–4900.

Briggs, C. A., and Cooper, J. R. (1981). A synaptosomal preparation from the guinea pig ileum myenteric plexus. *J. Neurochem.* **36**, 1097–1108.

Brown, P., and Dale, N. (2002). Modulation of K(+) currents in Xenopus spinal neurons by p2y receptors: a role for ATP and ADP in motor pattern generation. *J. Physiol.* **540**, 843–850.

Buell, G., Lewis, C., Collo, G., North, R. A., and Surprenant, A. (1996). An antagonist-insensitive P2X receptor expressed in epithelia and brain. *EMBO J.* **15**, 55–62.

Buller, K. M., Khanna, S., Sibbald, J. R., and Day, T. A. (1996). Central noradrenergic neurons signal via ATP to elicit vasopressin responses to haemorrhage. *Neuroscience* **73**, 637–642.

Burgard, E. C., Niforatos, W., van Biesen, T., Lynch, K. J., Touma, E., Metzger, R. E., Kowaluk, E. A., and Jarvis, M. F. (1999). P2X receptor-mediated ionic currents in dorsal root ganglion neurons. *J. Neurophysiol.* **82**, 1590–1598.

Burnstock, G. (1972). Purinergic nerves. *Pharmacol. Rev.* **24**, 509–581.

Burnstock, G. (1976). Do some nerve cells release more than one transmitter? *Neuroscience* **1**, 239–248.

Burnstock, G. (1977). Purine nucleotides and nucleosides as neurotransmitters or neuro-modulators in the central nervous system. In "Neuroregulators and Psychiatric Disorders" (E. Usdin, D. A. Hamburg, and J. D. Barchas, eds.), pp. 470–477. Oxford University Press, New York.

Burnstock, G. (1978). A basis for distinguishing two types of purinergic receptor. In "Cell Membrane Receptors for Drugs and Hormones: A Multidisciplinary Approach" (R. W. Straub and L. Bolis, eds.), pp. 107–118. Raven Press, New York.

Burnstock, G. (1988). Sympathetic purinergic transmission in small blood vessels. *Trends Pharmacol. Sci.* **9**, 116–117.

Burnstock, G. (1990a). Co-transmission. The fifth Heymans memorial lecture—Ghent, February 17, 1990. *Arch. Int. Pharmacodyn. Ther.* **304**, 7–33.

Burnstock, G. (1990b). Dual control of local blood flow by purines. *Ann. N. Y. Acad. Sci.* **603**, 31–44.

Burnstock, G. (1990c). Local mechanisms of blood flow control by perivascular nerves and endothelium. *J. Hypertens.* **8**(Suppl. 7), S95–S106.

Burnstock, G. (1993a). Introduction: Changing face of autonomic and sensory nerves in the circulation. *In* "Vascular Innervation and Receptor Mechanisms: New Perspectives" (L. Edvinsson and R. Uddman, eds.), pp. 1–22. Academic Press, San Diego, CA.

Burnstock, G. (Series Editor) (1993b). "The Autonomic Nervous System. Volume 3. Nervous Control of the Urogenital System."Harwood Academic Publishers, Chur, Switzerland.

Burnstock, G. (Guest Editor) (1996). Purinergic neurotransmission. *Semin. Neurosci.* **8**, 171–257.

Burnstock, G. (1997). The past, present and future of purine nucleotides as signalling molecules. *Neuropharmacology* **36**, 1127–1139.

Burnstock, G. (1999a). Current status of purinergic signalling in the nervous system. *Prog. Brain Res.* **120**, 3–10.

Burnstock, G. (1999b). Purinergic cotransmission. *Brain Res. Bull.* **50**, 355–357.

Burnstock, G. (2000). P2X receptors in sensory neurones. *Br. J. Anaesthesiol.* **84**, 476–488.

Burnstock, G. (2001a). Purine-mediated signalling in pain and visceral perception. *Trends Pharmacol. Sci.* **22**, 182–188.

Burnstock, G. (2001b). Purinergic signalling in gut. *In* "Handbook of Experimental Pharmacology. Purinergic and Pyrimidinergic Signalling II—Cardiovascular, Respiratory, Immune, Metabolic and Gastrointestinal Tract Function" (M. P. Abbracchio and M. Williams, eds.), Vol. 151/II, pp. 141–238. Springer-Verlag, Berlin.

Burnstock, G. (2001c). Purinergic signalling in lower urinary tract. *In* "Handbook of Experimental Pharmacology. Purinergic and Pyrimidinergic Signalling I—Molecular, Nervous and Urinogenitary System Function" (M. P. Abbracchio and M. Williams, eds.), Vol. 151/I, pp. 423–515. Springer-Verlag, Berlin.

Burnstock, G., and Holman, M. E. (1962). Effect of denervation and of reserpine treatment on transmission at sympathetic nerve endings. *J. Physiol.* **160**, 461–469.

Burnstock, G., and Wood, J. N. (1996). Purinergic receptors: Their role in nociception and primary afferent neurotransmission. *Curr. Opin. Neurobiol.* **6**, 526–532.

Burnstock, G., and Lavin, S. (2002). Interstitial cells of Cajal and purinergic signalling. *Auton. Neurosci.* **97**, 68–72.

Burnstock, G., Campbell, G., Satchell, D., and Smythe, A. (1970). Evidence that adenosine triphosphate or a related nucleotide is the transmitter substance released by non-adrenergic inhibitory nerves in the gut. *Br. J. Pharmacol.* **40**, 668–688.

Burnstock, G., Satchell, D. G., and Smythe, A. (1972). A comparison of the excitatory and inhibitory effects of non-adrenergic, non-cholinergic nerve stimulation and exogenously applied ATP on a variety of smooth muscle preparations from different vertebrate species. *Br. J. Pharmacol.* **46**, 234–242.

Burnstock, G., Cocks, T., Crowe, R., and Kasakov, L. (1978). Purinergic innervation of the guinea-pig urinary bladder. *Br. J. Pharmacol.* **63**, 125–138.

Burnstock, G., Allen, T. G. J., Hassall, C. J. S., and Pittam, B. S. (1987). Properties of intramural neurones cultured from the heart and bladder. *In* "Histochemistry and Cell Biology of Autonomic Neurons and Paraganglia. Experimental Brain Research Series 16" (C. Heym, ed.), pp. 323–328. Springer-Verlag, Heidelberg.

Burrows, G. H., and Barnea, A. (1982). Comparison of the effects of ATP, $Mg^{2+}$, and MgATP on the release of luteinizing hormone-releasing hormone from isolated hypothalamic granules. *J. Neurochem.* **38,** 569–573.

Carey, P. A., and Jordan, D. (1998). Effects of central $A_2$ receptors on aortic baroreceptor and cardiopulmonary receptor reflexes in anaesthetized rabbits. *J. Physiol.* **513,** 79P.

Castelucci, P., Robbins, H. L., Poole, D. P., and Furness, J. B. (2002). The distribution of purine P2X(2) receptors in the guinea-pig enteric nervous system. *Histochem. Cell Biol.* **117,** 415–422.

Chagoya, d.S. V (1995). Circadian variations of adenosine and of its metabolism. Could adenosine be a molecular oscillator for circadian rhythms? *Can. J. Physiol. Pharmacol.* **73,** 339–355.

Chen, C., Parker, M. S., Barnes, A. P., Deininger, P., and Bobbin, R. P. (2000). Functional expression of three $P2X_2$ receptor splice variants from guinea pig cochlea. *J. Neurophysiol.* **83,** 1502–1509.

Chen, C. C., Akopian, A. N., Sivilotti, L., Colquhoun, D., Burnstock, G., and Wood, J. N. (1995). A P2X purinoceptor expressed by a subset of sensory neurons. *Nature* **377,** 428–431.

Chen, G., and van den Pol, A. N. (1997). Adenosine modulation of calcium currents and presynaptic inhibition of GABA release in suprachiasmatic and arcuate nucleus neurons. *J. Neurophysiol.* **77,** 3035–3047.

Chen, T. H., Huang, H. P., Matsumoto, Y., Wu, S. H., Wang, M. F., Chung, S. Y., Uezu, K., Moriyama, T., Uezu, E., Korin, T., Sato, S., and Yamamoto, S. (1996). Effects of dietary nucleoside-nucleotide mixture on memory in aged and young memory deficient mice. *Life Sci.* **59,** L325–L330.

Chen, Z. P., Levy, A., and Lightman, S. L. (1994). Activation of specific ATP receptors induces a rapid increase in intracellular calcium ions in rat hypothalamic neurons. *Brain Res.* **641,** 249–256.

Cheng, J.-T., Liu, I.-M., Juang, S.-W., and Jou, S.-B. (2000). Decrease of adenosine A-1 receptor gene expression in cerebral cortex of aged rats. *Neurosci. Lett.* **283,** 227–229.

Chessell, I. P., Michel, A. D., and Humphrey, P. P. (1997). Functional evidence for multiple purinoceptor subtypes in the rat medial vestibular nucleus. *Neuroscience* **77,** 783–791.

Cheung, K.-K., and Burnstock, G. (2002). Localisation of $P2X_3$ and co-expression with $P2X_2$ receptors during rat embryonic neurogenesis. *J. Comp. Neurol.* **443,** 368–382.

Choca, J. I., Proudfit, H. K., and Green, R. D. (1987). Identification of $A_1$ and $A_2$ adenosine receptors in the rat spinal cord. *J. Pharmacol. Exp. Ther.* **242,** 905–910.

Chowdhury, M., and Fillenz, M. (1991). Presynaptic adenosine A2 and N-methyl-D-aspartate receptors regulate dopamine synthesis in rat striatal synaptosomes. *J. Neurochem.* **56,** 1783–1788.

Christofi, F. L., Guan, Z., Lucas, J. H., Rosenberg Schaffer, L. J., and Stokes, B. T. (1996). Responsiveness to ATP with an increase in intracellular free $Ca^{2+}$ is not a distinctive feature of calbindin-D28 immunoreactive neurons in myenteric ganglia. *Brain Res.* **725,** 241–246.

Christofi, F. L., Guan, Z., Wood, J. D., Baidan, L. V., and Stokes, B. T. (1997). Purinergic $Ca^{2+}$ signaling in myenteric neurons via P2 purinoceptors. *Am. J. Physiol.* **272,** G463–G473.

Clark, S. R., Costa, M., Tonini, M., and Brookes, S. J. (1996). Purinergic transmission is involved in a descending excitatory reflex in the guinea-pig small intestine. *Proc. Aust. Neurosci. Soc.* **7,** 176.

Cloues, R., Jones, S., and Brown, D. A. (1993). $Zn^{2+}$ potentiates ATP-activated currents in rat sympathetic neurons. *Pflugers Arch* **424,** 152–158.

Cockayne, D. A., Hamilton, S. G., Zhu, Q.-M., Dunn, P. M., Zhong, Y., Novakovic, S., Malmberg, A. B., Cain, G., Berson, A., Kassotakis, L., Hedley, L., Lachnit, W. G., Burnstock, G., McMahon, S. B., and Ford, A. P. D. W. (2000). Urinary bladder hyporeflexia and reduced pain-related behaviour in P2X$_3$-deficient mice. *Nature* **407**, 1011–1015.

Collo, G., North, R. A., Kawashima, E., Merlo-Pich, E., Neidhart, S., Surprenant, A., and Buell, G. (1996). Cloning of P2X$_5$ and P2X$_6$ receptors and the distribution and properties of an extended family of ATP-gated ion channels. *J. Neurosci.* **16**, 2495–2507.

Collo, G., Neidhart, S., Kawashima, E., Kosco-Vilbois, M., North, R. A., and Buell, G. (1997). Tissue distribution of the P2X$_7$ receptor. *Neuropharmacology* **36**, 1277–1283.

Cook, S. P., Vulchanova, L., Hargreaves, K. M., Elde, R., and McCleskey, E. W. (1997). Distinct ATP receptors on pain-sensing and stretch-sensing neurons. *Nature* **387**, 505–508.

Costenla, A. R., de Mendonca, A., and Ribeiro, J. A. (1999). Adenosine modulates synaptic plasticity in hippocampal slices from aged rats. *Brain Res.* **851**, 228–234.

Coutts, A. A., Jorizzo, J. L., Eady, R. A. J., Greaves, M. W., and Burnstock, G. (1981). Adenosine triphosphate-evoked vascular changes in human skin: Mechanism of action. *Eur. J.Pharmacol.* **76**, 391–401.

Crowe, R., Haven, A. J., and Burnstock, G. (1986). Intramural neurons of the guinea-pig urinary bladder: Histochemical localization of putative neurotransmitters in cultures and newborn animals. *J. Auton. Nerv. Syst.* **15**, 319–339.

Cunha, R. A., and Ribeiro, J. A. (2000a). ATP as a presynaptic modulator. *Life Sci.* **68**, 119–137.

Cunha, R. A., and Ribeiro, J. A. (2000b). Purinergic modulation of [$^3$H]GABA release from rat hippocampal nerve terminals. *Neuropharmacology* **39**, 1156–1167.

Daly, J. W., Bruns, R. F., and Snyder, S. H. (1981). Adenosine receptors in the central nervous system: Relationship to the central actions of methylxanthines. *Life Sci.* **28**, 2083–2097.

Dawid-Milner, M. S., Silva-Carvalho, L., Goldsmith, G. R., and Spyer, K. M. (1994). A potential role of central A$_1$ adenosine receptors in the responses to hypothalamic stimulation in the anaesthetized cat. *J. Auton. Nerv. Syst.* **49**, 15–19.

Day, T. A., Renaud, L. P., and Sibbald, J. R. (1990). Excitation of supraoptic vasopressin cells by stimulation of the A1 noradrenaline cell group: Failure to demonstrate role for established adrenergic or amino acid receptors. *Brain Res.* **516**, 91–98.

Day, T. A., Sibbald, J. R., and Khanna, S. (1993). ATP mediates an excitatory noradrenergic neuron input to supraoptic vasopressin cells. *Brain Res.* **607**, 341–344.

De Koninck, Y., Salter, M. W., and Henry, J. L. (1994). Substance P released endogenously by high-intensity sensory stimulation potentiates purinergic inhibition of nociceptive dorsal horn neurons induced by peripheral vibration. *Neurosci. Lett.* **176**, 128–132.

Deng, G., Matute, C., Kumar, C. K., Fogarty, D. J., and Miledi, R. (1998). Cloning and expression of a P$_{2y}$ purinoceptor from the adult bovine corpus callosum. *Neurobiol. Dis* **5**, 259–270.

Diaz-Hernandez, M., Pintor, J., Castro, E., and Miras-Portugal, M. T. (2002). Co-localisation of functional nicotinic and ionotropic nucleotide receptors in isolated cholinergic synaptic terminals. *Neuropharmacology* **42**, 20–33.

Di Cori, S., and Henry, J. L. (1984). Effects of ATP and AMP on hippocampal neurones of the rat *in vitro*. *Brain Res. Bull.* **13**, 199–201.

Dolphin, A. C., and Prestwich, S. A. (1985). Pertussis toxin reverses adenosine inhibition of neuronal glutamate release. *Nature* **316**, 148–150.

Dowd, E., McQueen, D. S., Chessell, I. P., and Humphrey, P. P. A. (1998). P2X receptor-mediated excitation of nociceptive afferents in the normal and arthritic rat knee joint. *Br. J. Pharmacol.* **125**, 341–346.

Dowe, G. H., Kilbinger, H., and Whittaker, V. P. (1980). Isolation of cholinergic synaptic vesicles from the myenteric plexus of guinea-pig small intestine. *J. Neurochem.* **35**, 993–1003.

Dragunow, M. (1988). Purinergic mechanisms in epilepsy. *Prog. Neurobiol.* **31**, 85–108.

Driessen, B., Reimann, W., Selve, N., Friderichs, E., and Bültmann, R. (1994). Antinociceptive effect of intrathecally administered P2-purinoceptor antagonists in rats. *Brain Res.* **666**, 182–188.

Driessen, B., Bültmann, R., Jurna, I., and Baldauf, J. (1998). Depression of C fiber-evoked activity by intrathecally administered reactive red 2 in rat thalamic neurons. *Brain Res.* **796**, 284–290.

Dubyak, G. R. (1991). Signal transduction by $P_2$-purinergic receptors for extracellular ATP. *Am. J. Respir. Cell Mol. Biol.* **4**, 295–300.

Dunn, P. M., Liu, M., Zhong, Y., King, B. F., and Burnstock, G. (2000). Diinosine pentaphosphate: An antagonist which discriminates between recombinant $P2X_3$ and $P2X_{2/3}$ receptors and between two P2X receptors in rat sensory neurones. *Br. J. Pharmacol.* **130**, 1378–1384.

Dunn, P. M., Zhong, Y., and Burnstock, G. (2001). P2X receptors in peripheral neurones. *Prog. Neurobiol.* **65**, 107–134.

Dunwiddie, T. V. (1985). The physiological role of adenosine in the central nervous system. *Int. Rev. Neurobiol.* **27**, 63–139.

Dunwiddie, T. V., Abbracchio, M. P., Bischofberger, N., Brundege, J. M., Buell, G., Collo, G., Corsi, C., Diao, L., Kawashima, E., Jacobson, K. A., Latini, S., Lin, R. C. S., North, R. A., Pazzagli, M., Pedata, F., Pepeu, G. C., Proctor, W. R., Rassendren, F., Surprenant, A., and Cattabeni, F. (1996). Purinoceptors in the central nervous system. *Drug Dev. Res.* **39**, 361–370.

Dunwiddie, T. V., Diao, L., and Proctor, W. R. (1997). Adenine nucleotides undergo rapid, quantitative conversion to adenosine in the extracellular space in rat hippocampus. *J. Neurosci.* **17**, 7673–7682.

Edwards, F. A., Gibb, A. J., and Colquhoun, D. (1992). ATP receptor-mediated synaptic currents in the central nervous system. *Nature* **359**, 144–147.

Ekonomou, A., Angelatou, F., Vergnes, M., and Kostopoulos, G. (1998). Lower density of A1 adenosine receptors in nucleus reticularis thalami in rats with genetic absence epilepsy. *Neuroreport* **9**, 2135–2140.

Ergene, E., Dunbar, J. C., O'Leary, D. S., and Barraco, R. A. (1994). Activation of $P_2$-purinoceptors in the nucleus tractus solitarius mediate depressor responses. *Neurosci. Lett.* **174**, 188–192.

Evans, R. J., and Surprenant, A. (1992). Vasoconstriction of guinea-pig submucosal arterioles following sympathetic nerve stimulation is mediated by the release of ATP. *Br. J. Pharmacol.* **106**, 242–249.

Evans, R. J., Derkach, V., and Surprenant, A. (1992). ATP mediates fast synaptic transmission in mammalian neurons. *Nature* **357**, 503–505.

Feldberg, W., and Hebb, C. (1948). The stimulating action of phosphate compounds on the perfused superior cervical ganglion of the cat. *J. Physiol.* **107**, 210–221.

Feldberg, W., and Sherwood, S. L. (1954). Injection of drugs into the lateral ventricle of the cat. *J. Physiol.* **123**, 148–167.

Ferguson, D. R., Kennedy, I., and Burton, T. J. (1997). ATP is released from rabbit urinary bladder epithelial cells by hydrostatic pressure changes—a possible sensory mechanism? *J. Physiol.* **505**, 503–511.

Ferre, S., Fredholm, B. B., Morelli, M., Popoli, P., and Fuxe, K. (1997). Adenosine-dopamine receptor-receptor interactions as an integrative mechanism in the basal ganglia. *Trends Neurosci.* **20**, 482–487.

Fieber, L. A., and Adams, D. J. (1991). Adenosine triphosphate-evoked currents in cultured neurones dissociated from rat parasympathetic cardiac ganglia. *J. Physiol.* **434**, 239–256.

Fields, R. D., and Stevens, B. (2000). ATP: An extracellular signaling molecule between neurons and glia. *Trends Neurosci.* **23**, 625–633.

Florenzano, F., Viscomi, M. T., Cavaliere, F., Volonte, C., and Molinari, M. (2002). Cerebellar lesion up-regulates P2X1 and P2X2 purinergic receptors in precerebellar nuclei. *Neuroscience* **115**, 425–434.

Fong, A. Y., Krstew, E. V., Barden, J., and Lawrence, A. J. (2002). Immunoreactive localisation of P2Y1 receptors within the rat and human nodose ganglia and rat brainstem: comparison with [alpha 33P]deoxyadenosine 5′-triphosphate autoradiography. *Neuroscience* **113**, 809–823.

Fredholm, B. B. (1995). Adenosine receptors in the central nervous system. *News Physiol. Sci* **10**, 122–128.

Fredholm, B. B., and Vernet, L. (1979). Release of $^3$H-nucleosides from $^3$H-adenine labelled hypothalamic synaptosomes. *Acta Physiol. Scand.* **106**, 97–107.

Frohlich, R., Boehm, S., and Illes, P. (1996). Pharmacological characterization of $P_2$ purinoceptor types in rat locus coeruleus neurons. *Eur. J. Pharmacol.* **315**, 255–261.

Fujii, S., Kato, H., Furuse, H., Ito, K., Osada, H., Hamaguchi, T., and Kuroda, Y. (1995). The mechanism of ATP-induced long-term potentiation involves extracellular phosphorylation of membrane proteins in guinea-pig hippocampal CA1 neurons. *Neurosci. Lett.* **187**, 130–132.

Fujii, S., Kato, H., and Kuroda, Y. (1999). Extracellular adenosine 5′-triphosphate plus activation of glutamatergic receptors induces long-term potentiation in CA1 neurons of guinea pig hippocampal slices. *Neurosci. Lett.* **276**, 21–24.

Fujii, S., Kuroda, Y., Ito, K., and Kato, H. (2000). Long-term potentiation—induction a synaptic catch mechanism released by extracellular phosphorylation. *Neuroscience* **96**, 259–266.

Fujii, S., Kato, H., and Kuroda, Y. (2002). Cooperativity between extracellular adenosine 5′-triphosphate and activation of N-methyl-D-aspartate receptors in long-term potentiation induction in hippocampal CA1 neurons. *Neuroscience* **113**, 617–628.

Funk, G. D., Kanjhan, R., Walsh, C., Lipski, J., Comer, A. M., Parkis, M. A., and Housley, G. D. (1997). P2 receptor excitation of rodent hypoglossal motoneuron activity in vitro and in vivo: A molecular physiological analysis. *J. Neurosci.* **17**, 6325–6337.

Furness, J. B., Kunze, W. A., Bertrand, P. P., Clerc, N., and Bornstein, J. C. (1998). Intrinsic primary afferent neurons of the intestine. *Prog. Neurobiol.* **54**, 1–18.

Furukawa, K., Ishibashi, H., and Akaike, N. (1994). ATP-induced inward current in neurons freshly dissociated from the tuberomammillary nucleus. *J. Neurophysiol.* **71**, 868–873.

Fuxe, K., Ferre, S., Zoli, M., and Agnati, L. F. (1998). Integrated events in central dopamine transmission as analyzed at multiple levels. Evidence for intramembrane adenosine A2A/dopamine D2 and adenosine A1/dopamine D1 receptor interactions in the basal ganglia. *Brain Res. Rev* **26**, 258–273.

Fyffe, R. E. W., and Perl, E. R. (1984). Is ATP a central synaptic mediator for certain primary afferent fibres from mammalian skin? *Proc. Natl. Acad. Sci. USA* **81**, 6890–6893.

Galindo, A., Krnjevic, K., and Schwartz, S. (1967). Micro-iontophoretic studies on neurones in the cuneate nucleus. *J. Physiol.* **192**, 359–377.

Galligan, J. J. (2002). Pharmacology of synaptic transmission in the enteric nervous system. *Curr. Opin. Pharmacol.* **2**, 623–629.

Garcia-Lecea, M., Delicado, E. G., Miras-Portugal, M. T., and Castro, E. (1999). $P2X_2$ characteristics of the ATP receptor coupled to $[Ca^{2+}]_i$ increases in cultured Purkinje neurons from neonatal rat cerebellum. *Neuropharmacology* **38**, 699–706.

Garcia-Lecea, M., Sen, R. P., Soto, F., Miras-Portugal, M. T., and Castro, E. (2001). P2 receptors in cerebellar neurons: Molecular diversity of ionotropic ATP receptors in Purkinje cells. *Drug Dev. Res.* **52,** 104–113.

Geiger, J. D., LaBella, F. S., and Nagy, J. I. (1984). Characterization and localization of adenosine receptors in rat spinal cord. *J. Neurosci.* **4,** 2303–2310.

Gibb, A. J., and Halliday, F. C. (1996). Fast purinergic transmission in the central nervous system. *Semin. Neurosci.* **8,** 225–232.

Giovannelli, L., Giovannini, M. G., Pedata, F., and Pepeu, G. (1988). Purinergic modulation of cortical acetylcholine release is decreased in aging rats. *Exp. Gerontol.* **23,** 175–181.

Giraldez, L., Diaz-Hernandez, M., Gomez-Villafuertes, R., Pintor, J., Castro, E., and Miras-Portugal, M. T. (2001). Adenosine triphosphate and diadenosine pentaphosphate induce [Ca(2+)](i) increase in rat basal ganglia aminergic terminals. *J. Neurosci. Research* **64,** 174–182.

Glushakov, A. V., Melishchuk, A. I., and Skok, V. I. (1996). ATP-induced currents in submucous plexus neurons of the guinea-pig small intestine. *Neurophysiology (Moscow)* **28,** 77–85.

Glushakov, A. V., Glushakova, H. Y., and Skok, V. I. (1998). Two types of $P_{2x}$-purinoceptors in neurons of the guinea pig ileum submucous plexus. *Neurophysiology (Moscow)* **30,** 242–245.

Gomez-Villafuertes, R., Gualix, J., and Miras-Portugal, M. T. (2001). Single GABAergic synaptic terminals from rat midbrain exhibit functional P2X and dinucleotide receptors, able to induce GABA secretion. *J. Neurochem.* **77,** 84–93.

Gomez-Villafuertes, R., Pintor, J., Gualix, J., and Miras-Portugal, M. T. (2003). GABA(B) receptor-mediated presynaptic potentiation of ATP ionotropic receptors in rat midbrain synaptosomes. *Neuropharmacology* **44,** 311–323.

Gordon, J. L. (1986). Extracellular ATP: Effects, sources and fate. *Biochem. J.* **233,** 309–319.

Gourine, A. V., Melenchuk, E. V., Poputnikov, D. M., Gourine, V. N., and Spyer, K. M. (2002). Involvement of purinergic signalling in central mechanisms of body temperature regulation in rats. *Br. J. Pharmacol.* **135,** 2047–2055.

Grondin, R., Bedard, P. J., Hadj, T. A., Gregoire, L., Mori, A., and Kase, H. (1999). Antiparkinsonian effect of a new selective adenosine A2A receptor antagonist in MPTP-treated monkeys. *Neurology* **52,** 1673–1677.

Grubb, B. D., and Evans, R. J. (1999). Characterization of cultured dorsal root ganglion neuron P2X receptors. *Eur. J. Neurosci.* **11,** 149–154.

Gu, J. G., and MacDermott, A. B. (1997). Activation of ATP P2X receptors elicits glutamate release from sensory neuron synapses. *Nature* **389,** 749–753.

Guieu, R., Couraud, F., Pouget, J., Sampieri, F., Bechis, G., and Rochat, H. (1996). Adenosine and the nervous system: Clinical implications. *Clin. Neuropharmacol.* **19,** 459–474.

Guo, A., Vulchanova, L., Wang, J., Li, X., and Elde, R. (1999). Immunocytochemical localization of the vanilloid receptor 1 (VR1): Relationship to neuropeptides, the $P2X_3$ purinoceptor and IB4 binding sites. *Eur. J. Neurosci.* **11,** 946–958.

Häggblad, J., and Heilbronn, E. (1988). $P_2$-purinoceptor-stimulated phosphoinositide turnover in chick myotubes. Calcium mobilization and the role of guanyl nucleotide-binding proteins. *FEBS Lett.* **235,** 133–136.

Halliday, F. C., and Gibb, A. J. (1997). Single-channel properties of P2X ATP receptors in rat cerebellar slices. *J. Physiol.* **504,** 51.

Hamilton, B. R., and Smith, D. O. (1991). Autoreceptor-mediated purinergic and cholinergic inhibition of motor nerve terminal calcium currents in the rat. *J. Physiol.* **432,** 327–341.

Hamilton, S. G., and McMahon, S. B. (2000). ATP as a peripheral mediator of pain. *J. Auton. Nerv. Syst.* **81,** 187–194.

Hamilton, S. G., Wade, A., and McMahon, S. B. (1999). The effects of inflammation and inflammatory mediators on nociceptive behaviour induced by ATP analogues in the rat. *Br. J. Pharmacol.* **126,** 326–332.

Hamilton, S. G., Warburton, J., Bhattachajee, A., Ward, J., and McMahon, S. B. (2000). ATP in human skin elicits a dose-related pain response which is potentiated under conditions of hyperalgesia. *Brain* **123,** 1238–1246.

Harms, H. H., Wardeh, G., and Mulder, A. H. (1979). Effects of adenosine on depolarization-induced release of various radiolabelled neurotransmitters from slices of rat corpus striatum. *Neuropharmacology* **18,** 577–580.

Harms, L., Finta, E. P., Tschöpl, M., and Illes, P. (1992). Depolarization of rat locus coeruleus neurons by adenosine 5′-triphosphate. *Neuroscience* **48,** 941–952.

Heinemann, A., Shahbazian, A., Barthó, L., and Holzer, P. (1999). Different receptors mediating the inhibitory action of exogenous ATP and endogenously released purines on guinea-pig intestinal peristalsis. *Br. J. Pharmacol.* **128,** 313–320.

Hiruma, H., and Bourque, C. W. (1995). $P_2$ purinoceptor-mediated depolarization of rat supraoptic neurosecretory cells *in vitro. J. Physiol.* **489,** 805–811.

Hollins, C., and Stone, T. W. (1980). Adenosine inhibition of $\gamma$-aminobutyric acid release from slices of rat cerebral cortex. *Br. J. Pharmacol.* **69,** 107–112.

Holton, P. (1959). The liberation of adenosine triphosphate on antidromic stimulation of sensory nerves. *J. Physiol.* **145,** 494–504.

Horiuchi, J., Potts, P. D., Tagawa, T., and Dampney, R. A. L. (1999). Effects of activation and blockade of $P_{2x}$ receptors in the ventrolateral medulla on arterial pressure and sympathetic activity. *J. Auton. Nerv. Syst.* **76,** 118–126.

Housley, G. D. (1997). Extracellular nucleotide signaling in the inner ear. *Mol. Neurobiol.* **16,** 21–48.

Housley, G. D. (2001). Nucleoside and nucleotide transmission in sensory systems. *In* "Handbook of Experimental Pharmacology. Purinergic and Pyrimidinergic SignallingI—Molecular, Nervous and Urinogenitary System Function" (M. P. Abbracchio and M. Williams, eds.), Vol. 151/I, pp. 339–369. Springer-Verlag, Berlin.

Hoyle, C. H. V., and Burnstock, G. (1985). Atropine-resistant excitatory junction potentials in rabbit bladder are blocked by $\alpha,\beta$-methylene ATP. *Eur. J. Pharmacol.* **114,** 239–240.

Hu, H. Z., Gao, N., Lin, Z., Gao, C., Liu, S., Ren, J., Xia, Y., and Wood, J. D. (2001). P2X(7) receptors in the enteric nervous system of guinea-pig small intestine. *J. Comp Neurol.* **440,** 299–310.

Hugel, S., and Schlichter, R. (2000). Presynaptic P2X receptors facilitate inhibitory GABAergic transmission between cultured rat spinal cord dorsal horn neurons. *J. Neurosci.* **20,** 2121–2130.

Ikeuchi, Y., and Nishizaki, T. (1995). ATP-evoked potassium currents in rat striatal neurons are mediated by a P2 purinergic receptor. *Neurosci. Lett.* **190,** 89–92.

Ikeuchi, Y., and Nishizaki, T. (1996). $P_2$ purinoceptor-operated potassium channel in rat cerebellar neurons. *Biochem. Biophys. Res. Commun.* **218,** 67–71.

Ikeuchi, Y., Nishizaki, T., and Okada, Y. (1995). A $P_2$ purinoceptor activated by ADP in rat medullar neurons. *Neurosci. Lett.* **198,** 71–74.

Ikeuchi, Y., Nishizaki, T., Mori, M., and Okada, Y. (1996). Regulation of the potassium current and cytosolic $Ca^{2+}$ release induced by 2-methylthio ATP in hippocampal neurons. *Biochem. Biophys. Res. Commun.* **218,** 428–433.

Illes, P., and Zimmermann, H. (1999). Nucleotides and their receptors in the nervous system. *Prog. Brain Res.* **120,** 1–432.

Illes, P., Sevcik, J., Finta, E. P., Frohlich, R., Nieber, K., and Norenberg, W. (1994). Modulation of locus coeruleus neurons by extra- and intracellular adenosine 5'-triphosphate. *Brain Res. Bull.* **35,** 513–519.

Illes, P., Wirkner, K., Nörenberg, W., Masino, S. A., and Dunwiddie, T. V. (2001). Interaction between the transmitters ATP and glutamate in the central nervous system. *Drug Dev. Res.* **52,** 76–82.

Inoue, K. (1998). ATP receptors for the protection of hippocampal functions. *Jpn. J. Pharmacol.* **78,** 405–410.

Inoue, T., and Kannan, M. S. (1988). Nonadrenergic and noncholinergic excitatory neurotransmission in rat intrapulmonary artery. *Am. J. Physiol.* **254,** H1142–H1148.

Inoue, K., and Koizumi, S. (2001). Mechanism of the inhibitory action of ATP in rat hippocampus. *Drug Dev. Res.* **52,** 95–103.

Inoue, K., Nakazawa, K., Fujimori, K., Watano, T., and Takanaka, A. (1992a). Extracellular adenosine 5'-triphosphate-evoked glutamate release in cultured hippocampal neurons. *Neurosci. Lett.* **134,** 215–218.

Inoue, K., Nakazawa, K., Watano, T., Ohara-Imaizumi, M., Fujimori, K., and Takanaka, A. (1992b). Dopamine receptor agonists and antagonists enhance ATP-activated currents. *Eur. J. Pharmacol.* **215,** 321–324.

Inoue, K., Koizumi, S., and Nakazawa, K. (1995). Glutamate-evoked release of adenosine 5'-triphosphate causing an increase in intracellular calcium in hippocampal neurons. *Neuroreport* **6,** 437–440.

Inoue, K., Koizumi, S., and Ueno, S. (1996). Implication of ATP receptors in brain functions. *Prog. Neurobiol.* **50,** 483–492.

Jahr, C. E., and Jessell, T. M. (1983). ATP excites a subpopulation of rat dorsal horn neurones. *Nature* **304,** 730–733.

Jang, I. S., Rhee, J. S., Kubota, H., and Akaike, N. (2001). Developmental changes in P2X purinoceptors on glycinergic presynaptic nerve terminals projecting to rat substantia gelatinosa neurones. *J. Physiol.* **536,** 505–519.

James, S., and Richardson, P. J. (1993). Production of adenosine from extracellular ATP at the striatal cholinergic synapse. *J. Neurochem.* **60,** 219–227.

Jarvis, M. F., Wismer, C. T., Schweitzer, E., Yu, H., van Biesen, T., Lynch, K. J., Burgard, E. C., and Kowaluk, E. A. (2001). Modulation of BzATP and formalin induced nociception: Attenuation by the P2X receptor antagonist, TNP-ATP and enhancement by the P2X$_3$ allosteric modulator, Cibacron blue. *Br. J. Pharmacol.* **132,** 259–269.

Jarvis, M. F., Burgard, E. C., McGaraughty, S., Honore, P., Lynch, K., Brennan, T. J., Subieta, A., van Biesen, T., Cartmell, J., Bianchi, B., Niforatos, W., Kage, K., Yu, H., Mikusa, J., Wismer, C. T., Zhu, C. Z., Chu, K., Lee, C. H., Stewart, A. O., Polakowski, J., Cox, B. F., Kowaluk, E., Williams, M., Sullivan, J., and Faltynek, C. (2002). A-317491, a novel potent and selective non-nucleotide antagonist of P2X3 and P2X2/3 receptors, reduces chronic inflammatory and neuropathic pain in the rat. *Proc. Natl. Acad. Sci. USA* **99,** 17179–17184.

Jo, Y. H., and Role, L. W. (2000). Purinergic and GABAergic co-transmission in lateral hypothalamus. *Soc. Neurosci. Abstr.* **716,** 19.

Jo, Y. H., and Schlichter, R. (1999). Synaptic corelease of ATP and GABA in cultured spinal neurons. *Nat. Neurosci.* **2,** 241–245.

Jo, Y. H., and Role, L. W. (2002a). Coordinate release of ATP and GABA at in vitro synapses of lateral hypothalamic neurons. *J. Neurosci.* **22,** 4794–4804.

Jo, Y. H., and Role, L. W. (2002b). Cholinergic modulation of purinergic and GABAergic co-transmission at in vitro hypothalamic synapses. *J. Neurophysiol.* **88,** 2501–2508.

John, G. R., Simpson, J. E., Woodroofe, M. N., Lee, S. C., and Brosnan, C. F. (2001). Extracellular nucleotides differentially regulate interleukin-1beta signaling in primary human astrocytes: implications for inflammatory gene expression. *J. Neurosci.* **21**, 4134–4142.

Jonzon, B., and Fredholm, B. B. (1985). Release of purines, noradrenaline, and GABA from rat hippocampal slices by field stimulation. *J. Neurochem.* **44**, 217–224.

Kamiji, T., Morita, K., and Katayama, Y. (1994). ATP regulates synaptic transmission by pre- and postsynaptic mechanisms in guinea-pig myenteric neurons. *Neuroscience* **59**, 165–174.

Kanjhan, R., Housley, G. D., Thorne, P. R., Christie, D. L., Palmer, D. J., Luo, L., and Ryan, A. F. (1996). Localization of ATP-gated ion channels in cerebellum using P2x$_2$R subunit-specific antisera. *Neuroreport* **7**, 2665–2669.

Kanjhan, R., Housley, G. D., Burton, L. D., Christie, D. L., Kippenberger, A., Thorne, P. R., Luo, L., and Ryan, A. F. (1999). Distribution of the P2X$_2$ receptor subunit of the ATP-gated ion channels in the rat central nervous system. *J. Comp. Neurol.* **407**, 11–32.

Kapoor, J. R., and Sladek, C. D. (2000). Purinergic and adrenergic agonists synergize in stimulating vasopressin and oxytocin release. *J. Neurosci.* **20**, 8868–8875.

Kapoor, J. R., and Sladek, C. D. (2001). Substance P and NPY differentially potentiate ATP and adrenergic stimulated vasopressin and oxytocin release. *Am. J. Physiol.* **280**, R69–R78.

Katayama, Y., and Morita, K. (1989). Adenosine 5'-triphosphate modulates membrane potassium conductance in guinea-pig myenteric neurones. *J. Physiol.* **408**, 373–390.

Kato, F., and Shigetomi, E. (2001). Distinct modulation of evoked and spontaneous EPSCs by purinoceptors in the nucleus tractus solitarii of the rat. *J. Physiol.* **530**, 469–486.

Keast, J. R. (1995). Visualization and immunohistochemical characterization of sympathetic and parasympathetic neurons in the male rat major pelvic ganglion. *Neuroscience* **66**, 655–662.

Kennedy, C. (2001). The role of purines in the peripheral nervous system. *In* "Handbook of Experimental Pharmacology. Purinergic and Pyrimidinergic Signalling I—Molecular, Nervous and Urinogenitary System Function" (M. P. Abbracchio and M. Williams, eds.), Vol. 151/I, pp. 289–304. Springer-Verlag, Berlin.

Khakh, B. S. (2001). Molecular physiology of P2X receptors and ATP signalling at synapses. *Nat. Rev. Neurosci.* **2**, 165–174.

Khakh, B. S., and Henderson, G. (1998). ATP receptor-mediated enhancement of fast excitatory neurotransmitter release in the brain. *Mol. Pharmacol.* **54**, 372–378.

Khakh, B. S., and Henderson, G. (2000). Modulation of fast synaptic transmission by presynaptic ligand-gated cation channels. *J. Auton. Nerv. Syst.* **81**, 110–121.

Khakh, B. S., Humphrey, P. P. A., and Surprenant, A. (1995). Electrophysiological properties of P2$_x$-purinoceptors in rat superior cervical, nodose and guinea-pig coeliac neurones. *J.Physiol.* **484**, 385–395.

Khakh, B. S., Humphrey, P. P., and Henderson, G. (1997). ATP-gated cation channels (P2X purinoceptors) in trigeminal mesencephalic nucleus neurons of the rat. *J. Physiol.* **498**, 709–715.

Khakh, B. S., Smith, W. B., Chiu, C. S., Ju, D., Davidson, N., and Lester, H. A. (2001). Activation-dependent changes in receptor distribution and dendritic morphology in hippocampal neurons expressing P2X$_2$-green fluorescent protein receptors. *Proc. Natl. Acad. Sci. USA* **98**, 5288–5293.

Khalil, Z., Ralevic, V., Bassirat, M., Dusting, G. J., and Helme, R. D. (1994). Effects of ageing on sensory nerve function in rat skin. *Brain Res.* **641**, 265–272.

Kidd, E. J., Grahames, C. B. A., Simon, J., Michel, A. D., Barnard, E. A., and Humphrey, P. P. A. (1995). Localization of P2X purinoceptor transcripts in the rat nervous system. *Mol. Pharmacol.* **48**, 569–573.

Kimball, B. C., and Mulholland, M. W. (1995). Neuroligands evoke calcium signaling in cultured myenteric neurons. *Surgery* 118, 162–169.

Kimball, B. C., Yule, D. I., and Mulholland, M. W. (1996). Extracellular ATP mediates $Ca^{2+}$ signaling in cultured myenteric neurons via a PLC-dependent mechanism. *Am. J. Physiol.* 270, G587–G593.

King, B. F., and North, R. A. (Guest Editors). (2000). Purines and the autonomic nervous system: From controversy to clinic. *J. Auton. Nerv. Syst.* 81, 1–298.

King, B. F., Townsend-Nicholson, A., Wildman, S. S., Thomas, T., Spyer, K. M., and Burnstock, G. (2000). Coexpression of rat $P2X_2$ and $P2X_6$ subunits in *Xenopus* oocytes. *J. Neurosci.* 20, 4871–4877.

Kirischuk, S., Matiash, V., Kulik, A., Voitenko, N., Kostyuk, P., and Verkhratsky, A. (1996). Activation of $P_2$-purino-, $\alpha_1$-adreno and $H_1$-histamine receptors triggers cytoplasmic calcium signalling in cerebellar Purkinje neurons. *Neuroscience* 73, 643–647.

Kirkpatrick, K., and Burnstock, G. (1987). Sympathetic nerve-mediated release of ATP from the guinea-pig vas deferens is unaffected by reserpine. *Eur. J. Pharmacol.* 138, 207–214.

Kirkup, A. J., Booth, C. E., Chessell, I. P., Humphrey, P. P., and Grundy, D. (1999). Excitatory effect of P2X receptor activation on mesenteric afferent nerves in the anaesthetised rat. *J. Physiol.* 520, 551–563.

Kitchen, A. M., Collins, H. L., DiCarlo, S. E., Scislo, T. J., and O'Leary, D. S. (2001). Mechanisms mediating NTS P2x receptor-evoked hypotension: cardiac output vs. total peripheral resistance. *Am. J. Physiol. Heart Circ. Physiol.* 281, H2198–H2203.

Kittner, H., Krugel, U., Hoffmann, E., and Illes, P. (2000). Effects of intra-accumbens injection of 2-methylthio ATP: A combined open field and electroencephalographic study in rats. *Psychopharmacology (Berl.)* 150, 123–131.

Knight, G. E., Bodin, P., de Groat, W. C., and Burnstock, G. (1999). Distension of the guinea pig ureter releases ATP from the epithelium. *Soc. Neurosci. Abstr.* 25, 1171.

Koch, H., von Kügelgen, I., and Starke, K. (1997). P2-receptor-mediated inhibition of noradrenaline release in the rat hippocampus. *Naunyn Schmiedebergs Arch. Pharmacol.* 355, 707–715.

Kocsis, J. D., Eng, D. L., and Bhisitkul, R. B. (1984). Adenosine selectively blocks parallel-fiber-mediated synaptic potentials in rat cerebellar cortex. *Proc. Natl. Acad. Sci. USA* 81, 6531–6534.

Kogure, K., and Alonso, O. F. (1978). A pictorial representation of endogenous brain ATP by a bioluminescent method. *Brain Res.* 154, 273–284.

Koizumi, S., and Inoue, K. (1997). Inhibition by ATP of calcium oscillations in rat cultured hippocampal neurones. *Br. J. Pharmacol.* 122, 51–58.

Kolb, H. A., and Wakelam, M. J. O. (1983). Transmitter-like action of ATP on patched membranes of cultured myoblasts and myotubes. *Nature* 303, 621–623.

Kowaluk, E. A., and Jarvis, M. F. (2000). Therapeutic potential of adenosine kinase inhibitors. *Exp. Opin. Investig. Drugs* 9, 551–564.

Krishtal, O. A., Marchenko, S. M., and Pidoplichko, V. I. (1983). Receptor for ATP in the membrane of mammalian sensory neurones. *Neurosci. Lett.* 35, 41–45.

Krügel, U., Kittner, H., and Illes, P. (1999). Adenosine 5'-triphosphate-induced dopamine release in the rat nucleus accumbens in vivo. *Neurosci. Lett.* 265, 49–52.

Krügel, U., Kittner, H., and Illes, P. (2001a). Mechanisms of adenosine 5'-triphosphate-induced dopamine release in the rat nucleus accumbens in vivo. *Synapse* 39, 222–232.

Krügel, U., Kittner, H., Franke, H., and Illes, P. (2001b). Stimulation of P2 receptors in the ventral tegmental area enhances dopaminergic mechanisms in vivo. *Neuropharmacology* 40, 1084–1093.

Geoffrey Burnstock

Kuroda, Y. (1989). "Tracing circuit" model for the memory process in human brain: Roles of ATP and adenosine derivatives for dynamic change of synaptic connections. *Neurochem. Int.* **14,** 309–319.

Kuroda, Y., and McIlwain, H. (1973). Subcellular localization of (14C)adenine derivatives newly-formed in cerebral tissues and the effects of electrical excitation. *J. Neurochem.* **21,** 889–900.

Kuroda, Y., and McIlwain, H. (1974). Uptake and relase of [$^{14}$C]adenine derivatives at beds of mammalian cortical synaptosomes in a superfusion system. *J. Neurochem.* **22,** 691–699.

Labrakakis, C., Gerstner, E., and MacDermott, A. B. (2000). Adenosine triphosphate-evoked currents in cultured dorsal root ganglion neurons obtained from rat embryos: Desensitization kinetics and modulation of glutamate release. *Neuroscience* **101,** 1117–1126.

Laitinen, J. T., Uri, A., Raidaru, G., and Miettinen, R. (2001). [(35)S]GTPgammaS autoradiography reveals a wide distribution of G(i/o)-linked ADP receptors in the nervous system: close similarities with the platelet P2Y (ADP) receptor. *J. Neurochem.* **77,** 505–518.

Lalo, U., and Kostyuk, P. (1998). Developmental changes in purinergic calcium signalling in rat neocortical neurones. *Brain Res. Dev. Brain Res.* **111,** 43–50.

Lalo, U., Voitenko, N., and Kostyuk, P. (1998). Iono- and metabotropically induced purinergic calcium signalling in rat neocortical neurons. *Brain Res.* **799,** 285–291.

Lara, D. R., and Souza, D. O. (2000). Schizophrenia: A purinergic hypothesis. *Med. Hypotheses* **54,** 157–166.

Laube, B. (2002). Potentiation of inhibitory glycinergic neurotransmission by Zn2+: a synergistic interplay between presynaptic P2X2 and postsynaptic glycine receptors. *European J. Neurosci.* **16,** 1025–1036.

Lê, K. T., Villeneuve, P., Ramjaun, A. R., McPherson, P. S., Beaudet, A., and Séguéla, P. (1998). Sensory presynaptic and widespread somatodendritic immunolocalization of central ionotropic P2X ATP receptors. *Neuroscience* **83,** 177–190.

Lee, H.-Y., Bardini, M., and Burnstock, G. (2000). Distribution of P2X receptors in the urinary bladder and ureter of the rat. *J. Urol.* **163,** 2002–2007.

Lee, K. S., Schubert, P., Emmert, H., and Kreutzberg, G. W. (1981). Effect of adenosine versus adenine nucleotides on evoked potentials in a rat hippocampal slice preparation. *Neurosci. Lett.* **23,** 309–314.

Lee, K. S., Schubert, P., Reddington, M., and Kreutzberg, G. W. (1986). The distribution of adenosine Al receptors and 5'-nucleotidase in the hippocampal formation of several mammalian species. *J. Comp. Neurol.* **246,** 427–434.

Lee, D. K., George, S. R., Evans, J. F., Lynch, K. R., and O'Dowd, B. F. (2001). Orphan G protein-coupled receptors in the CNS. *Curr. Opin. Pharmacol.* **1,** 31–39.

Lemos, J. R., and Wang, G. (2000). Excitatory versus inhibitory modulation by ATP of neurohypophysial terminal activity in the rat. *Exp. Physiol.* **85,** 67S–74S.

LePard, K. J., Messori, E., and Galligan, J. J. (1997). Purinergic fast excitatory postsynaptic potentials in myenteric neurons of guinea pig: Distribution and pharmacology. *Gastroenterology* **113,** 1522–1534.

Lewis, C., Neidhart, S., Holy, C., North, R. A., Buell, G., and Surprenant, A. (1995). Coexpression of P2X$_2$ and P2X$_3$ receptor subunits can account for ATP-gated currents in sensory neurons. *Nature* **377,** 432–435.

Lewis, J. (1927). "The Blood Vessels of the Human Skin and Their Responses." Shaw & Sons, London.

Li, C., Peoples, R. W., Li, Z., and Weight, F. F. (1993). Zn$^{2+}$ potentiates excitatory action of ATP on mammalian neurons. *Proc. Natl. Acad. Sci. USA* **90,** 8264–8267.

Li, C., Peoples, R. W., and Weight, F. F. (1997). Inhibition of ATP-activated current by zinc in dorsal root ganglion neurones of bullfrog. *J. Physiol.* **505,** 641–653.

Li, C., Peoples, R. W., Lanthorn, T. H., Li, Z. W., and Weight, F. F. (1999). Distinct ATP-activated currents in different types of neurons dissociated from rat dorsal root ganglion. *Neurosci. Lett.* **263**, 57–60.

Li, G. H., Lee, E. M., Blair, D., Holding, C., Poronnik, P., Cook, D. C., Barden, J. A., and Bennett, M. R. (2000). The disposition of P2x receptor clusters on individual neurons in sympathetic ganglia and their redistribution on agonist activation. *J. Biol. Chem.* **275**, 29107–29112.

Li, J., and Perl, E. R. (1994). Adenosine inhibition of synaptic transmission in the substantia gelatinosa. *J. Neurophysiol.* **72**, 1611–1621.

Li, J., and Perl, E. R. (1995). ATP modulation of synaptic transmission in the spinal substantia gelatinosa. *J. Neurosci.* **15**, 3357–3365.

Li, P., Calejesan, A. A., and Zhou, M. (1998). ATP $P_{2x}$ receptors and sensory synaptic transmission between primary afferent fibers and spinal dorsal horn neurons in rats. *J. Neurophysiol.* **80**, 3356–3360.

Li, G. H., Lee, E. M., Blair, D., Holding, C., Poronnik, P., Cook, D. I., Barden, J. A., and Bennett, M. R. (2000). The distribution of P2X receptor clusters on individual neurons in sympathetic ganglia and their redistribution on agonist activation. *J. Biol. Chem.* **275**, 29107–29112.

Liem, R. S., Copray, J. C., and van Willigen, J. D. (1991). Ultrastructure of the rat mesencephalic trigeminal nucleus. *Acta Anat. (Basel)* **140**, 112–119.

Linden, D. R., Sharkey, K. A., and Mawe, G. M. (2003). Enhanced excitability of myenteric AH neurones in the inflamed guinea-pig distal colon. *J. Physiol.* **547**, 589–601.

Lindsay, R. M. (1996). Role of neurotrophins and trk receptors in the development and maintenance of sensory neurons: An overview. *Philos. Trans. R. Soc. Lond. B Biol. Sci.* **351**, 365–373.

Llewellyn-Smith, I. J., and Burnstock, G. (1998). Ultrastructural localization of P2X$_3$ receptors in rat sensory neurons. *Neuroreport* **9**, 2245–2250.

Loesch, A., and Burnstock, G. (1998). Electron-immunocytochemical localization of the P2X$_1$ receptors in the rat cerebellum. *Cell Tissue Res.* **294**, 253–260.

Loesch, A., and Burnstock, G. (2001). ATP-gated P2X$_6$ receptors in the rat hypothalamo-neurohypophysial system: An ultrastructural study with peroxidase and gold-silver immunolabelling. *Neuroscience* **106**, 621–631.

Loesch, A., Miah, S., and Burnstock, G. (1999). Ultrastructural localisation of ATP-gated P2X$_2$ receptor immunoreactivity in the rat hypothalamo-neurohypophysial system. *J. Neurocytol.* **28**, 495–504.

Lu, Z., and Smith, D. O. (1998). Adenosine 5'-triphosphatase increases acetylcholine channel opening frequency in rat skeletal muscle. *J. Physiol.* **436**, 45–56.

Lundy, P. M., Hamilton, M. G., Mi, L., Gong, W., Vair, C., Sawyer, T. W., and Frew, R. (2002). Stimulation of Ca(2+) influx through ATP receptors on rat brain synaptosomes: identification of functional P2X(7) receptor subtypes. *Br. J. Pharmacol.* **135**, 1616–1626.

Lustig, K. D., Shiau, A. K., Brake, A. J., and Julius, D. (1993). Expression cloning of an ATP receptor from mouse neuroblastoma cells. *Proc. Natl. Acad. Sci. USA* **90**, 5113–5117.

Luthardt, J., Borvendeg, S. J., Sperlagh, B., Poelchen, W., Wirkner, K., and Illes, P. (2003). P2Y(1) receptor activation inhibits NMDA receptor-channels in layer V pyramidal neurons of the rat prefrontal and parietal cortex. *Neurochem. Int.* **42**, 161–172.

Maggi, C. A., and Meli, A. (1988). The sensory-efferent function of capsaicin-sensitive sensory nerves. *Gen. Pharmacol.* **19**, 1–43.

Maienschein, V., and Zimmermann, H. (1996). Immunocytochemical localization of ecto-5'-nucleotidase in cultures of cerebellar granule cells. *Neuroscience* **70**, 429–438.

Mally, J., and Stone, T. W. (1996). Potential role of adenosine antagonist therapy in pathological tremor disorders. *Pharmacol. Ther.* **72**, 243–250.

Marangos, P. J., and Boulenger, J. P. (1985). Basic and clinical aspects of adenosinergic neuromodulation. *Neurosci. Biobehav. Rev.* **9**, 421–430.

Marcaida, G., Minana, M. D., Grisolia, S., and Felipo, V. (1995). Lack of correlation between glutamate-induced depletion of ATP and neuronal death in primary cultures of cerebellum. *Brain Res.* **695**, 146–150.

Martinez-Mir, M. I., Probst, A., and Palacios, J. M. (1991). Adenosine A2 receptors: Selective localization in the human basal ganglia and alterations with disease. *Neuroscience* **42**, 697–706.

Masaki, E., Kawamura, M., and Kato, F. (2001). Reduction by sevoflurane of adenosine 5'-triphosphate-activated inward current of locus coeruleus neurons in pontine slices of rats. *Brain Res.* **921**, 226–232.

Masino, S. A., and Dunwiddie, T. V. (2001). Role of purines and pyrimidines in the central nervous system. *In* "Handbook of Experimental Pharmacology. Purinergic and Pyrimidinergic Signalling I—Molecular, Nervous and Urinogenitary System Function" (M. P. Abbracchio and M. Williams, eds.), Vol. 151/I, pp. 251–288. Springer-Verlag, Berlin.

Mateo, J., Garcia-Lecea, M., Miras-Portugal, M. T., and Castro, E. (1998). $Ca^{2+}$ signals mediated by P2X-type purinoceptors in cultured cerebellar Purkinje cells. *J. Neurosci.* **18**, 1704–1712.

Matsuka, Y., Neubert, J. K., Maidment, N. T., and Spigelman, I. (2001). Concurrent release of ATP and substance P within guinea pig trigeminal ganglia in vivo. *Brain Res.* **915**, 248–255.

Mayfield, R. D., Suzuki, F., and Zahniser, N. R. (1993). Adenosine $A_{2a}$ receptor modulation of electrically evoked endogenous GABA release from slices of rat globus pallidus. *J.Neurochem.* **60**, 2334–2337.

McQueen, D. S., Bond, S. M., Moores, C., Chessell, I., Humphrey, P. P. A., and Dowd, E. (1998). Activation of P2X receptors for adenosine triphosphate evokes cardiorespiratory reflexes in anaesthetized rats. *J. Physiol.* **507**, 843–855.

Mendoza-Fernandez, V., Andrew, D., and Barajas-López, C. (2000). ATP inhibits glutamate synaptic release by acting at P2Y receptors in pyramidal neurons of hippocampal slices. *J.Pharmacol. Exp. Ther.* **293**, 172–179.

Merlo, D., and Volonté, C. (1996). Binding and functions of extracellular ATP in cultured cerebellar granule neurons. *Biochem. Biophys. Res. Commun.* **225**, 907–914.

Mironov, S. L. (1994). Metabotropic ATP receptor in hippocampal and thalamic neurones: Pharmacology and modulation of $Ca^{2+}$ mobilizing mechanisms. *Neuropharmacology* **33**, 1–13.

Molliver, D. C., Cook, S. P., Carlsten, J. A., Wright, D. E., and McCleskey, E. W. (2002). ATP and UTP excite sensory neurons and induce CREB phosphorylation through the metabotropic receptor, P2Y2. *European J. Neurosci.* **16**, 1850–1860.

Moore, D., Chambers, J., Waldvogel, H., Faull, R., and Emson, P. (2000a). Regional and cellular distribution of the $P2Y_1$ purinergic receptor in the human brain: Striking neuronal localisation. *J. Comp. Neurol.* **421**, 374–384.

Moore, D., Iritani, S., Chambers, J., and Emson, P. (2000b). Immunohistochemical localization of the $P2Y_1$ purinergic receptor in Alzheimer's disease. *Neuroreport* **11**, 3799–3803.

Morán-Jiménez, M., and Matute, C. (2000). Immunohistochemical localization of the $P2Y_1$ purinergic receptor in neurons and glial cells of the central nervous system. *Brain Res. Mol. Brain Res.* **78**, 50–58.

Mori, M., Tsushima, H., and Matsuda, T. (1992). Antidiuretic effects of purinoceptor agonists injected into the hypothalamic paraventricular nucleus of water-loaded, ethanol-anesthetized rats. *Neuropharmacology* **31**, 585–592.

Mori, M., Tsushima, H., and Matsuda, T. (1994). Antidiuretic effects of ATP induced by microinjection into the hypothalamic supraoptic nucleus in water-loaded and ethanol-anesthetized rats. *Jpn. J. Pharmacol.* **66**, 445–450.

Mori, M., Heuss, C., Gahwiler, B. H., and Gerber, U. (2001). Fast synaptic transmission mediated by P2X receptors in CA3 pyramidal cells of rat hippocampal slice cultures. *J. Physiol.* **535**, 115–123.

Mosqueda-Garcia, R., Tseng, C. J., Appalsamy, M., Beck, C., and Robertson, D. (1991). Cardiovascular excitatory effects of adenosine in the nucleus of the solitary tract. *Hypertension* **18**, 494–502.

Motin, L., and Bennett, M. R. (1995). Effect of $P_2$-purinoceptor antagonists on glutamatergic transmission in the rat hippocampus. *Br. J. Pharmacol.* **115**, 1276–1280.

Müller, C. E. (1997). $A_1$-adenosine receptor antagonists. *Exp. Opin. Ther. Patents* **7**, 419–440.

Müller, J., Rocha, J. B. T., Battastini, A. M., Sarkis, J. J. F., and Dias, R. D. (1990). Ontogeny of ATP and ADP hydrolysis by cerebral cortex synaptosomes from rats. *Braz. J. Med. Biol. Res.* **23**, 935–939.

Nabekura, J., Ueno, S., Ogawa, T., and Akaike, N. (1995). Colocalization of ATP and nicotinic ACh receptors in the identified vagal preganglionic neurone of rat. *J. Physiol.* **489**, (pt. 2), 519–527.

Nagy, J. I., Buss, M., and Daddona, P. E. (1986). On the innervation of trigeminal mesencephalic primary afferent neurons by adenosine deaminase-containing projections from the hypothalamus in the rat. *Neuroscience* **17**, 141–156.

Nakatsuka, T., and Gu, J. G. (2001). ATP P2X receptor-mediated enhancement of glutamate release and evoked EPSCs in dorsal horn neurons of the rat spinal cord. *J. Neurosci.* **21**, 6522–6531.

Nakazawa, K., Inoue, K., and Inoue, K. (1994). ATP reduces voltage-activated $K^+$ current in cultured rat hippocampal neurons. *Pflugers Arch.* **429**, 143–145.

Nassauw, L. V., Brouns, I., Adriaensen, D., Burnstock, G., and Timmermans, J.-P. (2002). Neurochemical identification of enteric neurons expressing $P2X_3$ receptors in the guinea-pig ileum. *Histochem. Cell Biol.* **118**, 193–203.

Naylor, G. J., Worrall, E. P., Peet, M., and Dick, P. (1976). Whole blood adenosine triphosphate in manic-depressive illness. *Br. J. Psychiatry* **129**, 233–235.

Neary, J. T., Rathbone, M. P., Cattabeni, F., Abbracchio, M. P., and Burnstock, G. (1996). Trophic actions of extracellular nucleotides and nucleosides on glial and neuronal cells. *Trends Neurosci.* **19**, 13–18.

Nieber, K., Poelchen, W., and Illes, P. (1997). Role of ATP in fast excitatory synaptic potentials in locus coeruleus neurones of the rat. *Br. J. Pharmacol.* **122**, 423–430.

Nikbakht, M. R., and Stone, T. W. (2000). Suramin-sensitive suppression of paired-pulse inhibition by adenine nucleotides in rat hippocampal slices. *Neurosci. Lett.* **278**, 45–48.

Nishimura, T., and Tokimasa, T. (1996). Purinergic cation channels in neurons of rabbit vesical parasympathetic ganglia. *Neurosci. Lett.* **212**, 215–217.

Nishizaki, T., and Mori, M. (1998). Diverse signal transduction pathways mediated by endogenous $P_2$ receptors in cultured rat cerebral cortical neurons. *J. Neurophysiol.* **79**, 2513–2521.

Noguchi, J., and Yamashita, H. (2000). Adenosine inhibits voltage-dependent $Ca^{2+}$ currents in rat dissociated supraoptic neurones via $A_1$ receptors. *J. Physiol.* **526**, 313–326.

Novakovic, S. D., Kassotakis, L. C., Oglesby, I. B., Smith, J. A. M., Eglen, R. M., Ford, A. P. D. W., and Hunter, J. C. (1999). Immunocytochemical localization of $P2X_3$ purinoceptors in sensory neurons in naive rats and following neuropathic injury. *Pain* **80**, 273–282.

Nörenberg, W., and Illes, P. (2000). Neuronal P2X receptors: Localisation and functional properties. *Naunyn Schmiedebergs Arch. Pharmacol.* **362**, 324–339.

O'Connor, S. E., Dainty, I. A., and Leff, P. (1991). Further subclassification of ATP receptors based on agonist studies. *Trends Pharmacol. Sci.* **12**, 137–141.

Okada, M., Kawata, Y., Murakami, T., Wada, K., Mizuno, K., and Kaneko, S. (1999). Interaction between purinoceptor subtypes on hippocampal serotonergic transmission using in vivo microdialysis. *Neuropharmacology* **38**, 707–715.

Okada, Y., and Kuroda, Y. (1975). Inhibitory action of adenosine and adenine nucleotides on the postsynaptic potential of olfactory cortex slices of the guinea pig. *Proc. Jpn. Acad.* **51**, 491–494.

Oliet, S. H. R., and Poulain, D. A. (1999). Adenosine-induced presynaptic inhibition of IPSCs and EPSCs in rat hypothalamic supraoptic nucleus neurones. *J. Physiol.* **520**, 815–825.

Panchenko, V. A., Pintor, J., Tsyndrenko, A. Y., Miras-Portugal, M. T., and Krishtal, O. A. (1996). Diadenosine polyphosphates selectively potentiate N-type $Ca^{2+}$ channels in rat central neurons. *Neuroscience* **70**, 353–360.

Pankratov, Y., Castro, E., Miras-Portugal, M. T., and Krishtal, O. (1998). A purinergic component of the excitatory postsynaptic current mediated by P2X receptors in the CA1 neurons of the rat hippocampus. *Eur. J. Neurosci.* **10**, 3898–3902.

Pankratov, Y., Lalo, U., Castro, E., Miras-Portugal, M. T., and Krishtal, O. (1999). ATP receptor-mediated component of the excitatory synaptic transmission in the hippocampus. *Prog. Brain Res.* **120**, 237–249.

Pankratov, Y., Lalo, U., Krishtal, O., and Verkhratsky, A. (2002a). Ionotropic P2X purinoreceptors mediate synaptic transmission in rat pyramidal neurones of layer II/III of somato-sensory cortex. *J. Physiol.* **542**, 529–536.

Pankratov, Y. V., Lalo, U. V., and Krishtal, O. A. (2002b). Role for P2X receptors in long-term potentiation. *J. Neurosci.* **22**, 8363–8369.

Pannicke, T., Fischer, W., Biedermann, B., Schädlich, H., Grosche, J., Faude, F., Wiedemann, P., Allgaier, C., Illes, P., Burnstock, G., and Reichenbach, A. (2000). $P2X_7$ receptors in Müller glial cells from the human retina. *J. Neurosci.* **20**, 5965–5972.

Pape, H.-C. (1992). Adenosine promotes burst activity in guinea-pig geniculocortical neurones through two different ionic mechanisms. *J. Physiol.* **447**, 729–753.

Paton, J. F., De Paula, P. M., Spyer, K. M., Machado, B. H., and Boscan, P. (2002). Sensory afferent selective role of P2 receptors in the nucleus tractus solitarii for mediating the cardiac component of the peripheral chemoreceptor reflex in rats. *J. Physiol.* **543**, 995–1005.

Patel, M. K., Khakh, B. S., and Henderson, G. (2001). Properties of native P2X receptors in rat trigeminal mesencephalic nucleus neurones: Lack of correlation with known, heterologously expressed P2X receptors. *Neuropharmacology* **40**, 96–105.

Pelleg, A., and Hurt, C. M. (1996). Mechanism of action of ATP on canine pulmonary vagal C fibre nerve terminals. *J. Physiol.* **490**, 265–275.

Petruska, J. C., Cooper, B. Y., Gu, J. G., Rau, K. K., and Johnson, R. D. (2000a). Distribution of $P2X_1$, $P2X_2$, and $P2X_3$ receptor subunits in rat primary afferents: Relation to population markers and specific cell types. *J. Chem. Neuroanat.* **20**, 141–162.

Petruska, J. C., Cooper, B. Y., Johnson, R. D., and Gu, J. G. (2000b). Distribution patterns of different P2x receptor phenotypes in acutely dissociated dorsal root ganglion neurons of adult rats. *Exp. Brain Res.* **134**, 126–132.

Phillis, J. W. (1984). Potentiation of the action of adenosine on cerebral cortical neurones by the tricyclic antidepressants. *Br. J. Pharmacol.* **83**, 567–575.

Phillis, J. W., and Wu, P. H. (1981). The role of adenosine and its nucleotides in central synaptic transmission. *Prog. Neurobiol.* **16**, 187–239.

Phillis, J. W., Kostopoulos, G. K., and Limacher, J. J. (1974). Depression of corticospinal cells by various purines and pyrimidines. *Can. J. Physiol. Pharmacol.* **52**, 1226–1229.

Phillis, J. W., Edstrom, J. P., Kostopoulos, G. K., and Kirkpatrick, J. R. (1979a). Effects of adenosine and adenine nucleotides on synaptic transmission in the cerebral cortex. *Can. J.Physiol. Pharmacol.* **57**, 1289–1312.

Phillis, J. W., Jiang, Z. G., Chelack, B. J., and Wu, P. H. (1979b). Morphine enhances adenosine release from the in vivo rat cerebral cortex. *Eur. J. Pharmacol.* **65**, 97–100.

Phillis, J. W., Jiang, Z. G., Chelack, B. J., and Wu, P. H. (1980). The effect of morphine on purine and acetylcholine release from rat cerebral cortex: Evidence for a purinergic component in morphine's action. *Pharmacol. Biochem. Behav.* **13**, 421–427.

Phillis, J. W., Barraco, R. A., DeLong, R. E., and Washington, D. O. (1986). Behavioral characteristics of centrally administered adenosine analogs. *Pharmacol. Biochem. Behav.* **24**, 263–270.

Phillis, J. W., Scislo, T. J., and O'Leary, D. (1997). Purines and the nucleus of the solitary tract: Effects on cardiovascular and respiratory function. *Proc. Aust. Physiol. Pharmacol. Soc.* **28**, 14–21.

Pintor, J. (1999). Purinergic signalling in the eye. *In* "Nervous Control of the Eye" (G. Burnstock and A. M. Sillito, eds.), pp. 171–210. Harwood Academic Publishers, Amsterdam.

Pintor, J., Díaz-Hernández, M., Bustamante, C., Gualix, J., de Terreros, F. J., and Miras-Portugal, M. T. (1999). Presence of dinucleotide and ATP receptors in human cerebrocortical synaptic terminals. *Eur. J. Pharmacol.* **366**, 159–165.

Pintor, J., Díaz-Hernández, M., Gualix, J., Gómez-Villafuertes, R., Hernando, F., and Miras-Portugal, M. T. (2000). Diadenosine polyphosphate receptors, from rat and guinea-pig brain to human nervous system. *Pharmacol. Ther.* **87**, 103–115.

Poelchen, W., Sieler, D., Inoue, K., and Illes, P. (1998). Effect of extracellular adenosine 5'-triphosphate on principal neurons of the rat ventral tegmental area. *Brain Res.* **800**, 170–173.

Poelchen, W., Sieler, D., Wirkner, K., and Illes, P. (2001). Co-transmitter function of ATP in central catecholaminergic neurons of the rat. *Neuroscience* **102**, 593–602.

Poole, D. P., Castelucci, P., Robbins, H. L., Chiocchetti, R., and Furness, J. B. (2002). The distribution of P2X3 purine receptor subunits in the guinea pig enteric nervous system. *Auton. Neurosci.* **101**, 39–47.

Potter, P., and White, T. D. (1980). Release of adenosine 5'-triphosphate from synaptosomes from different regions of rat brain. *Neuroscience* **5**, 1351–1356.

Price, G. D., Robertson, S. J., and Edwards, F. A. (2003). Long-term potentiation of glutamatergic synaptic transmission induced by activation of presynaptic P2Y receptors in the rat medial habenula nucleus. *Eur. J. Neurosci.* **17**, 844–850.

Prothero, L. S., Mathie, A., and Richards, C. D. (2000). Purinergic and muscarinic receptor activation activates a common calcium entry pathway in rat neocortical neurons and glial cells. *Neuropharmacology* **39**, 1768–1778.

Pull, I., and McIlwain, H. (1972a). Adenine derivatives as neurohumoral agents in the brain: The quantities liberated on excitation of superfused cerebral tissues. *Biochem. J.* **130**, 975–981.

Pull, I., and McIlwain, H. (1972b). Metabolism of (14 C) adenine and derivatives by cerebral tissues, superfused and electrically stimulated. *Biochem. J.* **126**, 965–973.

Purinton, P. T., Fletcher, T. F., and Bradley, W. E. (1973). Gross and light microscopic features of the pelvic plexus in the rat. *Anat. Rec.* **175**, 697–705.

Rae, M. G., Rowan, E. G., and Kennedy, C. (1998). Pharmacological properties of P2X3-receptors present in neurones of the rat dorsal root ganglia. *Br. J. Pharmacol.* **124**, 176–180.

Ralevic, V., and Burnstock, G. (1998). Receptors for purines and pyrimidines. *Pharmacol. Rev.* **50,** 413–492.

Ralevic, V., Thomas, T., Burnstock, G., and Spyer, K. M. (1996). P2-Purinoceptor-mediated changes in activity of neurones recorded extracellularly from the rostral ventrolateral medulla of the rat. *Drug Dev. Res.* **37,** 157.

Ramme, D., Regenold, J. T., Starke, K., Busse, R., and Illes, P. (1987). Identification of the neuroeffector transmitter in jejunal branches of the rabbit mesenteric artery. *Naunyn Schmiedebergs Arch. Pharmacol.* **336,** 267–273.

Reddington, M., Lee, K. S., Schubert, P., and Kreutzberg, G. W. (1983). Biochemical and physiological characterization of adenosine receptors in rat brain. *Adv. Biochem. Psychopharmacol.* **37,** 465–476.

Reekie, F. M., and Burnstock, G. (1994). Some effects of purines on neurones of guinea pig superior cervical ganglia. *Gen. Pharmacol.* **25,** 143–148.

Reese, J. H., and Cooper, J. R. (1982). Modulation of the release of acetylcholine from ileal synaptosomes by adenosine and adenosine 5'-triphosphate. *J. Pharmacol. Exp. Ther.* **223,** 612–616.

Regenold, J. T., and Illes, P. (1990). Inhibitory adenosine $A_1$-receptors on rat locus coeruleus neurones. An intracellular electrophysiological study. *Naunyn Schmiedebergs Arch. Pharmacol.* **341,** 225–231.

Rhee, J. S., Wang, Z. M., Nabekura, J., Inoue, K., and Akaike, N. (2000). ATP facilitates spontaneous glycinergic IPSC frequency at dissociated rat dorsal horn interneuron synapses. *J. Physiol.* **524,** 471–483.

Ribeiro, J. A. (1995). Purinergic inhibition of neurotransmitter release in the central nervous system. *Pharmacol. Toxicol.* **77,** 299–305.

Richards, C. D., Snell, C. R., and Snell, P. H. (1983). Nicotinamide adenine dinucleotide depresses synaptic transmission in the hippocampus and has specific binding sites on the synaptic membranes. *Br. J. Pharmacol.* **79,** 553–564.

Richardson, P. J., and Brown, S. J. (1987). ATP release from affinity-purified rat cholinergic nerve terminals. *J. Neurochem.* **48,** 622–630.

Richardson, P. J., Kase, H., and Jenner, P. G. (1997). Adenosine A2A receptor antagonists as new agents for the treatment of Parkinson's disease. *Trends Pharmacol. Sci.* **18,** 338–344.

Robertson, S. J. (1998). Tissue distribution and functional contribution of P2X receptors in the CNS. *Drug Dev. Res.* **45,** 336–341.

Robertson, S. J., and Edwards, F. A. (1998). ATP and glutamate are released from separate neurones in the rat medial habenula nucleus: Frequency dependence and adenosine-mediated inhibition of release. *J. Physiol.* **508,** (pt. 3), 691–701.

Robertson, S. J., Rae, M. G., Rowan, E. G., and Kennedy, C. (1996). Characterization of a P2X-purinoceptor in cultured neurones of the rat dorsal root ganglia. *Br. J. Pharmacol.* **118,** 951–956.

Robertson, S. J., Burnashev, N., and Edwards, F. A. (1999). Ca2+ permeability and kinetics of glutamate receptors in rat medial habenula neurones: Implications for purinergic transmission in this nucleus. *J. Physiol.* **518,** (pt. 2), 539–549.

Rocha, I., Burnstock, G., and Spyer, K. M. (2001). Effect on urinary bladder function and arterial blood pressure of the activation of putative purine receptors in brainstem areas. *Auton. Neurosci.* **88,** 6–15.

Rogers, M., Colquhoun, L. M., Patrick, J. W., and Dani, J. A. (1997). Calcium flux through predominantly independent purinergic ATP and nicotinic acetylcholine receptors. *J.Neurophysiol.* **77,** 1407–1417.

Rong, W., Burnstock, G., and Spyer, K. M. (2000). P2X purinoceptor-mediated excitation of trigeminal lingual nerve terminals in an *in vitro* intra-arterially perfused rat tongue preparation. *J. Physiol.* **524**, 891–902.

Rong, W., Spyer, M., and Burnstock, G. (2002). Activation and sensitisation of low and high threshold afferent fibres mediated by P2X receptors in the mouse urinary bladder. *J. Physiol.* **541**, 591–600.

Ross, F. M., Brodie, M. J., and Stone, T. W. (1998). Modulation by adenine nucleotides of epileptiform activity in the CA3 region of rat hippocampal slices. *Br. J. Pharmacol.* **123**, 71–80.

Rubino, A., and Burnstock, G. (1996). Capsaicin-sensitive sensory-motor neurotransmission in the peripheral control of cardiovascular function. *Cardiovasc. Res.* **31**, 467–479.

Rubio, M. E., and Soto, F. (2001). Distinct localization of P2X receptors at excitatory postsynaptic specializations. *J. Neurosci.* **21**, 641–653.

Rudolphi, K. A., Schubert, P., Parkinson, F. E., and Fredholm, B. B. (1992). Neuroprotective role of adenosine in cerebral ischaemia. *Trends Pharmacol. Sci.* **13**, 439–445.

Rybakowski, J., Potok, E., and Strzyzewski, W. (1981). Erythrocyte membrane adenosine triphosphatase activities in patients with endogenous depression and healthy subjects. *Eur. J. Clin. Invest.* **11**, 61–64.

Ryten, M., Hoebertz, A., and Burnstock, G. (2001). Sequential expression of three receptor subtypes for extracellular ATP in developing rat skeletal muscle. *Dev. Dyn.* **221**, 331–341.

Ryu, J. K., Kim, J., Choi, S. H., Oh, Y. J., Lee, Y. B., Kim, S. U., and Jin, B. K. (2002). ATP-induced in vivo neurotoxicity in the rat striatum via P2 receptors. *Neuroreport* **13**, 1611–1615.

Saffrey, M. J., Hassall, C. J. S., Allen, T. G. J., and Burnstock, G. (1992). Ganglia within the gut, heart, urinary bladder and airways: Studies in tissue culture. *Int. Rev. Cytol.* **136**, 93–144.

Salgado, A. H., Gomez, M. V., Romano-Silva, M. A., and Prado, M. A. (1996). Effect of vesamicol on the release of ATP from cortical synaptosomes. *Neurosci. Lett.* **204**, 37–40.

Salt, T. E., and Hill, R. G. (1983). Excitation of single sensory neurones in the rat caudal trigeminal nucleus by iontophoretically applied adenosine 5′-triphosphate. *Neurosci. Lett.* **35**, 53–57.

Salter, M. W., and Henry, J. L. (1985). Effects of adenosine 5′-monophosphate and adenosine 5′-triphosphate on functionally identified units in the cat spinal dorsal horn. Evidence for a differential effect of adenosine 5′-triphosphate on nociceptive vs non-nociceptive units. *Neuroscience* **15**, 815–825.

Salter, M. W., and Hicks, J. L. (1994). ATP-evoked increases in intracellular calcium in neurons and glia from the dorsal spinal cord. *J. Neurosci.* **14**, 1563–1575.

Salter, M. W., and Hicks, J. L. (1995). ATP causes release of intracellular $Ca^{2+}$ via the phospholipase $C\beta/IP_3$ pathway in astrocytes from the dorsal spinal cord. *J. Neurosci.* **15**, 2961–2971.

Salter, M. W., and Sollevi, A. (2001). Roles of purines in nociception and pain. *In* "Handbook of Experimental Pharmacology. Purinergic and Pyrimidinergic Signalling I—Molecular, Nervous and Urinogenitary System Function" (M. P. Abbracchio and M. Williams, eds.), Vol. 151/I, pp. 371–401. Springer-Verlag, Berlin.

Salter, M. W., De Koninck, Y., and Henry, J. L. (1993). Physiological roles for adenosine and ATP in synaptic transmission in the spinal dorsal horn. *Prog. Neurobiol.* **41**, 125–156.

Sansum, A. J., Chessell, I. P., Hicks, G. A., Trezise, D. J., and Humphrey, P. P. A. (1998). Evidence that P2X purinoceptors mediate the excitatory effects of $\alpha\beta$methylene-ADP in rat locus coeruleus neurones. *Neuropharmacology* **37**, 875–885.

Santicioli, P., Del Bianco, E., and Maggi, C. A. (1993). Adenosine $A_1$ receptors mediate the presynaptic inhibition of calcitonin gene-related peptide release by adenosine in the rat spinal cord. *Eur. J. Pharmacol.* **231**, 139–142.

Sanz, J. M., Vendite, D., Fernandez, M., Andres, A., and Ros, M. (1996). Adenosine $A_1$ receptors in cultured cerebellar granule cells: Role of endogenous adenosine. *J. Neurochem.* **67**, 1469–1477.

Sarosi, G. A., Barnhart, D. C., Turner, D. J., and Mulholland, M. W. (1998). Capacitative $Ca^{2+}$ entry in enteric glia induced by thapsigargin and extracellular ATP. *Am. J. Physiol.* **275**, G550–G555.

Sattin, A., and Rall, T. W. (1970). The effect of adenosine and adenine nucleotides on the cyclic adenosine $3',5'$-phosphate content of guinea-pig cerebral cortex slices. *Mol. Pharmacol.* **6**, 13–23.

Sawynok, J. (1998). Adenosine receptor activation and nociception. *Eur. J. Pharmacol.* **317**, 1–11.

Sawynok, J., and Sweeney, M. I. (1989). The role of purines in nociception. *Neuroscience* **32**, 557–569.

Sawynok, J., Downie, J. W., Reid, A. R., Cahill, C. M., and White, T. D. (1993). ATP release from dorsal spinal cord synaptosomes: Characterization and neuronal origin. *Brain Res.* **610**, 32–38.

Schafer, R., and Reiser, G. (1999). ATPαS is a ligand for P2Y receptors in synaptosomal membranes: Solubilization of [$^{35}$S]ATPαS binding proteins associated with G-proteins. *Neurochem. Int.* **34**, 303–317.

Scholfield, C. N. (1978). Depression of evoked potentials in brain slices by adenosine compounds. *Br. J. Pharmacol.* **63**, 239–244.

Schrader, J., Wahl, M., Kuschinsky, W., and Kreutzberg, G. W. (1980). Increase of adenosine content in cerebral cortex of the cat during bicuculline-induced seizure. *Pflugers Arch.* **387**, 245–251.

Schubert, P., and Kreutzberg, G. W. (1974). Axonal transport of adenosine and uridine derivatives and transfer to postsynaptic neurons. *Brain Res.* **76**, 526–530.

Schubert, P., Lee, K., West, M., Deadwyler, S., and Lynch, G. (1976). Stimulation-dependent release of 3H-adenosine derivatives from central axon terminals to target neurones. *Nature* **260**, 541–542.

Scislo, T. J., and O'Leary, D. S. (1998). Differential control of renal vs. adrenal sympathetic nerve activity by NTS $A_{2a}$ and $P_{2x}$ purinoceptors. *Am. J. Physiol.* **275**, H2130–H2139.

Scislo, T. J., and O'Leary, D. S. (2000). Differential role of ionotropic glutamatergic mechanisms in responses to NTS $P_{2x}$ and $A_{2a}$ receptor stimulation. *Am. J. Physiol. Heart Circ. Physiol.* **278**, H2057–H2068.

Scislo, T. J., Augustyniak, R. A., Barraco, R. A., Woodbury, D. J., and O'Leary, D. S. (1997). Activation of $P_{2x}$-purinoceptors in the nucleus tractus solitarius elicits differential inhibition of lumbar and renal sympathetic nerve activity. *J. Auton. Nerv. Syst.* **62**, 103–110.

Scislo, T. J., Kitchen, A. M., Augustyniak, R. A., and O'Leary, D. S. (2001). Differential patterns of sympathetic responses to selective stimulation of nucleus tractus solitarius purinergic receptor subtypes. *Clin. Exp. Pharmacol. Physiol.* **28**, 120–124.

Sebastiao, A. M., and Ribeiro, J. A. (1996). Adenosine A2 receptor-mediated excitatory actions on the nervous system. *Prog. Neurobiol.* **48**, 167–189.

Shefner, S. A., and Chiu, T. H. (1986). Adenosine inhibits locus coeruleus neurons: An intracellular study in a rat brain slice preparation. *Brain Res.* **366**, 364–368.

Shen, K. Z., and North, R. A. (1993). Excitation of rat locus coeruleus neurons by adenosine $5'$-triphosphate: Ionic mechanism and receptor characterization. *J. Neurosci.* **13**, 894–899.

Shibuya, I., Tanaka, K., Hattori, Y., Uezono, Y., Harayama, N., Noguchi, J., Ueta, Y., Izumi, F., and Yamashita, H. (1999). Evidence that multiple P2X purinoceptors are functionally expressed in rat supraoptic neurones. *J. Physiol.* **514**, 351–367.

Siggins, G. R., Gruol, D. L., Padjen, A. L., and Formans, D. S. (1977). Purine and pyrimidine mononucleotides depolarise neurones of explanted amphibian sympathetic ganglia. *Nature* **270**, 263–265.

Silinsky, E. M. (1984). On the mechanism by which adenosine receptor activation inhibits the release of acetylcholine from motor nerve endings. *J. Physiol.* **346**, 243–256.

Silinsky, E. M., and Gerzanich, V. (1993). On the excitatory effects of ATP and its role as a neurotransmitter in coeliac neurons of the guinea-pig. *J. Physiol.* **464**, 197–212.

Simon, J., Kidd, E. J., Smith, F. M., Chessell, I. P., Murrell-Lagnado, R., Humphrey, P. P. A., and Barnard, E. A. (1997a). Localization and functional expression of splice variants of the P2X$_2$ receptor. *Mol. Pharmacol.* **52**, 237–248.

Simon, J., Webb, T. E., and Barnard, E. A. (1997b). Distribution of [$^{35}$S]dATP$\alpha$S binding sites in the adult rat neuraxis. *Neuropharmacology* **36**, 1243–1251.

Simpson, R. E., O'Regan, M. H., Perkins, L. M., and Phillis, J. W. (1992). Excitatory transmitter amino acid release from the ischemic rat cerebral cortex: Effects of adenosine receptor agonists and antagonists. *J. Neurochem.* **58**, 1683–1690.

Skladchikova, G., Rnn, L. C., Berezin, V., and Bock, E. (1999). Extracellular adenosine triphosphate affects neural cell adhesion molecule (NCAM)-mediated cell adhesion and neurite outgrowth. *J. Neurosci. Res.* **57**, 207–218.

Sneddon, P., and Burnstock, G. (1984). Inhibition of excitatory junction potentials in guinea-pig vas deferens by $\alpha,\beta$-methylene-ATP: Further evidence for ATP and noradrenaline as cotransmitters. *Eur. J. Pharmacol.* **100**, 85–90.

Sneddon, P., and Westfall, D. P. (1984). Pharmacological evidence that adenosine triphosphate and noradrenaline are co-transmitters in the guinea-pig vas deferens. *J. Physiol.* **347**, 561–580.

Snyder, S. H. (1985). Adenosine as a neuromodulator. *Annu. Rev. Neurosci.* **8**, 103–124.

Sokolova, E., Nistri, A., and Giniatullin, R. (2001). Negative cross talk between anionic GABAA and cationic P2X ionotropic receptors of rat dorsal root ganglion neurons. *J. Neurosci.* **21**, 4958–4968.

Sorimachi, M., Moritoyo, T., and Yamagami, K. (2000). Activation of ATP receptor increases the cytosolic Ca$^{2+}$ concentration in nucleus accumbens neurons of rat brain. *Brain Res.* **882**, 217–220.

Sorimachi, M., Ishibashi, H., Moritoyo, T., and Akaike, N. (2001). Excitatory effect of ATP on acutely dissociated ventromedial hypothalamic neurons of the rat. *Neuroscience* **105**, 393–401.

Sorimachi, M., Yamagami, K., and Wakomori, M. (2002). Activation of ATP receptor increases the cytosolic Ca(2+) concentration in ventral tegmental area neurons of rat brain. *Brain Res.* **935**, 129–133.

Soto, F., and Rubio, M. E. (2001). Cloned P2X receptor sub-units in cerebellum and hippocampus. *Drug Dev. Res.* **52**, 133–139.

Souslova, V., Cesare, P., Ding, Y., Akopian, A. N., Stanfa, L., Suzuki, R., Carpenter, K., Dickenson, A., Boyce, S., Hill, R., Nebenius-Oosthuizen, D., Smith, A. J. H., Kidd, E., and Wood, J. N. (2000). Warm-coding deficits and aberrant inflammatory pain in mice lacking P2X$_3$ receptors. *Nature* **407**, 1015–1017.

Spencer, N. J., Walsh, M., and Smith, T. K. (2000). Purinergic and cholinergic neuro-neuronal transmission underlying reflexes activated by mucosal stimulation in the isolated guinea-pig ileum. *J. Physiol.* **522**, 321–331.

364                                                                    Geoffrey Burnstock

Sperlágh, B., Kittel, A., Lajtha, A., and Vizi, E. S. (1995). ATP acts as fast neurotransmitter in rat habenula: Neurochemical and enzymecytochemical evidence. *Neuroscience* **66,** 915–920.

Sperlágh, B., Magloczky, Z., Vizi, E. S., and Freund, T. F. (1998a). The triangular septal nucleus as the major source of ATP release in the rat habenula: A combined neurochemical and morphological study. *Neuroscience* **86,** 1195–1207.

Sperlágh, B., Sershen, H., Lajtha, A., and Vizi, E. S. (1998b). Co-release of endogenous ATP and [³H]noradrenaline from rat hypothalamic slices: Origin and modulation by $\alpha_2$-adrenoceptors. *Neuroscience* **82,** 511–520.

Sperlágh, B., Erdelyi, F., Szabo, G., and Vizi, E. S. (2000). Local regulation of [³H]-noradrenaline release from the isolated guinea-pig right atrium by $P_{2x}$-receptors located on axon terminals. *Br. J. Pharmacol.* **131,** 1775–1783.

Sperlágh, B., Kofalvi, A., Deuchars, J., Atkinson, L., Milligan, C. J., Buckley, N. J., and Vizi, E. S. (2002). Involvement of P2X7 receptors in the regulation of neurotransmitter release in the rat hippocampus. *J. Neurochem.* **81,** 1196–1211.

Stanfa, L. C., Kontinen, V. K., and Dickenson, A. H. (2000). Effects of spinally administered P2X receptor agonists and antagonists on the responses of dorsal horn neurones recorded in normal, carrageenan-inflamed and neuropathic rats. *Br. J. Pharmacol.* **129,** 351–359.

Stone, T. W. (1981). The effects of morphine and methionine-enkephalin on the release of purines from cerebral cortex slices of rats and mice. *Br. J. Pharmacol.* **74,** 171–176.

Stone, T. W., and Taylor, D. A. (1978). Antagonist by clonidine of neuronal depressant responses to adenosine, adenosine-5′-monophosphate and adenosine triphosphate. *Br. J. Pharmacol.* **64,** 369–374.

Sun, M. K., Wahlestedt, C., and Reis, D. J. (1992). Action of externally applied ATP on rat reticulospinal vasomotor neurons. *Eur. J. Pharmacol.* **224,** 93–96.

Suran, A. A. (1974). 5′-Nucleotidase and an acid phosphatase of spinal cord. Comparative histochemistry and specificity of the enzymes in mouse and cat spinal cords. Cytologic localization in mouse substantia gelatinosa. *J. Histochem. Cytochem.* **22,** 802–811.

Surprenant, A. (1994). Control of the gastrointestinal tract by enteric neurons. *Annu. Rev. Physiol.* **56,** 117–140.

Svenningsson, P., Le Moine, C., Kull, B., Sunahara, R., Bloch, B., and Fredholm, B. B. (1997). Cellular expression of adenosine $A_{2A}$ receptor messenger RNA in the rat central nervous system with special reference to dopamine innervated areas. *Neuroscience* **80,** 1171–1185.

Svenningsson, P., Le Moine, C., Fisone, G., and Fredholm, B. B. (1999). Distribution, biochemistry and function of striatal adenosine $A_{2A}$ receptors. *Prog. Neurobiol.* **59,** 355–396.

Sweeney, M. I., White, T. D., and Sawynok, J. (1989). Morphine, capsaicin and $K^+$ release purines from capsaicin-sensitive primary afferent nerve terminals in the spinal cord. *J.Pharmacol. Exp. Ther.* **248,** 447–454.

Tanaka, J., Murate, M., Wang, C. Z., Seino, S., and Iwanaga, T. (1996). Cellular distribution of the $P2X_4$ ATP receptor mRNA in the brain and non-neuronal organs of rats. *Arch. Histol. Cytol.* **59,** 485–490.

Tao, S., and Abdel-Rahman, A. A. (1993). Neuronal and cardiovascular responses to adenosine microinjection into the nucleus tractus solitarius. *Brain Res. Bull.* **32,** 407–417.

Taylor, D. A., and Stone, T. W. (1980). The action of adenosine on noradrenergic neuronal inhibition induced by stimulation of locus coeruleus. *Brain Res.* **183,** 367–376.

Terrian, D. M., Hernandez, P. G., Rea, M. A., and Peters, R. I. (1989). ATP release, adenosine formation, and modulation of dynorphin and glutamic acid release by adenosine analogues in rat hippocampal mossy fiber synaptosomes. *J. Neurochem.* **53,** 1390–1399.

Theobald, R. J. Jr., and de Groat, W. C. (1977). The effects of ATP and related adenine derivatives on the superior cervical ganglion (SCG) of the rat. *Fed. Proc.* **36,** 290.

Theobald, R. J. Jr., and de Groat, W. C. (1989). The effects of purine nucleotides on transmission in vesical parasympathetic ganglia of the cat. *J. Auton. Pharmacol.* **9,** 167–182.

Thomas, S., Virginio, C., North, R. A., and Surprenant, A. (1998). The antagonist trinitrophenyl-ATP reveals co-existence of distinct P2X receptor channels in rat nodose neurones. *J. Physiol.* **509,** 411–417.

Thomas, T., and Spyer, K. M. (1999). A novel influence of adenosine on ongoing activity in rat rostral ventrolateral medulla. *Neuroscience* **88,** 1213–1223.

Thomas, T., and Spyer, K. M. (2000). ATP as a mediator of mammalian central $CO_2$ chemoreception. *J. Physiol.* **523,** 441–447.

Thomas, T., Ralevic, V., Gadd, C. A., and Spyer, K. M. (1999). Central $CO_2$ chemoreception: A mechanism involving P2 purinoceptors localized in the ventrolateral medulla of the anaesthetized rat. *J. Physiol.* **517,** 899–905.

Thomas, T., St Lambert, J. H., Dashwood, M. R., and Spyer, K. M. (2000). Localization and action of adenosine $A_{2a}$ receptors in regions of the brainstem important in cardiovascular control. *Neuroscience* **95,** 513–518.

Torres, I. L., Buffon, A., Dantas, G., Furstenau, C. R., Bohmer, A. E., Battastini, A. M., Sarkis, J. J., Dalmaz, C., and Ferreira, M. B. (2002). Chronic stress effects on adenine nucleotide hydrolysis in the blood serum and brain structures of rats. *Pharmacol. Biochem. Behav.* **74,** 181–186.

Trendelenburg, A. U., and Bultmann, R. (2000). P2 receptor-mediated inhibition of dopamine release in rat neostriatum. *Neuroscience* **96,** 249–252.

Trezise, D. J., Black, M. D., and Grahames, C. B. A. (1996). Novel pharmacological characteristics of $P_2$ purinoceptors mediating increases in firing rate of rat locus coeruleus neurones in vitro. *Br. J. Pharmacol.* **117,** 319P.

Troadec, J. D., and Thirion, S. (2002). Multifaceted purinergic regulation of stimulus-secretion coupling in the neurohypophysis. *Neuroendocrinol. Lett.* **23,** 273–280.

Troadec, J. D., Thirion, S., Nicaise, G., Lemos, J. R., and Dayanithi, G. (1998). ATP-evoked increases in $[Ca^{2+}]_i$ and peptide release from rat isolated neurohypophysical terminals via a $P_{2x2}$ purinoceptor. *J. Physiol.* **511,** 89–103.

Tseng, C. J., Biaggioni, I., Appalsamy, M., and Robertson, D. (1988). Purinergic receptors in the brainstem mediate hypotension and bradycardia. *Hypertension* **11,** 191–197.

Tsuda, M., Ueno, S., and Inoue, K. (1999a). Evidence for the involvement of spinal endogenous ATP and P2X receptors in nociceptive responses caused by formalin and capsaicin in mice. *Br. J. Pharmacol.* **128,** 1497–1504.

Tsuda, M., Ueno, S., and Inoue, K. (1999b). In vivo pathway of thermal hyperalgesia by intrathecal administration of $\alpha,\beta$-methylene ATP in mouse spinal cord: Involvement of the glutamate-NMDA receptor system. *Br. J. Pharmacol.* **127,** 449–456.

Ueno, S., Harata, N., Inoue, K., and Akaike, N. (1992). ATP-gated current in dissociated rat nucleus solitarii neurons. *J. Neurophysiol.* **68,** 778–785.

Ueno, S., Tsuda, M., Iwanaga, T., and Inoue, K. (1999). Cell type-specific ATP-activated responses in rat dorsal root ganglion neurons. *Br. J. Pharmacol.* **126,** 429–436.

Ueno, T., Ueno, S., Akaike, N., and Nabekura, J. (2000). Gene expression of P2X receptor on rat dorsal motor nucleus of vagus neurons and its functional property. *Soc. Neurosci. Abstr.* **39,** 15.

Ueno, S., Yamada, H., Moriyama, T., Honda, K., Takano, Y., Kamiya, H. O., and Katsuragi, T. (2002). Measurement of dorsal root ganglion P2X mRNA by SYBR Green fluorescence. *Brain Res. Brain Res. Protoc.* **10,** 95–101.

Ueno, T., Ueno, S., Kakazu, Y., Akaike, N., and Nabekura, J. (2001). Bidirectional modulation of P2X receptor-mediated response by divalent cations in rat dorsal motor nucleus of the vagus neurons. *J. Neurochem.* **78,** 1009–1018.

Ulrich, D., and Huguenard, J. R. (1995). Purinergic inhibition of GABA and glutamate release in the thalamus: Implications for thalamic network activity. *Neuron* **15**, 909–918.

Valera, S., Hussy, N., Evans, R. J., Adani, N., North, R. A., Surprenant, A., and Buell, G. (1994). A new class of ligand-gated ion channel defined by $P_{2x}$ receptor for extra-cellular ATP. *Nature* **371**, 516–519.

van den Pol, A. N., Finkbeiner, S. M., and Cornell-Bell, A. H. (1992). Calcium excitability and oscillations in suprachiasmatic nucleus neurons and glia *in vitro*. *J. Neurosci.* **12**, 2648–2664.

Virginio, C., Robertson, G., Surprenant, A., and North, R. A. (1998). Trinitrophenyl-substituted nucleotides are potent antagonists selective for $P2X_1$, $P2X_3$, and heteromeric $P2X_{2/3}$ receptors. *Mol. Pharmacol.* **53**, 969–973.

Vitolo, O. V., Ciotti, M. T., Galli, C., Borsello, T., and Calissano, P. (1998). Adenosine and ADP prevent apoptosis in cultured rat cerebellar granule cells. *Brain Res.* **809**, 297–301.

Vizi, E. S., Liang, S. D., Sperlágh, B., Kittel, A., and Juranyi, Z. (1997). Studies on the release and extracellular metabolism of endogenous ATP in rat superior cervical ganglion: Support for neurotransmitter role of ATP. *Neuroscience* **79**, 893–903.

Vlaskovska, M., Kasakov, L., Rong, W., Bodin, P., Bardini, M., Koch, B., Cockayne, D. A., Ford, A. P. D. W., and Burnstock, G. (2000). $P2X_3$ deficient mice reveal a major sensory role for urothelial released ATP. *Soc. Neurosci. Abstr.* **353**, 2.

Vlaskovska, M., Kasakov, L., Rong, W., Bodin, P., Bardini, M., Cockayne, D. A., Ford, A. P. D. W., and Burnstock, G. (2001). $P2X_3$ knockout mice reveal a major sensory role for urothelially released ATP. *J. Neurosci.* **21**, 5670–5677.

Volonté, C., and Merlo, D. (1996). Selected $P_2$ purinoceptor modulators prevent glutamate-evoked cytotoxicity in cultured cerebellar granule neurons. *J. Neurosci. Res.* **45**, 183–193.

Volonté, C., Ciotti, M. T., D'Ambrosi, N., Lockhart, B., and Spedding, M. (1999). Neuroprotective effects of modulators of P2 receptors in primary culture of CNS neurones. *Neuropharmacology* **38**, 1335–1342.

von Kügelgen, I., Spath, L., and Starke, K. (1992). Adenosine but not an adenine nucleotide mediates tonic purinergic inhibition, as well as inhibition by glutamate, of noradrenaline release in rabbit brain cortex slices. *Naunyn Schmiedebergs Arch. Parmacol.* **346**, 677–684.

von Kügelgen, I., Spath, L., and Starke, K. (1994). Evidence for P2-purinoceptor-mediated inhibition of noradrenaline release in rat brain cortex. *Br. J. Pharmacol.* **113**, 815–822.

Vulchanova, L., Arvidsson, U., Riedl, M., Wang, J., Buell, G., Surprenant, A., North, R. A., and Elde, R. (1996). Differential distribution of two ATP-gated channels (P2X receptors) determined by immunocytochemistry. *Proc. Natl. Acad. Sci. USA* **93**, 8063–8067.

Vulchanova, L., Riedl, M., Shuster, S. J., Buell, G., Surprenant, A., North, R. A., and Elde, R. (1997). Immunocytochemical study of the $P2X_2$ and $P2X_3$ receptor subunits in rat and monkey sensory neurons and their central terminals. *Neuropharmacology* **36**, 1229–1242.

Vulchanova, L., Riedl, M. S., Shuster, S. J., Stone, L. S., Hargreaves, K. M., Buell, G., Surprenant, A., North, R. A., and Elde, R. (1998). $P2X_3$ is expressed by DRG neurons that terminate in inner lamina II. *Eur. J. Neurosci.* **10**, 3470–3478.

Wardas, J. (2002). Neuroprotective role of adenosine in the CNS. *Pol. J. Pharmacol.* **54**, 313–326.

Webb, T. E., Simon, J., Krishek, B. J., Bateson, A. N., Smart, T. G., King, B. F., Burnstock, G., and Barnard, E. A. (1993). Cloning and functional expression of a brain G-protein-coupled ATP receptor. *FEBS Lett.* **324**, 219–225.

Wells, D. G., Zawisa, M. J., and Hume, R. I. (1995). Changes in responsiveness to extracellular ATP in chick skeletal muscle during development and upon denervation. *Dev. Biol.* **172**, 585–590.

Whalley, L. J., Scott, M., Reading, H. W., and Christie, J. E. (1980). Effect of electroconvulsive therapy on erythrocyte adenosine triphosphatase activity in depressive illness. *Br. J. Psychiatry* **137**, 343–345.

White, T. D. (1978). Release of ATP from a synaptosomal preparation by elevated extracellular $K^+$ and by veratridine. *J. Neurochem.* **30**, 329–336.

White, T., Potter, P., and Wonnacott, S. (1980). Depolarisation-induced release of ATP from cortical synaptosomes is not associated with acetylcholine release. *J. Neurochem.* **34**, 1109–1112.

Wieraszko, A. (1996). Extracellular ATP as a neurotransmitter: Its role in synaptic plasticity in the hippocampus. *Acta Neurobiol. Exp. (Warsz.)* **56**, 637–648.

Wieraszko, A., and Ehrlich, Y. H. (1994). On the role of extracellular ATP in the induction of long-term potentiation in the hippocampus. *J. Neurochem.* **63**, 1731–1738.

Wieraszko, A., and Seyfried, T. N. (1989a). ATP-induced synaptic potentiation in hippocampal slices. *Brain Res.* **491**, 356–359.

Wieraszko, A., and Seyfried, T. N. (1989b). Increased amount of extracellular ATP in stimulated hippocampal slices of seizure prone mice. *Neurosci. Lett.* **106**, 287–293.

Williams, M. (1984). Mammalian central adenosine receptors. *In* "Handbook of Neurochemistry" (A. Lajtha, ed.), Vol. 6, pp. 1–25. Plenum, New York.

Williams, M. (1987). Purinergic receptors and central nervous system function. *In* "Psychopharmacology: The Third Generation of Progress" (H. Y. Meltzer ed.), pp. 289–301. Raven Press, New York.

Williams, M., and Burnstock, G. (1997). Purinergic neurotransmission and neuromodulation: A historical perspective. *In* "Purinergic Approaches in Experimental Therapeutics" (K. A. Jacobson and M. F. Jarvis eds.), pp. 3–26. Wiley-Liss, New York.

Wirkner, K., Koles, L., Thummler, S., Luthardt, J., Poelchen, W., Franke, H., Furst, S., and Illes, P. (2002). Interaction between P2Y and NMDA receptors in layer V pyramidal neurons of the rat prefrontal cortex. *Neuropharmacology* **42**, 476–488.

Wirkner, K., Koles, L., Thummler, S., Luthardt, J., Poelchen, W., Franke, H., Furst, S., and Illes, P. (2002). Interaction between P2Y and NMDA receptors in layer V pyramidal neurons of the rat pretrontal cortex. *Neuropharmacology* **42**, 476–488.

Wise, S. P., and Jones, E. G. (1976). Transneuronal or retrograde transport of [³H]adenosine in the rat somatic sensory system. *Brain Res.* **107**, 127–131.

Wojcik, W. J., and Neff, N. H. (1983). Adenosine A1 receptors are associated with cerebellar granule cells. *J. Neurochem.* **41**, 759–763.

Wong, A. Y. C., Burnstock, G., and Gibb, A. J. (2000). Single channel properties of P2X ATP receptors in outside-out patches from rat hippocampal granule cells. *J. Physiol.* **527**, 529–547.

Xiang, Z., Bo, X., and Burnstock, G. (1998a). Localization of ATP-gated P2X receptor immunoreactivity in rat sensory and sympathetic ganglia. *Neurosci. Lett.* **256**, 105–108.

Xiang, Z., Bo, X., Oglesby, I. B., Ford, A. P. D. W., and Burnstock, G. (1998b). Localization of ATP-gated $P2X_2$ receptor immunoreactivity in the rat hypothalamus. *Brain Res.* **813**, 390–397.

Yamazaki, Y., Fujii, S., Nakamura, T., Miyakawa, H., Kudo, Y., Kato, H., and Ito, K. (2002). Changes in [Ca2+](i) during adenosine triphosphate-induced synaptic plasticity in hippocampal CA1 neurons of the guinea pig. *Neuroscience Letters* **324**, 65–68.

Yao, S. T., Barden, J. A., Finkelstein, D. I., Bennett, M. R., and Lawrence, A. J. (2000). Comparative study on the distribution patterns of P2X(1)-P2X(6) receptor immunoreactivity in the brainstem of the rat and the common marmoset *(Callithrix jacchus):* Association with catecholamine cell groups. *J. Comp. Neurol.* **427**, 485–507.

Yao, S. T., Barden, J. A., and Lawrence, A. J. (2001). On the immunohistochemical distribution of ionotropic P2X receptors in the nucleus tractus solitarius of the rat. *Neuroscience* **108**, 673–685.

Yoshida, K., Nakagawa, T., Kaneko, S., Akaike, A., and Satoh, M. (2002). Adenosine 5′-triphosphate inhibits slow depolarization induced by repetitive dorsal root stimulation via P2Y purinoceptors in substantia gelatinosa neurons of the adult rat spinal cord slices with the dorsal root attached. *Neuroscience Lett.* **320**, 121–124.

Yoshioka, K., Hosoda, R., Kuroda, Y., and Nakata, H. (2002). Hetero-oligomerization of adenosine A1 receptors with P2Y1 receptors in rat brains. *FEBS Lett.* **531**, 299–303.

Zhang, J., Kornecki, E., Jackman, J., and Ehrlich, Y. H. (1988). ATP secretion and extracellular protein phosphorylation by CNS neurons in primary culture. *Brain Res. Bull.* **21**, 459–464.

Zhang, M., Zhong, H., Vollmer, C., and Nurse, C. A. (2000). Co-release of ATP and ACh mediates hypoxic signalling at rat carotid body chemoreceptors. *J. Physiol.* **525**, 143–158.

Zhang, Y. X., Yamashita, H., Ohshita, T., Sawamoto, N., and Nakamura, S. (1995). ATP increases extracellular dopamine level through stimulation of P2Y purinoceptors in the rat striatum. *Brain Res.* **691**, 205–212.

Zhong, Y., Dunn, P. M., Xiang, Z., Bo, X., and Burnstock, G. (1998). Pharmacological and molecular characterisation of P2X purinoceptors in rat pelvic ganglion neurons. *Br. J. Pharmacol.* **125**, 771–781.

Zhong, Y., Dunn, P. M., and Burnstock, G. (2000a). Guinea-pig sympathetic neurons express varying proportions of two distinct P2X receptors. *J. Physiol.* **523**, 391–402.

Zhong, Y., Dunn, P. M., and Burnstock, G. (2000b). Pharmacological comparison of P2X receptors on rat coeliac, mouse coeliac and mouse pelvic ganglion neurons. *Neuropharmacology* **39**, 172–180.

Zhong, Y., Dunn, P. M., and Burnstock, G. (2001). Multiple P2X receptors on guinea pig pelvic ganglion neurons exhibit novel pharmacological properties. *Br. J. Pharmacol.* **132**, 221–233.

Zhou, X., and Galligan, J. J. (1996). P2X purinoceptors in cultured myenteric neurons of guinea-pig small intestine. *J. Physiol.* **496**, 719–729.

Zhou, X., and Galligan, J. J. (1998). Non-additive interaction between nicotinic cholinergic and P2X purine receptors in guinea-pig enteric neurons in culture. *J. Physiol.* **513**, 685–697.

Zinchuk, V. S., Okada, T., Kobayashi, T., and Seguchi, H. (1999). Ecto-ATPase activity in cerebellum: Implication to the function of synaptic transmission. *Brain Res.* **815**, 111–115.

Zona, C., Marchetti, C., Volontè, C., Mercuri, N. B., and Bernardi, G. (2000). Effect of P2 purinoceptor antagonists on kainate-induced currents in rat cultured neurons. *Brain Res.* **882**, 26–35.

# CHAPTER 11

# Purinergic (P2) Receptors in the Kidney

**Jens Leipziger,\* Matthew A. Bailey,† and Robert J. Unwin†**
\*Institute of Physiology, Åarhus University, DK-8000 Åarhus, Denmark
†Centre for Nephrology and Department of Physiology, Royal Free and University College
Medical School, University College London, London NW3 2QG, United Kingdom

## I. INTRODUCTION

There has been a growing interest in describing the effects of adenosine on renal function, the distribution of its P1 ($A_{1-3}$ subtypes) receptors in kidney tissue (blood vessels and tubules), and the potential of this receptor system for therapeutic intervention (1,2). Progress in research on the renal effects of ATP via interaction with P2 cell surface receptors has been much slower. Curiously, although adenosine acting as a hormone or paracrine factor is now widely accepted, the concept that its precursor, extracellular ATP, could fulfill a similar and, perhaps, even more important physiological role is still met with some resistance and skepticism. After all, it is reasoned that ATP is the universal currency of energy exchange. The idea that it could be released in a controlled way to act as a selective messenger between cells is too far-fetched for something so abundant and ubiquitous.

*Current Topics in Membranes, Volume 54*

It has been said that teleology is "the mistress biologists need but don't like to be seen with." As a primitive bioactive molecule, however, it is likely that ATP has (and had) many important and diverse functions throughout evolution (3), and it is this aspect of ATP biology that has expanded rapidly (4).

Because of this conceptual difficulty, attention on the renal effects of ATP has focused for many years on extracellular ATP as a supplement to maintain and restore its intracellular levels following acute renal injury, especially after ischemia (5). Although there have been reports going back more than 20 years concerning the effects of ATP, both on isolated renal tissue and on cultured renal cells, these studies have been limited by the poor specificity of both agonists and antagonists for P2 receptors (Fig. 1). Moreover, these studies have been complicated by interspecies differences in receptors and agonist/antagonist sensitivities, and they have been confounded by the transient nature of intact ATP, associated with the intrinsic activity of its various degradation products, including adenosine (6) (Fig. 2).

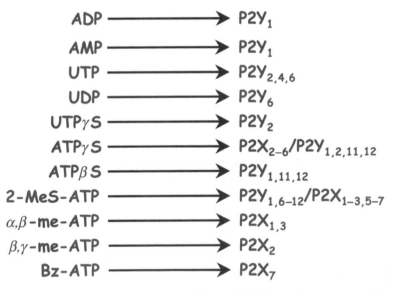

## P2 receptors and (relative) agonist selectivity

### (Nonselective blockers: PPADS and Suramin)

$$ADP \longrightarrow P2Y_1$$
$$AMP \longrightarrow P2Y_1$$
$$UTP \longrightarrow P2Y_{2,4,6}$$
$$UDP \longrightarrow P2Y_6$$
$$UTP\gamma S \longrightarrow P2Y_2$$
$$ATP\gamma S \longrightarrow P2X_{2-6}/P2Y_{1,2,11,12}$$
$$ATP\beta S \longrightarrow P2Y_{1,11,12}$$
$$2\text{-MeS-ATP} \longrightarrow P2Y_{1,6-12}/P2X_{1-3,5-7}$$
$$\alpha,\beta\text{-me-ATP} \longrightarrow P2X_{1,3}$$
$$\beta,\gamma\text{-me-ATP} \longrightarrow P2X_2$$
$$Bz\text{-ATP} \longrightarrow P2X_7$$

**FIGURE 1**   A summary of ATP analogs and their preferred P2 receptor subtypes, used to define ("profile") P2 receptors *in vitro*. (Nonselective blockers: PPADS and suramin.) Based on references 4 and 133.

# ATP, adenosine and ATP breakdown

$$ATP \Rightarrow ADP + P_i$$

$$ATP \Rightarrow AMP + P_i$$

$$ADP \Rightarrow AMP + P_i$$

$$cAMP \Rightarrow AMP$$

$$AMP \Rightarrow ADENOSINE + P_i$$

$$ADENOSINE + ATP \Rightarrow AMP + ADP$$

$$AMP \Leftrightarrow ADENOSINE \Rightarrow INOSINE$$

**FIGURE 2**    A summary and scheme of ATP breakdown and its conversion to adenosine (see text for relevance).

This chapter aims to summarize our current state of knowledge of renal P2 receptors and their relationship to the renal effects of ATP. It focuses on the nonvascular P2 receptors of renal epithelia, since their potential renal vascular effects are dealt by Inscho (see elsewhere in this volume).

P2 purinoceptors now consist of many subtypes within both the ionotropic P2X (1–7 and M) and metabotropic (G protein-coupled) P2Y (1,2,4,6,11,12,13) structural subgroups (Fig. 1). Studies of these receptors in the kidney are still at an essentially descriptive and taxonomic stage (13). With these receptors being identified at the molecular level, it is much easier to detect and identify them with specific (mostly polyclonal) antibodies for immunohistochemistry and Western blotting, and to localize their messenger RNA (mRNA) by Northern analysis, reverse transcriptase-polymerase chain reaction (RT-PCR), and *in situ* hybridization. Yet, despite the availability of these tools to detect and define P2 receptors, there is still no consensus as to their renal distribution or function. It is clear that they are widespread throughout the nephron (9) and that each cell may express more than one P2X and/or P2Y subtype (13). Part of the reason for this seemingly complex pattern of renal P2 receptors is that much of the published work to date is based on studies of immortalized renal cell lines, such as MDCK (14–21), MDCT (22), mlMCD-K2 (23), M-1 (24,25), and LLC-PK1 (7,26,27), or on primary renal cell cultures from various species (28–30), rather than on intact renal tissue. Of course, the main advantage of such systems is that downstream signaling and functional effects of ATP can be explored more easily. However, extrapolating the findings in such cell lines to normal tissue is fraught with difficulties. A major difficulty is that of relating the various subtypes of P2 receptor and their functional properties, as defined in any isolated cell, to their distribution and function in native renal tissue. Further complications arise with apparent differences between species, the paucity of human data, and the limited functional studies of these receptors *in vitro* and *in vivo*.

A second troublesome issue concerning ATP is that of the likely source(s) of extracellular ATP: peritubular capillary blood, glomerular filtrate/tubular fluid, renal nerve endings, or the renal epithelium itself. It is also unclear whether these sources can result in a sustained presence of ATP locally at concentrations sufficient to activate P2 receptors (31). Attempts have been made to measure ambient concentrations of extracellular ATP and the luciferase–luciferin system is one of the most sensitive and widely used assays for this purpose (32). However, the use of this assay, and that of several ingenious and elegant biodetectors for ATP (31,33), has so far been confined to *in vitro* systems. At the very least, some investigators have performed these luciferase–luciferin assays on primary cultures of human renal epithelial cells grown as polarized monolayers (30,32,34), showing predominantly lumenal release under basal and stimulated conditions. The issue of cellular mechanisms of ATP release is addressed more specifically by Schwiebert, Zsembery, and Geibel (see elsewhere in this volume). Having introduced the potential as well as the pitfalls in this field, we proceed to review what is currently known of renal (nonvascular) P2 receptors.

## II. THE KIDNEY

It was initially thought that the P2X receptors were confined largely to blood vessels, whereas the presence of P2Y receptors had been demonstrated on the endothelium lining the vasculature (35), including the glomerulus (36). P2X receptors were thought to be involved in the control of intrarenal blood flow and glomerular filtration (37); whereas the P2Y receptors were present mainly on tubular epithelial cells and were thought to affect tubular fluid composition (6). However, this broad division in terms of localization and function has proved far too simple. As already mentioned, both P2X and P2Y receptor subtypes are found along the entire length of the nephron, probably in the same cell and in the same apical or basolateral membrane domain. The same appears true of the vascular endothelium (38–40). Indeed, at the time of writing, a large number of P2Y and P2X receptor subtypes have been reported to be present in mammalian renal

**TABLE I**

Currently Identified P2Y Receptors along Freshly Isolated Mammalian Nephron Segments[a]

|  | PT | Loop of Henle TAL (DTL, ATL, TAL) | CCD | OMCD | IMCD |
|---|---|---|---|---|---|
| P2Y$_1$ | 9[b],71[b] | | 98[c] | 9,[b]71[b] | |
| P2Y$_2$ | 9[b] | 9,[b] 85[c] | 98,[c] 102,[d] 105[c] | 99,[b] 71[b] | 99,[b] 100,[b] 101[b] |
| P2Y$_4$ | ? | ? | ? | ? | ? |
| P2Y$_6$ | 76[b] | | | | |

*Abbreviations:* ATL, Ascending thin limb of Henle's loop; CCD, cortical collecting duct; DTL, descending thin limb of Henle's loop; IMCD, inner medullary collecting duct; OMCD, outer medullary collecting duct; PT, proximal tubule; TAL, thick ascending limb.

[a]Renal P2Y receptor distributions along the mammalian nephron (rat, mouse, and rabbit). Given the difficulty in pharmacologically discriminating between the P2Y$_2$ and P2Y$_4$ receptors, it is still uncertain which of these two receptors has been identified. This is indicated with a question mark. There is evidence to date of the expression of the P2Y$_{11}$ receptor in intact renal tubules (4) and the P2Y$_{12}$ receptor is expressed exclusively on thrombocytes (132).

[b]Rat.

[c]Mouse.

[d]Rabbit.

epithelia, suggesting a local regulatory or "housekeeping" role for these receptors with perhaps overlapping functions and redundancy as a fail-safe mechanism. Table I summarizes our current knowledge of P2Y receptor distribution along the intact mammalian nephron and Table II lists the main sites of kidney immunolocalization for both P2Y and P2X subtypes.

## III. THE GLOMERULUS

A search of the literature reveals limited information on the nature and role of P2 receptors in the mammalian glomerulus. This probably reflects the complex and structurally heterogeneous architecture of the intact glomerulus, which consists of a central tuft of endothelial cells and mesangial cells, encapsulated by a double layer of visceral and parietal epithelial cells. Nevertheless, calcium transients in response to extracellular ATP have been described for the glomerulus as a whole (8,41), mediated through distinct receptors displaying the characteristics of P2Y$_1$ and either P2Y$_2$ or P2Y$_4$ (41) (Fig. 3). Immunohistochemical analysis supports these findings, but clearly show that P2Y$_2$, and not P2Y$_4$, is present, despite the reported detection of mRNA encoding the P2Y$_4$ receptor (42,43). P2X receptors appear to be expressed only at low levels in nonvascular structures of the intact glomerulus (43) (Table II).

Most of the detailed information relating to glomerular P2 receptors comes from studies of cells in culture. In this regard, most P2Y and some

# P2-receptor(s) profile in isolated rat glomeruli

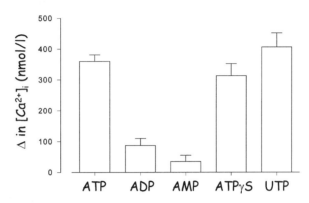

**FIGURE 3** The change in intracellular calcium concentration ($[Ca^{2+}]_i$) in rat isolated whole glomeruli in response to various P2 agonists, including ATP. The profile suggests a predominant $P2Y_2$ or $P2Y_4$ subtype response.

P2X subtypes have been identified in glomerular mesangial cells (44), glomerular endothelial cells (36,45), and podocytes (46,47). It is apparent from these studies, as indeed it is in other cell models, that P2Y receptors couple to multiple signaling pathways. Discussion of these pathways lies beyond the scope of this chapter and the interested reader is referred to the chapter by Inscho (elsewhere in this volume) and to a review (48).

Although the presence of P2 receptors in all cell types of the glomerulus is now clear, their function remains unclear. It has been known for many years

**TABLE II**

Immunolocalization of P2 Receptor Distribution in Normal Rat Kidney

| P2 receptor subtype | Immunolocalization |
|---|---|
| $P2X_1$ | Vascular smooth muscle |
| $P2X_2$ | Vascular smooth muscle |
| $P2X_3$ | Vasculature associated with nerve terminals |
| $P2X_4$ | Tubular staining in the cortex |
| $P2X_5$ | Collecting ducts |
| $P2X_6$ | Low levels throughout the tubule |
| $P2X_7$ | Podocytes—low levels in control tissue |
| $P2Y_1$ | Mesangial cells and vascular smooth muscle |
| $P2Y_2$ | Mesangial cells |
| $P2Y_4$ | Proximal tubules |

that infusion of ATP into the renal artery causes vasoconstriction (6). This appears to be mediated through P2X activation, since selective activation of P2Y receptors is vasodilatory (45). At this point, it becomes important to distinguish between the nonvascular cells of the glomerulus and renal microvasculature. In elegant studies by Inscho and colleagues (48), the function of both P2X and P2Y receptors is relatively well understood. In the intact glomerulus, devoid of renal arterioles, extracellular purines, probably via a $P2Y_1$ receptor, relieve angiotensin II-mediated constriction through the production of nitric oxide (NO) and cyclic GMP (49). $P2Y_1$-mediated induction of NO production has been reported in glomerular endothelial cells (45). However, mechanisms other than NO production also need to be considered. P2Y activation may also facilitate mesangial cell contraction by inhibiting adenyl cyclase (50), which would oppose any relaxant effect from local prostaglandin release (51). Indeed, P2Y activation by ATP has been shown to induce contraction of other glomerular cells containing myofibrils, such as the epithelial parietal sheet (52), podocytes (47), mesangial cells (53), and the glomerulus as a whole (54). Although these findings of relaxation and contraction of the intact glomerulus may seem a little confusing, such studies suggest a physiological role for extracellular ATP in controlling formation of the glomerular filtrate (55), which includes changes in hydraulic permeability that are independent of any effects on preglomerular vessels. It is tempting to speculate that local release of ATP, whether it is from red blood cells squeezed during their passage through the glomerular capillary bed (56) or from renal sympathetic nerve endings as a cotransmitter with norepinephrine (57), may regulate glomerular filtration, either via a direct action or by modulation of other locally produced vasoactive factors (58).

Returning to the signaling pathways of P2Y activation in glomerular cells, one theme is readily apparent. There are several reports in the literature of extracellular ATP-activating pathways associated with inflammation, such as the classic mitogen-activated protein kinase/extracellular signal-regulated kinase (59,60) and the stress-activated protein kinase cascade (61,62). Reports indicate that growth hormones can potentiate the proinflammatory effects of nucleotides (63). When one considers that the early response genes, c-*fos* and c-*jun*, are activated in mesangial cells following exposure to extracellular ATP (60), it is not unreasonable to expect that P2 receptor expression and activation may be involved in the response to some forms of glomerular injury. Indeed, activation of P2Y receptors reportedly aggravates the course of glomerulonephritis in some experimental models (60). Another feature of glomerulonephritis, mesangial cell proliferation, might also involve P2Y activation, since extracellular ATP is a potent mitogen for these cells (43,64).

# The P2 'death' receptor – P2X$_7$ (P2Z)

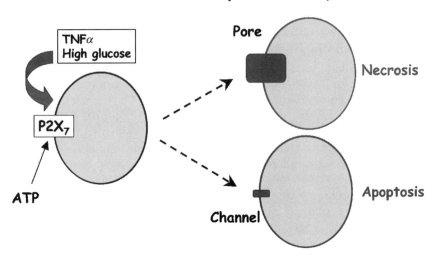

**FIGURE 4**  Surface expression and activation of the low-affinity P2X$_7$ receptor may lead to cell necrosis, associated with formation of a large membrane pore, or apoptosis.

Also of potential relevance to glomerular pathology is the P2X receptor, P2X$_7$, activation of which can cause cell death by necrosis or apoptosis (Fig. 4). It is expressed by mesangial cells and, when activated in these cells, it causes apoptosis (65). Its expression can also be induced by exposure to tumor necrosis factor $\alpha$ (TNF-$\alpha$) (43) and, in at least some cultured nonrenal cells (skin fibroblasts), by a high extracellular glucose concentration (66), as well as in the diabetic glomerulus (67). Apoptosis appears to be a major repair mechanism in several models of glomerulonephritis (68), although uncontrolled activity of this process can also be detrimental, causing progression to sclerosis in some other forms of glomerulonephritis (69).

That the same ligand, ATP, should be able to produce such contrasting effects in the same mesangial cell is intriguing, and how this may be regulated is completely unknown. Within these cells, it clearly must depend on different signal transduction pathways, but also on different receptor affinities for ATP (e.g., P2X$_{1,3}$ > P2X$_{2,4,5,6}$ > P2X$_7$) (4), associated ATPase activity (70), and local ATP concentrations. For example, stimulation of the P2Y$_2$ or P2Y$_4$ subtype is likely to be responsible for the proliferative response to ATP. The low-affinity P2X$_7$ receptor is probably the cause of cell death, triggered by high local concentrations of ATP in the setting of tissue damage, associated with inflammation and platelet aggregation (70).

## IV. THE PROXIMAL TUBULE

The proximal tubule is a highly active component of the kidney, accounting for reabsorption of ~70% of the filtered sodium and chloride, ~100% of the filtered glucose, amino acids, and small "tubular proteins," and ~90% of the filtered bicarbonate. In contrast to the glomerulus, the proximal tubule is a relatively homogeneous collection of cells. Although this segment is divided into three portions, S1, S2, and S3, there is in reality only one general cell type. It is largely the rates at which transport processes occur in each part that differs, rather than the underlying mechanisms of transport. Indeed, since the proximal tubule is a polarized epithelium, there may be distinct receptor subtypes in each membrane domain, apical and basolateral, which can influence local transport rates.

Most of the studies performed in native tissue have focused on the basolateral membrane of the proximal tubule. Initially, they have identified $P2Y_1$ as the dominant receptor subtype, both on the basis of measurements of calcium transients (9,12,71) (Fig. 5) and identification of receptor mRNA (9). The presence of $P2Y_1$-like receptors was also confirmed in primary cultures of rat (72) and rabbit proximal tubule (28), clonal kidney cells

## Profile of P2Y receptor [Ca²⁺]ᵢ response to ATP along the rat nephron

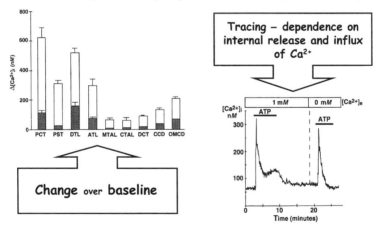

**FIGURE 5** *Left:* Response of different renal tubular segments to ATP as measured by the change in intracellular calcium concentration ($[Ca^{2+}]_i$). ATL, Ascending thin limb; CCD, cortical collecting duct; CTAL, cortical thick ascending limb; DCT, distal convoluted tubule; DTL, descending thin limb; MTAL, medullary thick ascending limb; OMCD, outer medullary collecting duct; PCT, proximal collecting tubule; PST, proximal straight tubule. *Right:* A characteristic PY receptor response, showing its time course and dependence of its plateau on extracellular $Ca^{2+}$ influx. Adapted from reference 9.

(73,74), and immortalized cell lines with a proximal phenotype (7,75). One study also provided evidence for $P2Y_2$ and $P2Y_4$ receptors in the rat proximal tubule (9). Even more recently, it has been observed that the pyrimidine-specific $P2Y_6$ receptor is also present in the basolateral membrane of this segment (76).

To date, there is limited information about the presence of specific subtypes in the lumenal membrane, although $P2Y_1$ receptors have been identified in this membrane in polarized LLC-PK1 cells (75). P2X receptors ($P2X_1$ and $P2X_4$) have, at the time of writing, been described only in established renal cell lines (26,74) but they may also be present in the intact tubule (77). Schwiebert and colleagues have undertaken the major task of establishing the quantitative distribution of mRNA for all seven P2X subtypes in primary cell cultures derived from specific regions of the mammalian nephron. Although detailed results are not yet published, it seems that the proximal tubule expresses $P2X_4$ and $P2X_5$ mRNAs abundantly (13). However, while our own immunohistochemical studies using polyclonal antibodies against these receptors in intact rat and human kidney tissue confirm the presence of $P2X_4$ in the proximal tubule (C. Turner, M. A. Bailey, and R. J. Unwin, unpublished observation), $P2X_5$ protein cannot be detected in this segment and appears to be confined to more distal segments, particularly the collecting duct (78) (Fig. 6). There is a low level of expression for $P2X_7$, which is perhaps not surprising if its cell surface expression risks autodestruction. Systematic studies of P2X receptor mRNA and protein distribution are underway in native real tissue; however, a complete and detailed picture has yet to fully emerge.

In marked contrast to the collecting duct (see Section VI), there is scant information concerning a physiological role for P2 receptor activation in the proximal tubule. In renal cell lines, ATP is known to activate "maxi-K" potassium channels (73) and induce acute inhibition of $Na^+$, $K^+$-ATPase (75). In addition, basolateral chloride channels are reportedly activated in the *Necturus* proximal tubule (79), which would also point to a modulatory role for extracellular ATP in vectorial transport along the proximal tubule. It is certainly the case that ATP-induced calcium transients are stronger in this segment than in any other, comparable to those induced by known regulators of proximal cell function, such as norepinephrine (9) (Fig. 5). Nevertheless, with the exception of one report in which a P2 receptor was shown to stimulate indirectly sodium-dependent phosphate uptake (77), convincing evidence of a regulatory role is still lacking.

It is most likely that extracellular ATP exerts autocrine/paracrine control of renal epithelial cell function, since tightly regulated, physiological release of nucleotides probably occurs in a variety of cell types (80,81). It is certain that ATP is released across both apical and basolateral membranes and

## P2X₅ and P2X₆ receptor localization by immunostaining

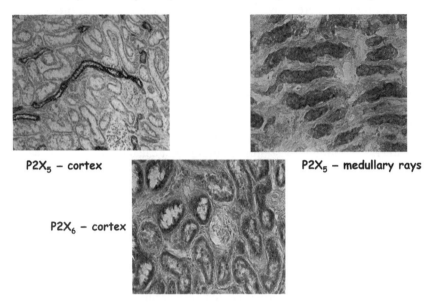

P2X₅ – cortex

P2X₅ – medullary rays

P2X₆ – cortex

**FIGURE 6** Positive immunostaining with polyclonal antibodies against P2X₅ and P2X₆ receptors in frozen sections of rat kidney tissue (original magnification, × 40).

local concentrations in the proximal tubule may reach the micromolar range (13). With this in mind, researchers are beginning to uncover a subtler role for P2 receptor activation, demonstrating modulation of hormonal action (7) and effects on cell metabolism, such as stimulation of gluconeogenesis (82) via $P2Y_1$ activation. More recently, elegant double-microperfusion experiments involving the simultaneous perfusion of both the tubular lumen and the peritubular capillaries uncovered P2Y-mediated control of acidification in the rat proximal tubule (83). In these experiments, increasing the viscosity (and thus perhaps the shear force) of the peritubular fluid resulted in an NO-dependent acidification. Because the increase in peritubular fluid viscosity and associated rise in shear stress may promote the release of ATP from capillary endothelial cells (84), the effect of P2 receptor activation was then investigated. It was found that ATP mimicked the action of increasing fluid viscosity and that this effect was also NO dependent. Although the source of the ATP and NO remains uncertain (endothelial or tubular), the effect of receptor activation is clear and implicates ATP (and therefore a P2 receptor) in the control of $Na^+–H^+$ exchange (NHE-3), the major mechanism of proton secretion in the proximal tubule.

In summary, purinoceptors of both types are clearly present in the proximal tubule. There is increasing evidence that these P2 receptors could exert a local effect on electrolyte transport in this nephron segment, and would as well interact with other circulating and paracrine messengers.

## V. THE LOOP OF HENLE

Until now, the loop of Henle has received little attention with respect to P2 receptors. In nonperfused mouse thick ascending limb (TAL) segments, basolateral ATP and UTP stimulate intracellular calcium ($[Ca^{2+}]_i$) transients (85). However, without further characterization, the authors suggested that a P2Y receptor mediated this effect. Given the expression of multiple P2 receptors in numerous epithelial cell preparations, it is most likely that, in the TAL, a number of different subtypes are also present. In addition, there is evidence for expression of multiple P2 receptors in both the descending thin limb (DTL) and ascending thin limb (ATL) of Henle's loop (9).

In the TAL, its "pharmacological fingerprint" suggests the presence of a $P2Y_2$ or $P2Y_4$ receptor. On the basis of a number of studies of the distal nephron (see below), we proposed that extracellular ATP could serve a common purpose along the nephron, namely to inhibit active solute transport. As indicated above, ATP can be released from epithelial cells and cell swelling appears to be a crucial activating mechanism (13). Therefore, ATP could function as a mechanism to protect tubular cells from ischemic injury by reducing the energy consumption used for ion transport. However, in the TAL, the opposite appears to be true. Our own preliminary studies, and those by A. Di Stefano and M. Wittner (unpublished observations), show that basolateral ATP increased transepithelial voltage across the TAL from some +8 to +13 mV, suggesting a stimulation of ion transport. In addition to the elevation of $[Ca^{2+}]_i$, ATP also produced a small but significant increase of cAMP. However, it must be remembered that released ATP will serve as a precursor for adenosine and that adenosine release per se during hypoxia (86,87) can, via $A_1$ receptor stimulation, inhibit NaCl absorption by up to 50%. The authors propose an antiischemic effect of the adenosine released (or formed) in this way (88).

In summary, the situation may be quite complex (Fig. 7). The source of basolateral ATP could be from sympathetic nerve endings. $\beta$-Adrenergic receptor stimulation is known to increase cAMP and activate transport in the TAL (10), and ATP is often coreleased from sympathetic nerve endings (4,58). Thus, ATP may function as one of several basolateral stimuli to the multihormonally activated NaCl absorption in the TAL (89). Yet, another mechanism could also come into play, again under ischemic conditions.

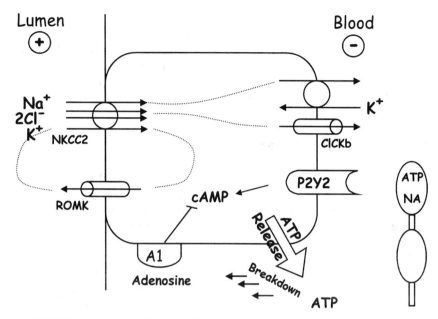

**FIGURE 7**   A thick ascending limb (TAL) cell showing the key transport proteins and their potential relationship to a P2Y$_2$ receptor, ATP, and adenosine; see text for details. ClCKb, Basolateral renal ClC-chloride channel, NKCC2, Na$^+$, K$^+$, 2Cl$^-$ symporter; ROMK, renal outer medullary potassium channel.

After autocrine release of ATP, its rapid breakdown by ecto-ATPases (90) will occur, leading to the formation of adenosine, which can then activate A$_1$ receptors and cause inhibition of NaCl transport (88,91,92). Figure 7 represents a possible cell model and role for P2 receptors in regulating TAL function. It is worth nothing that our own preliminary data could not demonstrate nucleotide-stimulated [Ca$^{2+}$]$_i$ elevations when applied to the lumenal side of isolated perfused TAL segments, which is against the expression of lumenal P2 receptors in this segment.

The loop of Henle terminates at the macula densa, an epithelial "sensor" responsible for balancing the rate of glomerular filtration with tubular reabsorption (tubuloglomerular feedback, TGF). The critical mediator of TGF appears to be adenosine, as was proposed previously (93). The strongest evidence for this has come from an A$_1$ receptor knockout mouse, in which normal TGF was absent (94). Importantly, TGF depends on the Na$^+$, 2Cl$^-$, K$^+$ cotransporter mediating the "sensed" change in early distal tubular NaCl load (95). However, the link between loop diuretic sensitive NaCl absorption and interstitial (basolateral) adenosine release, or production, is still uncertain. Preliminary work suggests that it is ATP release from macula densa cells that is the functional link. Lumenal cotransporter-mediated NaCl uptake results in an elevated cell volume (96), leading to ATP release (13); breakdown of ATP then forms adenosine, which reduces GFR (97). Although not directly relevant to P2 receptors, these novel findings highlight the close interrelationship between adenosine and ATP as extracellular signaling molecules.

## VI. THE DISTAL TUBULE AND COLLECTING DUCT

A number of studies have addressed the effect of extracellular nucleotides on epithelial cells of distal tubular origin. Also, in this part of the nephron, a comprehensive register of P2 receptors is still pending (Table I). Current data indicate strongly that, along the entire distal nephron, P2Y$_1$ and P2Y$_2$ receptors are coexpressed. It is believed currently that the same epithelial cell expresses both these receptors and, perhaps, even more subtypes. Given the continuous uncertainty in discriminating between the P2Y$_2$ and P2Y$_4$ receptors, the role of the P2Y$_4$ receptor remains to be determined. No firm evidence for the expression of a P2Y$_6$ receptor is present. Along the distal tubule, the P2Y$_2$ receptor is also expressed in the lumenal membrane (98,99). Functionally basolateral ATP was shown to elevate [Ca$^{2+}$]$_i$, in rat inner medullary collecting duct (IMCD) (100), and it was subsequently shown in rat that basolateral ATP inhibited antidiuretic hormone (ADH)-stimulated H$_2$O transport, possibly by a decrease in cAMP (101). This effect was also found in the rabbit cortical collecting duct (CCD) (102). Pharmacological profiling suggested a basolateral P2Y$_2$ receptor (102). A number of studies using permanent distal tubular cell lines in culture (e.g., MDCK and M-1) demonstrated that extracellular ATP activated chloride secretion and inhibited electrogenic sodium absorption (19,21,23–25). Thus, cultured distal tubular epithelial cells appear functionally similar to respiratory epithelium, where the cystic fibrosis transmembrane regulator (CFTR) and lumenal calcium-activated chloride channels provide the apical chloride exist pathway for NaCl secretion (103); however, the intact distal nephron does not seem to transport chloride transcellularly (104). Also, evidence indicates that the isolated perfused CCD does not show nucleotide-mediated activation of a lumenal calcium-activated chloride channel (105). A similar phenomenon has been observed in colonic secretory epithelium. In the intact colon, the CFTR chloride channel is the exclusive chloride exist pathway in the lumenal membrane (106). However, cell culture systems commonly show the expression of calcium-activated chloride channels that serve as an alternative chloride secretory pathway (106). We suspect that the expression of lumenal calcium-activated chloride channels in distal renal

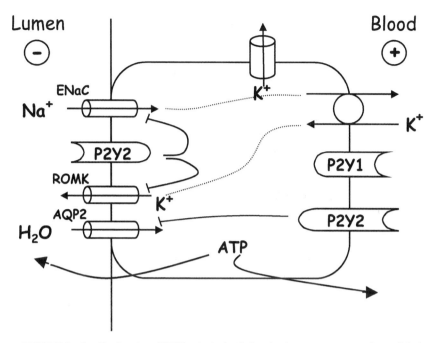

**FIGURE 8** A collecting duct (CCD) principal cell showing key transport proteins and their potential relationship to ATP and P2Y receptor activation. AQP2, Aquaporin 2; ENaC, epithelial sodium channel.

tubular cells in culture is possibly the result of culture conditions and dedifferentiation.

Nucleotide-mediated inhibition of electrogenic sodium absorption was found in the isolated perfused mouse CCD (105) and in a number of distal tubular cell lines (23,25,29). Similarly, calcium absorption was also inhibited by stimulation of a P2Y receptor (29). Finally, renal outer medullary potassium (ROMK) channels, which mediate potassium secretion, can be inhibited by UTP and ATP when studied in a split open collecting duct preparation (107). Figure 8 is a principal cell model of P2Y receptor localization and its functional effects on transport. It is becoming clear that all relevant transcellular transport properties are inhibited by P2Y receptor stimulation.

P2 receptors are frequently also expressed in the lumenal membrane of epithelia. This has been found for lung, intestine, and exocrine glands and also in a variety of cell cultures derived from distal tubular epithelia (21,25,108,109). Lumenal expression of a $P2Y_2$ receptor was also found in the intact isolated perfused collecting duct (98,99); its activation led to inhibition of electrogenic sodium absorption (105). The same lumenal receptor is also thought to mediate ATP-induced inhibition of potassium secretion in principal cells (107).

ATP released from epithelial cells appears to be predominantly transported into the lumenal compartment. The mechanism of ATP release is still obscure, but cell volume changes appear to be critical. It has been suggested that ATP secretion, or release, into the tubular lumen serves to regulate ion transport distal to its site of release. The issue of cellular mechanisms of ATP release is addressed more specifically by Schwiebert, Zsembery, and Geibel (see elsewhere in this volume).

## VII. FUNCTIONAL IMPLICATIONS OF RENAL EPITHELIAL P2 RECEPTORS

An exciting area in this rapidly expanding and promising field of P2 receptor research is the potential role for extracellular nucleotides in renal pathophysiology and, therefore, therapeutics. It is anticipated that by determining the function of ATP in situations of abnormal function, a clue as to its role in normal function will emerge.

### A. Role for ATP Receptors in Polycystic Kidney Disease

One interesting example is a potential action of extracellular ATP in the progression of autosomal-dominant polycystic disease (ADPKD), the fourth leading cause of end-stage renal failure worldwide (110). ADPKD is associated with the formation of (and accumulation of fluid in) cysts derived from renal tubular epithelia. It is thought that the expansion of cysts contributes to the progression of disease and development of renal failure, and that it is driven by both cellular proliferation and fluid secretion, leading to progressive cyst enlargement. Cysts can arise from any portion of the renal nephron in ADPKD in mice (111), rats (112), and humans (13,30,31,34); however, in ARPKD, dilated tubules rather than completely formed cysts are found and primarily in the distal nephron (34,113).

There is a growing belief that, although there is an underlying and specific gene defect causing cyst formation in ADPKD, the progress of this disorder is to some extent determined by locally released factors acting as autocrine or paracrine regulators. Highly suggestive circumstantial evidence comes from the observation that progression of ADPKD to end-stage renal disease is highly variable among individuals, that the cysts are focal and asymmetric (between the two kidneys) in their distribution, and that the expansion of cysts follows no consistent pattern. It was demonstrated that cyst fluid contains high concentrations of ATP (30) and that this level is maintained by cellular secretion of ATP into the cyst lumen (34). In renal (11,24,114) nonrenal tissues (109, 115–123), ATP is a proven secretagogue. In renal epithelia, it is also mitogenic (74). Therefore, it is possible that the presence of "trapped" extracellular purines bathing the lumenal surface of the cyst-lining epithelium could augment cyst expansion by a direct stimulating effect on fluid secretion and cell proliferation. This may also arise through an interaction with other agents, such as the implicated cytokines epidermal growth factor (EGF) and TNF-$\alpha$ (124,125). Another interesting development is the observation that cysts may expand through the loss of adjacent cells, perhaps mediated by apoptotic cell death (110). This process can be triggered by ATP acting on cell surface-expressed P2X$_7$ receptors (65), which can be induced by TNF-$\alpha$ (43). However, if this process is important, it seems to be confined to cells between cysts in the interstitium rather than to cells lining the cysts.

## B. P2 Receptors and the Regulation of Cell Volume

The transcellular transport of fluids and electrolytes and the osmotic stresses within the kidney make it a major site of epithelial cell volume regulation. A role for extracellular ATP and P2 receptors has been demonstrated in hepatocytes (126) and bile duct epithelia (126). In both cell types, cell swelling was shown to release ATP and to activate P2 receptors and subsequently calcium-activated potassium and chloride conductances. The resulting cellular KCl loss mediates regulatory volume decrease (RVD). As referred to above, activation of potassium and chloride channels is also seen in cultured renal epithelia (11,24,114). A similar cell volume regulatory mechanism appears to exist in the intact renal epithelium (13). However, so far, there is no evidence for calcium-activated chloride channels in intact nephron segments. Therefore, it is questionable whether this mechanism exists in the native duct or tubule.

Cell volume regulation is important in intact renal epithelia and is necessary to adjust for rapid changes in transcellular flux of fluid and solutes and, therefore, must depend on transcellular ion transport mechanisms, as shown in macula densa cells (96). Reduction of apical ion influx reduces cell volume and vice versa. A common functional consequence of extracellular ATP action in the distal tubule is inhibition of the major transport processes. Thus, it is not unreasonable to speculate that small increases in cell volume during increased transcellular transport would elicit "regulated" ATP release and a consequent negative feedback regulation of fluid and electrolyte transport.

## C. ATP and Ischemic Protection

In the intact distal nephron, a consistent finding has been that all major transporting activity is downregulated when lumenal and/or basolateral P2Y receptors are activated (29,101,105,107). For more proximal tubular segments, this has not been studied in any detail. In the distal tubule, we propose that extracellular (lumenal) ATP acts as an endotubular signaling molecule and serves to protect the tubular epithelium under ischemic conditions. The sequence of events could be as follows. First, ATP is released from epithelial cells by mechanical stimulation or cell swelling in physiological states (32,127,128). In ischemia, epithelial cells will suffer energy depletion and also swell. Second, this, in turn, will trigger "regulated" ATP release, and the released ATP will enter the tubular fluid. This charged molecule at physiologic pH will also be trapped in the tubular lumen, where it can then act as an autocrine or paracrine agonist (limited only by the amount of ecto-ATPase activity present). Third, the ATP will bind to P2 receptors and "autoinhibit" energy-consuming transport processes in more distal tubular segments and so "protect" them. Basolateral P2 receptors could mediate a similar process, although the ATP trapped in the interstitium may be more limited in its diffusion.

In renal ischemia, increased amounts of adenosine are released into the urine (87). However, it may well be that ATP is the primary metabolite released from ischemic cells and that it is then rapidly converted to adenosine. In addition, extracellular ATP acting via P2 receptors has been shown to stimulate cell growth and division in a number of renal (43,59,129) and nonrenal tissues (130). An intriguing preliminary study suggests that shortly after renal ischemia, ion transport processes are down-regulated, whereas expression of the $P2Y_2$ receptor itself is upregulated (131). Thus, it is tempting to regard this ubiquitous compound, ATP, as an "intelligent" extracellular signaling molecule that can prevent ischemic damage, almost before it occurs, and can also assist in cell recovery and regeneration after damage has occurred.

### Acknowledgments
M. A. B. and R. J. U. are grateful to the Wellcome Trust and the St. Peter's Trust for research support. Work by J. L. was made possible by the Deutsche Forschungsgemeinschaft.

### References
1. Jackson, E. K., and Dubey, R. K. (2001). Role of the extracellular cAMP-adenosine pathway in renal physiology. *Am. J. Physiol. Renal Physiol.* **281**, F597–F612.
2. Williams, M., and Jarvis, M. F. (2000). Purinergic and pyrimidinergic receptors as potential drug targets. *Biochem. Pharmacol.* **59**, 1173–1185.

3. Burnstock, G. (1996). Purinoceptors: Ontogeny and phylogeny. *Drug Dev. Res.* **39,** 204–242.
4. Ralevic, V., and Burnstock, G. (1998). Receptors for purines and pyrimidines. *Pharmacol. Rev.* **50,** 413–492.
5. Sumpio, B. E., Hull, M. J., Baue, A. E., and Chaudry, I. H. (1987). Comparison of effects of ATP-MgCl₂ and adenosine-MgCl₂ on renal function following ischemia. *Am. J. Physiol.* **252,** R388–R393.
6. Chan, C. M., Unwin, R. J., and Burnstock, G. (1998). Potential functional roles of extracellular ATP in kidney and urinary tract. *Exp. Nephrol.* **6,** 200–207.
7. Anderson, R. J., Breckon, R., and Dixon, B. S. (1991). ATP receptor regulation of adenylate cyclase and protein kinase C activity in cultured renal LLC-PK1 cells. *J. Clin. Invest.* **87,** 1732–1738.
8. Bailey, M. A., Hillman, K. A., and Unwin, R. J. (2000). P2 receptors in the kidney. *J. Auton. Nerv. Syst.* **81,** 264–270.
9. Bailey, M. A., Imbert-Teboul, M., Turner, C., Marsy, S., Srai, K., Burnstock, G., and Unwin, R. J. (2000). Axial distribution and characterization of basolateral P2Y receptors along the rat renal tubule. *Kidney Int.* **58,** 1893–1901.
10. Bailly, C., Imbert-Teboul, M., Roinel, N., and Amiel, C. (1990). Isoproterenol increases Ca, Mg, and NaCl reabsorption in mouse thick ascending limb. *Am. J. Physiol.* **258,** F1224–F1231.
11. Boese, S. H., Glanville, M., Aziz, O., Gray, M. A., and Simmons, N. L. (2000). Ca²⁺ and cAMP-activated Cl⁻ conductances mediate Cl⁻ secretion in a mouse renal inner medullary collecting duct cell line. *J. Physiol.* **523,** 325–338.
12. Bouyer, P., Paulais, M., Cougnon, M., Hulin, P., Anagnostopoulos, T., and Planelles, G. (1998). Extracellular ATP raises cytosolic calcium and activates basolateral chloride conductance in *Necturus* proximal tubule. *J. Physiol.* **510,** 535–548.
13. Schwiebert, E. M., and Kishore, B. K. (2001). Extracellular nucleotide signaling along the renal epithelium. *Am. J. Physiol. Renal Physiol.* **280,** F945–F963.
14. Firestein, B. L., Xing, M., Hughes, R. J., Corvera, C. U., and Insel, P. A. (1996). Heterogeneity of P2u and P2Y purinergic receptor regulation of phospholipases in MDCK cells. *Am. J. Physiol.* **271,** F610–F618.
15. Friedrich, F., Weiss, H., Paulmichl, M., and Lang, F. (1989). Activation of potassium channels in renal epithelioid cells (MDCK) by extracellular ATP. *Am. J. Physiol.* **256,** C1016–C1021.
16. Insel, P. A., Ostrom, R. S., Zambon, A. C., Hughes, R. J., Balboa, M. A., Shehnaz, D., Gregorian, C., Torres, B., Firestein, B. L., Xing, M., and Post, S. R. (2001). P2Y receptors of MDCK cells: Epithelial cell regulation by extracellular nucleotides. *Clin. Exp. Pharmacol. Physiol.* **28,** 351–354.
17. Lang, F., and Paulmichl, M. (1995). Properties and regulation of ion channels in MDCK cells. *Kidney Int.* **48,** 1200–1205.
18. Orlov, S. N., Dulin, N. O., Gagnon, F., Gekle, M., Douglas, J. G., Schwartz, J. H., and Hamet, P. (1999). Purinergic modulation of Na⁺, K⁺, Cl⁻ contransport and MAP kinases is limited to C11-MDCK cells resembling intercalated cells from collecting ducts. *J. Membr. Biol.* **172,** 225–234.
19. Simmons, N. L. (1981). Identification of a purine (P₂) receptor linked to ion transport in a cultured renal (MDCK) epithelium. *Br. J. Pharmacol.* **73,** 379–384.
20. Woo, J. S., Inoue, C. N., Hanaoka, K., Schwiebert, E. M., Guggino, S. E., and Guggino, W. B. (1998). Adenylyl cyclase is involved in desensitization and recovery of ATP-stimulated Cl⁻ secretion in MDCK cells. *Am. J. Physiol.* **274,** C371–C378.

21. Zegarra-Moran, O., Romeo, G., and Galietta, L. J. V. (1995). Regulation of transepithelial ion transport by two different purinoceptors in the apical membrane of canine kidney (MDCK) cells. *Br. J. Pharmacol.* **114**, 1052–1056.

22. Dai, L. J., Kang, H. S., Kerstan, D., Ritchie, G., and Quamme, G. A. (2001). ATP inhibits $Mg^{2+}$ uptake in mouse distal convoluted tubule cells via P2X purinoceptors. *Am. J. Physiol.* **281**, F833–F840.

23. McCoy, D. E., Taylor, A. L., Kudlow, B. A., Karlson, K., Slattery, M. J., Schwiebert, L. M., Schwiebert, E. M., and Stanton, B. A. (1999). Nucleotides regulate NaCl transport in mIMCD-K2 cells via P2X and P2Y purinergic receptors. *Am. J. Physiol.* **277**, F552–F559.

24. Cuffe, J. E., Bielfeld-Ackermann, A., Thomas, J., Leipziger, J., and Korbmacher, C. (2000). ATP stimulates $Cl^-$ secretion and reduces amiloride-sensitive $Na^+$ absorption in M-1 mouse cortical collecting duct cells. *J. Physiol.* **524**, 77–90.

25. Thomas, J., Deetjen, P., Ko, W. H., Jacobi, C., and Leipziger, J. (2001). $P2Y_2$ receptor-mediated inhibition of amiloride-sensitive short circuit current in M-1 mouse cortical collecting duct cells. *J. Membr. Biol.* **183**, 115–124.

26. Filipovic, D. M., Adebanjo, O. A., Zaidi, M., and Reeves, W. B. (1998). Functional and molecular evidence for P2X receptors in LLC-PK1 cells. *Am. J. Physiol.* **274**, F1070–F1077.

27. Harada, H., Kanai, Y., Tsuji, Y., and Suketa, Y. (1991). P2-purinoceptors in a renal epithelial cell line (LLC-PK1). *Biochem. Pharmacol.* **42**, 1495–1497.

28. Cejka, J. C., Le Maout, S., Bidet, M., Tauc, M., and Poujeol, P. (1994). Activation of calcium influx by ATP and store depletion in primary cultures of renal proximal cells. *Pflugers Arch.* **427**, 33–41.

29. Koster, H. P. G., Hartog, A., van Os, C. H., and Bindels, R. J. M. (1996). Inhibition of $Na^+$ and $Ca^{2+}$ reabsorption by P2U purinoceptors requires PKC but not $Ca^{2+}$ signaling. *Am. J. Physiol.* **270**, F53–F60.

30. Wilson, P. D., Hovater, J. S., Casey, C. C., Fortenberry, J. A., and Schwiebert, E. M. (1999). ATP release mechanisms in primary cultures of epithelia derived from the cysts of polycystic kidneys. *J. Am. Soc. Nephrol.* **10**, 218–229.

31. Schwiebert, E. M. (2001). ATP release mechanisms, ATP receptors and purinergic signalling along the nephron. *Clin. Exp. Pharmacol. Physiol.* **28**, 340–350.

32. Taylor, A. L., Kudlow, B. A., Marrs, K. L., Gruenert, D. C., Guggino, W. B., and Schwiebert, E. M. (1998). Bioluminescence detection of ATP release mechanisms in epithelia. *Am. J. Physiol.* **275**, C1391–C1406.

33. Hazama, A., Hayashi, S., and Okada, Y. (1998). Cell surface measurements of ATP release from single pancreatic beta cells using a novel biosensor technique. *Pflugers Arch.* **437**, 31–35.

34. Schwiebert, E. M., Wallace, D. P., Braunstein, G. M., King, S. R., Peti-Peterdi, J., Hanaoka, K., Guggino, W. B., Guay-Woodford, L. M., Bell, P. D., Sullivan, L. P., Grantham, J. J., and Taylor, A. L. (2002). Autocrine extracellular purinergic signaling in epithelial cells derived from polycystic kidneys. *Am. J. Physiol. Renal Physiol.* **282**, F763–775.

35. Eltze, M., and Ullrich, B. (1996). Characterization of vascular P2 purinoceptors in the rat isolated perfused kidney. *Eur. J. Pharmacol.* **306**, 139–152.

36. Briner, V. A., and Kern, F. (1994). ATP stimulates $Ca^{2+}$ mobilization by a nucleotide receptor in glomerular endothelial cells. *Am. J. Physiol.* **266**, F210–F217.

37. Chan, C. M., Unwin, R. J., Bardini, M., Oglesby, I. B., Ford, A. P., Townsend-Nicholson, A., and Burnstock, G. (1998). Localization of P2X1 purinoceptors by autoradiography and immunohistochemistry in rat kidneys. *Am. J. Physiol.* **274**, F799–F804.

38. Glass, R., and Burnstock, G. (2001). Immunohistochemical identification of cells expressing ATP-gated cation channels (P2X receptors) in the adult rat thyroid. *J. Anat.* **198**, 569–579.

39. Marrelli, S. P. (2001). Mechanisms of endothelial $P2Y_1$- and $P2Y_2$-mediated vasodilation involve differential $[Ca^{2+}]_i$ responses. *Am. J. Physiol Heart Circ. Physiol.* **281**, H1759–H1766.

40. Yamamoto, K., Korenaga, R., Kamiya, A., Qi, Z., Sokabe, M., and Ando, J. (2000). $P2X_4$ receptors mediate ATP-induced calcium influx in human vascular endothelial cells. *Am. J. Physiol. Heart Circ. Physiol.* **279**, H285–H292.

41. Bailey, M. A., Imbert-Teboul, M., Burnstock, G., and Unwin, R. J. (2000). ATP-stimulated phosphoinositide metabolism in rat glomeruli is not stimulated by acute sympathectomy [abstract]. *J. Physiol.* **78**, 523P.

42. Chan, C. M., Unwin, R. J., and Burnstock, G. (1997). RT-PCR micro-localization of P2Y2 purinoceptor mRNA along the rat nephron [abstract]. *J. Am. Soc. Nephrol.* **8**, 434A.

43. Harada, H., Chan, C. M., Loesch, A., Unwin, R. J., and Burnstock, G. (2000). Induction of proliferation and apoptotic cell death via P2Y and P2X receptors, respectively, in rat glomerular mesangial cells 1. *Kidney Int.* **57**, 949–958.

44. Huber-Lang, M., Fischer, K. G., Gloy, J., Schollmeyer, P., Kramer-Guth, A., Greger, R., and Pavenstädt, H. (1997). UTP and ATP induce different membrane voltage responses in rat mesangial cells. *Am. J. Physiol.* **272**, F704–F711.

45. Churchill, P. C., and Ellis, V. R. (1993). Pharmacological characterization of the renovascular P2 purinergic receptors. *J. Pharmacol. Exp. Ther.* **265**, 334–338.

46. Fischer, K. G., Saueressig, U., Jacobshagen, C., Wichelmann, A., and Pavenstädt, H. (2001). Extracellular nucleotides regulate cellular functions of podocytes in culture. *Am. J. Physiol. Renal Physiol.* **281**, F1075–F1081.

47. Pavenstädt, H. (2000). Roles of the podocyte in glomerular function. *Am. J. Physiol. Renal Physiol.* **278**, F173–F179.

48. Inscho, E. W. (2001). P2 receptors in regulation of renal microvascular function. *Am. J. Physiol. Renal Physiol.* **280**, F927–F944.

49. Jankowski, M., Szczepanska-Konkel, M., Kalinowski, L., and Angielski, S. (2001). Cyclic GMP-dependent relaxation of isolated rat renal glomeruli induced by extracellular ATP. *J. Physiol.* **530**, 123–130.

50. Schulze-Lohoff, E., Bitzer, M., Ogilvie, A., and Sterzel, R. B. (1995). P2U-purinergic receptor activation mediates inhibition of cAMP accumulation in cultured renal mesangial cells. *Renal Physiol. Biochem.* **18**, 219–230.

51. Mene, P., and Dunn, M. J. (1988). Eicosanoids and control of mesangial cell contraction. *Circ. Res.* **62**, 916–925.

52. Hus-Citharel, A., Marchetti, J., Meneton, P., and Morelli, A. (1994). Exogenous ATP induces calcium transients and contraction in the Bowman's parietal capsule of the rat glomerulus [abstract]. *J. Am. Soc. Nephrol.* **3**, 717A.

53. Pavenstädt, H., Gloy, J., Leipziger, J., Pfeilschifter, J., Schollmeyer, P., and Greger, R. (1993). Effect of extracellular ATP on contraction, cytosolic calcium activity, membrane voltage and ion currents of rat mesangial cell in primary culture. *Br. J. Pharmacol.* **109**, 953–959.

54. Jankowski, M., Szczepanska-Konkel, M., Kalinowski, L., and Angielski, S. (2000). Bidirectional action of extracellular ATP on intracapillary volume of isolated rat renal glomeruli. *J. Physiol. Pharmacol.* **51**, 491–496.

55. Jankowski, M., Szczepanska-Konkel, M., Kalinowski, L., and Angielski, S. (2001). The role of P2Y-receptors in the regulation of glomerular volume. *Med. Sci. Monit.* **7**, 635–640.

56. Sprague, R. S., Ellsworth, M. L., Stephenson, A. H., and Lonigro, A. J. (1996). ATP: The red blood cell link to NO and local control of the pulmonary circulation. *Am. J. Physiol.* **271,** H2717–H2722.

57. Schwartz, D. D., and Malik, K. U. (1989). Renal periarterial nerve stimulation-induced vasoconstriction at low frequencies is primarily due to release of a purinergic transmitter in the rat. *J. Pharmacol. Exp. Ther.* **250,** 764–771.

58. Oberhauser, V., Vonend, O., and Rump, L. C. (1999). Neuropeptide Y and ATP interact to control renovascular resistance in the rat. *J. Am. Soc. Nephrol.* **10,** 1179–1185.

59. Huwiler, A., and Pfeilschifter, J. (1994). Stimulation by extracellular ATP and UTP of the mitogen-activated protein kinase cascade and proliferation of rat renal mesangial cells. *Br. J. Pharmacol.* **113,** 1455–1463.

60. Schulze-Lohoff, E., Ogilvie, A., and Sterzel, R. B. (1996). Extracellular nucleotides as signalling molecules for renal mesangial cells. *J. Auton. Pharmacol.* **16,** 381–384.

61. Huwiler, A., van Rossum, G., Wartmann, M., and Pfeilschifter, J. (1997). Stimulation by extracellular ATP and UTP of the stress-activated protein kinase cascade in rat renal mesangial cells. *Br. J. Pharmacol.* **120,** 807–812.

62. Huwiler, A., Wartmann, M., van den Bosch, H., and Pfeilschifter, J. (2000). Extracellular nucleotides activate the p38-stress-activated protein kinase cascade in glomerular mesangial cells. *Br. J. Pharmacol.* **129,** 612–618.

63. Gutierrez, A. M., Lou, X., Erik, A., Persson, G., and Ring, A. (2000). Growth hormones reverse desensitization of $P2Y_2$ receptors in rat mesangial cells. *Biochem. Biophys. Res. Commun.* **270,** 594–599.

64. Schulze-Lohoff, E., Zanner, S., Ogilvie, A., and Sterzel, R. B. (1992). Extracellular ATP stimulates proliferation of cultured mesangial cells via P2-purinergic receptors. *Am. J. Physiol.* **263,** F374–F383.

65. Schulze-Lohoff, E., Hugo, C., Rost, S., Arnold, S., Gruber, A., Brune, B., and Sterzel, R. B. (1998). Extracellular ATP causes apoptosis and necrosis of cultured mesangial cells via P2Z/P2X7 receptors. *Am. J. Physiol.* **275,** F962–F971.

66. Solini, A., Chiozzi, P., Falzoni, S., Morelli, A., Fellin, R., and Di Virgilio, F. (2000). High glucose modulates P2X7 receptor-mediated function in human primary fibroblasts. *Diabetologia* **43,** 1248–1256.

67. Vonend, O., Turner, C., Loesch, A., Srai, K., Burnstock, G., and Unwin, R. (2001). P2X7 receptor expression in diabetic nephropathy [abstract]. *J. Am. Soc. Nephrol.* **12,** 849A.

68. Baker, A. J., Mooney, A., Hughes, J., Lombardi, D., Johnson, R. J., and Savill, J. (1994). Mesangial cell apoptosis: The major mechanism for resolution of glomerular hypercellularity in experimental mesangial proliferative nephritis. *J. Clin. Invest.* **94,** 2105–2116.

69. Sugiyama, H., Kashihara, N., Makino, H., Yamasaki, Y., and Ota, A. (1996). Apoptosis in glomerular sclerosis. *Kidney Int.* **49,** 103–111.

70. Poelstra, K., Heynen, E. R., Baller, J. F., Hardonk, M. J., and Bakker, W. W. (1992). Modulation of anti-Thy1 nephritis in the rat by adenine nucleotides: Evidence for an anti-inflammatory role for nucleotidases. *Lab. Invest.* **66,** 555–563.

71. Cha, S. H., Sekine, T., and Endou, H. (1998). P2 purinoceptor localization along rat nephron and evidence suggesting existence of subtypes $P2Y_1$ and $P2Y_2$. *Am. J. Physiol.* **274,** F1006–F1014.

72. Dockrell, M. E., Noor, M. I., James, A. F., and Hendry, B. M. (2001). Heterogeneous calcium responses to extracellular ATP in cultured rat renal tubule cells. *Clin. Chim. Acta* **303,** 133–138.

73. Hafting, T., and Sand, O. (2000). Purinergic activation of BK channels in clonal kidney cells (Vero cells). *Acta Physiol. Scand.* **170,** 99–109.

74. Takeda, M., Kobayashi, M., and Endou, H. (1998). Establishment of a mouse clonal early proximal tubule cell line and outer medullary collecting duct cells expressing P2 purinoceptors. *Biochem. Mol. Biol. Int.* **44,** 657–664.

75. Jin, W., and Hopfer, U. (1997). Purinergic-mediated inhibition of $Na^+$-$K^+$-ATPase in proximal tubule cells: Elevated cytosolic $Ca^{2+}$ is not required. *Am. J. Physiol.* **272,** C1169–C1177.

76. Bailey, M. A., Imbert-Teboul, M., Turner, C., Srai, S. K., Burnstock, G., and Unwin, R. J. (2001). Evidence for basolateral $P2Y_6$ receptors along the rat proximal tubule: Functional and molecular characterization. *J. Am. Soc. Nephrol.* **12,** 1640–1647.

77. Lederer, E. D., and McLeish, K. R. (1995). P2 purinoceptor stimulation attenuates PTH inhibition of phosphate uptake by a G protein-dependent mechanism. *Am. J. Physiol.* **269,** F309–F316.

78. Chan, C. M., Unwin, R. J., and Burnstock, G. (1998). Immunohistochemical localization of P2X5 and P2X6 receptors along the normal rat nephron [abstract]. *J. Am. Soc. Nephrol.* **9,** 420A.

79. Bouyer, P., Paulais, M., Cougnon, M., Hulin, P., Anagnostopoulos, T., and Planelles, G. (1998). Extracellular ATP raises cytosolic calcium and activates basolateral chloride conductance in *Necturus* proximal tubule. *J. Physiol.* **510,** 535–548.

80. Homolya, L., Steinberg, T. H., and Boucher, R. C. (2000). Cell to cell communication in response to mechanical stress via bilateral release of ATP and UTP in polarized epithelia. *J. Cell Biol.* **150,** 1349–1360.

81. Mitchell, C. H., Carre, D. A., McGlinn, A. M., Stone, R. A., and Civan, M. M. (1998). A release mechanism for stored ATP in ocular ciliary epithelial cells. *Proc. Natl. Acad. Sci. USA* **95,** 7174–7178.

82. Cha, S. H., Jung, K. Y., and Endou, H. (1995). Effect of P2Y-purinoceptor stimulation on renal gluconeogenesis in rats. *Biochem. Biophys. Res. Commun.* **211,** 454–461.

83. Diaz-Sylvester, P., Mac, L. M., and Amorena, C. (2001). Peritubular fluid viscosity modulates $H^+$ flux in proximal tubules through NO release. *Am. J. Physiol. Renal Physiol.* **280,** F239–F243.

84. Saiag, B., Bodin, P., Shacoori, V., Catheline, M., Rault, B., and Burnstock, G. (1995). Uptake and flow-induced release of uridine nucleotides from isolated vascular endothelial cells. *Endothelium* **2,** 279–285.

85. Paulais, M., Baudouin-Legros, M., and Teulon, J. (1995). Extracellular ATP and UTP trigger calcium entry in mouse cortical thick ascending limbs. *Am. J. Physiol.* **268,** F496–F502.

86. Beach, R. E., Watts, B. A., III, Good, D. W., Benedict, C. R., and DuBose, T. D., Jr. (1991). Effects of graded oxygen tension on adenosine release by renal medullary and thick ascending limb suspensions. *Kidney Int.* **39,** 836–842.

87. Osswald, H., Schmitz, H. J., and Kemper, R. (1977). Tissue content of adenosine, inosine and hypoxanthine in the rat kidney after ischemia and postischemic recirculation. *Pflügers Arch.* **371,** 45–49.

88. Beach, R. E., and Good, D. W. (1992). Effects of adenosine on ion transport in rat medullary thick ascending limb. *Am. J. Physiol.* **263,** F482–F487.

89. Feraille, E., and Doucet, A. (2001). Sodium–potassium–adenosine triphosphatase-dependent sodium transport in the kidney: Hormonal control. *Physiol. Rev.* **81,** 345–418.

90. Rorive, G., and Kleinzeller, A. (1972). The effect of ATP and $Ca^{2+}$ on the cell volume in isolated kidney tubules. *Biochim. Biophys. Acta* **274,** 226–239.

91. Lemmens, R., Kupers, L., Sevigny, J., Beaudoin, A. R., Grondin, G., Kittel, A., Waelkens, E., and Vanduffel, L. (2000). Purification, characterization, and localization of an ATP diphosphohydrolase in porcine kidney. *Am. J. Physiol. Renal Physiol.* **278,** F978–F988.

92. Sabolic, I., Culic, O., Lin, S. H., and Brown, D. (1992). Localization of ecto-ATPase in rat kidney and isolated renal cortical membrane vesicles. *Am. J. Physiol.* **262,** F217–F228.

93. Osswald, H., Muhlbauer, B., and Schenk, F. (1991). Adenosine mediates tubuloglomerular feedback response: An element of metabolic control of kidney function. *Kidney Int. Suppl.* **32,** S128–S131.

94. Schnermann, J., Sun, D., Samuelson, L. C., Yang, T., Huang, Y. G., Paliege, A., Saunders, T., and Briggs, J. P. (2001). Absence of tubuloglomerular feedback in adenosine 1 receptor (A1AR) knockout mice [abstract]. *Kidney Blood Press. Res.* **24,** 241–242.

95. Persson, A. E. G., Salomonsson, M., Westerlund, P., Greger, R., Schlatter, E., and Gonzalez, E. (1991). Macula densa cell function. *Kidney Int.* **39,** S39–S44.

96. Gonzalez, E., Salomonsson, M., Muller-Suur, C., and Persson, A. E. (1988). Measurements of macula densa cell volume changes in isolated and perfused rabbit cortical thick ascending limb. II. Apical and basolateral cell osmotic water permeabilities. *Acta Physiol. Scand.* **133,** 159–166.

97. Le Hir, M., and Kaissling, B. (1993). Distribution and regulation of renal ecto-5′-nucleotidase: Implications for physiological functions of adenosine. *Am. J. Physiol.* **264,** F377–F387.

98. Deetjen, P., Thomas, J., Lehrmann, H., Kim, S. J., and Leipziger, J. (2000). The luminal P2Y receptor in the isolated perfused mouse cortical collecting duct. *J. Am. Soc. Nephrol.* **11,** 1798–1806.

99. Kishore, B. K., Ginns, S. N., Krane, C. M., Nielsen, S., and Knepper, M. A. (2000). Cellular localization of P2Y2 purinoceptor in rat renal medulla and lung. *Am. J. Physiol.* **278,** F43–F51.

100. Ecelbarger, C. A., Maeda, Y., Gibson, C. C., and Knepper, M. A. (1994). Extracellular ATP increases intracellular calcium in rat terminal collecting duct via a nucleotide receptor. *Am. J. Physiol.* **267,** F998–F1006.

101. Kishore, B. K., Chou, C.-L., and Knepper, M. A. (1995). Extracellular nucleotide receptor inhibits AVP-stimulated water permeability in inner medullary collecting duct. *Am. J. Physiol.* **269,** F863–F869.

102. Rouse, D., Leite, M., and Suki, W. N. (1994). ATP inhibits the hydrostatic effect of AVP in rabbit CCT: Evidence for a nucleotide P2U receptor. *Am. J. Physiol.* **267,** F289–F295.

103. Pilewski, J. M., and Frizzell, R. A. (1999). Role of CFTR in airway disease. *Physiol. Rev.* **79,** S215–S255.

104. Schlatter, E., Greger, R., and Schafer, J. A. (1990). Principal cells of cortical collecting ducts of the rat are not a route of transepithelial Cl⁻ transport. *Pflügers Arch. Eur. J. Physiol.* **417,** 317–323.

105. Lehrmann, H., Thomas, J., Kim, S. J., Jacobi, C., and Leipziger, J. (2002). Luminal P2Y$_2$ receptor-mediated inhibition of Na$^+$ absorption in isolated perfused mouse CCD. *J. Am. Soc. Nephrol.* **13,** 10–18.

106. Greger, R. (2000). Role of CFTR in the colon. *Annu. Rev. Physiol.* **62,** 467–491.

107. Lu, M., MacGregor, G. G., Wang, W., and Giebisch, G. H. (2000). Extracellular ATP inhibits small-conductance K channels on the apical membrane of the cortical collecting duct from mouse kidney. *J. Gen. Physiol.* **116,** 299–310.

108. Gordjani, N., Nitschke, R., Greger, R., and Leipziger, J. (1997). Capacitative Ca$^{2+}$ entry (CCE) induced by luminal and basolateral ATP in polarized MDCK-C7 cells is restricted to the basolateral membrane. *Cell Calcium* **22,** 121–128.

109. Taylor, A. L., Schwiebert, L. M., Smith, J. J., King, C., Jones, J. R., Sorscher, E. J., and Schwiebert, E. M. (1999). Epithelial P2X purinergic receptor channel expression and function. *J. Clin. Invest.* **104,** 875–884.

110. Grantham, J. J., Cowley, B. D., and Torres, V. E. (2000). Progression of autosomal dominant polycystic kidney disease. In "The Kidney: Physiology and Pathophysiology" (D. W. Seldin and G. H. Giebisch, eds.), pp. 2513–2536. Lippincott Williams & Wilkins, Philadelphia, PA.

111. Lu, W., Peissel, B., Babakhanlou, H., Pavlova, A., Geng, L., Fan, X., Larson, C., Brent, G., and Zhou, J. (1997). Perinatal lethality with kidney and pancreas defects in mice with a targeted Pkd1 mutation. *Nat. Genet.* **17,** 179–181.

112. Cowley, B. D., Jr., Gudapaty, S., Kraybill, A. L., Barash, B. D., Harding, M. A., Calvet, J. P., and Gattone, V. H. (1993). Autosomal-dominant polycystic kidney disease in the rat. *Kidney Int.* **43,** 522–534.

113. Mandell, J., Koch, W. K., Nidess, R., Preminger, G. M., and McFarland, E. (1983). Congenital polycystic kidney disease: Genetically transmitted infantile polycystic kidney disease in C57BL/6J mice. *Am. J. Pathol.* **113,** 112–114.

114. Banderali, U., Brochiero, E., Lindenthal, S., Raschi, C., Bogliolo, S., and Ehrenfeld, J. (1999). Control of apical membrane chloride permeability in the renal A6 cell line by nucleotides. *J. Physiol.* **519,** 737–751.

115. Inglis, S. K., Collett, A., McAlroy, H. L., Wilson, S. M., and Olver, R. E. (1999). Effect of luminal nucleotides on $Cl^-$ secretion and $Na^+$ absorption in distal bronchi. *Pflügers Arch.* **438,** 621–627.

116. Chan, H. C., Liu, C. Q., Fong, S. K., Law, S. H., Wu, L. J., So, E., Chung, Y. W., Ko, W. H., and Wong, P. Y. D. (1997). Regulation of $Cl^-$ secretion by extracellular ATP in cultured mouse endometrial epithelium. *J. Membr. Biol.* **156,** 45–52.

117. Clarke, L. L., Harline, M. C., Otero, M. A., Glover, G. G., Garrad, R. C., Krugh, B., Walker, N. M., Gonzalez, F. A., Turner, J. T., and Weisman, G. A. (1999). Desensitization of P2Y2 receptor-activated transepithelial anion secretion. *Am. J. Physiol.* **276,** C777–C787.

118. Furukawa, M., Ikeda, K., Yamaya, M., Oshima, T., Sasaki, H., and Takasaka, T. (1997). Effects of extracellular ATP on ion transport and $[Ca^{2+}]_i$ in Mongolian gerbil middle ear epithelium. *Am. J. Physiol.* **272,** C827–C836.

119. Kerstan, D., Gordjani, N., Nitschke, R., Greger, R., and Leipziger, J. (1998). Luminal ATP induces $K^+$ secretion via a P2Y2 receptor in rat distal colonic mucosa. *Pflügers Arch. Eur. J. Physiol.* **436,** 712–716.

120. Leipziger, J., Kerstan, D., Nitschke, R., and Greger, R. (1997). ATP increases $[Ca^{2+}]_i$, and ion secretion via a basolateral P2Y receptor in rat distal colonic mucosa. *Pflügers Arch. Eur. J. Physiol.* **434,** 77–83.

121. Guo, X., Merlin, D., Harvey, R. D., Laboisse, C., and Hopfer, U. (1995). Stimulation of $Cl^-$ secretion by extracellular ATP does not depend on increased cytosolic $Ca^{2+}$ in HT-29.cl16E. *Am. J. Physiol.* **269,** C1457–C1463.

122. Inoue, C. N., Woo, J. S., Schwiebert, E. M., Morita, T., Hanaoka, K., Guggino, S. E., and Guggino, W. B. (1997). Role of purinergic receptors in chloride secretion in Caco-2 cells. *Am. J. Physiol.* **272,** C1862–C1870.

123. Yamaya, M., Sekizawa, K., Kakuta, Y., Ohrui, T., Sawai, T., and Sasaki, H. (1996). P2u-purinoceptor regulation of chloride secretion in cultured human tracheal submucosal glands. *Am. J. Physiol.* **270,** L979–L984.

124. Gardner, K. D., Jr., Burnside, J. S., Elzinga, L. W., and Locksley, R. M. (1991). Cytokines in fluids from polycystic kidneys. *Kidney Int.* **39,** 718–724.

125. Grantham, J. J. (1997). Mechanisms of progression in autosomal dominant polycystic kidney disease. *Kidney Int. Suppl.* **63,** S93–S97.

126. Wang, Y., Roman, R. M., Lidofsky, S. D., and Fitz, J. G. (1996). Autocrine signaling through ATP release represents a novel mechanism for cell volume regulation. *Proc. Natl. Acad. Sci. USA* **93,** 12020–12025.

127. Lazarowski, E. R., Boucher, R. C., and Harden, T. K. (2000). Constitutive release of ATP and evidence for major contribution of ecto-nucleotide pyrophosphatase and nucleoside diphosphokinase to extracellular nucleotide concentrations. *J. Biol. Chem.* **275,** 31061–31068.
128. Koyama, T., Oike, M., and Ito, Y. (2001). Involvement of Rho-kinase and tyrosine kinase in hypotonic stress-induced ATP release in bovine aortic endothelial cells. *J. Physiol.* **532,** 759–769.
129. Paller, M. S., Schnaith, E. J., and Rosenberg, M. E. (1998). Purinergic receptors mediate cell proliferation and enhanced recovery from renal ischemia by adenosine triphosphate. *J. Lab. Clin. Med.* **131,** 174–183.
130. Hou, M., Moller, S., Edvinsson, L., and Erlinge, D. (2000). Cytokines induce upregulation of vascular $P2Y_2$ receptors and increased mitogenic responses to UTP and ATP. *Arterioscler. Thromb. Vasc. Biol.* **20,** 2064–2069.
131. Kishore, B. K., Wang, Z., Rabb, H., Haq, M., and Soleimani, M. (1997). Upregulation of P2Y2 purinoceptor during ischemic reperfusion injury (IRI): Possible relevance to diuresis of IRI [abstract]. *J. Am. Soc. Nephrol.* **9,** 580A.
132. Hollopeter, G., Jantzen, H. M., Vincent, D., Li, G., England, L., Ramakrishnan, V., Yang, R. B., Nurden, P., Nurden, A., Julius, D., and Conley, P. B. (2001). Identification of the platelet ADP receptor targeted by antithrombotic drugs. *Nature* **409,** 202–207.
133. Khakh, B. S., Burnstock, G., Kennedy, C., King, B. F., North, R. A., Seguela, P., Voigt, M., and Humphrey, P. P. (2001). International union of pharmacology. XXIV. Current status of the nomenclature and properties of P2X receptors and their subunits. *Pharmacol. Rev.* **53,** 107–118.

# CHAPTER 12

# Purinergic Receptors and Hepatobiliary Function

**Andrew P. Feranchak\* and J. Gregory Fitz[†]**
\*Department of Pediatrics and [†]Department of Medicine, University of Colorado Health Sciences Center, Denver, Colorado 80262

## I. INTRODUCTION

While purinergic signaling has been well characterized in neuronal models, evidence that epithelial cells also utilize extracellular nucleotides to regulate a broad range of cell and organ level functions has emerged only more recently. In liver, purinergic receptors have been identified on the plasma membrane of the two principal epithelial cell types that form the bile secretory unit, including *hepatocytes*, which constitute the liver parenchymal cells, and *cholangiocytes*, which line the lumen of intrahepatic bile ducts. Activation of purinergic receptors by selective binding of nucleotide or nucleoside agonists has been linked to several fundamental responses important to cellular metabolism, ion channel activation, cell volume regulation, and bile formation. Accordingly, the first portion of this review describes the basic mechanisms that contribute to purinergic signaling in liver cells, with special attention to (1) a molecular and pharmacologic description of the receptors present, (2) mechanisms and regulation of ATP release, and (3) nucleotide/nucleoside degradation and reuptake pathways. The second portion of this review addresses current concepts regarding the potential roles of purinergic signaling in the regulation of liver-specific functions, including (1) hepatic metabolism/glycogenolysis, (2) cell volume regulation, and (3) secretion and bile formation.

## II. HEPATIC PURINERGIC RECEPTOR EXPRESSION

In liver, there is evidence that both hepatocytes and cholangiocytes express adenosine receptors as well as one or more P2 receptor types (Harden *et al.*, 1997; Schlenker et al., 1997). A cDNA encoding a functional $P2Y_2$ receptor of 374 amino acids was cloned from a rat liver cDNA library. When expressed in *Xenopus laevis* oocytes, the agonist properties (ATP = UTP > ADP ≫ adenosine) are highly homologous to $P2Y_2$ receptors from brain and other organs. In addition, pharmacologic studies suggest the presence of one or more novel P2Y receptors in hepatocytes that do not conform to the known agonist potencies of the cloned receptors. In rat liver plasma membrane preparations, binding of $\beta,\gamma$-methylene-ATP stimulates phospholipase A in the absence of effects on phospholipase C (Malcolm *et al.*, 1995). Definition of the number and types of P2Y receptors and their localization within the hepatic lobule is likely to provide interesting insights into the structural anatomy of this signaling complex.

ATP and UTP in nanomolar concentrations also mobilize intracellular calcium in cholangiocytes, and mRNA encoding the same $P2Y_2$ receptor found in hepatocytes is detectable as a 33-kDa band by Northern analysis (Schlenker *et al.*, 1997). Moreover, stimulation of the apical membrane of normal rat cholangiocytes (NRCs) in monolayer culture results in large increases in $Cl^-$ secretion with an agonist specificity of ATP = UTP > ADP ≫ adenosine, properties consistent with expression in the apical domain of

P2Y$_2$ receptors (Salter *et al.*, 2000a). In addition, polymerase chain reaction (PCR) products corresponding to P2Y$_1$, P2Y$_2$, P2Y$_4$, and P2Y$_6$ have been identified in primary rat cholangiocytes as well (Dranoff *et al.*, 2001). In polarized cholangiocyte monolayers, the apical and basolateral domains exhibit distinct pharmacologic profiles, consistent with regional differences in the type(s) of receptor expressed (Salter *et al.*, 2000a). The *apical* response to nucleotides (as measured by changes in short-circuit current, $\Delta I_{sc}$) is dominated by expression of purinergic receptors of the P2Y$_2$ subtype: ATP, UTP, and ATPγS are equally effective at stimulation of transepithelial currents with half-maximal responses at UTP concentrations near 300 n$M$. ADP, AMP, and adenosine are without effect. In contrast, the *basolateral* response shows a different agonist preference of ADP $\geq$ 2-methylthio-ATP (2MeS-ATP) $\geq$ ATP $>$ adenosine 5′-$O$-3-thiotriphosphate (ATPγS) $\geq$ UTP $\gg$ AMP. Adenosine, adenylyl ($\beta$,γ-methylene)-diphosphonate (AMP-PCP), and $\alpha$,$\beta$-methylene ATP (AMP-CPP) have no effect. The maximal current response of the basolateral membrane to ATP occurs at concentrations near 500 n$M$; half-maximal increases occur at ATP concentrations near 50 n$M$ (Salter *et al.*, 2000a). These basolateral responses are not consistent with any of the cloned P2Y receptor subtypes. Thus, the basolateral membrane of NRC monolayers shows a pattern of response distinctly different from the apical membrane, and more than one receptor subtype is involved.

Overall, it is clear that small concentrations of nucleotides can exert potent regulatory effects on both liver and biliary cells. The response, however, can be complex based on the types of nucleotides and nucleosides available at the cell surface as well as the repertoire of receptors expressed by target cells. Both hepatocytes and cholangiocytes express P2Y$_2$ receptors, but each cell type also expresses one or more additional receptor subtypes that have not yet been defined but contribute importantly to the overall response.

## III. ATP RELEASE MECHANISMS

In intact liver, ATP is present in both venous effluent and bile (Nukina *et al.*, 1994; Chari *et al.*, 1996). In cultured hepatocytes and cholangiocytes, nanomolar concentrations of ATP are always detectable in the supernatant fluid because of constitutive release of cellular ATP. Moreover, the apparent rate of ATP release is closely regulated in response to changing physiologic demands, consistent with a role for extracellular ATP as a secondary signaling molecule. At present, the molecular basis of epithelial ATP release is unknown. Theoretically, epithelial cell ATP release could be mediated through vesicular transport, membrane diffusion, or conductive movement through membrane channels or pores. Using electrophysiologic techniques, it has been demonstrated that charged ATP molecules are capable of traversing cell membranes (Wang *et al.*, 1996; Taylor *et al.*, 1998). This observation suggests that ATP release is electrogenic and is mediated, at least in part, through the opening of membrane channels. Definition of the molecular basis of this conductance and its relationship to other channel proteins represents an imperative issue for defining the cellular strategies regulating purinergic signaling.

## A. Hepatocyte ATP Release

ATP release has been demonstrated in a number of liver preparations, including primary rat hepatocytes, hepatoma cells, and isolated perfused rat liver (Wang *et al.*, 1996; Roman *et al.*, 1997). Importantly, primary human hepatocytes also demonstrate constitutive, mechanical, and volume-stimulated ATP release (Fig. 1) (Feranchak *et al.*, 2000), suggesting that these observations are relevant *in vivo* as well. To date, increases in cell volume represent the most potent stimulus for ATP release yet defined. Consequently, hypotonic exposure (to increase cell volume) has been widely utilized as a model for investigation of the mechanisms involved. In each of these models, increases in cell volume stimulate 3- to 10-fold increases in extracellular ATP concentrations that peak within 1 min and then gradually decrease toward basal levels as cell volume is restored toward basal levels. To date there has been no clear localization of the cellular site(s) (e.g., apical/canalicular versus basolateral) of ATP release. Our bias, described below, is that ATP release occurs at both sites. However, the physiologic roles and cellular mechanisms involved are likely to be distinct. If ATP signals to cells further down the bile secretory unit, then an apical/canalicular orientation of release into the bile lumen would be anticipated. At present, there is no direct evidence in support of this because of the difficulty in sampling from the small canalicular space between cells, which measures less than 1 $\mu$m in diameter. However, ATP is present in bile collected downstream in the common bile duct, and there is considerable evidence for a heavy cellular investment in the enzymes and transporters that contribute to ATP availability in the canalicular space. For example, the ATP channel regulator MDR-1 (multidrug resistance P-glycoprotein) is present in the apical/canalicular membrane, and ATP released into the canalicular lumen is subject to degradation by resident ectonucleotidase activity (Che *et al.*, 1997). Because of the anatomical relationship between the canalicular space and the lumen of intrahepatic bile ducts, any ATP that escapes local degradation within the canaliculus would have direct access to

P2Y receptors located further down the secretory unit in the apical membrane of cholangiocytes.

## B. Cholangiocyte ATP Release

On the basis of studies of cholangiocytes in culture, regulated release of ATP from cholangiocytes also is likely to contribute to the ATP detected in bile. Normal rat cholangiocytes (NRCs) grow in monolayer culture and the apical and basolateral compartments are isolated by high-resistance junctions ($>1000$ $\Omega \cdot cm^2$) between cells, so the different compartments can be studied in isolation (Vroman and LaRusso, 1996; Salter et al., 2000b). Interestingly, both membrane domains exhibit constitutive ATP efflux and express P2 receptors. However, ATP release is polarized, with concentrations in the apical (lumenal) chamber ($\sim$250 n$M$) that are consistently 5-fold greater than those in the basolateral chamber ($\sim$50 n$M$) (Salter et al., 2000a). Notably, extracellular ATP is derived from constitutive release of $<1\%$ of total cellular stores, and results in local extracellular ATP concentrations that are close to those required for half-maximal stimulation of $Cl^-$ secretion.

Taken together, these findings indicate that the majority of ATP in bile is likely to be derived from release across the canalicular membrane of hepatocytes and the apical membrane of cholangiocytes. At present, the relative contribution of the two cell types is unknown because of the inability to make measurements within the lumen of the canalicular space. Maintenance of local ATP concentrations within the lumen is likely to be complex because of changes in bile flow, changes in the rates of ATP release, and potent mechanisms for nucleotide degradation (described in more detail below) (Che et al., 1997). However, even small changes in ATP availability would be expected to influence cholangiocyte $P2Y_2$ receptor activity and $Cl^-$ secretion since the best evidence suggests that local concentrations are within the range required for receptor binding (Chari et al., 1996).

## IV. REGULATION OF NUCLEOTIDE RELEASE

### A. Cell Volume

Increases in cell volume result in a rapid increase in membrane ATP permeability that parallels changes in the transmembrane osmolar gradient; and decreases in cell volume have opposite effects. This relationship is likely to be physiologically relevant since in liver epithelia cell volume is not a

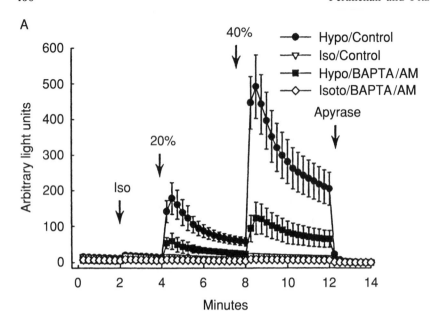

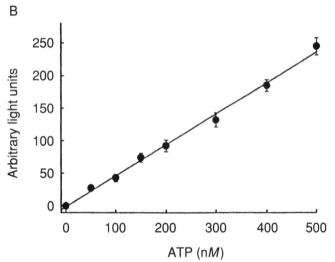

**FIGURE 1**   Primary human hepatocytes release ATP. (A) A luciferase–luciferin assay was
utilized to detect extracellular ATP and values are reported as arbitrary light units. Basal ATP
efflux is always detected. Addition of isotonic medium to cells leads to a small increase in ATP
release secondary to mechanical stimulation. Increases in cell volume stimulated by dilution of
medium by 20 and 40% increases ATP release (Hypo/Control) much more than is observed with
equal additions of isotonic medium (Iso/Control). Preincubation with BAPTA-AM (50 $\mu M$)
decreases both basal and hypotonically induced ATP release. Addition of apyrase to scavenge

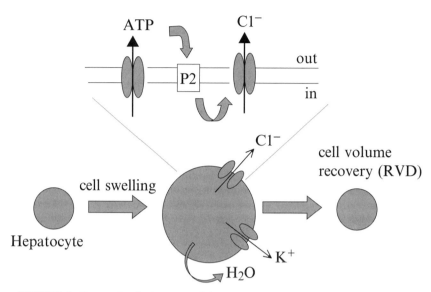

**FIGURE 2** Proposed role for extracellular ATP in regulation of hepatic cell volume. Increase in hepatocyte cell volume (secondary to uptake of organic solutes, exposure to insulin, or hypotonicity) result in ATP release, P2 receptor stimulation, and $Cl^-$ and $K^+$ channel activation. Efflux of ions is followed by water and subsequent cell volume recovery. This process is known as regulatory volume decrease (RVD).

static parameter. Rather, there are regulated changes that occur in response to hormonal signals and changes in solute transport (Haussinger and Schliess, 1999). Moreover, exposure to exogenous ATP causes a rapid decrease in cell volume due to activation of $K^+$ and $Cl^-$ channels; and findings in different liver and biliary cell models suggest that extracellular ATP plays a key role in recovery from cell swelling (Fig. 2) (Wang *et al.*, 1996; Roman *et al.*, 1997). Interestingly, volume-sensitive changes in ATP release are not limited to liver cells, but have been demonstrated in some (but not all) other cell types as well (Taylor *et al.*, 1998; Schwiebert *et al.*, 1998; Musante *et al.*, 1999; Light *et al.*, 1999; Mitchell *et al.*, 1998). Accordingly, considerable attention has been focused on defining the cellular mechanisms that sense these changes in cell volume and coupling them to changes in membrane ATP permeability. One proposed model

---

extracellular ATP abolishes bioluminescence. (B) ATP standard curve. Bioluminescence was determined by adding standard amounts of ATP to OptiMEM-1 containing luciferase–luciferin in the absence of cells. Adapted with permission from Feranchak, A. P., Fitz, J. G., and Roman, R. M. (2000). *J. Hepatol.* **33**, 174–182.

implicating a role for purinergic signaling in cell volume regulation is described in more detail below.

## B. ABC Proteins

In the search to identify the molecular basis for channel-mediated ATP transport, particular attention has been devoted to the potential role(s) of ATP-binding cassette (ABC) proteins, such as the $Cl^-$ channel protein encoded by the cystic fibrosis transmembrane conductance regulator *(CFTR)* gene and the multidrug resistance P-glycoprotein encoded by *MDR1*. Native expression, upregulation, and heterologous expression of these proteins are associated with enhanced ATP permeability and ATP channel activity in some but not all model systems (Taylor *et al.*, 1998; Roman *et al.*, 1997; Bosch *et al.*, 1996). However, reconstituted membrane vesicles expressing CFTR do not transport ATP (Al-Awqati, 1995). In addition, many cells exhibit regulated ATP release in the absence of apparent CFTR or MDR-1 expression, and studies support the concept that CFTR and MDR are more likely to serve as ATP channel regulators, increasing the amount of ATP released through an as-yet-unidentified ATP channel (Pasyk and Foskett, 1997; Braunstein *et al.*, 2001). Notably, however, modulation of CFTR or MDR function is capable of modulating ATP release.

In hepatocytes, which do not express CFTR, MDR-1 appears to play an important regulatory role (Roman *et al.*, 2001). In HTC hepatoma cells, overexpression of MDR proteins increases both constitutive and volume-stimulated ATP release (Roman *et al.*, 1997). In addition, volume-stimulated ATP release is inhibited by the MDR-1 transport inhibitors verapamil and cyclosporin (Roman *et al.*, 2001). Since volume-sensitive ATP release leads to receptor binding and activation of $Cl^-$ channels in liver and biliary cells, measurement of volume-sensitive $Cl^-$ currents provides a convenient bioassay for ATP available at the cell surface. Intracellular dialysis with antibodies to MDR proteins (C219 antibodies) inhibits volume-stimulated $Cl^-$ currents. Last, overexpression of MDR-1 in NIH 3T3 cells increases ATP release rates. However, the transport activity of MDR-1 can be dissociated from its role in regulated ATP release. $Gd^{3+}$, a known inhibitor of ATP release (Roman *et al.*, 1999), has no effect on the transport of rhodamine-123. Conversely, alteration of MDR-1 substrate selectivity, by mutation of specific protein sequences, has no effect on ATP release. Taken together, these studies demonstrate that MDR-1 plays an important role in regulation of hepatocyte ATP permeability, but it is probably not the ATP channel itself.

Cholangiocytes are unique in that they express both MDR-1 and CFTR proteins (Scoazec *et al.*, 1997). CFTR has been shown to regulate ATP release and cell volume regulation in a manner analogous to MDR (Braunstein *et al.*, 2001). CFTR expression stimulated both constitutive and volume-sensitive ATP release, and enhanced recovery from cell swelling. The relationship between CFTR, MDR-1, and ATP release in cholangiocytes is presently unknown, but is likely to be complex. It seems likely that multiple ATP transport pathways will be defined in future, and mechanisms of regulation are likely to be cell type specific in order to respond to different physiologic demands. This represents an exciting area for future investigation.

## C. Calcium

Intracellular $Ca^{2+}$ is required for ATP release, but it is not clear that it serves a primary regulatory role. These findings are of interest since intracellular $Ca^{2+}$ also modulates responses to cell volume increases and mediates adaptive opening of $K^+$ and $Cl^-$ channels, solute efflux, and restoration of cell volume toward basal values (Roman *et al.*, 1996, 1998). In human hepatocytes, chelation of intracellular $Ca^{2+}$ by exposure to 1,2-bis(2-aminophenoxy)ethane-$N,N,N',N'$-tetraacetic acid (BAPTA-AM) inhibits both constitutive and volume-sensitive ATP release by ~60 and ~90%, respectively, and prevents $Cl^-$ channel opening and recovery from cell swelling (Feranchak *et al.*, 2000). Similarly, BAPTA-AM inhibits ATP release from biliary cells (Salter *et al.*, 2000a). In contrast, neither increases in cAMP levels nor inhibition of cAMP-dependent signaling has any substantial regulatory effect. It is notable that increases in cell volume have only small effects on intracellular $Ca^{2+}$ in these cells, and that increases in intracellular $Ca^{2+}$ caused by exposure to ionomycin (2 $\mu M$) or thapsigargin (1 $\mu M$) stimulate ATP release but to a much smaller extent (Salter *et al.*, 2000a). Thus, intracellular $Ca^{2+}$ appears to be necessary for ATP release, but may not be the primary signaling pathway involved.

## D. Kinase-Dependent Signaling Pathways

Increases in cell volume have been shown to cause parallel activation of a number of kinases, including protein kinase C and phosphoinositol 3-kinase (PI 3-kinase) (Krause *et al.*, 1996). PI 3-kinase is a heterodimer that on activation phosphorylates phosphatidylinositol and generates at least three lipid mediators, including phosphatidylinositol 3-phosphate (PtdIns-3-P), phosphatidylinositol 3,4-diphosphite (PtdIns-3,4-$P_2$), and phosphatidylinositol 3,4,5-triphosphate (PtdIns-3,4,5-P$_3$). In liver cells, there is increasing evidence that these phospholipids play a major role in vesicular trafficking, movement of bile acid transporters to the canalicular membrane, and bile formation (Folli et al., 1997; Misra et al., 1998, 1999).

Several observations indicate that PI 3-kinase also plays a central role in the regulation of hepatic ATP release (Feranchak et al., 1998, 1999). First, exposure to the PI 3-kinase inhibitors wortmannin and LY294002 sharply decreases cellular ATP release and prevents volume-sensitive activation of Cl$^-$ currents, a convenient bioassay for ATP available at the cell surface, in both hepatocytes and cholangiocytes (Fig. 3). Second, the same inhibitors also prevent recovery from swelling in isolated cells and $\Delta I_{sc}$ in NRC monolayers (Feranchak et al., 1999). Each effect is attributable to impaired delivery of ATP to the cell surface since the Cl$^-$ secretory response to exogenous ATP is retained. Third, intracellular dialysis with the synthetic products of PI 3-kinase leads to activation of Cl$^-$ currents in cholangiocytes even in the absence of cell volume changes (Feranchak et al., 1999). Intracellular delivery of these lipids through the patch pipette permits careful comparison among cells under conditions in which key components of the membrane regulatory apparatus (i.e., receptors, cytoskeleton, and channels) remain largely intact. Intracellular dialysis with PtdIns-3,4-P$_2$ or PtdIns-3,4,5-P$_3$ results in spontaneous activation of currents over several minutes as the lipids diffuse into the cell interior. Moreover, the response to intracellular phospholipids is eliminated by removal of extracellular ATP, consistent with a model wherein the lipid products of PI 3-kinase directly modulate membrane ATP permeability (Feranchak et al., 1999).

This represents an attractive mechanism for site-specific regulation of membrane ATP permeability by a membrane-derived phospholipid. On the basis of the established role of PI 3-kinase in the regulation of vesicular trafficking, including other transporters involved in bile formation (Misra et al., 1998, 1999), it is possible that PI 3-kinase signaling leads to insertion of new ATP channels or to the opening of preexisting ATP channels. However, the molecular basis of ATP release has not been defined, so the specific site of action of these lipids is not known.

## V. DEGRADATION AND REUPTAKE PATHWAYS

### A. ATPase Activity

Degradation of extracellular nucleotides is essential for termination of purinergic responses and for generation of purine and pyrimidine bases that can then be taken up by the cell and recycled into nucleotides. The precise mechanisms involved are not fully defined in liver. Ectonucleotidase activity has been well documented in the hepatocellular canalicular membrane (Che et al., 1997). At present, however, the definitive molecular identities of ectonucleotidase in liver are unknown. Transcripts encoding an ectoapyrase (CD39) isolated from rat brain are abundant in liver (Wang et al., 1997). In addition, there is a potent ecto-5′-nucleotidase in the same domain, and membrane turnover of the enzyme is rapid (Misumi et al., 1990). Little is presently known about the presence and type of ecto-ATP/Dases on biliary epithelial cells. However, in vitro studies demonstrate substantial differences in the kinetics of ATP degradation between apical and basolateral surfaces. In the apical compartment, the time course of nucleotide degradation can be described by a single exponential resulting in the clearance of 13% of ATP in the first minute. In the basolateral compartment, the time course for clearance is faster and more complex, suggesting the presence of more than one degradation pathway. These differences in degradation kinetics are likely due to location-specific biliary cell ecto-ATPases (Salter et al., 2000a). Clearly, the regulation of nucleotide degradation is an important factor in the modulation of nucleotide concentrations in bile and hence bile formation.

## B. Salvage Pathways

Cellular uptake of nucleosides attenuates adenosine-mediated effects and provides substrates for the regeneration of intracellular nucleotides and important metabolic pathways (Griffith and Jarvis, 1996). The liver represents a key source for purine bases and nucleosides that are used by other tissues, and hepatocytes are the primary scavengers of nucleosides from the extrahepatic circulation. Several equilibrative nucleoside transporters (ENTs) have been identified in liver, including rENT1 (rat equilabrative nucleoside transporter) (Yao et al., 1997). Equilibrative transporters are notably bidirectional and therefore also represent potential efflux pathways in hepatocytes. In addition, concentrative nucleoside transporters (CNTs) are also present. Both CNT-2 (SPNT, sodium-dependent purine nucleoside transporter) and CNT-1 cDNAs have been isolated from liver (Che et al., 1995; Felipe et al., 1998). These transporters are electrogenic, with net movement of positive charge with each transport cycle. Together, equilibrative and $Na^+$-dependent transporters account for avid cellular uptake of adenosine and uridine from the extracellular space.

Taken together, these findings support the presence of an extensive network of enzymes and tranporters that function to regulate the fate and concentration of nucleotides released into the extracellular space. Because

A

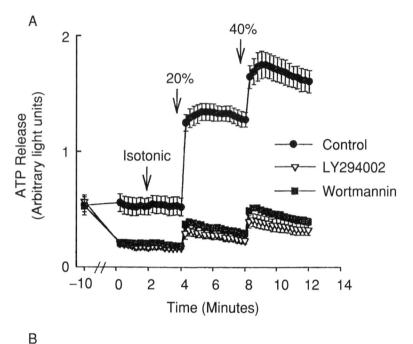

B

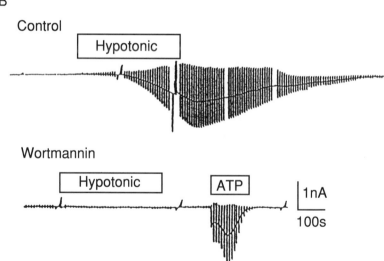

**FIGURE 3** ATP release and activation of volume-sensitive Cl⁻ channels is modulated by PI
3-kinase in cholangiocytes. (A) Volume-stimulated ATP release. Increases in cell volume
(stimulated by 20 and 40% hypotonicity) enhance ATP release from normal rat cholangiocyte
(NRC) monolayers. Inhibition of PI 3-kinase with wortmannin (50 n$M$ ) or LY294002 (10 $\mu M$ )
attenuates basal and volume-stimulated ATP efflux. (B) Inhibition of PI 3-kinase prevents

the product of one reaction is often the substrate for another reaction, elucidation of specific enzyme kinetics, regulation, and tissue distribution is likely to provide important insights into local agonist availability. The overall kinetics of ATP release, receptor binding, and nucleotide degradation are yet to be defined.

## VI. POTENTIAL PHYSIOLOGICAL ROLES FOR HEPATIC PURINERGIC SIGNALING

Liver cells possess a broad range of purinergic receptor subtypes. The presence and roles of adenosine and P2X receptors are not well defined. However, studies from a variety of laboratories are beginning to define distinct roles for P2Y signaling in hepatic and biliary function. In this section, emerging evidence for regulatory effects of extracellular ATP on liver cell metabolism, cell volume regulation, paracrine signaling, and bile formation are described.

## VII. HEPATIC GLUCOSE METABOLISM

Purinergic stimulation of liver cells is capable of influencing an array of metabolic processes including hepatic protein synthesis, gluconeogenesis, and ureagenesis. Among these, the effects on glucose metabolism have been best defined. Purinergic stimulation in isolated, perfused rat liver elicits a potent glucagon-like glycolytic response mediated by activation of glycogen phosphorylase and stimulation of hepatic glucose output (Keppens and De Wulf, 1991; Haussinger et al., 1987; Keppens et al., 1993). The signals that couple P2 receptor stimulation to the glycolytic response are presently unknown. However, both ATP and UTP stimulate hepatic glucose output through activation of separate receptors and signaling pathways. The hepatocyte response to ATP involves stimulation of phospholipase C, generation of inositol phosphates, and increases in cytosolic $Ca^{2+}$

---

activation of volume-sensitive $Cl^-$ currents. Whole cell $Cl^-$ currents (measured at $-80$-mV test pulse at 10-s intervals) were measured under basal conditions and following exposure to hypotonic medium (20% decrease in NaCl). Incubation with wortmannin to inhibit PI 3-kinase (lower tracing) inhibited the $Cl^-$ current response. However, the inhibition was overcome by subsequent addition of ATP (10 $\mu M$) to the medium. These results suggest that extracellular ATP is essential for volume-stimulated $Cl^-$ channel activation and that ATP release is mediated in part by PI 3-kinase. Adapted with permission from Feranchak, A. P., Roman, R. M., Salter, K. D., Toker, A., and Fitz, J. G. (1999). *J. Biol. Chem.* **274**, 30979–30986.

concentrations (Fitz and Sostman, 1994). Notably, the $P2Y_1$-preferring agonist 2MeS-ATP is $\sim$25 times more potent than ATP, activating glycogen phosphorylase in the apparent absence of significant effects on inositol phosphates (Keppens and De Wulf, 1991). While ATP exerts potent regulatory control over hepatic glucose metabolism, the mechanisms involved are complex and characterization of the receptors and signals involved will require further study.

## VIII. AUTOCRINE REGULATION OF CELL VOLUME

Purinergic signaling has been shown to play a fundamental role in autocrine regulation of liver and biliary cell volume. Cell volume regulation is of special importance in liver since these cells receive dual perfusion from both the systemic and portal circulations and are exposed to unusually large changes in the concentrations of solutes such as amino acids, bile acids, and glucose between the fed and fasted states. Since these solutes are concentrated intracellularly, and osmolar gradients as low as $\sim$1 mOsm $\cdot$ liter$^{-1}$ are sufficient to induce transmembrane water movement, liver cell volume can increase by 5–10% in response to increased solute uptake (Haussinger, 1996; Haussinger and Schliess, 1995). These primary changes in volume are followed after a delay by compensatory increases in membrane ion permeability of 20-fold or more (opening of $K^+$ and $Cl^-$ channels), water efflux, and restoration of cell volume toward basal values (Fig. 2). This process is referred to as regulatory volume decrease (RVD) (Sarkadi and Parker, 1991; Strange et al., 1996). Several lines of evidence suggest that extracellular ATP plays a key role in RVD (Wang et al., 1996; Roman et al., 1997). First, as described previously, increases in cell volume represent a potent stimulus for ATP release, increasing local concentrations of extracellular ATP through effects on membrane ATP permeaility (Wang et al., 1996; Feranchak et al., 1998, 2000). Second, removal of extracellular ATP with apyrase (which hydrolyzes ATP) abolishes both $Cl^-$ channel activation and cell volume recovery from swelling. However, ATPγS, a nonhydrolyzable ATP analog, restores volume recovery in the presence of apyrase. Finally, volume-activated $Cl^-$ currents and RVD are inhibited by P2Y receptor blockers, such as suramin and Reactive Blue 2. These findings are consistent with a model whereby ATP released from liver cells serves as an autocrine signal controlling cell volume recovery through activation of membrane $Cl^-$ channels coupled to P2 receptors.

It is notable that volume-sensitive ATP release is not unique to liver cells, but is detectable in many epithelia. Thus, regulated release of ATP represents a general property of epithelial cells. Consequently, it is attractive to speculate that epithelial ATP release and P2 receptor binding may play a central role in coupling physiologic changes in the cellular hydration state to activation of cell-specific effector pathways in cells appropriately endowed with these signaling elements.

## IX. PARACRINE COORDINATION OF CALCIUM SIGNALING

In liver, ATP is detectable in blood, bile, and interstitial fluids under normal conditions. The amount of ATP released is small, representing <1% of total cellular stores. However, nanomolar concentrations of ATP at levels 10,000-fold below those found intracellularly are sufficient to activate local purinergic receptors. What is the purpose of ongoing ATP release? Evidence for a role for ATP as a paracrine molecule responsible for coordination of calcium signaling between cells has been presented, suggesting a role for ATP in cell-to-cell coupling along the sinusoid (Dranoff et al., 2001; Frame and de Feijter, 1997). Specifically, ATP released by one cell binds to P2 receptors on neighboring cells, stimulating a rise in cytosolic calcium. Thus, ATP released into the sinusoidal space by periportal cells may be carried by sinusoidal flow to bind to receptors on more pericentral cells, and coordinate their transport activities through effects on calcium signaling.

## X. HEPATOBILIARY COUPLING AND BILE FORMATION

The formation of bile is an essential function of the liver that depends on complementary interactions between hepatocytes and cholangiocytes that together constitute the bile secretory unit (Fitz, 1996). Secretion is initiated by transport of bile acids and other organic solutes into the canalicular space between hepatocytes. As bile exits the canalicular space it enters the lumen of intrahepatic ducts, where it undergoes alkalinization and dilution. Despite the comparatively small number of cholangiocytes, the network of intrahepatic ducts is extensive, and ductular secretion is thought to account for up to 40% of bile volume in humans. Increases in the apical membrane $Cl^-$ permeability permit movement of $Cl^-$ ions out of the duct cell and into the duct lumen (Fitz, 1996). Consequently, opening or closing of $Cl^-$ channels in the apical membrane of cholangiocytes represents an important site for modulation of transepithelial movement of water and regulation of ductular bile formation. There is evidence to suggest that ATP may serve as a signaling molecule that is released by hepatocytes and stimulates cholangiocyte $Cl^-$ secretion through activation of P2Y2 receptors in the apical membrane, consistent with a role for ATP in coordination of

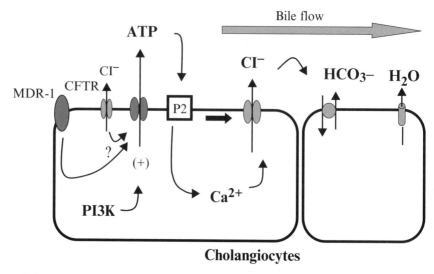

**Cholangiocytes**

**FIGURE 4** Proposed model of purinergic signaling in the regulation of bile flow. ATP in bile, derived from release across the canalicular domain of hepatocytes and the apical domain of cholangiocytes, stimulates P2 receptors on the apical membrane of cholangiocytes. Receptor binding results in Cl$^-$ channel activation through activation of Ca$^{2+}$ or other signaling pathways. Cl$^-$ secretion from the apical membrane serves to drive Cl$^-$/HCO$_3^-$ exchange and stimulate water efflux through membrane aquaporins, resulting in alkalinization and dilution of bile. ATP release from cholangiocytes may be modulated by phosphatidylinositol 3-kinase (PI3K) as well as other membrane proteins such as CFTR and MDR-1.

the separate hepatocyte and cholangiocyte contributions to bile formation (Fig. 4).

It is difficult to estimate the overall contribution of purinergic signaling to the regulation of bile formation. In humans, secretin acts on cholangiocytes to increase bile flow from 0.67 to 1.54 ml·min$^{-1}$, indicating that ductular secretion contributes importantly to the volume and composition of bile (Lenzen *et al.*, 1997). In NRC in culture, the amplitude of the $\Delta I_{sc}$ response to apical ATP exceeds that of secretin. While there are obvious limitations that preclude extrapolation of these observations *in vitro* to intact liver, the robust responses observed in isolated cells are sufficient to suggest that ATP represents a good candidate for local regulation of constitutive and volume-stimulated cholangiocyte Cl$^-$ secretion.

A role for ATP in hepatobiliary coupling and bile formation is supported by the observations that (1) ATP is released by liver and biliary cells, and is present in mammalian bile in concentrations sufficient to activate P2 receptors (Chari *et al.*, 1996), and (2) P2Y$_2$ receptor binding stimulates Cl$^-$ efflux from cholangiocytes (McGill *et al.*, 1994), similar to observations in airway epithelia (Knowles *et al.*, 1991). For example, in voltage-clamp studies of polarized NRCs apical exposure to ATP (or UTP) stimulates a robust Cl⁻ secretory response (Schlenker *et al.*, 1997; Salter *et al.*, 2000a). Together, these findings suggest that ATP in bile is derived from both hepatocytes and cholangiocytes and regulates cholangiocyte secretion through activation of $P2Y_2$ receptors in the apical membrane. The apical orientation of signaling is intriguing, representing an anatomic structure well suited for hepatocyte-to-cholangiocyte or cholangiocyte-to-cholangiocyte signaling by release of ATP into bile.

## XI. CONCLUSION

The progress in identification in liver of the many contributors to purinergic signaling, including ATP channels, purinergic receptors, P-glycoproteins, and nucleotidases, suggests that extracellular nucleotides are likely to contribute to local regulation of liver cell and organ function. Studies to date support roles for extracellular ATP in regulation of metabolism, cell volume, hepatobiliary coupling, and possibly paracrine regulation of bile formation. Moreover, it is attractive to speculate that pharmacologic modulation of ATP release and purinergic signaling might provide novel strategies for the management of cholestasis and other disorders characterized by impaired bile flow (Parr *et al.*, 1994; Clarke *et al.*, 1994; Bennett *et al.*, 1996).

### References

Al-Awqati, Q. (1995). Regulation of ion channels by ABC transporters that secrete ATP. *Science* **269**, 805–806.

Bennett, W. D., Olivier, K. N., Zeman, K. L., Hohneker, K. W., Boucher, R. C., and Knowles, M. R. (1996). Effect of uridine 5′-triphosphate plus amiloride on mucociliary clearance in adult cystic fibrosis. *Am. J. Respir. Crit. Care Med.* **153**, 1796–1801.

Bosch, I., Jackson, G., Croop, J. M., and Cantiello, H. F. (1996). Expression of *Drosophilia melanogaster* P-glycoproteins is associated with ATP channel activity. *Am. J. Physiol.* **271**, C1527–C1538.

Braunstein, G. M., Roman, R. M., Clancy, J. P., Kudlow, B. A., Taylor, A. L., Shylonsky, V. G., Jovov, B., Peter, K., Jilling, T., Ismailov, I. I. *et al.* (2001). Cystic fibrosis transmembrane conductance regulator facilitates ATP release by stimulating a separate ATP release channel for autocrine control of cell volume regulation. *J. Biol. Chem.* **276**, 6621–6630.

Chari, R. S., Schutz, S. M., Haebig, J. A., Shimokura, G. H., Cotton, P. B., Fitz, J. G., and Meyers, W. C. (1996). Adenosine nucleotides in bile. *Am. J. Physiol.* **270**, G246–G252.

Che, M., Ortiz, D. F., and Arias, I. M. (1995). Primary structure and functional expression of a cDNA encoding the bile canalicular purine-specific Na⁺-nucleoside cotransporter. *J. Biol. Chem.* **270**, 13596–13599.

Che, M., Gatmaitan, Z., and Arias, I. M. (1997). Ectonucleotidases, purine nucleoside transporter, and function of the bile canalicular plasma membrane of the hepatocyte. *FASEB J.* **11,** 101–108.

Clarke, L. L., Grubb, B. R., Yankaskas, J., Cotton, C. U., McKenzie, A., and Boucher, R. C. (1994). Relationship of a non-cystic fibrosis transmembrane conductance regulator-mediated chloride conductance to organ-level disease in $cftr^{-/-}$ mice. *Proc. Natl. Acad. Sci. USA* **91,** 479–483.

Dranoff, J. A., Masyuk, A. I., Kruglov, E. A., LaRusso, N. F., and Nathanson, M. H. (2001). Polarized expression and function of P2Y ATP receptors in rat bile duct epithelia. *Am. J. Physiol. Gastrointest. Liver Physiol.* **281,** G1059–G1067.

Felipe, A., Valdes, R., Santo, B., Lloberas, J., Casado, J., and Pastor-Anglada, M. (1998). $Na^+$-dependent nucleoside transport in liver: Two different isoforms from the same gene family are expressed in liver cells. *Biochem. J.* **330,** 997–1001.

Feranchak, A. P., Roman, R. M., Schwiebert, E. M., and Fitz, J. G. (1998). Phosphatidyl inositol 3-kinase represents a novel signal regulating cell volume through effects on ATP release. *J. Biol. Chem.* **273,** 14906–14911.

Feranchak, A. P., Roman, R. M., Doctor, R. B., Salter, K. D., Toker, A., and Fitz, J. G. (1999). The lipid products of phosphoinositide 3-kinase contribute to regulation of cholangiocyte ATP and chloride transport. *J. Biol. Chem.* **274,** 30979–30986.

Feranchak, A. P., Fitz, J. G., and Roman, R. M. (2000). Volume-sensitive purinergic signaling in human hepatocytes. *J. Hepatol.* **33,** 174–182.

Fitz, J. G. (1996). Cellular mechanisms of bile secretion. *In* "Hepatology" (D. Zakim and T. D. Boyer, eds.), pp. 362–376. W.B. Saunders, Philadelphia, PA.

Fitz, J. G., and Sostman, A. (1994). Nucleotide receptors activate cation, potassium, and chloride currents in a liver cell line. *Am. J. Physiol.* **266,** G544–G553.

Folli, F., Alvaro, D., Gigliozzi, A., Bassotti, C., Kahn, C. R., Pontiroli, A. E., Capocaccia, L., Jezequel, A. M., and Benedetti, A. (1997). Regulation of endocytic–transcytotic pathways and bile secretion by phosphatidylinositol 3-kinase in rats. *Gastroenterology* **113,** 954–965.

Frame, M. K., and de Feijter, A. W. (1997). Propagation of mechanically induced intercellular calcium waves via gap junctions and ATP receptors in rat liver epithelial cells. *Exp. Cell Res.* **230,** 197–207.

Griffith, D. A., and Jarvis, S. M. (1996). Nucleoside and nucleobase transport systems of mammalian cells. *Biochim. Biophys. Acta* **1286,** 153–181.

Harden, T. K., Boyer, J. L., and Nicholas, R. A. (1997). $P_2$-purinergic receptors: Subtype-associated signaling responses and structure. *Annu. Rev. Pharmacol. Toxicol.* **35,** 541–579.

Haussinger, D. (1996). Regulation and functional significance of liver cell volume. *In* "Progress in Liver Disease" (J. L. Boyer and R. K. Ockner, eds.), pp. 29–53. W.B. Saunders, Philadelphia, PA.

Haussinger, D., and Schliess, F. (1995). Cell volume and hepatocellular function. *J. Hepatol.* **22,** 94–100.

Haussinger, D., and Schliess, F. (1999). Osmotic induction of signaling cascades: Role in regulation of cell function. *Biochem. Biophys. Res. Commun.* **255,** 551–555.

Haussinger, D., Stehle, T., and Gerok, W. (1987). Actions of extracellular ATP and UTP in perfused rat liver. *Eur. J. Biochem.* **167,** 65–71.

Keppens, S., and De Wulf, H. (1991). Characterization of the biological effects of 2-methylthio-ATP on rat hepatocytes: Clear-cut differences with ATP. *Br. J. Pharmacol.* **104,** 301–304.

Keppens, S., Vandekerchove, A., Moshage, H., Yap, S. H., Aerts, R., and De Wulf, H. (1993). Regulation of glycogen phosphorylase activity in isolated human hepatocytes. *Hepatology* **17,** 610–614.

Knowles, M. R., Clarke, L. L., and Boucher, R. C. (1991). Activation by extracellular nucleotides of chloride secretion in the airway epithelia of patients with cystic fibrosis. *N. Engl. J. Med.* **325**, 533–538.

Krause, U., Rider, M. H., and Hue, L. (1996). Protein kinase signalling pathway triggered by cell swelling and involved in the activation of glycogen synthase and acetyl-CoA carboxylase in isolated rat hepatocytes. *J. Biol. Chem.* **271**, 16668–16673.

Lenzen, R., Elster, J., Behrend, C., Hampel, K.-E., Bechstein, W.-O., and Neuhaus, P. (1997). Bile acid-independent bile flow is differentially regulated by glucagon and secretin in humans after orthotopic liver transplantation. *Hepatology* **26**, 1272–1281.

Light, D. B., Capes, T. L., Gronau, R. T., and Adler, M. R. (1999). Extracellular ATP stimulates volume decrease in *Necturus* red blood cells. *Am. J. Physiol.* **277**, C480–C491.

Malcolm, K. C., Trammell, S. E., and Exton, J. H. (1995). Purinergic agonist and G protein stimulation of phospholipase D in rat liver plasma membranes: Independence from phospholipase C activation. *Biochim. Biophys. Acta* **1268**, 152–158.

McGill, J., Basavappa, S., Shimokura, G. H., Middleton, J. P., and Fitz, J. G. (1994). Adenosine triphosphate activates ion permeabilities in biliary epithelial cells. *Gastroenterology* **107**, 236–243.

Misra, S., Ujhazy, P., Gatmaitan, Z., Varticovski, L., and Arias, I. M. (1998). The role of phosphoinositide 3-kinase in taurocholate-induced trafficking of ATP-dependent canalicular transporters in rat liver. *J. Biol. Chem.* **273**, 26638–26644.

Misra, S., Ujhazy, P., Varticovski, L., and Arias, I. M. (1999). Phosphoinositide 3-kinase lipid products regulate ATP-dependent transport by sister of P-glycoprotein and multidrug resistance associated protein 2 in bile canalicular membrane vesicles. *Proc. Natl. Acad. Sci. USA* **96**, 5814–5819.

Misumi, Y., Ogata, S., Hirose, S., and Ikehara, Y. (1990). Primary structure of rat liver 5′-nucleotidase deduced from the cDNA. *J. Biol. Chem.* **265**, 2178–2183.

Mitchell, C. H., Carre, D. A., McGlinn, A. M., Stone, R. A., and Civan, M. M. (1998). A release mechanism for stored ATP in ocular ciliary epithelial cells. *Proc. Natl. Acad. Sci. USA* **95**, 7174–7178.

Musante, L., Zegarra-Moran, O., Montaldo, P. G., Ponzoni, M., and Galietta, L. J. (1999). Autocrine regulation of volume-sensitive anion channels in airway epithelial cells by adenosine. *J. Biol. Chem.* **274**, 11701–11707.

Nukina, S., Fusaoka, T., and Thurman, R. G. (1994). Gylcogenolytic effect of adenosine involves ATP from hepatocytes and eicosanoids from Kupffer cells. *Am. J. Physiol.* **266**, G99–G105.

Parr, C. E., Sullivan, D. M., Paradiso, A. M., Lazarowski, E. R., Burch, L. H., Olsen, J. C., Erb, L., Weisman, G. A., Boucher, R. C., and Turner, J. T. (1994). Cloning and expression of a human P2u nucleotide receptor, a target for cystic fibrosis pharmacotherapy. *Proc. Natl. Acad. Sci. USA* **91**, 3275–3279.

Pasyk, E. A., and Foskett, J. K. (1997). Cystic fibrosis transmembrane conductance regulator-associated ATP and adenosine 3′-phosphate 5′-phosphosulfate channels in endoplasmic reticulum and plasma membranes. *J. Biol. Chem.* **272**, 7746–7751.

Roman, R. M., Wang, Y., and Fitz, J. G. (1996). Regulation of cell volume in a human biliary cell line: Calcium-dependent activation of $K^+$ and $Cl^-$ currents. *Am. J. Physiol.* **271**, G239–G248.

Roman, R. M., Wang, Y., Lidofsky, S. D., Feranchak, A. P., Lomri, N., Scharschmidt, B. F., and Fitz, J. G. (1997). Hepatocellular ATP-binding cassette protein expression enhances ATP release and autocrine regulation of cell volume. *J. Biol. Chem.* **272**, 21970–21976.

Roman, R. M., Bodily, K., Wang, Y., Raymond, J. R., and Fitz, J. G. (1998). Activation of protein kinase C couples cell volume to membrane Cl⁻ permeability in HTC hepatoma and Mz-ChA-1 cholangiocarcinoma cells. *Hepatology* **28**, 1073–1080.

Roman, R. M., Feranchak, A. P., Davison, A. K., Schwiebert, E. M., and Fitz, J. G. (1999). Evidence for $Gd^{3+}$ Inhibition of membrane ATP permeability and purinergic signaling. *Am. J. Physiol.* **277**, G1222–G1230.

Roman, R. M., Lomri, N., Braunstein, G., Feranchak, A. P., Simeoni, L. A., Davison, A. K., Mechetner, E., Schwiebert, E. M., and Fitz, J. G. (2001). Evidence for multidrug resistance-1 P-glycoprotein-dependent regulation of cellular ATP permeability. *J. Membr. Biol.* **183**, 165–173.

Salter, K. D., Fitz, J. G., and Roman, R. M. (2000a). Domain-specific purinergic signaling in polarized rat cholangiocytes. *Am. J. Physiol. Gastrointest. Liver Physiol.* **31**, 1045–1054.

Salter, K. D., Roman, R. M., LaRusso, N. R., Fitz, J. G., and Doctor, R. B. (2000b). Modified culture conditions enhance expression of differentiated phenotypic properties of normal rat cholangiocytes. *Lab. Invest.* **80**, 1775–1778.

Sarkadi, B., and Parker, J. C. (1991). Activation of ion transport pathways by changes in cell volume. *Biochim. Biophys. Acta* **1071**, 407–427.

Schlenker, T., Romac, J. M. J., Sharara, A., Roman, R. M., Kim, S., LaRusso, N., Liddle, R., and Fitz, J. G. (1997). Regulation of biliary secretion through apical purinergic receptors in cultured rat cholangiocytes. *Am. J. Physiol.* **273**, G1108–G1117.

Scoazec, J. Y., Bringuier, A. F., Medina, J. F., Martinez-Anso, E., Veissiere, D., Feldmann, G., and Housset, C. (1997). The plasma membrane polarity of human biliary epithelial cells: In situ immunohistochemical analysis and functional implications. *J. Hepatol.* **26**, 543–553.

Schwiebert, E. M., Egan, M. E., and Guggino, W. B. (1998). Assays of dynamics, mechanisms, and regulation of ATP transport and release: Implications for study of ABC transporter function. *Methods Enzymol.* **292**, 664–675.

Strange, K., Emma, F., and Jackson, P. S. (1996). Cellular and molecular physiology of volume-sensitive ion channels. *Am. J. Physiol.* **270**, C711–C730.

Taylor, A. L., Kudlow, B. A., Marrs, K. L., Gruenert, D., Guggino, W. B., and Schwiebert, E. M. (1998). Bioluminescence detection of ATP release mechanisms in epithelia. *Am. J. Physiol.* **275**, C1391–C1406.

Vroman, B., and LaRusso, N. (1996). Development and characterization of polarized primary cultures of rat intrahepatic bile duct epithelial cells. *Lab. Invest.* **74**, 303–313.

Wang, T. F., Rosenberg, P. A., and Guidotti, G. (1997). Characterization of a brain ecto-apyrase: Evidence for only one ecto-apyrase (CD39) gene. *Mol. Brain Res.* **47**, 295–302.

Wang, Y., Roman, R. M., Lidofsky, S. D., and Fitz, J. G. (1996). Autocrine signaling through ATP release represents a novel mechanism for cell volume regulation. *Proc. Natl. Acad. Sci. USA* **93**, 12020–12025.

Yao, S. Y., Ng, A. M., Muzyka, W. R., Griffiths, M., Cass, C. E., Baldwin, S. A., and Young, J. D. (1997). Molecular cloning and functional characterization of nitrobenzylthioinosine (NBMPR)-sensitive (*es*) and NBMPR-insensitive (*ei*) equilibrative nucleoside transporter proteins (rENT1 and rENT2) from rat tissues. *J. Biol. Chem.* **272**, 28423–28430.

# CHAPTER 13

# ATP in the Treatment of Advanced Cancer

**Edward H. Abraham, Anna Y. Salikhova, and Eliezer Rapaport**
Dartmouth Medical School, Dartmouth-Hitchcock Medical Center, Lebanon,
New Hampshire 03756

## I. INTRODUCTION

The roles of adenosine $5'$-triphosphate (ATP) as the major cellular energy source, as a phosphate group donor in phosphorylation reactions, and as an allosteric regulator of the activities of many cellular proteins have been well established. In addition to its role in cellular metabolism, extracellular ATP

*Current Topics in Membranes, Volume 54*

⇓  ⇓  ⇓

| MOLECULAR LEVEL | CELLULAR LEVEL: VIA P2 ACTIVATION | MACROCELLULAR LEVEL |
|---|---|---|
| • Phosphate group donor in phosphorylation reactions | • Cell growth inhibition | • Induction of hypotension |
| • Allosteric regulator of the activities of cellular proteins | • Apoptosis or cell lysis | • Survival benefit after various stresses |
|  | • Cell volume decrease | • Vasodilation of pulmonary vasculature |
|  | • Ca++ mobilization | • Bone marrow radiation tolerance |
|  | • Membrane permeability | • Inhibition of weigh loss |
|  | • Stimulation of inositol phosphate metabolism |  |

**FIGURE 1**    Extracellular ATP/adenylate and tumor cellular interaction.

and its breakdown product adenosine exert pronounced effects in a variety of biological processes including neurotransmission, muscle contraction, cardiac function, platelet function, vasodilatation, and liver glycogen metabolism. Both P1 (adenosine) and P2 (ATP) receptors in untransformed cells mediate these effects. Tumor cells express functional P2-purinergic receptors linked to phospholipase C, and ATP and other agonists of these receptors are markedly growth inhibitory, suggesting a novel therapeutic approach to cancer treatment. Figures 1 and 4 show possible tumor–extracellular ATP interactions.

## II. PRONOUNCED DEPLETION OF PURINES IN AGING AND ADVANCED DISEASE

### A. Endogenous ATP Levels

**1. Extracellular and Intracellular ATP: Balance between ATP Release from Cells and Its Extracellular Degradation**
Extracellular ATP reflects the balance between ATP release from cells and its extracellular degradation. Extracellular ATP is available to interact with cell surface P2 purinergic receptors. The release of ATP from cells to the extracellular compartment was shown to occur in association with the presence of so-called membrane ATP-binding cassette (ABC) proteins such as P-glycoprotein and the cystic fibrosis transmembrane conductance regulator (CFTR) (1–6). A scenario is that ATP release occurs through membrane complexes. Support for the view comes from findings that ectoapyrases, also referred to as ecto-ATPases, are found in plasma membranes and associate with ABC proteins. The level of ecto-ATPase

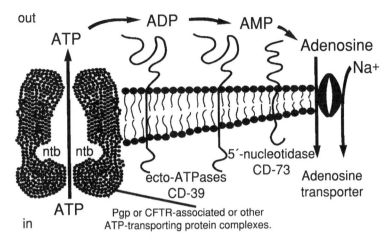

**FIGURE 2**  Models for role of ABC family member and allied membrane proteins in transmembrane ATP cycling. Interplay between proteins that release ATP and proteins responsible for the extracellular degradation determines the level of extracellular ATP. The resulting amount of extracellular ATP is available to interact with ATP receptors (P2, etc.). Intravenous administration of ATP is one approach to elevate the systemic extracellular ATP levels (20,21). ntb, Nucleotide-binding fold; Pgp, P-glycoprotein.

expression appears to be correlated with ABC protein expression (7–11). This is also the case for ecto-5′-nucleotidase (12,13). Ecto-ATPases hydrolyze the terminal phosphate residues of nucleoside triphosphates and diphosphates on the extracellular surface of cell. In concert, ecto-5′-nucleotidases (ecto-5′-NT) hydrolyze nucleotide monophosphates to their respective nucleosides. This latter enzyme is bound to the external surface of cell by glycosylphophatidylinositol anchoring. Ectoapyrase, shown to be the known marker CD39, and ecto-5′-nucleotidase, identified as CD73, can be viewed as converting nontransportable nucleotides to transportable nucleosides (12,14–19). This sequential adenylate scavenger pathway can supply cells with required sources of purines when internal sources are depleted. When ABC proteins are overexpressed, cells appear to modulate levels of ectoapyrase (CD39) and ecto-5′-NT (CD73), and also nucleoside transporters. This provides a mechanism for retrieving the ATP adenylate moieties (salvage precursors) accumulating on the cell surface in the vicinity of the ABC proteins, as shown in Fig. 2 (20,21).

## 2. ATP and Age

Blood plasma levels of purines (ATP and its catabolic metabolites) were shown to decline progressively on aging in experimental animals (22–24). Claims were made as to the causality of low blood plasma purine levels in the spontaneous increases in blood pressure in old rats. Rabini *et al.* (25) have established significant declines (about 50%) in human blood ATP as compared with younger adult populations. A whole body $^{31}$P magnetic resonance spectroscopy (MRS) study showed a decline in myocardial ATP on aging in humans as well as further significant declines in individuals suffering from hypertrophic cardiomyopathy or hypertension (26).

## B. Physiological Role of ATP in Tumor Development

### 1. Blood ATP and Cancer

Laciak and Witkowski (27) observed diminished adenine compound composition in the erythrocytes obtained from patients with laryngeal cancer. Both Wand and Rieche (28) and De la Morena Garcia *et al.* (29) showed decreased levels of adenine nucleotide in erythrocytes and thrombocytes obtained from patients with various malignancies. Another study reported a 27% decline in erythrocyte ATP level in subjects with gastrointestinal tumors in comparison with control subjects. This observation was interpreted to result from enzymatic and metabolic modifications previously observed in red blood cells (30,31). Pilot studies by A.Y. Salikhova and E. H. Abraham (heretofore unpublished) have shown that patients with more advanced cancers have diminished blood ATP levels and plasma ATP and ATP RBC release rates in comparison with control subjects (Tables I and II). The presence of decreased ATP pools correlates with the increased probability of tumor. The observations of Rabini *et al.* (25) showed that human blood ATP levels decline as individuals of the population age. These findings are both in agreement with the various studies that have recognized a strong relationship between aging and the development of cancer (32–36). Hence our premise that there is a connection between diminished blood ATP levels and the increased risk of some types of cancer. As was proposed previously, the risk of the developing cancer demonstrates a tendency to increase with age while blood ATP values decrease with age. The genetic disease cystic fibrosis is associated with elevation in blood ATP. Age-adjusted rates of prostate, breast, and other tumors were shown to decrease in individuals homozygous for the cystic fibrosis (CF) defect, who have elevations in blood ATP levels (37–41). These findings would be consistent with our idea that there is a causal relationship between higher extracellular ATP levels and suppression of certain tumor types.

The biological significance of extracellular ATP and other adenylates in regulating tumor growth has been emphasized by reports regarding the inverse relationship between exercise and tumor development in experimental animals (42–45). In exercising rats, tumor volumes were reduced

**TABLE I**
Comparison of Total Blood ATP, Plasma ATP, and ATP Release Rates for Prostate Cancer and Control Subjects

| Subjects | Total blood ATP (mole/RBC) | | | Plasma ATP (mole/RBC) | | | RBC ATP release rate (mole/RBC/min) | | |
| | Value | SE | $n$ | Value | SE | $n$ | Value | SE | $n$ |
|---|---|---|---|---|---|---|---|---|---|
| Control | $1.6 \times 10^{-16}$ | $1.29 \times 10^{-17}$ | 12 | $1.04 \times 10^{-19}$ | $6.41 \times 10^{-21}$ | 10 | $6.83 \times 10^{-21}$ | $1.76 \times 10^{-21}$ | 8 |
| Stage 4 | $1.31 \times 10^{-16}$ | $1.1 \times 10^{-17}$ | 5 | $3.77 \times 10^{-21}$ | $2.05 \times 10^{-21}$ | 4 | $2.68 \times 10^{-21}$ | $0.91 \times 10^{-21}$ | 4 |
| $p$ Value vs. control | 0.0484 | | | 0.0001 | | | 0.030 | | |

*Abbreviation:* RBC, Red blood cell.

**TABLE II**
Total Blood and Plasma ATP Levels in Breast Cancer versus Control Subjects

| Subjects (average age, in years, ± SE) | Total blood ATP (mole/RBC) | | | | Plasma ATP (mole/RBC) | | | |
|---|---|---|---|---|---|---|---|---|
| | Value | SE | $n$ | $p$ Value vs. control | Value | SE | $n$ | $p$ Value vs. control |
| Control (52.1 ± 3.75) | $1.95 \times 10^{-16}$ | $0.28 \times 10^{-16}$ | 9 | | $5.77 \times 10^{-20}$ | $1.18 \times 10^{-20}$ | 8 | |
| Stage 1 or 2 (50.3 ± 0.81) | $1.29 \times 10^{-16}$ | $0.25 \times 10^{-16}$ | 3 | 0.081 | Not done | | | |
| Stage 4 (55.0 ± 5.56) | $1.13 \times 10^{-16}$ | $0.15 \times 10^{-16}$ | 6 | 0.009 | $4.15 \times 10^{-20}$ | $0.71 \times 10^{-20}$ | 4 | 0.089 |

*Abbreviation*: RBC, Red blood cell.

significantly in a manner inversely related to tumor ATP pools. Exercising human muscles release significant amounts of adenosine and ATP (46,47). It was found that tumor metastases to striated muscle are clinically rare. The elevations in ATP and its catabolic products ADP, AMP, and adenosine were directly analyzed in interstitial exercising skeletal muscle and these elevations were directly related to the level of work performed by the muscle. One study concluded that the low molecular weight factors, which are released by muscle cells and are inhibitory to tumor development and growth, were found to be adenosine and related compounds (45). These observations indicate that the extracellular purine levels are important modulators of tumor growth.

### 2. ATP and Cancer Cachexia

Cancer cachexia in humans as well as in experimental animals not only results in host weight loss but also anorexia, hormonal aberrations, and depletions and redistribution of host metabolic factors, all of which lead to a progressive decline in vital host functions (48,49). With improvements in the capabilities of accurately determining tissue ATP levels, much of the attention in cancer cachexia has focused on alterations in energy metabolism in cancer patients and cachexia tumor models (Fig. 3).

Gluconeogenesis from lactate, three-carbon amino acids, and glycerol, which are the respective products of tumor, muscle, and adipose tissue catabolism, has been considered the source of the enormous energy expenditures that occur in the liver during cancer cachexia (50–52). Gluconeogenesis is costly in terms of the number of ATP molecules required for the synthesis of one molecule of glucose (six ATP molecules). Glycolysis in tumors yields fewer molecules of ATP (two ATP molecules per molecule of glucose). This type of host–tumor interplay is the main reason

TUMOR GROWTH
⇓
↑Gluconeogenesis (↑ATP consumption)
↓Glycolysis (↓ATP synthesis)
⇓
Depletion of host visceral energy stores
⇓
CANCER CACHEXIA

**FIGURE 3** Alterations in energy metabolism in cancer patients with cachexia.

for the depletion of energy stores and progressive weight loss in advanced cancer patients. These type of aberrations in glucose metabolism have been demonstrated with radioactively labeled precursors in human cancer patients suffering from progressive weight loss (53–57). Furthermore, the significant depletion of host visceral energy stores by a growing tumor was demonstrated in experimental animals (58–60). Along with the decline in hepatic ATP pools in cachexia tumor animal models, severe declines in skeletal muscle ATP pools were demonstrated (42,44,58–60).

### 3. ATP and Glutathione Level

*aNontumor Cells.*   Glutathione (GSH) is a sulfur-containing tripeptide consisting of the amino acids cysteine, glutamic acid, and glycine. Low intracellular glutathione levels in antigen-presenting cells correlated with defective processing of antigen with disulfide bonds, indicating that this thiol may be a critical factor in regulating productive antigen processing. The study by Short *et al.* (61) showed that decreased cellular ATP and depletion of the antioxidant L-glutathione are two factors that adversely affect antigen processing and presentation. It was found that there is a correlation between diminished blood, hepatocellular ATP level, and glutathione content under stress-induced conditions in non-tumor-bearing animals (62–67). The detailed mechanism of this correlation is not clear yet.

*bTumor Cells.*   Multidrug resistance (MDR) in cancer cells is often associated with an elevation in the concentration of glutathione and the expression of $\gamma$-glutamylcysteine synthetase ($\gamma$-GCS), a rate-limiting enzyme for GSH (68, 69). The ABC protein MRP1 (multiple drug resistance protein 1) mediates the active transport of numerous GSH-, sulfate-, and glucuronide-conjugated organic anions and can transport certain xeno-biotics by a mechanism that may involve cotransport with GSH (70). Increased MRP expression also is associated with resistance to radiation, anthracyclines, and etoposide in cells treated with fractionated $\gamma$ radiation. Treatment of the human T cell leukemia cell line (CEM cells) with 2.0 Gy of $\gamma$ radiation caused an increase in MRP mRNA within 4 h, which, by 24 h, was greater than 5-fold that of the untreated CEM cells. Harvie *et al.* (71) concluded that MRP is involved in the immediate response to radiation and it may account for the drug resistance that often develops following radiation treatment.

ATP administration was shown effective in sensitizing a preclinical mouse tumor model to radiation therapy and producing "cures" when other commonly utilized sensitizers yielded only minor effects (72–75). The mechanism of this particular activity was attributed to the ability of exogenously administered ATP to selectively reduce intracellular glutathione levels in the tumor but not in host tissues (72–75). The ability of ATP to effect a selective and significant reduction of tumor glutathione levels *in vivo* suggests that the use of ATP in the treatment of refractory patients who have failed prior chemo- and/or radiation therapy may be beneficial. This mechanistically includes the well-known role of ATP in the function of MDR/MRP1 in removing toxins, including chemotherapeutic agents from resistant cells. The protection of host functions by ATP is a powerful contributor to the incentive for utilizing combination therapy with mainstay cytotoxic drugs and/or radiation therapy. ATP synergistically improves the efficacy of certain cytotoxic agents in model systems (76,77). ATP has been shown to enhance doxorubicin (Adriamycin) (77) or vincristine (76) entry into tumor cells, significantly enhancing cell killing.

## C. Why Are P2 Receptors Considered a Family of Novel Therapeutic Targets by the Pharmaceutical Industry?

### 1. Cell Line Studies: ATP Cell Growth Inhibition

Early experiments have demonstrated that relatively low levels of extracellular ATP inhibit the growth of a variety of human tumor cells and subsequently yield substantial increases in cell killing in *in vitro* systems (78,79). The mechanism of tumor cell killing was attributed to the permeation into tumor cell membranes of extracellular ATP as well as the arrest of tumor cells in S phase followed by cell death (78–83). The presence of extracellular ATP has also been correlated with increased membrane permeability by non-receptor-mediated $Na^+$ channels (84–88). Inhibitory effects of extracellular ATP was noted more than 30 years ago but the elucidation of various specific P2 receptors (purinoreceptors) involved in the process began about 10 years ago (89–91).

The physiologic effects of extracellular ATP include the following: apoptosis or cell lysis, cell growth inhibition, cell shrinkage, $Ca^{2+}$ mobilization, membrane permeability, enhancement of penetration of chemotherapeutic agents, and stimulation of inositol phosphate metabolism (some detailed literature for these effects is included in Table III). As seen from Table III, three types of P2 receptors are involved in tumor cell growth inhibition: P2X (ion channel gated), P2Y (G protein coupled), and P2Z or pore-forming ATP receptor type [3'-*O*-(4-benzoyl)benzoyl-ATP (BzATP) as a selective and potent agonist]. Cells of the monocytic/macrophage lineage express P2Z receptors, more recently identified as a $P2X_7$ receptor by Suprenant *et al.* (92). These receptors form large pores across the cytoplasmic membrane after ligand binding (93–96).

**TABLE III**

Interaction of Extracellular Adenylates with Purinergic Receptors and Effects on Cellular Metabolism

| Specific receptor | Cell/tissue type | Extracellular nucleotide effect on cell growth and related events | Rank order of potency/dose/conclusion | Ref. |
|---|---|---|---|---|
| | | **Solid Tumor Cells: Apoptosis and Growth Inhibition** | | |
| Subtype not specified | Melanoma cell line B16 | ATP-dependent cell lysis was observed only in several transformed cell lines, and not in untransformed mouse fibroblasts | Effect of ATP on the membrane permeability in transformed cells is elicited under physiological conditions and this would be useful in some limited way for cancer chemotherapy management | 166 |
| P2X | Mouse tumor cells: 1. YAC-1 2. P-815 | Extracellular ATP ($ATP_o$) caused dose-dependent lysis tumor cells and DNA fragmentation | Plasma membrane depolarization and $Ca^{2+}$ influx occurred within a few seconds of $ATP_o$ addition | 167 |
| $P2Y_2$ | 1. Human esophageal cancers 2. Squamous esophageal cancer cell line Kyse-140 | ATP and various nucleotides revealed dose-dependent inhibition of cell proliferation. This was due to both an induction of apoptosis and cell cycle arrest. | The rank order of potency was ATP = UTP > ATPγS > ADP = UDP | 168 |
| $P2Y_2$ | Colorectal cancer cells: 1. Ht29 2. Colo320 DM | Extracellular nucleotides induce apoptosis and inhibit cell growth. Receptor-induced [$Ca^{2+}$]i] signaling appears to play a major role in the observed antiproliferative and apoptosis-inducing effects | Short-term stimulation of $P2Y_2$ receptors (6 h, 500 $\mu M$ ATP or ATPγS) caused $Ca^{2+}$ mobilization from intracellular stores and a subsequent transmembrane $Ca^{2+}$ influx. Prolonged application (24–72 h) induced a time-dependent increase in apoptosis (up to 50% above control values) in both cell lines and caused dose-dependent inhibition of cell proliferation of more then 65% | 169 |

| | | | |
|---|---|---|---|
| P2U*[a] | 1. Colorectal tumor tissues<br>2. Colorectal cancer cell line Ht29 | In HT29 cells, the hydrolysis-resistant ATP analog ATP$\gamma$S inhibited cell proliferation and, also, induced apoptosis in a dose-dependent manner. Stimulation with either extracellular ATP or UTP caused a biphasic rise of $[Ca^{2+}]_i$ in a dose-dependent manner and cross-desensitization between both nucleotides was observed | The rank order of potency was ATP $\geq$ UTP > ATP$\gamma$S > ADP > adenosine, which is characteristic for a P2U receptor subtype. Thus, human colorectal cancer cells express functional P2U receptors, which may play a role in the regulation of cell proliferation and apoptosis | 169 |
| P2X | Human colon carcinoma cells:<br>1. LoVo-Dx<br>2. Multidrug-resistant (MDR-1[+]) derivative | LoVo-Dx cells were sensitive to both LAK (interleukin 2-activated peripheral blood lymphoid ells) and 100 $\mu M$ ATP | ATP acts as a natural amplifier of physical, or immune cytotoxic damage since it may be released in large amounts from target cells injured by several cytotoxic mediators secreted by LAK effectors | 103 |
| P2, subtype not specified | Human bladder carcinoma detrusor strips | Contractile effects of ATP and of its slowly hydrolyzable analogs were investigated | The rank order of potency was $\alpha$, $\beta$-MeATP > $\beta$, $\gamma$-MeATP > ATP. Evidence exists for the presence of both pre- and postjunctional P2-purinoceptor subtypes in human isolated urinary bladder | 170 |

(continued)

425

**TABLE III**

Interaction of Extracellular Adenylates with Purinergic Receptors and Effects on Cellular Metabolism

| Specific receptor | Cell/tissue type | Extracellular nucleotide effect on cell growth and related events | Rank order of potency/dose/conclusion | Ref. |
|---|---|---|---|---|
| 1. P2Y$_2$<br>2. P2Y$_6$, P2X$_4$, P2X$_5$, P2X$_7$<br>3. P2Y$_{11}$, P2X$_4$ and P2X$_5$ | Human prostate carcinoma cell lines:<br>1. LNCaP<br>2. PC-3<br>3. DU145 | ATP inhibited the growth of PC-3 and DU145 cells. On the other hand, in PC-3 cells, BzATP reproduced the effect of ATP, which was associated with a moderate decrease in proliferation and with an increase in apoptosis. In DU145 cells, ATP was more potent than BzATP and growth inhibition was associated mainly with necrosis | In both PC-3 and DU145 cells, ATP and UTP stimulated inositol phosphate accumulation in an equipotent, equiactive, and nonaddictive way, suggesting the involvement of P2Y$_2$ receptors. The calcium influx induced by BzATP confirmed the functional expression of P2X receptors. So P2X receptors might be involved in the inhibition by nucleotides of prostate carcinoma cell growth | 97 |
| P2, subtype not specified | Androgen-independent prostate carcinoma cells | ATP and certain hydrolysis-resistant adenine nucleotides induced a rapid, transient increase in cytoplasmic free Ca$^{2+}$ that was detectable at 50 to 100 nM ATP, was maximal at 100 μM ATP, and was inhibited approximately 50% by chelation of extracellular Ca$^{2+}$. In addition to stimulating phosphatidylinositol turnover and Ca$^{2+}$ mobilization. ATP and hydrolysis-resistant ATP analogs induced greater than 90% inhibition of the growth of all lines tested | Cells express functional P2-purinergic receptors linked to phospholipase C, and agonists of this receptor are markedly growth inhibitory, suggesting a novel therapeutic approach to this common adult neoplasm | 171 |

**Solid Tumor Cells: Ca$^{2+}$ Mobilization**

| | | | |
|---|---|---|---|
| P2Z | Glioma hybrid cell line NG108-15 | Suggested that ATP$^{4-}$-promoted pores are antagonized by Na$^+$ and Mg$^{2+}$ in dibutyryl cyclic AMP-differentiated NG108-15 cells | 107 |
| P2Y$_2$ | Human ovarian carcinoma cells: 1. EFO-21 2. EFO-27 | Extracellular ATP induced a concentration-dependent rise in intracellular calcium concentration ([Ca$^{2+}$]$_i$), suggesting the expression of a purinoreceptor | 108 |
| P2Y$_2$ | 1. Endometrial carcinoma HEC-1A 2. Ishikawa cells | ATP induced a rapid and extracellular Ca$^{2+}$-independent rise in cytosolic Ca$^{2+}$ concentration ([Ca$^{2+}$]$_i$) in a dose-dependent manner, with an ED$_{50}$ of about 10 $\mu M$ | 109 |
| P2Y$_1$ and P2Y$_2$ | Ehrlich ascites tumor cells | Mechanical stimulation of Ehrlich cells (centrifugation at 700 × g for 45 s) leads to release of 2.6 ± 0.2 $\mu M$ ATP which in turn stimulates both P2Y$_1$ and P2Y$_2$ receptors, resulting in Ca$^{2+}$ influx as well as release and activation of an outwardly rectifying whole-cell current | 172 |

*(continued)*

The order of agonist potency for this receptor was ATP = UTP > ATPαS >> ADP. Indicates that P2Y$_2$ receptors may participate in control of the cell cycle of endometrial carcinoma cells

Indicated that ATP may act as an extracellular messenger in controlling the ovarian epithelial cell cycle through P2Y$_2$ receptors

Extracellular ATP induced a suramin-sensitive, transient, concentration-dependent increase in [Ca$^{2+}$]$_i$ activation of an outwardly rectifying whole-cell current, and a hyperpolarization of the plasma membrane

427

**TABLE III**

Interaction of Extracellular Adenylates with Purinergic Receptors and Effects on Cellular Metabolism

| Specific receptor | Cell/tissue type | Extracellular nucleotide effect on cell growth and related events | Rank order of potency/dose/conclusion | Ref. |
|---|---|---|---|---|
| P2, subtype not specified | Ehrlich ascites tumor cells | The increase in $[Ca^{2+}]_i$ evoked by ATP or UTP is abolished after depletion of intracellular $Ca^{2+}$ stores with thapsigargin in $Ca^{2+}$-free medium, and is inhibited by U73122, an inhibitor of phospholipase C (PLC), indicating that the increase in $[Ca^{2+}]_i$ is primarily due to release from intracellular, $Ins(1,4,5)P_3$-sensitive $Ca^{2+}$ stores | ATP and UTP elicit rapid cell shrinkage, presumably due to activation of $Ca^{2+}$-sensitive $K^+$ and $Cl^-$ efflux pathways | 110 |
| P2U*[a] | Human lung small cell adenocarcinoma cell line A549 | Cells possess a functional P2U receptor that, on activation, causes $Ca^{2+}$ mobilization from $Ca^{2+}$-ATPase inhibitor-insensitive stores followed by protein kinase C-regulated $Ca^{2+}$ influx | The cells show $Ca^{2+}$ mobilization on addition of various nucleotides, with an order of agonist potency: UTP $\geq$ ATP > ADP > ADP$\beta$S > AMP; adenosine is ineffective. The $EC_{50}$ values for UTP and ATP are $12.5 \pm 0.4$ $\mu M$ and $18.9 \pm 0.5$ $\mu M$, respectively | 111 |
| P2, subtype not specified | Human epidermoid carcinoma cell line A-431 | Stimulation of P2-purinergic receptors with extracellular ATP caused production of inositol 1,4,5-trisphosphate ($InsP_3$), followed by mobilization of $Ca^{2+}$ from intracellular stores; $Ca^{2+}$ influx from the extracellular fluid and breakdown of phosphatidylcholine (PtdCho) also accompanied this $InsP_3/Ca^{2+}$ signaling | | 112 |

| P2, subtype not specified | Human epidermoid carcinoma cell line A431 | Extracellular ATP (5–100 $\mu M$) caused the rapid formation of inositol trisphosphate and later its metabolites, inositol bisphosphate and inositol monophosphate. ATP also induced the efflux of $^{45}Ca^{2+}$ from pre-loaded cells. In addition, an increase in the rate of influx of $^{45}Ca^{2+}$ stimulated by extracellular ATP was detected | 113 |
|---|---|---|---|

**Endothelial Cells and Connective Tissue: Apoptosis and Growth Inhibition**

| | | | |
|---|---|---|---|
| P2Y | Endothelial cells in *cd39*-null mice | Deletion of CD39 (the major ectonucleotidase of endothelial cells and monocytes, catalyzes phosphohydrolysis of extracellular nucleotides) causes perturbations in the hydrolysis of ADP, ATP, and UTP in the vasculature. Activation of P2 receptors appears to influence endothelial cell chemotactic and mitogenic responses *in vitro*. Aberrant regulation of nucleotide P2 receptors may influence angiogenesis in *cd39*-null mice. In *cd39*-null mice, absolute failure of new vessel ingrowth was consistently observed | 14 |
| | | Demonstrated a role for CD39 and phosphohydrolysis of extracellular nucleotides in the regulation of the cellular infiltration and new vessel growth in a model of angiogenesis | |
| P2Z | Mouse microglial cell lines: 1. N9 2. N13 | Exposure of microglial cells to 3 m$M$ ATP resulted in activation of NF-$\kappa$B, a transcription factor controlling cytokine expression and apoptosis | 173 |
| | | ATP may control the expression of a subset of NF-$\kappa$B target genes distinct from those activated by classic proinflammatory mediators | |
| P2Z | Fibroblast cell line L929 | ATP-induced cell death occurred by two alternative mechanisms: colloido-osmotic lysis or apoptosis | 96 |

*(continued)*

429

**TABLE III**

Interaction of Extracellular Adenylates with Purinergic Receptors and Effects on Cellular Metabolism

| Specific receptor | Cell/tissue type | Extracellular nucleotide effect on cell growth and related events | Rank order of potency/dose/conclusion | Ref. |
|---|---|---|---|---|
| | | **Endothelial Cells and Connective Tissue: $Ca^{2+}$ Mobilization** | | |
| $P2Y_{12}$ | 1. Blood platelets<br>2. C6-2B glioma cells<br>3. B10 microvascular endothelial cells | P2Y receptor in blood platelets, C6-2B glioma cells, and in B10 microvascular endothelial cells, a P2Y receptor subtype, which couples to inhibition of adenylyl cyclase, historically termed $P2Y_{AC}$, ($P2T_{AC}$ or $P2_T$ in platelets, equal to $P2Y_{12}$ in P2 receptor nomenclature) has been identified | Demonstrated that the $P2Y_{AC}$ receptor in C6-2B cells is pharmacologically identical to the $P2T_{AC}$ receptor in rat platelets | 174 |
| $P2Y_2$ | Human umbilicalve in endothelial cell culture, ECV304 | ATP and uridine triphosphate (UTP) stimulated inositol phosphate metabolism in ECV304 cells without alteration of cAMP levels. Selective P2Y receptor agonists indicated that this response, leading to calcium mobilization from intracellular stores, was predominantly mediated by the activation of $P2Y_2$ receptors | Similarity in functional responses and genetic structure between ECV304 and T24/83, a human bladder cancer cell line | 175 |
| $P2X_7$ | Rat cultured brain microglia | Extracellular ATP triggers TNF-α release in rat microglia via $P2X_7$, by a mechanism that is dependent on both the sustained $Ca^{2+}$ influx and ERK/p38 cascade, regulated independently of $Ca^{2+}$ influx | ATP (1 m$M$ ATP) potently stimulates TNF-α release, resulting from TNF-α mRNA expression | 176 |

**Immune cells**

| | | | |
|---|---|---|---|
| P2X | Human leukaemic cells, HL-60 | A single addition of ATP (20–1000 $\mu M$) to cultures resulted here in permanent, $Ca^{2+}$-independent inhibition of cellular proliferation, evident 48 h following treatment. ATP was maximally effective at 250 $\mu M$, giving $90 \pm 1.5\%$ growth inhibition. Up to a concentration of 250 $\mu M$ $ATP_o$, growth inhibition is solely attributable to $ATP_o$, while at higher $ATP_o$ concentrations adenosine generated from $ATP_o$ hydrolysis contributes to this effect | The order of potency for growth inhibition was $ATP = ADP > AMP >$ adenosine | 177 |
| P2U*[a] | Lymphocytes from patients with chronic lymphocytic leukemia | Agonists for endothelial P2 purinoceptors trigger a signaling pathway producing $Ca^{2+}$ responses in lymphocytes adherent to endothelial cells | Stimulation of endothelial P2U purinoceptors triggers an endothelial–lymphocyte signaling pathway that releases internal $Ca^{2+}$ in adherent lymphocytes | 178 |
| 1. $P2Y_1$, $P2Y_2$, $P2Y_4$, $P2Y_6$, $P2Y_{11}$<br>2. $P2X_1$, $P2X_2$, $P2X_4$, $P2X_5$, $P2X_7$ | Human dendritic cells, DCs | Signaling through both receptor families by selected nucleotides induces different aspects of DC activation. ATP, which can be released by damaged cells and has been demonstrated to synergize with TNF-$\alpha$ in the activation of DCs (a), acts through $P2Y_{11}$ receptor signaling. This signaling pathway leads to the generation of cAMP. The $P2X_7$ receptor is important in cytokine secretion in human DCs (b) and antigen presentation in murine DCs (c). In addition to DC activation, it | In Ref. 179:<br>(a) 180;<br>(b) 181;<br>(c) 102;<br>(d) 182 | |

*(continued)*

**TABLE III**

Interaction of Extracellular Adenylates with Purinergic Receptors and Effects on Cellular Metabolism

| Specific receptor | Cell/tissue type | Extracellular nucleotide effect on cell growth and related events | Rank order of potency/dose/conclusion | Ref. |
|---|---|---|---|---|
| | | has also been observed that signaling through nucleotide receptors by ATP leads to DC apoptosis (d) and aberrant DC activation. DCs pretreated with low, non toxic doses of ATP produced lower amounts of IL-1, IL-1, TNF-$\alpha$, IL-6, and IL-12 after subsequent stimulation with LPS or CD40L (e) | | |
| P2, subtype not specified | Human monocyte-derived DCs | In the presence of extracellular ATP, DCs transiently increase their endocytotic activity. Subsequently, DCs upregulate CD86, CD54, and MHC-II; secrete IL-12; and exhibit an improved stimulatory capacity for allogeneic T cells | Extracellular ATP may play an important immunomodulatory role by activating DCs and by skewing the immune reaction toward a Th1 response through the induction of IL-12 secretion | 183 |
| Subtype not specified | Thymocytes | Extracellular ATP is shown here to induce apoptosis in thymocytes and certain tumor cell lines. These changes include DNA fragmentation or lysis without DNA fragmentation. The changes caused by ATP are preceded by a rapid increase in the cytoplasmic calcium fluxes, which by themselves, however, are not sufficient to cause apoptosis | Proposed two independent mechanism of cell death, one of which, requiring active participation on the part of the cell, takes place through apoptosis | 184 |

| P2X | 1. Mouse macro-phages | Extracellular ATP$^{4-}$ opens pores in the | 185 |
| | 2. J774 macrophage-like | plasma membrane of mouse macrophages | |
| | cell line | and the J774 macrophage-like cell line, | |
| | | which leads to apoptosis | Demonstrated that mouse macrophages express the connexin-43 gap junction mRNA and protein and strongly suggested that in these cells connexin-43 forms "half-gap junctions" in response to extracellular ATP$^{4-}$ |

[a]P2U*, Old named for receptor; equal to P2Y in current receptor classification.

433

The above-mentioned experiments (Table III) have demonstrated that extracellular ATP inhibits the growth of a variety of human tumors including prostate, breast, colon, colorectal, esophageal, and melanoma cancer cells. ATP receptor polymorphisms have been implicated in the pathogenesis of prostate cancer (97) and of malignant melanoma and breast cancer (98).

Extracellular ATP is a potent trigger of apoptosis not only for tumor cells but also of the associated tumor vasculature/endothelial cells. The role of ATP and its metabolite adenosine in regulating tumor angiogeneis is currently under investigation. Absence of CD39 in CD39 knockout mice leads to decreased angiogenesis and can be viewed as the result of either increased extracellular ATP or decreased production of extracellular adenosine (14). It has been shown (99,100) that extracellular ATP can lead to endothelial apoptosis. This has been postulated to be the result of intracellular uptake of ATP hydrolysis products.

It has been shown that extracellular ATP activates the transcription factor NF-$\kappa$B through activation of P2Z receptors. The ATP-induced generation of NF-$\kappa$B leads to endothelial cell apoptotic death (16,101). Extracellular ATP has also been proposed as a mediator of the activities of human dendritic cells (102) and natural killer cells (103–106).

It should be noted that the presence of extracellular ATP does not result in cell death in all tumor cell systems studied. For example, cell killing in response to extracellular ATP is not observed in some glioma cells, endometrial carcinoma cells, Ehrlich acites lines, ovarian cancer cells, lung small cell cancer cells, and epidermoid adenocarcinoma cells (107–113). Some authors have indicated that ATP may act as an extracellular messenger in controlling ovarian epithelial and endometrial growth through modulation of the cell cycle via signaling through P2Y$_2$ receptors (108,109). It is not reported whether the absence of ATP-induced cell killing in these systems might be the result of rapid extracellular ATP hydrolysis. In these experiments extracellular ATP induced pore formation, concentration-dependent increases in [Ca$^{2+}$], hypolarization of the plasma membrane, cell shrinkage, and production of inositol triphosphate and later metabolites.

This chapter focuses on the therapeutic role of ATP and P2 receptor interactions. Receptor-mediated processes have been proposed as potential drug targets in the treatment of cancer and other diseases (114–116). *In vitro* studies have indicated that various nonhydrolyzable analogs of ATP could also be used in place of ATP but *in vivo* studies are required (including eventually clinical phase I studies) to demonstrate whether therapeutic relevant concentrations can be achieved without significant toxicities.

## III. ABILITY OF EXOGENOUSLY ADMINISTERED ATP TO EXPAND ORGAN, BLOOD, AND PLASMA (EXTRACELLULAR) POOLS OF ATP

*In vivo*, red blood cells (RBCs) can modulate many physiological functions (20,21,117–119). In addition to acting as transporters of oxygen and carbon dioxide, RBCs function to carry ATP from the liver to peripheral organs (Fig. 4). In the rapid proliferative phase of tumor development total [$^{14}$C] adenylate incorporation into purine pools is much faster than in the resting phase. Intracellular ATP turnover as well as purine breakdown to hypoxanthine and uric acid production are increased. This corresponds to previous findings of increased nucleotide pool sizes as well as elevations in ATP-consuming processes in this growth period (120,121). The net effect is of sequestration of adenylates within the tumor cells and reduction of the extracellular ATP available to interact with surface purinergic (P2) receptors.

ATP release from RBCs appears to be triggered by factors encountered as erythrocytes pass through the tumor vasculature. The tumor environment has been characterized as hypoxic (low oxygen tension) and acidotic

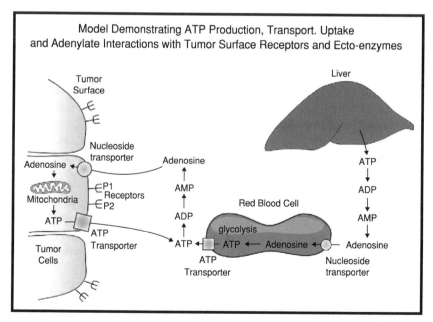

**FIGURE 4**  Model demonstrating ATP production, transport, and uptake, and adenylate interactions with tumor surface receptors and ectoenzymes.

(low pH) (121). These conditions have been shown to be potent stimuli for release of ATP from erythrocytes (122–125). Large increases in red blood cell ATP pools cannot be achieved under any *in vitro* condition. Therefore the intravenous administration of ATP offers a novel approach to the treatment of advanced cancer. ATP released from RBCs passing through tumors can thus inhibit tumor growth through direct interaction with the tumor purine receptors as well as through ATP-mediated antiangiogenic effects, in this fashion blocking potential tumor angiogenesis.

## A. Mechanism of ATP Expansion

Administration of ATP or other hydrolyzable adenine nucleotides into the systemic circulation or extravascular sites results in the immediate rapid degradation of the nucleotides to adenosine and inorganic phosphate, as the result of the powerful activities of ectoenzymes and soluble enzymes (12, 126–128). The adenosine and inorganic phosphate are then incorporated into liver ATP pools via an adenosine–phosphate cotransport mechanism. Detailed radioactive labeling kinetics experiments have established that the turnover of liver ATP pools supplies the adenosine precursors for red blood cell ATP synthesis and that expansion of hepatic ATP pools yields via the same mechanism an expansion of red blood cell ATP pools (127,128). Red blood cells utilize only salvage precursors for ATP synthesis, mostly adenosine. Red blood cell ATP is continuously released in a nonhemolytic fashion, resulting in elevated pools of ATP plasma levels. The ATP released by the red blood cells produces the precursor for salvage purine synthesis in all extrahepatic sites.

It was concluded that the supply of purines (adenosine) precursors for red blood cell ATP synthesis is generated in the hepatic sinusoids (127,128). The generation of increases in red blood cell (total blood) ATP pools immediately after administration of ATP has now been confirmed in a human clinical trial (129). Intravenous administration of ATP in advanced cancer patients resulted in immediate (3–4 h) increases in total blood ATP levels. At that time point, 3–4 h, essentially all the administered ATP ended up in expanding blood ATP pools. These findings demonstrate the much lower catabolic activities in humans as compared with experimental animals and why animal studies underestimate the efficacies of administered ATP. The higher enzymatic catabolic activities in experimental animals as compared with humans are related to their much higher basal metabolic rates and their need for an increased supply of purine precursors for salvage nucleotide biosynthesis.

More than 20 years ago, Rapaport and Zamecnik demonstrated that administration of adenosine *in vivo*, at levels below those that produce significant cardiovascular side effects, can expand liver ATP pools (130). Adenosine became later the key component, along with inorganic phosphate, in the solution used for solid organ preservation by cold storage, after graft harvesting and ischemic damage and prior to graft transplantation (131). The successful clinical transplantation of livers and kidneys (131) and the success of experimental transplantation of pancreatic grafts (132) are dependent on the ability of adenosine and inorganic phosphate to expand the graft's ATP pools. The expanded ATP pools in the ischemically damaged graft are directly responsible for the initiation of metabolic processes vital to organ function. The relationship of graft ATP pools to the successful transplantation of the grafts was accurately established in experimental transplantation of the pancreas (132). Adenosine, rather than adenine, inosine, or hypoxanthine, was the only precursor that could be effectively utilized in the expansion of depleted ATP pools of pancreatic grafts (132), as was shown earlier for liver ATP pools *in vivo* (130). Expansions of organ ATP pools *in vivo* were directly linked to improvements in organ functions in liver (133,134), kidney (135), splenic immune functions (136), gut absorptive capacity (137), and pancreatic B cell function (138). Red blood cell function in delivering oxygen to peripheral sites was shown long ago to benefit from high cellular levels of ATP. Elevated red blood cell ATP pools yield elevated red blood cell 2,3-diphosphoglycerate (2,3-DPG) (139), which in turn causes a shift in the hemoglobin–oxygen affinity curve to the right, yielding enhanced oxygen delivery under normal nonhypoxic conditions.

## B. *Pharmacological Effects of Intravenous Infusion of ATP: Inhibition of Host Weight Loss and Protection of Nonmalignant Cells from Ionizing Radiation*

Therapeutic levels of ATP can be easily achieved in experimental animals through administration of ATP via either the intraperitoneal or intravenous route. The administration of ATP or other adenine nucleotides was shown to significantly elevate ATP plasma pools. For example, Rapaport and Fontaine (127,128) achieved plasma ATP levels of 3 to 5 $\mu M$, 1 to 5 h after intraperitoneal administration of ATP, and this prolonged increase in plasma ATP levels was subsequently shown to be due to an expansion of erythrocyte ATP levels. Similar elevations of mouse red blood cell ATP levels and blood plasma ATP levels after administration (intravenous or intraperitoneal) of ATP were reported by others (72).

Pharmacological effects of intravenous infusion of ATP have been established both in experimental animal systems (140–146) and in human subjects (147–151). In addition, studies in experimental animals have shown the efficacy of ATP administration in several disease models (152). ATP, as a sodium salt or metal chelate, when administered to murine hosts, was shown to possess antitumor activities *in vivo* in several experimental models (40,72,127,128,153–157). The administration of ATP to tumor-bearing murine hosts results in improvements in liver function as well as significant inhibition of weight loss (127,128,133,158).

ATP administration has also been shown to exert a radioprotective effect in radiation therapy (159,160). ATP administration, intravenously or intraperitoneally, in conjunction with radiation, produced cures of murine tumors refractory to other treatment (72,169). Administration of ATP as a plantinum chelate was demonstrated to be effective in murine tumor models without producing nephrotoxicity, myelosuppression, or hematologically adverse side effects (153). In comparison with similar doses of *cis*-dichlorodiammineplatinum(II) (DDP), Pt-ATP produced somewhat less tumor growth inhibition than DDP but yielded significantly greater increases in host life span in two mouse tumor models.

## IV. SUMMARY OF PREVIOUS ATP CANCER TRIALS

The value of ATP as a potential treatment in advanced cancers depends on three separate relationships that, in our opinion, have to a large extent been established.

1. The decline in muscle and liver ATP pools, resulting in lower levels of blood and plasma pools of ATP in advanced cancer patients
2. The ability to expand host ATP pools by administering exogenously supplied ATP in advanced disease patients
3. The ability of elevated pools of ATP to stabilize advanced disease by inhibiting tumor growth and favorably affecting host functions

As has been shown above, extracellular ATP exhibits anticancer activity *in vivo* when injected into tumor-bearing animals. These findings contribute to the rationale of the development of ATP as an anticancer agent, as a single agent or part of combination therapy. Three clinical trials have examined the effects of continuous intravenous infusions of ATP in advanced cancer patients. The published results of these trials provide information about the toxicity, pharmacokinetics, and anticachectic and antitumor actions of ATP.

## A. Study 1

In the initial phase I/II clinical trial (129), ATP was administered as a continuous intravenous infusion for 96 h, once every 4 weeks, at rates of 50, 75, or 100 $\mu$g/kg per minute (Table IV). The trial included 14 men with advanced, nonresectable cancer, 8 of whom suffered from stage IIIB/IV non-small cell lung cancer. Most of the patients were chemotherapy naive. One patient received one infusion, four received two infusions, six received three infusions, one received four infusions, and two received six infusions.

The dose-limiting toxicity seen in this study was a cardiopulmonary reaction characterized by tightness of the chest and dyspnea that resolved shortly after discontinuing the ATP infusion. This reaction was seen in all three patients (100%) infused at 100 $\mu$g/kg per minute; in 3 of 6 (50%) patients infused at 75 $\mu$g/kg per minute; and in 4 of 11 patients (36%) who received 50 $\mu$g/kg per minute. In some cases, this reaction was accompanied by electrocardiographic changes suggestive of myocardial ischemia. Less frequent or less prominent adverse effects that may, or may not, have been related to ATP treatment were injection site reactions ("local reactions," pain and phlebitis), hypoxia, hypotension, electrocardiogram abnormalities, nausea and/or emesis, abdominal pain, dizziness, headache, anxiety, back or neck pain, anemia, and leukopenia. In summary, 21% of all adverse events recorded in this phase I/II trial were grade 3 or 4 (defined as severe or potentially life threatening in nature). These included six events of dyspnea, two events of hypoxia, five events of chest pain, one event of hypotension, two events of nausia/emesis, one event of dizziness, one event of grade 3 headache, and one event of grade 3 anxiety. Of all adverse events recorded, 79% were of grade 1 or 2 (defined as mild or moderate). Patients required

**TABLE IV**
Clinical Trials of Infusional Therapy ATP in Patients with Advanced Cancer:
Range of Dose Rates Tolerated

| Ref. | Average weight (kg) | MTD ($\mu$g/kg/min) | Duration (h) | Total ATP dose (g) | Daily ATP dose (g) | No. of infusions |
|------|---------------------|---------------------|--------------|--------------------|--------------------|------------------|
| 129  | Assumed 70          | 50                  | 96           | 21.6               | 5.04               | >45              |
| 161  | 74.6 ($\pm$) 16.5   | 65–75               | 27[a]        | 8.5 ($\pm$) 1.9    | 7.1                | 77               |
| 129  | Assumed 70          | 75                  | 96           | 32.4               | 8.1                | 11               |
| 129  | Assumed 70          | 100                 | 96           | 43.2               | 14.4               | 3                |

*Abbreviation:* MTD, maximally tolerated dose.

[a]ATP was infused over 30 h. ATP infusions were started at 20 $\mu$g/kg/min and increased by increments of 10 $\mu$g/kg/min every 30 min until the maximum dose of 75 $\mu$g/kg/min had been reached.

hospitalization for at least 4 days once every 4 weeks for the purpose of the monitored ATP administration.

With respect to the pharmacokinetic properties of ATP, Haskell *et al.* measured the whole blood concentrations of ATP of 18 subjects before and 24, 48, 72, and 96 h during and after infusions at rates of 50, 75, or 100 $\mu$g/kg per minute. The ATP concentrations were found to vary widely but, in general, they increased 30–40% after 4 h and the highest blood concentrations were seen at 24 h of infusion. These concentrations averaged 63, 67, and 13%, respectively, higher than the pretreatment values of blood ATP levels. The blood concentrations of ATP were relatively constant or slightly declined during the interval between 24 and 96 h of infusion. Few data are available concerning the decay of ATP concentrations postinfusion (129). These authors concluded that prolonged infusions of ATP are feasible with acceptable toxicity and that 50 $\mu$g/kg per minute is both the maximally tolerated dose and the most appropriate dose rate for subsequent phase II testing of 96-h infusions of ATP in patients with advanced cancer.

## B. Study 2

The second human clinical trial was a phase II multicenter study of 15 chemotherapy-naive, stage IIIB/IV non-small cell lung cancer patients. These patients were continuously infused at rates of 50 or 65 $\mu$g/kg per minute of ATP for 96 h, once every 4 weeks (162,163). Two patients received one infusion, eight received two infusions; four received four infusions; and one received seven infusions. A large proportion of the patients in this study experienced a variety of adverse effects, including chest pain, dyspnea, coughing, anxiety, injection site pain, chest tightness, headache, insomnia, and hot flashes. Almost one-half of the patients exhibited abnormal electrocardiograms. In some patients, there were minor reductions in hematocrit, hemoglobin, total protein, albumin, sodium, and calcium and minor increases in serum glucose levels. No significant hematological, renal, hepatic, or gastrointestinal toxicity was noted. Six patients reported severe (grade 3) adverse effects and two patients had life-threatening dyspnea (grade 4). All patients were hospitalized for at least 4 days during each of the monitored 96-h infusion cycles.

Although no significant tumor shrinking was observed, the majority of patients exhibited stable disease after treatment with ATP. In addition, beneficial effects were seen in terms of weight gain, performance status, and the overall survival of patients with non-small cell lung cancer. The positive effects of ATP administration on median survival and quality of life in advanced cancer are the result of the ATP-promoted expansions of hepatic

ATP pools, red blood cell ATP pools, and blood plasma ATP pools, which in turn positively affect a variety of physiological functions. The most pronounced effects include improvements in liver functions, which are directly related to hepatic ATP pools in the advanced disease patient, such as the significant inhibition of hepatic gluconeogenesis, which is responsible for futile cycles and for the depletion of the body's energy stores (133).

## C. Study 3

The third trial (164) was a randomized, open label phase III study of 52 assessable, previously treated, refractory, stage IIIB/IV non-small cell lung cancer patients who failed previous chemotherapy and/or radiation therapy. The patients were randomized into two groups. One group of 25 patients received best supportive care and infusions of ATP while the other group of 27 patients received only best supportive care. The ATP-treated patients received ten 30-h infusions, the first seven at 2-week intervals and the last three at 4-week intervals. In each case, the infusion was started at 20 $\mu$g/kg per minute and increased every 30 min by 10 $\mu$g/kg per minute until adverse effects developed or a maximum dose of 75 $\mu$g/kg per minute was reached. If adverse effects developed, the dose was reduced stepwise until the adverse effects disappeared. Eleven patients received 1–3 infusions, 5 received 4–6 infusions, and 12 received 7–10 infusions of ATP.

The adverse experiences seen during treatment were generally mild or moderate and consisted of chest discomfort, an urge to take a deep breath, flushing, nausea, lightheadedness, headache, sweating, anxiety, and palpitations. More pronounced side effects were injection site reactions and dyspnea. In those patients experiencing chest discomfort, electrocardiograms did not exhibit changes suggestive of myocardial ischemia. All side effects resolved within minutes of lowering the rate of ATP infusion. Of all adverse events recorded in this trial, 81 were of grade 1 (mild), 5 were of grade 2 (moderate), none were of grade 3 (severe), and 5 were of grade 4 (life threatening). Patients were hospitalized for 1 or 2 days during the monitored infusions of ATP, which were administered at 2- or 4-week intervals.

In this study, ATP administration was associated with beneficial effects on body weight and voluntary muscle strength and with improvements in a variety of Quality of Life domains. Thus, the authors concluded that ATP shows promise as an agent for the palliation of cancer cachexia. Both preclinical (133) and clinical (164,165) data suggest that administration of ATP to mice or humans results in the induction of enzymatic activities that result in the degradation of the extracellular ATP. Future clinical trials will utilize schedules of administration of less than 48 h in induration. Detailed publications of three human clinical trials established a particular protocol of ATP administration in advanced cancer patients. Such a protocol includes continuous intravenous infusions of ATP for periods longer than 24 h during which time the patient is being hospitalized. The continuous infusions are administered for multiple treatment cycles every 2–4 weeks. These schedules would also significantly reduce many of the side effects that normally materialize after 48 h of continuous infusion. It is important to emphasize, however, that any side effect or discomfort to the patient is resolved within a few seconds to 2 min once the infusion is disconnected. During these infusion protocols some grade 3 and 4 (severe and life threatening) toxicity has been recorded.

### D. Other Studies

Measurable positive effects by ATP in the advanced disease patient include the lowering of pulmonary vascular resistance and systemic vascular resistance without any effect on arterial blood pressure (147,151). The increases in total blood (RBC) and blood plasma ATP pools provide a supply of salvage purines to peripheral sites as well as increasing oxygenation of peripheral sites. The *de novo* synthesis of purines is expensive in terms of endogenous ATP pools and in the advanced disease patient it significantly contributes to the overall decline in energy stores (133).

### V. CONCLUSIONS: ADMINISTRATION OF ATP IN THE TREATMENT OF ADVANCED CANCER

Three properties of intravenously administered ATP discussed in this review contribute to its efficacy, and are expected to significantly affect the outcomes of patients with advanced cancer:

*Cytolytic activities of ATP:* Many research groups have established cytolytic activities of ATP in preclinical studies. They are attributed to both purinergic receptors as well as non- receptor-mediated mechanisms.

*Induction of resistance by ATP:* ATP administration induces resistance to chemo- and radiation therapy in normal tissues.

*Anticachexia effects of ATP:* The third proven activity of ATP, its ability to effectively expand liver, red blood cell, and blood plasma ATP pools, is expected as well to significantly contribute to a favorable outcome in the treatment of advanced cachectic cancers especially in older patients, by delivering purines to purine-depleted peripheral sites. As discussed earlier, a large amount of data establishes that on aging and in cachectic disease models, a significant depletion of purines *in vivo* can be easily noted. The anticachexia effects of ATP translate into significant improvements in quality of life parameters in the advanced disease patient. As stated, the anticachexia effects of ATP, and its effectiveness in delivering purines to purine-depleted peripheral sites, act to protect the host from the otherwise devastating effects of high-dose cytotoxic and/or radiation therapy. All the large volumes of preclinical data suggesting the utilization of ATP in the treatment of advanced cancers, along with initial phase I/II and phase II human trials, indicate that ATP has a future place as a useful anticancer agent. Its effects are most noticeable in increasing median survival and improving a variety of quality of life parameters.

## VI. THE FUTURE

In human advanced cancers tumor growth-inhibitory and cytolytic activities of ATP are expected to yield stable disease in selected cancer patients. The combination of ATP administration with other anticancer modalities is an important area of research that is beginning to be explored.

### References

1. Abraham, E. H., *et al.* (1993). The multidrug resistance (*mdr1*) gene product functions as an ATP channel. *Proc. Natl. Acad. Sci. USA* **90**, 312–316.
2. Abraham, E. H., *et al.* (1997). Cystic fibrosis transmembrane conductance regulator and adenosine triphosphate. *Science* **275**, 1324–1326.
3. Cantiello, H. F. (2001). Electrodiffusional ATP movement through CFTR and other ABC transporters. *Pflugers Arch.* **443**(Suppl. 1), S22–S27.
4. Mitchell, C. H., *et al.* (1998). A release mechanism for stored ATP in ocular ciliary epithelial cells. *Proc. Natl. Acad. Sci. USA* **95**, 7174–7178.
5. Reisin, I. L., *et al.* (1994). The cystic fibrosis transmembrane conductance regulator is a dual ATP and chloride channel. *J. Biol. Chem.* **269**, 20584–20591.
6. Sugita, M., Yue, Y., and Foskett, J. K. (1998). CFTR Cl⁻ channel and CFTR-associated ATP channel: Distinct pores regulated by common gates. *EMBO J.* **17**, 898–908.
7. Handa, M., and Guidotti, G. (1996). Purification and cloning of a soluble ATP-diphosphohydrolase (apyrase) from potato tubers *(Solanum tuberosum)*. *Biochem. Biophys. Res. Commun.* **218**, 916–923.
8. Wang, T. F., and Guidotti, G. (1996). CD39 is an ecto-(Ca$^{2+}$, Mg$^{2+}$)-apyrase. *J. Biol. Chem.* **271**, 9898–9901.
9. Grinthal, A., and Guidotti, G. (2002). Transmembrane domains confer different substrate specificities and adenosine diphosphate hydrolysis mechanisms on CD39, CD39L1, and chimeras. *Biochemistry* **41**, 1947–1956.
10. Chen, W., and Guidotti, G. (2001). Soluble apyrases release ADP during ATP hydrolysis. *Biochem. Biophys. Res. Commun.* **282**, 90–95.
11. Wang, T. F., Ou, Y., and Guidotti, G. (1998). The transmembrane domains of ectoapyrase (CD39) affect its enzymatic activity and quaternary structure. *J. Biol. Chem.* **273**, 24814–24821.

12. Ujhazy, P., *et al.* (1996). Evidence for the involvement of ecto-5'-nucleotidase (CD73) in drug resistance. *Int. J. Cancer* **68**, 493–500.

13. Ujhazy, P., *et al.* (1994). Ecto-5'-nucleotidase (CD73) in multidrug-resistant cell lines generated by doxorubicin. *Int. J. Cancer* **59**, 83–93.

14. Goepfert, C., *et al.* (2000). CD39 modulates endothelial cell activation and apoptosis. *Mol. Med.* **6**, 591–603.

15. Dzhandzhugazyan, K. N., *et al.* (1998). Ecto-ATP diphosphohydrolase/CD39 is over-expressed in differentiated human melanomas. *FEBS Lett.* **430**, 227–230.

16. von Albertini, M., *et al.* (1998). Extracellular ATP and ADP activate transcription factor NF-$\kappa$B and induce endothelial cell apoptosis. *Biochem. Biophys. Res. Commun.* **248**, 822–829.

17. Schulte am Esch, J., II., *et al.* (1999). Structural elements and limited proteolysis of CD39 influence ATP diphosphohydrolase activity. *Biochemistry* **38**, 2248–2258.

18. Kittel, A., *et al.* (1999). CD39 as a caveolar-associated ectonucleotidase. *Biochem. Biophys. Res. Commun.* **262**, 596–599.

19. Imai, M., *et al.* (2000). CD39 modulates IL-1 release from activated endothelial cells. *Biochem. Biophys. Res. Commun.* **270**, 272–278.

20. Abraham, E. H., *et al.* (2001). Cellular and biophysical evidence for interactions between adenosine triphosphate and P-glycoprotein substrates: Functional implications for adenosine triphosphate/drug cotransport in P-glycoprotein overexpressing tumor cells and in P-glycoprotein low-level expressing erythrocytes. *Blood Cells Mol. Dis.* **27**, 181–200.

21. Abraham, E. H., *et al.* (2001). Erythrocyte membrane ATP binding cassette (ABC) proteins: MRP1 and CFTR as well as CD39 (ecto-apyrase) involved in RBC ATP transport and elevated blood plasma ATP of cystic fibrosis. *Blood Cells Mol. Dis.* **27**, 165–801.

22. Hashimoto, M., *et al.* (1999). Hypotension induced by exercise is associated with enhanced release of adenyl purines from aged rat artery. *Am. J. Physiol.* **276**, H970–H975.

23. Shinozuka, K., *et al.* (2001). Participation of ATP in cell volume regulation in the endothelium after hypotonic stress. *Clin. Exp. Pharmacol. Physiol.* **28**, 799–803.

24. Hashimoto, M., *et al.* (1995). The effects of age on the release of adenine nucleosides and nucleotides from rat caudal artery. *J. Physiol.* **489**, 841–848.

25. Rabini, R. A., *et al.* (1997). Diabetes mellitus and subjects' ageing: A study on the ATP content and ATP-related enzyme activities in human erythrocytes. *Eur. J. Clin. Invest.* **27**, 327–332.

26. Okada, M., *et al.* (1998). Influence of aging or left ventricular hypertrophy on the human heart: Contents of phosphorus metabolites measured by $^{31}$P MRS. *Magn. Reson. Med.* **39**, 772–782.

27. Laciak, J., and Witkowski, S. (1966). [Studies on the content of adenine compounds in the erythrocytes in laryngeal cancer patients]. *Otolaryngol. Pol.* **20**, 269–275.

28. Wand, H., and Rieche, K. (1972). [Content and liberation of adenine nucleotides from isolated thrombocytes of cancer patients prior and during treatment]. *Dtsch. Gesundheitsw* **27**, 1072–1076.

29. De la Morena Garcia, E., *et al.* (1987). [Levels of adenine nucleotide in erythrocytes of normal subjects and in neoplastic patients]. *Rev. Clin. Esp.* **146**, 221–223.

30. Stocchi, V., *et al.* (1987). Adenine and pyridine nucleotides in the red blood cells of subjects with solid tumors. *Tumori.* **73**, 25–28.

31. Stocchi, V., *et al.* (1987). Adenine and pyridine nucleotides during rabbit reticulocyte maturation and cell aging. *Mech. Ageing Dev.* **39**, 29–44.

32. Henson, D. E., and Tarone, R. E. (1994). Involution and the etiology of breast cancer. *Cancer* **74**(Suppl. 1), 424–429.

33. Braendle, W. (2000). [Estrogens—a breast cancer risk?]. *Ther. Umsch.* **57**, 646–650.

34. Kikuchi, S., *et al.* (2000). Effect of age on the relationship between gastric cancer and *Helicobacter pylori*. Tokyo Research Group of Prevention for Gastric Cancer. *Jpn. J. Cancer Res.* **91**, 774–779.

35. Cortopassi, G., Liu, Y., and Hutchin, T. (1996). Degeneration of human oncogenes and mitochondrial genes occurs in cells that exhibit age-related pathology. *Exp. Gerontol.* **31**, 253–265.

36. Liu, Y., *et al.* (1994). BCL2 translocation frequency rises with age in humans. *Proc. Natl. Acad. Sci. USA* **91**, 8910–8914.

37. Warren, N., *et al.* (1991). Frequency of carriers of cystic fibrosis gene among patients with myeloid malignancy and melanoma. *BMJ* **302**, 760–761.

38. Padua, R. A., *et al.* (1997). The cystic fibrosis ΔF508 gene mutation and cancer. *Hum. Mutat.* **10**, 45–48.

39. Neglia, J. P., *et al.* (1995). The risk of cancer among patients with cystic fibrosis. Cystic Fibrosis and Cancer Study Group. *N. Engl. J. Med.* **332**, 494–499.

40. Abraham, E. H., *et al.* (1996). Cystic fibrosis hetero- and homozygosity is associated with inhibition of breast cancer growth. *Nat. Med.* **2**, 593–596.

41. Miro, C., and Orecchia, R. (2002). Cystic fibrosis heterozygosity: Darwinian bet on cancer protection? *Lancet Oncol.* **3**, 395 [discussion on p. 396].

42. Daneryd, P., *et al.* (1998). Protection of metabolic and exercise capacity in unselected weight-losing cancer patients following treatment with recombinant erythropoietin: A randomized prospective study. *Cancer Res.* **58**, 5374–5379.

43. Daneryd, P. L., Hafstrom, L. R., and Karlberg, I. H. (1990). Effects of spontaneous physical exercise on experimental cancer anorexia and cachexia. *Eur. J. Cancer* **26**, 1083–1088.

44. Daneryd, P., *et al.* (1995). Tumour purine nucleotides and cell proliferation in response to exercise in rats. *Eur. J. Cancer* **31A**, 2309–2312.

45. Fishman, P., Bar-Yehuda, S., and Vagman, L. (1998). Adenosine and other low molecular weight factors released by muscle cells inhibit tumor cell growth. *Cancer Res.* **58**, 3181–3187.

46. Hellsten, Y., *et al.* (1998). Adenosine concentrations in the interstitium of resting and contracting human skeletal muscle. *Circulation* **98**, 6–8.

47. Hellsten, Y., *et al.* (1998). Urate uptake and lowered ATP levels in human muscle after high-intensity intermittent exercise. *Am. J. Physiol.* **274**, E600–E606.

48. Nelson, K. A. (2000). The cancer anorexia–cachexia syndrome. *Semin. Oncol.* **27**, 64–68.

49. Nelson, K. A. (2001). Modern management of the cancer anorexia–cachexia syndrome. *Curr. Pain Headache Rep.* **5**, 250–256.

50. Stein, T. P. (1978). Cachexia, gluconeogenesis and progressive weight loss in cancer patients. *J. Theor. Biol.* **73**, 51–59.

51. Torosian, M. H., *et al.* (1993). Effect of tumor burden on futile glucose and lipid cycling in tumor-bearing animals. *J. Surg. Res.* **55**, 68–73.

52. Bartlett, D. L., Stein, T. P., and Torosian, M. H. (1995). Effect of growth hormone and protein intake on tumor growth and host cachexia. *Surgery* **117**, 260–267.

53. Holroyde, C. P., *et al.* (1975). Altered glucose metabolism in metastatic carcinoma. *Cancer Res.* **35**, 3710–3714.

54. Holroyde, C. P., *et al.* (1977). Metabolic response to total parenteral nutrition in cancer patients. *Cancer Res.* **37**, 3109–3114.

55. Holroyde, C. P., and Reichard, G. A. (1981). Carbohydrate metabolism in cancer cachexia. *Cancer Treat. Rep.* **65**(Suppl. 5), 55–59.

56. Holroyde, C. P., *et al.* (1984). Glucose metabolism in cachectic patients with colorectal cancer. *Cancer Res.* **44**, 5910–5913.

57. Holroyde, C. P., and Reichard, G. A. Jr. (1986). General metabolic abnormalities in cancer patients: Anorexia and cachexia. *Surg. Clin. North Am.* **66**, 947–956.

58. Schneeberger, A. L., *et al.* (1989). Effect of cancer on the in vivo energy state of rat liver and skeletal muscle. *Cancer Res.* **49**, 1160–1164.

59. Peacock, J. L., *et al.* (1987). Resting energy expenditure and body cell mass alterations in noncachectic patients with sarcomas. *Surgery* **102**, 465–472.

60. Inculet, R. I., *et al.* (1987). Gluconeogenesis in the tumor-influenced rat hepatocyte: Importance of tumor burden, lactate, insulin, and glucagon. *J. Natl. Cancer Inst.* **79**, 1039–1046.

61. Short, S., *et al.* (1996). Defective antigen processing correlates with a low level of intracellular glutathione. *Eur. J. Immunol.* **26**, 3015–3020.

62. el Hag, I. A., Roos, G., and Stenram, U. (1990). Decrease of liver energy charge, ATP and glutathione at concomitant intraarterial administration of Adriamycin and degradable starch microspheres in rat. *Sel. Cancer Ther.* **6**, 135–144.

63. Luthen, R. E., Niederau, C., and Grendell, J. H. (1994). Glutathione and ATP levels, subcellular distribution of enzymes, and permeability of duct system in rabbit pancreas following intravenous administration of alcohol and cerulein. *Dig. Dis. Sci.* **39**, 871–879.

64. Luthen, R., Niederau, C., and Grendell, J. H. (1995). Intrapancreatic zymogen activation and levels of ATP and glutathione during caerulein pancreatitis in rats. *Am. J. Physiol.* **268**, G592–G604.

65. Kobayashi, T., *et al.* (1993). Glutathione depletion alters hepatocellular high-energy phosphate metabolism. *J. Surg. Res.* **54**, 189–195.

66. Kobayashi, K., *et al.* (1988). ATP stimulates the uptake of *S*-dinitrophenylglutathione by rat liver plasma membrane vesicles. *FEBS Lett.* **240**, 55–58.

67. Kobayashi, H., *et al.* (1992). Changes in the glutathione redox system during ischemia and reperfusion in rat liver. *Scand. J. Gastroenterol.* **27**, 711–716.

68. Iida, T., *et al.* (2001). Hammerhead ribozyme against γ-glutamylcysteine synthetase sensitizes human colonic cancer cells to cisplatin by down-regulating both the glutathione synthesis and the expression of multidrug resistance proteins. *Cancer Gene Ther.* **8**, 803–814.

69. Iida, M., *et al.* (1999). Effect of glutathione-modulating compounds on platinum compound-induced cytotoxicity in human glioma cell lines. *Anticancer Res.* **19**, 5383–5384.

70. Leslie, E. M., *et al.* (2001). Transport of the β-*O*-glucuronide conjugate of the tobacco-specific carcinogen 4-(methylnitrosamino)-1-(3-pyridyl)-1-butanol (NNAL) by the multidrug resistance protein 1 (MRP1). Requirement for glutathione or a non-sulfur-containing analog. *J. Biol. Chem.* **276**, 27846–27854.

71. Harvie, R. M., Davey, M. W., and Davey, R. A. (1997). Increased MRP expression is associated with resistance to radiation, anthracyclines and etoposide in cells treated with fractionated γ-radiation. *Int. J. Cancer* **73**, 164–167.

72. Estrela, J. M., *et al.* (1995). Elimination of Ehrlich tumours by ATP-induced growth inhibition, glutathione depletion and X-rays. *Nat. Med.* **1**, 84–88.

73. Lasso de la Vega, M. C., *et al.* (1994). Inhibition of cancer growth and selective glutathione depletion in Ehrlich tumour cells in vivo by extracellular ATP. *Biochem. J.* **298**, 99–105.

74. Navarro, J., *et al.* (1997). Blood glutathione as an index of radiation-induced oxidative stress in mice and humans. *Free Radic. Biol. Med.* **22**, 1203–1209.

75. Obrador, E., *et al.* (1997). Glutathione and the rate of cellular proliferation determine tumour cell sensitivity to tumour necrosis factor in vivo. *Biochem. J.* **325**, 183–189.

76. Mure, T., Sano, K., and Kitagawa, T. (1992). Modulation of membrane permeability, cell proliferation and cytotoxicity of antitumor agents by external ATP in mouse tumor cells. *Jpn. J. Cancer Res.* **83**, 121–126.

77. Maymon, R., *et al.* (1994). Enhancing effect of ATP on intracellular Adriamycin penetration in human ovarian cancer cell lines. *Biochim. Biophys. Acta* **1201**, 173–178.

78. Rapaport, E. (1983). Treatment of human tumor cells with ADP or ATP yields arrest of growth in the S phase of the cell cycle. *J. Cell. Physiol.* **114**, 279–283.

79. Rapaport, E., Fishman, R. F., and Gercel, C. (1983). Growth inhibition of human tumor cells in soft-agar cultures by treatment with low levels of adenosine 5'-triphosphate. *Cancer Res.* **43**, 4402–4406.

80. Schroder, E. W., and Rapaport, E. (1984). Retinoic acid alters subcellular compartmentalization of ATP pools in 3T3 cells but not in HeLa cells. *J. Cell. Physiol.* **120**, 204–210.

81. Rapaport, E., Schroder, E. W., and Black, P. H. (1982). Retinoic acid-promoted expansion of total cellular ATP pools in 3T3 cells can mediate its stimulatory and growth inhibitory effects. *J. Cell. Physiol.* **110**, 318–322.

82. Rapaport, E. (1980). Compartmentalized ATP pools produced from adenosine are nuclear pools. *J. Cell. Physiol.* **105**, 267–274.

83. Rapaport, E., Garcia-Blanco, M. A., and Zamecnik, P. C. (1979). Regulation of DNA replication in S phase nuclei by ATP and ADP pools. *Proc. Natl. Acad. Sci. USA* **76**, 1643–1647.

84. Wiley, J. S., *et al.* (1990). Extracellular ATP stimulates an amiloride-sensitive sodium influx in human lymphocytes. *Arch. Biochem. Biophys.* **280**, 263–268.

85. Wiley, J. S., *et al.* (1992). The $ATP^{4-}$ receptor-operated ion channel of human lymphocytes: Inhibition of ion fluxes by amiloride analogs and by extracellular sodium ions. *Arch. Biochem. Biophys.* **292**, 411–418.

86. Wiley, J. S., *et al.* (1994). The P2Z-purinoceptor of human lymphocytes: Actions of nucleotide agonists and irreversible inhibition by oxidized ATP. *Br. J. Pharmacol.* **112**, 946–950.

87. Wiley, J. S., *et al.* (1996). Transduction mechanisms of P2Z purinoceptors. *Ciba Found. Symp.* **198**, 149–160.

88. Wiley, J. S., and Dubyak, G. R. (1989). Extracellular adenosine triphosphate increases cation permeability of chronic lymphocytic leukemic lymphocytes. *Blood* **73**, 1316–1323.

89. Dubyak, G. R. (1991). Signal transduction by P2-purinergic receptors for extracellular ATP. *Am. J. Respir. Cell Mol. Biol.* **4**, 295–300.

90. Dubyak, G. R., and el-Moatassim, C. (1993). Signal transduction via P2-purinergic receptors for extracellular ATP and other nucleotides. *Am. J. Physiol.* **265**, C577–C606.

91. Dubyak, G. R. (2002). Focus on "Extracellular ATP Signaling and P2X Nucleotide Receptors in Monolayers of Primary Human Vascular Endothelial Cells." *Am. J. Physiol. Cell Physiol.* **282**, C242–C244.

92. Surprenant, A., *et al.* (1996). The cytolytic P2Z receptor for extracellular ATP identified as a P2X receptor (P2X7). *Science* **272**, 735–738.

93. Steinberg, T. H., and Di Virgilio, F. (1991). Cell-mediated cytotoxicity: ATP as an effector and the role of target cells. *Curr. Opin. Immunol.* **3**, 71–75.

94. Chiozzi, P., *et al.* (1996). Role of the purinergic P2Z receptor in spontaneous cell death in J774 macrophage cultures. *Biochem. Biophys. Res. Commun.* **218**, 176–181.

95. Murgia, M., *et al.* (1993). Oxidized ATP: An irreversible inhibitor of the macrophage purinergic P2Z receptor. *J. Biol. Chem.* **268**, 8199–8203.

96. Pizzo, P., *et al.* (1992). Role of P2z purinergic receptors in ATP-mediated killing of tumor necrosis factor (TNF)-sensitive and TNF-resistant L929 fibroblasts. *J. Immunol.* **149,** 3372–3378.

97. Janssens, R., and Boeynaems, J. M. (2001). Effects of extracellular nucleotides and nucleosides on prostate carcinoma cells. *Br. J. Pharmacol.* **132,** 536–546.

98. Csaszar, A., and Abel, T. (2001). Receptor polymorphisms and diseases. *Eur. J. Pharmacol.* **414,** 9–22.

99. Rounds, S., *et al.* (1998). Mechanism of extracellular ATP- and adenosine-induced apoptosis of cultured pulmonary artery endothelial cells. *Am. J. Physiol.* **275,** L379–L388.

100. Dawicki, D. D., *et al.* (1997). Extracellular ATP and adenosine cause apoptosis of pulmonary artery endothelial cells. *Am. J. Physiol.* **273,** L485–L494.

101. von Albertini, M. A., *et al.* (1997). Adenosine nucleotides induce E-selectin expression in porcine endothelial cells. *Transplant. Proc.* **29,** 1062.

102. Mutini, C., *et al.* (1999). Mouse dendritic cells express the P2X7 purinergic receptor: Characterization and possible participation in antigen presentation. *J. Immunol.* **163,** 1958–1965.

103. Correale, P., *et al.* (1997). Extracellular adenosine 5'-triphosphate involvement in the death of LAK-engaged human tumor cells via P2X-receptor activation. *Immunol. Lett.* **55,** 69–78.

104. Macino, B., *et al.* (1996). CD45 regulates apoptosis induced by extracellular adenosine triphosphate and cytotoxic T lymphocytes. *Biochem. Biophys. Res. Commun.* **226,** 769–776.

105. Di Virgilio, F., *et al.* (1990). Extracellular ATP as a possible mediator of cell-mediated cytotoxicity. *Immunol. Today* **11,** 274–277.

106. Correale, P., *et al.* (1995). Bryostatin 1 enhances lymphokine activated killer sensitivity and modulates the $\beta_1$ integrin profile of cultured human tumor cells. *Anticancer Drugs* **6,** 285–290.

107. Song, S. L., and Chueh, S. H. (1996). Antagonistic effect of $Na^+$ and $Mg^{2+}$ on P2z purinoceptor-associated pores in dibutyryl cyclic AMP-differentiated NG108–15 cells. *J. Neurochem.* **67,** 1694–1701.

108. Schultze-Mosgau, A., *et al.* (2000). Characterization of calcium-mobilizing, purinergic P2Y₂ receptors in human ovarian cancer cells. *Mol. Hum. Reprod.* **6,** 435–442.

109. Katzur, A. C., *et al.* (1999). Expression and responsiveness of P2Y2 receptors in human endometrial cancer cell lines. *J. Clin. Endocrinol Metab.* **84,** 4085–4091.

110. Pedersen, S. F., *et al.* (1998). P2 receptor-mediated signal transduction in Ehrlich ascites tumor cells. *Biochim. Biophys. Acta* **1374,** 94–106.

111. Clunes, M. T., and Kemp, P. J. (1996). P2u purinoceptor modulation of intracellular $Ca^{2+}$ in a human lung adenocarcinoma cell line: Down-regulation of $Ca^{2+}$ influx by protein kinase C. *Cell Calcium* **20,** 339–346.

112. Sugita, K., *et al.* (1994). Effects of pertussis toxin on signal transductions via P2-purinergic receptors in A-431 human epidermoidal carcinoma cells. *Enzyme Protein* **48,** 222–228.

113. Gonzalez, F. A., *et al.* (1989). Receptor specific for certain nucleotides stimulates inositol phosphate metabolism and $Ca^{2+}$ fluxes in A431 cells. *J. Cell. Physiol.* **141,** 606–617.

114. Cutler, C. S., Lewis, J. S., and Anderson, C. J. (1999). Utilization of metabolic, transport and receptor-mediated processes to deliver agents for cancer diagnosis. *Adv. Drug Deliv. Rev.* **37,** 189–211.

115. Dubowchik, G. M., and Walker, M. A. (1999). Receptor-mediated and enzyme-dependent targeting of cytotoxic anticancer drugs. *Pharmacol. Ther.* **83,** 67–123.

116. Williams, M., and Jarvis, M. F. (2000). Purinergic and pyrimidinergic receptors as potential drug targets. *Biochem. Pharmacol.* **59,** 1173–1185.

117. Konishi, Y., and Ichihara, A. (1979). Transfer of purines from liver to erythrocytes: In vivo and in vitro studies. *J. Biochem. (Tokyo)* **85,** 295–301.
118. Pritchard, J. B., Chavez-Peon, F., and Berlin, R. D. (1970). Purines: Supply by liver to tissues. *Am. J. Physiol.* **219,** 1263–1267.
119. Pritchard, J. B., *et al.* (1975). Uptake and supply of purine compounds by the liver. *Am. J. Physiol.* **229,** 967–972.
120. Grune, T., *et al.* (1992). Adenine metabolism of Ehrlich mouse ascites cells in proliferating and resting phase of tumor growth. *Biochem. Int.* **26,** 199–209.
121. Stubbs, M., *et al.* (1990). Monitoring tumor growth and regression by $^{31}$P magnetic resonance spectroscopy. *Adv. Enzyme Regul.* **30,** 217–230.
122. Forrester, T. (1990). Release of ATP from heart: Presentation of a release model using human erythrocyte. *Ann. N.Y. Acad. Sci.* **603,** 335–351.
123. Thiriot, C., *et al.* (1982). Effects of whole-body $\gamma$ irradiation on oxygen transport by rat erythrocytes. *Biochimic.* **64,** 79–83.
124. Bergfeld, G. R., and Forrester, T. (1992). Release of ATP from human erythrocytes in response to a brief period of hypoxia and hypercapnia. *Cardiovasc. Res.* **26,** 40–47.
125. Sprague, R. S., *et al.* (2001). Impaired release of ATP from red blood cells of humans with primary pulmonary hypertension. *Exp. Biol. Med. (Maywood)* **226,** 434–439.
126. Kaczmarek, E., *et al.* (1996). Identification and characterization of CD39/vascular ATP diphosphohydrolase. *J. Biol. Chem.* **271,** 33116–33122.
127. Rapaport, E., and Fontaine, J. (1989). Generation of extracellular ATP in blood and its mediated inhibition of host weight loss in tumor-bearing mice. *Biochem. Pharmacol* **38,** 4261–4266.
128. Rapaport, E., and Fontaine, J. (1989). Anticancer activities of adenine nucleotides in mice are mediated through expansion of erythrocyte ATP pools. *Proc. Natl. Acad. Sci. USA* **86,** 1662–1666.
129. Haskell, C. M., *et al.* (1996). Phase I trial of extracellular adenosine 5'-triphosphate in patients with advanced cancer. *Med. Pediatr. Oncol.* **27,** 165–173.
130. Rapaport, E., and Zamecnik, P. C. (1976). Incorporation of adenosine into ATP: Formation of compartmentalized ATP. *Proc. Natl. Acad. Sci. USA* **73,** 3122–3125.
131. Belzer, F. O., and Southard, J. H. (1988). Principles of solid-organ preservation by cold storage. *Transplantation* **45,** 673–676.
132. Fujino, Y., *et al.* (1993). The effect of fasting and exogenous adenosine on ATP tissue concentration and viability of canine pancreas grafts during preservation by the two-layer method. *Transplantation* **56,** 1083–1086.
133. Rapaport, E. (1990). Mechanisms of anticancer activities of adenine nucleotides in tumor-bearing hosts. *Ann. N.Y. Acad. Sci.* **603,** 142–149.
134. Soni, M. G., and Mehendale, H. M. (1994). Adenosine triphosphate protection of chlordecone-amplified $CCl_4$ hepatotoxicity and lethality. *J. Hepatol.* **20,** 267–274.
135. Kanwar, Y., *et al.* (1992). Biosynthetic regulation of proteoglycans by aldohexoses and ATP. *Proc. Natl. Acad. Sci. USA* **89,** 8621–8625.
136. Meldrum, D. R., *et al.* (1991). Association between decreased splenic ATP levels and immunodepression: Amelioration with ATP-$MgCl_2$. *Am. J. Physiol.* **261,** R351–R357.
137. Singh, G., Chaudry, K. I., and Chaudry, I. H. (1993). ATP-$MgCl_2$ restores gut absorptive capacity early after trauma-hemorrhagic shock. *Am. J. Physiol.* **264,** R977–R983.
138. Rapaport, E. (1995). Involvement of elevated intracellular and extracellular ATP in the regulation of insulin secretion: Therapeutic targets in non-insulin-dependent diabetes mellitus. *Am. J. Ther.* **2,** 283–289.
139. Brewer, G. J., and Eaton, J. W. (1971). Erythrocyte metabolism: Interaction with oxygen transport. *Science* **171,** 1205–1211.

140. Fukunaga, A. F. (1995). Intravenous administration of large dosages of adenosine or adenosine triphosphate with minimal blood pressure fluctuation. *Life Sci.* **56**, L209–L218.

141. Newberg, L. A., Milde, J. H., and Michenfelder, J. D. (1985). Cerebral and systemic effects of hypotension induced by adenosine or ATP in dogs. *Anesthesiology* **62**, 429–436.

142. Stange, K., *et al.* (1989). Effects of adenosine-induced hypotension on cerebral blood flow and metabolism in the pig. *Acta Anaesthesiol. Scand.* **33**, 199–203.

143. Sollevi, A., *et al.* (1984). Relationship between arterial and venous adenosine levels and vasodilatation during ATP- and adenosine-infusion in dogs. *Acta Physiol. Scand.* **120**, 171–176.

144. Fukunaga, A. F., Flacke, W. E., and Bloor, B. C. (1982). Hypotensive effects of adenosine and adenosine triphosphate compared with sodium nitroprusside. *Anesth. Analg.* **61**, 273–278.

145. Fineman, J. R., Crowley, M. R., and Soifer, S. J. (1990). Selective pulmonary vasodilation with ATP-MgCl$_2$ during pulmonary hypertension in lambs. *J. Appl. Physiol.* **69**, 1836–1842.

146. Paidas, C. N., *et al.* (1988). Adenosine triphosphate: A potential therapy for hypoxic pulmonary hypertension. *J. Pediatr. Surg.* **23**, 1154–1160.

147. Brook, M. M., *et al.* (1994). Use of ATP-MgCl$_2$ in the evaluation and treatment of children with pulmonary hypertension secondary to congenital heart defects. *Circulation* **90**, 1287–1293.

148. Hashimoto, K., *et al.* (1993). Perfusion pressure control by adenosine triphosphate given during cardiopulmonary bypass. *Ann. Thorac. Surg.* **55**, 123–126.

149. Gaba, S. J., and Prefaut, C. (1990). Comparison of pulmonary and systemic effects of adenosine triphosphate in chronic obstructive pulmonary disease—ATP: A pulmonary controlled vasoregulator? *Eur. Respir. J.* **3**, 450–455.

150. Kato, M., *et al.* (1999). Adenosine 5'-triphosphate induced dilation of human coronary microvessels in vivo. *Intern. Med.* **38**, 324–329.

151. Shiode, N., *et al.* (1998). Vasomotility and nitric oxide bioactivity of the bridging segments of the left anterior descending coronary artery. *Am. J. Cardiol.* **81**, 341–343.

152. Jacobson, G. R., Lodge, J., and Poy, F. (1989). Carbohydrate uptake in the oral pathogen *Streptococcus mutans*: Mechanisms and regulation by protein phosphorylation. *Biochimie* **71**, 997–1004.

153. Pal, S., *et al.* (1997). Pt-ATP as an antineoplastic agent in an experimental mice model system. *J. Exp. Clin. Cancer Res.* **16**, 255–260.

154. Nayak, K. K., *et al.* (1990). Antitumour activities of copper–ATP complex on transplantable murine lymphoma. *Pharmacology* **41**, 350–356.

155. Pal, S., Nayak, K. K., and Maity, P. (1991). Investigation on phosphate dependent glutaminase (EC 3.5.1.2) activity in host tissues of EAC-bearing mice and response of liver EC 3.5.1.2 on Cu-ATP therapy. *Cancer Lett.* **58**, 151–153.

156. Froio, J., *et al.* (1995). Effect of intraperitoneal ATP on tumor growth and bone marrow radiation tolerance. *Acta Oncol.* **34**, 419–422.

157. Pal, S., Ray, M. R., and Maity, P. (1993). Tumor inhibition and hematopoietic stimulation in mice by a synthetic copper-ATP complex. *Anticancer Drugs.* **4**, 505–510.

158. Rapaport, E., and Zamecnik, P. C. (1989). Similarities in ATP-binding domain between *E. coli* tyrosyl-tRNA synthetase and human estrogen receptor. *FASEB J.* **3**, 2554–2555.

159. Senagore, A. J., *et al.* (1992). Adenosine triphosphate-magnesium chloride in radiation injury. *Surgery* **112**, 933–939.

160. Szeinfeld, D., and De Villiers, N. (1992). Radioprotective properties of ATP and modification of acid phosphatase response after a lethal dose of whole body p(66MeV)/Be neutron radiation to BALB/c mice. *Cancer Biochem. Biophys.* **13**, 123–132.

161. Agteresch, H. J., *et al.* (2000). Pharmacokinetics of intravenous ATP in cancer patients. *Eur. J. Clin. Pharmacol.* **56**, 49–55.
162. Mendoza, E. F. F., Haskell, C., Pisters, K., Orlandi, C., Dixon, M., and Figlin, R. (1996). Adenosine triphosphate (ATP) for advanced non-small cell lung cancer (NSCLC): A phase II multicenter study. *In* "Proceedings of the American Society for Clinical Oncology".
163. Haskell, C. M., *et al.* (1998). Phase II study of intravenous adenosine 5'-triphosphate in patients with previously untreated stage IIIB and stage IV non-small cell lung cancer. *Invest. New Drugs* **16**, 81–85.
164. Agteresch, H. J., *et al.* (2000). Randomized clinical trial of adenosine 5'-triphosphate in patients with advanced non-small-cell lung cancer. *J. Natl. Cancer Inst.* **92**, 321–328.
165. Agteresch, H. J., *et al.* (1999). Adenosine triphosphate: Established and potential clinical applications. *Drugs* **58**, 211–232.
166. Kitagawa, T., Amano, F., and Akamatsu, Y. (1988). External ATP-induced passive permeability change and cell lysis of cultured transformed cells: Action in serum-containing growth media. *Biochim. Biophys. Acta* **941**, 257–263.
167. Zanovello, P., *et al.* (1990). Responses of mouse lymphocytes to extracellular ATP. II. Extracellular ATP causes cell type-dependent lysis and DNA fragmentation. *J. Immunol.* **145**, 1545–1550.
168. Maaser, K., *et al.* (2002). Extracellular nucleotides inhibit growth of human oesophageal cancer cells via P2Y$_2$-receptors. *Br. J. Cancer* **86**, 636–644.
169. Hopfner, M., *et al.* (2001). Growth inhibition and apoptosis induced by P2Y2 receptors in human colorectal carcinoma cells: Involvement of intracellular calcium and cyclic adenosine monophosphate. *Int. J. Colorectal Dis.* **16**, 154–166.
170. Palea, S., *et al.* (1995). Evidence for the presence of both pre- and postjunctional P2-purinoceptor subtypes in human isolated urinary bladder. *Br. J. Pharmacol.* **114**, 35–40.
171. Fang, W. G., *et al.* (1992). P2-purinergic receptor agonists inhibit the growth of androgen-independent prostate carcinoma cells. *J. Clin. Invest.* **89**, 191–196.
172. Pedersen, S., *et al.* (1999). Mechanical stress induces release of ATP from Ehrlich ascites tumor cells. *Biochim. Biophys. Acta* **1416**, 271–284.
173. Ferrari, D. M., *et al.* (1998). ERp28, a human endoplasmic-reticulum-lumenal protein, is a member of the protein disulfide isomerase family but lacks a CXXC thioredoxin-box motif. *Eur. J. Biochem.* **255**, 570–579.
174. Jin, J., *et al.* (2002). Adenosine diphosphate (ADP)-induced thromboxane A$_2$ generation in human platelets requires coordinated signaling through integrin $\alpha_{IIb}$ $\beta_3$ and ADP receptors. *Blood* **99**, 193–198.
175. Brown, J., *et al.* (2000). Critical evaluation of ECV304 as a human endothelial cell model defined by genetic analysis and functional responses: A comparison with the human bladder cancer derived epithelial cell line T24/83. *Lab. Invest* **80**, 37–45.
176. Hide, I., *et al.* (2000). Extracellular ATP triggers tumor necrosis factor-$\alpha$ release from rat microglia. *J. Neurochem.* **75**, 965–972.
177. Seetulsingh-Goorah, S. P., and Stewart, B. W. (1998). Growth inhibition of HL-60 cells by extracellular ATP: Concentration-dependent involvement of a P2 receptor and adenosine generation. *Biochem. Biophys. Res. Commun* **250**, 390–396.
178. Wiley, J. S., *et al.* (1995). Agonists for endothelial P2 purinoceptors trigger a signalling pathway producing Ca$^{2+}$ responses in lymphocytes adherent to endothelial cells. *Biochem. J.* **311**, 589–594.
179. Ni, H., *et al.* (2002). Extracellular mRNA induces dendritic cell activation by stimulating tumor necrosis factor-$\alpha$ secretion and signaling through a nucleotide receptor. *J. Biol. Chem.* **277**, 12689–12696.

180. Wilkin, F., *et al.* (2001). The P2Y11 receptor mediates the ATP-induced maturation of human monocyte-derived dendritic cells. *J. Immunol.* **166,** 7172–7177.
181. Ferrari, D., *et al.* (2000). The P2 purinergic receptors of human dendritic cells: Identification and coupling to cytokine release. *FASEB J.* **14,** 2466–2476.
182. Coutinho-Silva, R., *et al.* (1999). P2Z/P2X7 receptor-dependent apoptosis of dendritic cells. *Am. J. Physiol.* **276,** C1139–C1147.
183. Schnurr, M., *et al.* (2000). Extracellular ATP and TNF-α synergize in the activation and maturation of human dendritic cells. *J. Immunol.* **165,** 4704–4709.
184. Zheng, L. M., *et al.* (2000). Extracellular ATP as a trigger for apoptosis or programmed cell death. *J. Cell Biol.* **112,** 279–288.
185. Beyer, E. C., and Steinberg, T. H. (1991). Evidence that the gap junction protein connexin-43 is the ATP-induced pore of mouse macrophages. *J. Biol. Chem.* **266,** 7971–7974.

# CHAPTER 14

# Purinergic Receptors in the Glomerulus and Vasculature of the Kidney

**Edward W. Inscho**

Department of Physiology, Medical College of Georgia, Augusta, Georgia 30912

## I. INTRODUCTION

Renal microvascular control, through activation of purinergic receptors, has been an active area of research for many decades. Early interest began with investigations of the role adenosine plays in regulating renal vascular resistance, renal blood flow, glomerular filtration rate, renin secretion, and autoregulation. More recently, attention has been given to the role of ATP and

P2 receptor activation in those same renal hemodynamic and renal functional parameters. Although much has been learned in the ensuing years, the specific physiological roles that extracellular adenosine and ATP play in the regulation of renal hemodynamics remain unresolved. This chapter summarizes observations in the area of purinergic control of renal microvascular and glomerular function and tries to highlight important questions that remain to be resolved. Primary emphasis is placed on new developments related to intrarenal P2 receptors; however, some attention is directed toward more recent observations related to P1 or adenosine-sensitive purinergic receptors.

## II. P2 RECEPTOR DISTRIBUTION ALONG THE RENAL VASCULATURE AND GLOMERULUS

Numerous reports in the literature provide functional evidence in support of P2 receptor expression along the intrarenal microvasculature (Bo and Burnstock, 1993; Chan *et al.*, 1998a; Eltze and Ullrich, 1996; Gutiérrez *et al.*, 1999; Inscho *et al.*, 1992, 1994, 1996a,b; Majid *et al.*, 1999; Majid and Navar, 1992; Von Kügelgen *et al.*, 1995; Weihprecht *et al.*, 1992), mesangial cells (Gutierrez *et al.*, 1999a,b; Huber-Lang *et al.*, 1997; Huwiler *et al.*, 1997b; Kleta *et al.*, 1995; Schlatter *et al.*, 1995; Schulze-Lohoff *et al.*, 1998; Takeda *et al.*, 1996; Tepel *et al.*, 1996) and glomeruli (Briner and Kern, 1994; Huwiler *et al.*, 1997a; Pavenstädt *et al.*, 1996). Immunohistochemistry studies have clearly established the presence of $P2X_1$ receptors along the preglomerular vasculature, but reveal little evidence for the presence of these receptors on glomeruli, tubules, or postglomerular arterioles (Chan *et al.*, 1998a). While the physiological roles of intrarenal P2 receptors remain unclear, there is growing interest in P2 receptor activation by extracellular ATP, as an important mechanism involved in the regulation of renal hemodynamics under physiological or pathophysiological conditions (Chan *et al.*, 1998b; Gabriëls *et al.*, 2000; Inscho *et al.*, 1996b, 2001a,b; Inscho and Cook, 2002; Majid *et al.*, 1999; Mitchell and Navar, 1993; Navar *et al.*, 1996; Nishiyama *et al.*, 2000, 2001b; Van der Giet *et al.*, 1998, 1999; Zhao *et al.*, 2001).

## III. RENAL VASCULAR RESPONSE TO P2 RECEPTOR ACTIVATION

### A. Renal Vascular Response to Infused ATP

The ability of infused ATP to alter renal vascular resistance was reported by Drury and Szent-Györgyi in the late 1920s (Drury and Szent-Györgyi, 1929). Beyond this initial observation, little additional information about the renal vascular effects of ATP came to light until nearly 60 years later, when Schwartz and Malik (1989) suggested that ATP might play an important role in the regulation of renal vascular resistance by renal nerves. Since that time, interest in ATP as a regulator of renal and glomerular function has grown.

The effect of ATP infusion on renal blood flow or perfusion pressure varies according to the species and model being investigated. Intraarterial infusion of ATP into the rat kidney modulates renal vascular resistance in a tone-dependent manner. For example, in the isolated perfused rat kidney preparation, infusion of ATP stimulates renal vasoconstriction under basal tone conditions, or both vasoconstriction and vasodilation when renal vascular resistance is increased with norepinephrine (Eltze and Ullrich, 1996) or phenylephrine (Vargas *et al.*, 1996). Isolated perfused rabbit kidneys respond to ATP infusion with a modest but demonstrable vasoconstriction (Needleman *et al.*, 1974), whereas infusion of ATP into the dog kidney produces vasodilation mediated by the synthesis and release of endothelium-derived nitric oxide (Majid and Navar, 1992). Inhibition of nitric oxide synthase activity, by pretreating with $N^{\omega}$-nitro-L-arginine methyl ester, eliminates the vasodilator response to ATP and leads to a marked ATP-mediated renal vasoconstriction. Examination of responses of renal arteries from human kidneys presents a slightly different picture. Renal artery rings, preconstricted with either norepinephrine or prostagland in $F_{2\alpha}$, vasodilate in response to ATP; however, this vasodilation appears to be, in part, mediated by locally generated adenosine with the subsequent activation of P1 receptors (Rump *et al.*, 1998). It should be noted that a wide disparity can exist in the segmental responsiveness of renal vascular and microvascular segments to a given agonist and thus the responses observed with isolated renal arterial rings may not accurately represent intrarenal microvascular responses to the same agonists.

## B. Renal Vascular Response to Infused P2 Receptor Agonists

Characterization of P2 receptors present in various tissues has tradition-ally been surmised by evaluating the vascular response to ATP and a collection of "selective" P2 receptor agonists. On the basis of the pharmacological selectivity of these analogs, the P2 receptor composition of particular vascular tissues or vascular beds has been postulated. A similar approach has been used to construct a template of P2 receptor subtypes present along the renal vasculature (Churchill and Ellis, 1993; Eltze and Ullrich, 1996; Gabriëls *et al.*, 2000). Intrarenal infusion of 2-methylthio-ATP, UTP or ATPγS, which are considered to exhibit reasonable selectivity for the P2Y receptor family, stimulates vasodilation of isolated, perfused rat kidneys (Churchill and Ellis, 1993; Eltze and Ullrich, 1996) and that vasodilator response is largely dependent on nitric oxide generation. In contrast, infusion of $\alpha,\beta$-methylene-ATP or $\beta,\gamma$-methylene-ATP into the isolated perfused rat kidney elicits a prompt, concentration dependent vasoconstriction that is augmented by inhibition of nitric oxide synthesis (Churchill and Ellis, 1993; Eltze and Ullrich, 1996). These data suggest that the renal microcirculation expresses multiple P2 receptor subtypes that are accessible to lumenally delivered P2 agonists. The renal vascular response to P2 receptor activation depends, in part, on the receptor(s) being activated and the ambient vascular tone present in that vascular segment or vascular bed. Activation of P2Y receptors, under elevated tone conditions, commonly leads to nitric oxide-dependent renal vasodilation whereas activation of P2X receptors usually leads to renal vasoconstriction.

## C. Renal Vascular Response to Perivascular P2 Agonists

The source of extracellular nucleotides potentially involved in regulating renal vascular resistance remains an important and open question. Certainly, local release of ATP from renal nerves could deliver physiologically relevant concentrations of ATP directly to the adjacent microvascular smooth muscle cells (Schwartz and Malik, 1989; Sehic et al., 1994). Similarly, local release of ATP from vascular, endothelial, and renal epithelial cells could elevate the interstitial ATP concentration to levels that will modulate local renal vascular tone (Inscho, 2001a,b; Nishiyama et al., 2000, 2001b; Schwiebert, 2001; Schwiebert and Kishore, 2001). Therefore, assessment of the renal microvascular response to lumenally delivered ATP may not provide a complete picture of how ATP and P2 agonists can influence renal microvascular function.

The distribution of P2 receptors along renal vascular and microvascular tissue has been documented by biochemical and pharmacological approaches. Immunofluorescence and in situ hybridization techniques have clearly established the expression of multiple P2X receptor subtypes in large, medium, and small divisions of renal arteries (Bo and Burnstock, 1993; Lewis and Evans, 2001). Among the most prominent receptors found was evidence for $P2X_1$, $P2X_2$, $P2X_3$, $P2X_4$, and $P2X_6$. Differing profiles were found, depending on the techniques applied and the size of the arterial segment examined. The issue of P2 receptor expression by smooth muscle found in the renal microvasculature was addressed by Chan and co-workers (1998a). Immunohistochemical assessment for $P2X_1$ receptors, using a $P2X_1$ receptor-selective antibody, revealed strong positive staining along all segments of the preglomerular microvasculature. No visible staining was evident along the efferent arteriole or glomeruli. These data provide strong support for the expression of $P2X_1$ receptor protein by preglomerular vascular smooth muscle cells.

Functional evaluation of segment specific responsiveness reveals that only the arteries and arterioles comprising the preglomerular renal circulation respond to ATP administration (Inscho *et al.*, 1992). Moreover, as shown in Fig. 1, only the afferent arteriole exhibits a sustained vasoconstriction in response to ATP concentrations below 10 $\mu M$. Notably, the diameter of efferent arterioles exposed to ATP or to UTP remains unchanged during nucleotide administration. Weihprecht *et al.* (1992) also reported that ATP concentrations below 10 $\mu M$ evoked a significant vasoconstriction of

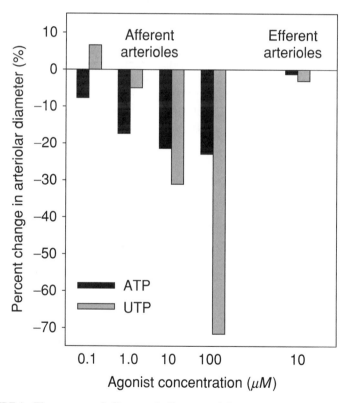

**FIGURE 1**    The response of afferent and efferent arterioles to exogenously applied ATP or UTP. Data are expressed as a percentage of the control diameter. The steady state responses to ATP are represented by solid bars. Steady state responses to UTP are represented by shaded bars. These data are modified from Inscho *et al.* (1992, 1998b). The responses of efferent arterioles to UTP are unpublished observations (E. Inscho).

isolated rabbit afferent arterioles but did not assess efferent arteriolar
responsiveness to ATP or UTP. The parallels between documented P2
receptor expression and the segment-specific microvascular responses
evoked by ATP administration are remarkably consistent. Essentially, only
the preglomerular microvascular segments show clear expression of P2
receptors, and these same microvascular segments are the only segments
documented to respond to ATP treatment.

Identification of the receptors responsible for the afferent arteriolar
response to ATP has not been clearly established. As mentioned previously,
Chan and co-workers determined that arcuate arteries, interlobular arteries,
and afferent arterioles strongly express P2X$_1$ receptors; however, a thorough
biochemical examination of receptor expression has not been completed.
Pharmacological assessment of rank order potencies to purportedly
receptor-selective P2 agonists on afferent arterioles has been performed
and the data suggest the possible presence of multiple P2 receptor subtypes
(Inscho et al., 1998b). The P2X receptor-selective agonists $\alpha,\beta$-methylene-
ATP and $\beta,\gamma$-methylene-ATP yielded rapid biphasic vasoconstrictor
responses consistent with the expression of P2X receptors on afferent
arterioles (Fig. 2) (Inscho et al., 1998b; Inscho and Cook, 2002). The rapid
initial vasoconstriction rapidly subsided to a stable and reversible response.
The afferent arteriolar vasoconstriction elicited by the endogenous ligand

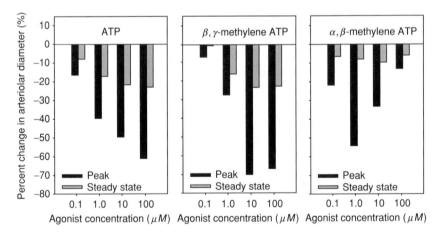

**FIGURE 2**   Peak and steady state changes in microvascular diameter in response to ATP
and the P2X agonists $\alpha,\beta$-methylene ATP and $\beta,\gamma$-methylene ATP. Data are expressed as the
percent change in afferent arteriolar diameter compared with the control diameter. Peak
responses are indicated by solid bars and steady state responses are indicated by shaded bars.
These data are modified from Inscho et al. (1998b).

ATP closely paralleled the response evoked by the P2X agonist $\beta,\gamma$-methylene-ATP (Inscho et al., 1998b; Weihprecht et al., 1992). Interestingly, sequential addition of increasing concentrations of $\alpha,\beta$-methylene-ATP resulted in a rapid desensitization of the afferent arterioles to this agonist (Inscho et al., 1996b). Rapid desensitization of P2X receptors by $\alpha,\beta$-methylene-ATP is a frequently described characteristic of $P2X_1$ receptors (Abbracchio and Burnstock, 1994; Dubyak and El-Moatassim, 1993; North and Surprenant, 2000; Ralevic and Burnstock, 1998).

Administration of the P2Y agonist 2-methylthio-ATP, UTP, UDP, or ATP$\gamma$S yielded markedly different results (Fig. 3) (Inscho et al., 1998b; Inscho and Cook, 2002). 2-Methylthio-ATP evoked substantial initial vasoconstrictions that resulted in only modest, steady state responses. UTP and ATP$\gamma$S were less potent than ATP or either of the P2X agonists tested. Nevertheless, when agonist concentrations reached 10 and 100 $\mu M$, these agents elicited significantly larger monophasic vasoconstrictions (Inscho et al., 1998b). Taken together, these experiments suggest that rat afferent arterioles express both P2X and P2Y receptors. Activation of either receptor family will evoke sustained, concentration-dependent afferent arteriolar vasoconstriction.

## D. Diadenosine Polyphosphates

Diadenosine polyphosphates represent a member of a larger group of naturally occurring, vasoactive nucleotide agonists that also activate P2 receptors (Flores et al., 1999; Ralevic and Burnstock, 1998; Schlüter et al., 1996). In 1975, Harrison and Brossmer reported the effects of $\alpha,\omega$-diadenosine polyphosphate to inhibit platelet aggregation (Harrison and Brossmer, 1975). Since that time, numerous additional reports have appeared in the literature describing broadly varied biological effects of diadenosine phosphates (Baxi and Vishwanatha, 1995; Flodgaard and Klenow, 1982; Flores et al., 1999; Luthje and Ogilvie, 1983; Miras-Portugal et al., 1999; Pintor et al., 1997; Schlüter et al., 1996). Within the family of diadenosine polyphosphates are many structurally distinct molecules that exhibit varying biological activity (Baxi and Vishwanatha, 1995; Flores et al., 1999; Schlüter et al., 1996; Tepel et al., 1997; Van der Giet et al., 1997, 1998, 1999). Diadenosine phosphate compounds exist as two adenosine molecules linked by a series of two to six phosphate groups. The number of phosphate groups linking the adenosine molecules together significantly alters the biological activity each diadenosine polyphosphate compound exhibits and influences the interaction of each compound with various purinoceptor subtypes.

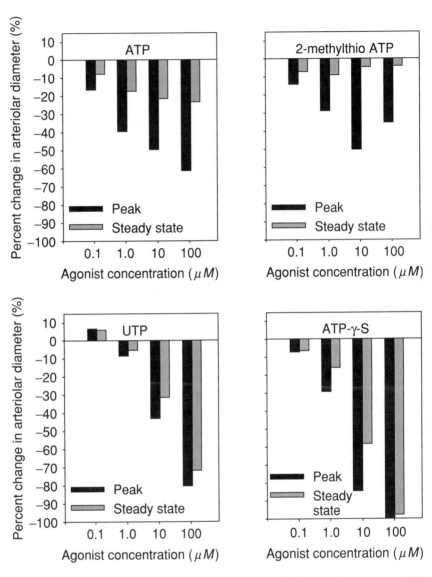

**FIGURE 3**  Peak and steady state changes in microvascular diameter in response to ATP and the P2Y agonists 2-methylthio-ATP, UTP, and ATP $\gamma$ S. Data are expressed as the percent change in afferent arteriolar diameter compared with the control diameter. Peak responses are indicated by solid bars and steady state responses are indicated by shaded bars. These data are modified from Inscho *et al.* (1998b).

Relevant to this review are several reports describing the effects of these agents on the renal microvasculature. Whole kidney studies reveal that intravenous administration of $Ap_4A$, $Ap_5A$, or $Ap_6A$ transiently reduces mean arterial blood pressure and cardiac output while increasing total peripheral resistance (Khattab *et al.*, 1998). Accompanying this systemic response was a marked decrease in renal blood flow and renal cortical blood flow and an increase in renal vascular resistance. These responses were transient and usually reversed within 2–3 min. It is difficult to determine the direct effects of the diadenosine polyphosphates on renal hemodynamics because of the pronounced systemic hemodynamic changes that accompanied agonist administration. In another study, $Ap_6A$ was found to induce variable fluctuations in mean arterial blood pressure and to increase urine flow and sodium excretion (Hohage *et al.*, 1996). These effects appeared to occur through mechanisms distinct from those of ATP or the P2X agonist $\alpha,\beta$-methylene ATP. $Ap_5A$ did not alter renal function, whereas $Ap_4A$ and $Ap_3A$ modestly reduced urine flow and sodium excretion. The effects of $Ap_6A$ on renal function occur without a change in glomerular filtration rate and can be blocked or blunted by indomethacin or bosentan (Hohage *et al.*, 1996).

In the isolated perfused rat kidney, infusion of $Ap_4A$, $Ap_5A$, or $Ap_6A$ increases renal perfusion pressure, consistent with intrarenal vasoconstriction (Van der Giet *et al.*, 1997, 1998, 1999). The vasoconstriction occurs at agonist concentrations as low as 1.0 n$M$ (Van der Giet *et al.*, 1999) and can be antagonized by the nonselective P2 receptor antagonists suramin or pyridoxal phosphate-6-azophenyl-2',4'-disulfonic acid (PPADS) (Van der Giet *et al.*, 1997, 1998, 1999). The vasoconstrictor response could be completely blocked by including the P1 receptor blocker 8-cyclopentyl-1,3-dipropylxanthine (DPCPX) with the P2 receptor blocker PPADS (Van der Giet *et al.*, 1998). This observation suggests that the vasoconstrictor response is mediated partly through activation of P2X receptors and partly through activation of adenosine-sensitive P1 receptors. Whether P1 receptor activation reflects direct activation of P1 receptors by diadenosine polyphosphate administration, or actually represents P1 receptor activation by breakdown of the polyphosphate agonist to the constituent adenosine, remains unclear.

Isolated renal resistance arteries, in the range of 200 to 250 $\mu$m, contract during exposure to $Ap_4A$, $Ap_5A$, and $Ap_6A$ at concentrations above 0.1 $\mu M$ (Tepel *et al.*, 1997). The estimated 50% effective concentration ($EC_{50}$) values were approximately 0.80, 0.80, and 0.56 $\mu M$, respectively. The magnitude of the response from renal resistance arteries is significantly greater than responses obtained from aortic strips. Removal of calcium from the extracellular bathing medium or blockade of calcium channels, with nifedipine, attenuated the contractile response to $Ap_4A$ whereas P2 receptor blockade with PPADS nearly abolished the response. These data indicate that $Ap_4A$-mediated contraction of isolated renal resistance arteries involves activation of PPADS-sensitive P2 receptors and that the vasoconstrictor response is largely dependent on the influx of extracellular calcium.

Gabriëls and co-workers examined the effects of $Ap_3A$ and $Ap_5A$ on the preglomerular arteries and arterioles and postglomerular arterioles of hydronephrotic kidneys (Gabriëls *et al.*, 2000). These agents induced transient vasoconstriction of interlobular arteries and afferent arterioles but had little effect on efferent arteriolar diameter. The preglomerular vasoconstrictor response to $Ap_3A$ was completely inhibited during adenosine $A_1$ receptor blockade with DPCPX and was attenuated during adenosine $A_2$ receptor blockade with DMPX or with P2 receptor blockade. Thus it appears that, in the hydronephrotic kidney model, adenosine-sensitive P1 receptors play a major role in the response to $Ap_3A$. Of the agents examined, only $Ap_5A$ produced a sustained vasoconstriction and then only of the afferent arteriole. Vasoconstrictor responses elicited by $Ap_5A$ were attenuated slightly by the adenosine receptor blocker DPCPX, but appeared more sensitive to P2 receptor blockade with PPADS. These data suggest that in the hydronephrotic kidney, $Ap_5A$-mediated vasoconstriction is predominantly mediated through P2 receptor activation (Gabriëls *et al.*, 2000).

On the basis of these findings, it appears that diadenosine polyphosphates can alter renal microvascular diameter and renal vascular resistance. The length of the polyphosphate chain is an important consideration in determining the mechanisms by which diadenosine polyphosphates influence microvascular function. There appears to be some disagreement about the nature of the systemic effects of these compounds, but these differences could be related to differences in the experimental conditions. Also to be considered are the potential interactions between diadenosine polyphosphates and P1 and P2 receptors, recognizing that the response profile may vary depending on the vascular tissue being examined. Further investigations will surely resolve some of these issues.

## IV. RENAL VASCULAR SIGNALING MECHANISMS

Calcium is a critical signaling element in agonist-mediated afferent arteriolar vasoconstrictor responses, autoregulatory responses, and tubuloglomerular feedback responses (Imig *et al.*, 2001; Inscho *et al.*, 1997, 1998a; Inscho, 2001a; Navar *et al.*, 1996; Schroeder *et al.*, 2000). Stimulation of P2 receptors leads to activation of multiple calcium signaling mechanisms in other cell types (Navar *et al.*, 1996). P2X receptor stimulation involves activation of nonselective cation channels that can directly increase intracellular calcium concentration (Ralevic and Burnstock, 1998). Furthermore, the net influx of cations can depolarize the plasma membrane and activate voltage-dependent calcium channels (Inscho, 2001a). Alternatively, P2Y receptors can be coupled to phospholipase C and generate inositol triphosphate ($IP_3$)-dependent mobilization of calcium from intracellular stores (Ralevic and Burnstock, 1998). This diversity in second-messenger pathways is also evident when the calcium signaling mechanisms utilized by afferent arteriolar smooth muscle in responding to P2 receptor stimulation are assessed. For example, L-type calcium channels are modestly involved in the initial vasoconstriction induced by P2X receptor activation but seem to account for all of the sustained vasoconstriction (Inscho *et al.*, 1995; Inscho, 2001a; Inscho and Cook, 2002; White *et al.*, 2001). Calcium channel blockade with diltiazem attenuated the initial response to $\alpha,\beta$-methylene-ATP by approximately 50% and abolished the sustained response (Inscho *et al.*, 1995; Inscho, 2001a; Inscho and Cook, 2002; Zhao *et al.*, 2001). Attempts to inhibit the activity of nonselective cations channels with mefenamic acid failed to alter the magnitude of the initial vasoconstriction to $\alpha,\beta$-methylene-ATP and only slightly attenuated the sustained vasoconstriction (Inscho and Cook, 2002). Stimulation of P2Y receptors with UTP evoked a more monophasic vasoconstriction that was only slightly attenuated by calcium channel blockade (Inscho and Cook, 2002). Low concentrations of ATP were almost completely abolished by calcium channel blockade; however, sustained vasoconstrictions were observed when ATP concentrations exceeded 1 $\mu M$ (Inscho and Cook, 2002). Similar results were obtained with the nonselective calcium channel blocker cadmium. These observations suggest that activation of P2 receptors on afferent arterioles induces vasoconstriction through distinct signal transduction pathways. Activation of P2X receptors relies largely on activation of voltage-dependent L-type calcium channels to evoke afferent arteriolar vasoconstriction. In contrast, P2Y receptor activation stimulates vasoconstriction of afferent arterioles largely through L-type calcium channel-independent mechanisms.

The distinct calcium signaling pathways found to be involved in the receptor-selective vasoconstrictor responses are also clearly apparent when evaluated at the cellular level. Direct assessment of the calcium signaling pathways invoked by P2 receptor activation using single microvascular smooth muscle cells yielded results that are remarkably consistent with the functional data. P2X receptor activation increases intracellular calcium concentration in preglomerular smooth muscle cells by stimulating calcium influx through voltage-dependent L type calcium channels (Inscho *et al.*, 1999; Inscho, 2001a; White *et al.*, 2001). Blockade of calcium channels with diltiazem or nickel eliminated the response. Similarly, removal of calcium from the bathing solution also eliminated the increase in cytosolic calcium induced by $\alpha,\beta$-methylene-ATP. Finally, the response was also blocked in the presence of the P2X receptor antagonist NF-279 (8,8'-[carbonylbis(imino-4,1-phenylenecarbonylimino-4,1-phenylenecarbonylimino)]bis-1,3,5-napthalenetrisulfonic acid hexasodium salt) (Inscho, 2001a; Lambrecht et al., 1999; Rettinger et al., 2000; White et al., 2001). Therefore, $\alpha,\beta$-methylene-ATP stimulates afferent arteriolar vasoconstriction by stimulating calcium influx through pathways that include voltage-dependent L-type calcium channels. In contrast, stimulation of P2Y receptors with UTP increases intracellular calcium concentration almost exclusively by mobilizing calcium from intracellular stores (Inscho et al., 1999). The effect of selective P2 receptor activation on intracellular calcium concentration corresponds with the calcium channel blocker experiments performed with intact afferent arterioles (Inscho et al., 1995; Inscho and Cook, 2002). These data establish a significant contribution of voltage-dependent calcium influx and calcium mobilization in the afferent arteriolar response to P2 receptor stimulation. In addition, they lend further support to the hypothesis that afferent arteriolar smooth muscle cells express both P2X and P2Y receptor subtypes that utilize distinct calcium signaling mechanisms to influence afferent arteriolar function.

As discussed, P2 receptor-mediated calcium signaling can occur through direct activation of ligand-gated ion channels or through stimulation of intracellular signaling pathways. Evidence also implicates a role for cytochrome P450 metabolites participating in calcium-dependent response of renal microvascular smooth muscle to P2 receptor activation. In studies by Zhao et al. (2001), the cytochrome P450 metabolite 20-hydroxyeicosatetraenoic acid (20-HETE) was found to participate in the voltage-dependent calcium response to P2 receptor activation. Administration of the 20-HETE antagonist 20-hydroxyeicosa-6(Z),15(Z)-dienoic acid (20-HEDE) attenuated the initial afferent arteriolar vasoconstriction induced by ATP and abolished the sustained vasoconstriction. Similar results were obtained when 20-HETE synthesis was inhibited with the cytochrome P450 hydroxylase inhibitor N-methylsulfonyl-12,12-dibromododec-11-enamide (DDMS). Efforts to differentiate the receptor subtype involved suggest that the 20-HETE effect occurs through activation of a P2X receptor, probably P2X$_1$. This conclusion stems from the observation that inhibition of the effects of 20-HETE with 20-HEDE or DDMS also attenuated the initial vasoconstriction and eliminated the sustained vasoconstriction elicited by the P2X agonist $\alpha,\beta$-methylene-ATP, whereas the afferent arteriolar vasoconstriction evoked by the P2Y agonist UTP was unchanged by 20-HEDE or DDMS treatment. When calcium signaling was directly assessed in freshly isolated preglomerular smooth muscle cells, 20-HEDE attenuated the peak and steady state increase in cytosolic calcium concentration (Zhao et al., 2001). The effect of 20-HETE inhibition on the response of afferent arterioles to P2 receptor activation parallels the blockade of L-type calcium channels with diltiazem (Inscho et al., 1995, 1999; Inscho and Cook, 2002; White et al., 2001; Zhao et al., 2001).

Accordingly, the effects of 20-HETE inhibition on depolarization-induced afferent arteriolar vasoconstriction were determined. Voltage-dependent afferent arteriolar vasoconstriction induced by 55 m$M$ KCl remained unchanged before and after inhibition of 20-HETE with 20-HEDE (Zhao et al., 2001). These data suggest that afferent arteriolar responses to ATP or P2X receptor activation involve the generation of 20-HETE, which subsequently influences calcium influx through voltage-gated L-type calcium channels. The nature of the role played by 20-HETE in the sequence of signaling events remains to be determined but the data strongly suggest that this cytochrome P450 metabolite may influence agonist-induced changes in membrane potential.

## V. RENAL VASCULAR RESPONSE TO P1 RECEPTOR ACTIVATION

P2 receptors represent only one side of the family of purinoceptors that potentially participate in the regulation of renal function and renal hemodynamics. P1 receptors have also been found to make important contributions in the kidney. P1 receptors are divided into four unique subtypes classified as $A_1$, $A_{2a}$, $A_{2b}$, and $A_3$ (Jackson, 2001; Jackson and Dubey, 2001; Klotz, 2000; Linden, 2001; Olsson, 1996; Ralevic and Burnstock, 1998). There is clear evidence for the involvement of $A_1$ and $A_{2a}$ receptors in renal function (Jackson, 2001; McCoy et al., 1993; Navar et al., 1996; Schnermann, 1998). Less is known about $A_{2b}$ and $A_3$ receptors with regard to their place in renal physiology.

The renal vasculature expresses both $A_1$ and $A_{2a}$ receptors and both likely play a role in regulating renal microvascular function (Jackson, 2001; McCoy et al., 1993; Navar et al., 1996; Schnermann, 1998). Adenosine receptors are also distributed along different tubular segments of the nephron, where they may influence regulation of tubular transport (Jackson, 2001; McCoy et al., 1993; Navar et al., 1996; Schnermann, 1998). The influence of adenosine on renal hemodynamics is complex. Typically, intrarenal infusion of adenosine results in a transient reduction in renal blood flow followed by a sustained increase in blood flow (Jackson, 2001; McCoy et al., 1993; Navar et al., 1996; Osswald et al., 1978b; Schnermann, 1998). It has been stated that the initial renal vasoconstriction involves activation of preglomerular $A_1$ receptors followed by a postglomerular vasodilation, mediated by $A_2$ receptors (Jackson, 2001). Indeed, the prevailing view is that vasodilation of the postglomerular circulation, particularly in the inner cortex, redistributes blood flow from the cortex to the medulla (Dinour and Brezis, 1991; Levens et al., 1991a,b; Spielman et al., 1980; Ueda, 1972). Direct assessment of microvascular function, however, only corroborates the preglomerular $A_1$ receptor-mediated vasoconstriction and does not substantiate the postulated vasodilation of postglomerular vessels. Two reports directly assessed efferent arteriolar responses to adenosine in the inner cortex (Carmines and Inscho, 1994; Nishiyama et al., 2001a). In both reports, application of 10 $\mu M$ adenosine reduced juxtamedullary afferent and efferent arteriolar diameter. Greater adenosine concentrations reversed the vasoconstriction and returned arteriole diameters to near control values. Selective blockade of $A_1$ receptors with KW-3902 resulted in a significant adenosine-mediated vasodilation of afferent and efferent arterioles, whereas selective blockade of $A_{2a}$ receptors with KW-17837 markedly augmented adenosine-mediated vasoconstrictor responses in both arteriolar segments (Nishiyama et al., 2001a). These data demonstrate that vasoconstriction is the predominant response observed in response to adventitial administration of adenosine to the juxtamedullary afferent and efferent arterioles of the inner cortex and conflict with the postulate that the inner cortex responds to adenosine with vasodilation. It should be noted, however, that adenosine may provide an important vasodilatory influence in the medullary circulation. Zou et al. (2001) reported that administration of the $A_2$ receptor antagonist 3,7-dimethyl-1-propargylxanthine (DMPX) decreased outer and inner medullary blood flow as measured by laser–Doppler flowmetry. Similar results were not obtained during $A_1$ receptor blockade with 8-cyclopentyl-1,3-dipropyl-xanthin (DPCPX) or during $A_3$ receptor blockade with $N^6$-benzyl-5′-($N$-ethylcarbonxamido)adenosine. These data suggest that while the inner cortical microvasculature does not exhibit adenosine-mediated vasodilation, it is possible that resistance elements of the medullary circulation respond to endogenous adenosine with vasodilation, mediated by $A_2$ receptor activation.

Investigations into the mechanisms by which adenosine influenced renal vascular resistance have led to the speculation that a significant interaction exists between adenosine and angiotensin II (Carmines and Inscho, 1994; Dietrich et al., 1991; Hall and Granger, 1986; Osswald et al., 1975; Spielman and Osswald, 1979; Traynor et al., 1998). Whole animal studies suggest that the ambient adenosine concentration presets renal microvascular reactivity and serves to modulate preglomerular responsiveness to angiotensin II (Hall and Granger, 1986). More recent studies, using mice deficient in $AT_{1A}$ receptors, reveal that adenosine-mediated preglomerular vasoconstriction is markedly attenuated compared with heterozygous or wild-type mice (Traynor *et al.*, 1998). These data are consistent with an interaction between adenosine and angiotensin to regulate preglomerular vascular resistance. However, not all studies support such an interaction. Direct assessment of microvascular responses to angiotensin II indicate that preglomerular vasoconstriction can occur independent of adenosine receptor activation (Dietrich *et al.*, 1991). Furthermore, evaluation of the role of endogenous adenosine to serve as an essential cofactor in facilitating angiotensin II-mediated vasoconstriction of juxtamedullary afferent or efferent arterioles revealed no significant interaction (Carmines and Inscho, 1994). In those studies, pharmacological blockade of adenosine receptors or exposure to exogenous adenosine did not alter angiotensin II-mediated vasoconstrictor responses compared with untreated control kidneys. Taken together, the issue of whether the renal microvascular response to angiotensin II and adenosine is significantly, or importantly, altered by an interaction between the angiotensin and purine nucleoside remains unresolved.

## VI. REGULATION OF RENIN SECRETION BY ADENOSINE

The regulation of renin secretion has been extensively reviewed (Bader and Ganten, 2000; Churchill and Churchill, 1988; Hackenthal *et al.*, 1990; Navar *et al.*, 1996; Schnermann, 1998; Spielman and Arend, 1991). Intrarenal infusion of adenosine inhibits renin secretion (Arend *et al.*, 1986; Deray *et al.*, 1989; Kuan *et al.*, 1989; Osswald *et al.*, 1978a; Spielman, 1984; Tagawa and Vander, 1970) and this effect involves activation of P1 receptors expressed by juxtaglomerular cells (Arend *et al.*, 1986; Churchill and Churchill, 1988). Studies using selective adenosine receptor agonists and antagonists support the argument that $A_1$ and $A_2$ receptors exert counterbalancing influences to regulate renin secretion. $A_1$ receptor activation inhibits renin secretion whereas A2 receptor activation stimulates renin secretion (Churchill and Bidani, 1987; Murray and Churchill, 1985; Webb *et al.*, 1990; Weihprecht *et al.*, 1990).

## VII. GENERAL HYPOTHESIS ON THE ROLE OF P1 AND P2 RECEPTORS IN THE PHYSIOLOGICAL REGULATION OF RENAL HEMODYNAMICS

Purinoceptors are found on renal microvascular smooth muscle cells, endothelial cells, glomerular epithelial cells, proximal tubular cells, cultured mesangial cells, and tubular epithelial cells but their physiological role is unknown. Therefore, it is reasonable to assume that purinoceptors serve numerous renal physiological functions. One possible function that has received considerable attention is the possibility that purinoceptors provide a mechanism for the autocrine and/or paracrine regulation of preglomerular resistance. Presently, there are two major schools of thought on this issue. One focuses on the involvement of P1 receptors and the other focuses on the involvement of the P2 receptors in regulating preglomerular resistance.

## A. Regulation of Tubuloglomerular Feedback by Adenosine

Tubuloglomerular feedback refers to the mechanism by which preglomerular resistance is regulated by signals originating from the macula densa (Harris, 1996; Imig, 2000; Layton et al., 1995; Navar et al., 1996; Osswald et al., 1991; Schnermann, 1998). The concept that the macula densa monitors distal tubular fluid composition and elicits compensatory adjustments in afferent arteriolar resistance is generally accepted; however, the mechanism by which macula densa cells signal for resistance adjustments has not been resolved. For some time, adenosine has been suggested as a viable candidate to mediate these changes in afferent arteriolar resistance (Harris, 1996; Imig, 2000; Layton et al., 1995; Navar et al., 1996; Osswald et al., 1991; Schnermann, 1998), although an alternative hypothesis that ATP may constitute the signaling molecule has also been proposed (Inscho et al., 1996b, 2001a; Mitchell and Navar, 1993; Navar et al., 1996).

In experiments utilizing gene knockout strategies, Sun et al. (2001) and Brown et al. (2001) have independently reported that mice lacking the adenosine $A_1$ receptor do not exhibit normal tubuloglomerular feedback responses. Although there are marked disparities in the physiological variables provided for these mouse populations, micropuncture studies, utilizing uniquely derived $A_1$ knockout mice, revealed that normal transforming growth factor (TGF)-mediated reductions in proximal tubular stop-flow pressure (an indicator of TGF-mediated preglomerular vasoconstriction) were abolished in the knockout mice (Brown et al., 2001; Sun et al., 2001). In the report by Sun et al. (2001), TGF responses were attenuated in heterozygous animals whereas in the report by Brown et al. (2001), heterozygous mice exhibited normal TGF responses. These observations are interesting and support an important role for an adenosine-dependent linkage between distal tubular flow and preglomerular resistance. These results must be viewed with appropriate caution, however, in view of earlier observations that knockout of angiotensin $AT_{1A}$ receptors also completely blocked TGF responses (Schnermann *et al.*, 1997) and attenuated renal microvascular responses to adenosine $A_1$ receptor activation (Traynor *et al.*, 1998).

## B. Autoregulation and ATP

An alternative hypothesis implicating P2 receptor activation in the mediation of autoregulatory and TGF-mediated adjustments in preglomerular resistance has been put forth (Inscho *et al.*, 1996b; Inscho, 2001a; Mitchell and Navar, 1993; Navar *et al.*, 1996). This hypothesis states that elevation of renal arterial perfusion pressure stimulates the release of ATP from microvascular cells or macular densa cells to effect autoregulatory increases in preglomerular resistance. In those studies, interventions designed to inactivate P2 receptors have consistently inhibited autoregulatory adjustments in afferent arteriolar diameter and changes in renal interstitial ATP concentration.

*In vitro* experiments were performed to directly assess pressure-mediated alterations in afferent arteriolar diameter before and after P2 receptor inactivation by receptor desensitization or by clamping the extracellular P2 agonist concentration at high levels. In those studies, pressure-mediated autoregulatory adjustments in afferent arteriolar diameter were blocked or attenuated without attenuating the vasoconstrictor responses to other P2 receptor-independent agonists such as KCl, angiotensin II, or norepinephrine (Inscho *et al.*, 1996b). Pharmacological blockade of P2 receptors with PPADS or suramin abolished or attenuated pressure-mediated vasoconstrictor responses (Fig. 4) (Inscho *et al.*, 1996b). Similar results were obtained using the more selective P2X receptor antagonist NF-279 (Inscho and Cook, 2000). In those preliminary experiments, NF-279 blocked pressure-mediated afferent arteriolar vasoconstriction and reversed established pressure-mediated vasoconstrictor responses that were already established (Inscho and Cook, 2000).

*In vivo* studies in dogs demonstrated that clamping extracellular ATP at high levels impaired the ability to autoregulate renal blood flow or the glomerular filtration rate, in response to acute changes in renal perfusion pressure (Majid *et al.*, 1999). Removal of the ATP-clamp restored autoregulatory capability to control levels. These data, collected in *in vitro* and *in vivo* settings from two different species, strongly support the hypothesis that renal microvascular P2 receptors play a central role in mediating pressure-dependent autoregulatory adjustments in afferent arteriolar diameter.

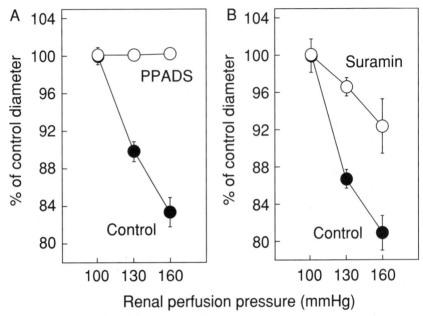

**FIGURE 4**   Effect of PPADS (A) or suramin (B) on the afferent arteriolar response to acute
increases in renal perfusion pressure. Afferent arteriolar diameter response to an increase in
perfusion pressure is shown during the control period (solid circles) and during exposure to
50 $\mu M$ PPADS ($n = 9$) or 500 $\mu M$ suramin ($n = 5$) (open circles). Responses are expressed as a
percentage of the control diameter. The data are modified from Inscho *et al.* (1996b).

## C. Tubuloglomerular Feedback and ATP

Micropuncture methods have played a crucial role in establishing our
ability to evaluate tubuloglomerular feedback responses. This approach has
also been used to evaluate the potential involvement of P2 receptors in
mediating tubuloglomerular feedback responses in the rat kidney. Experiments performed by Mitchell and Navar (1993) attempted to address this
issue by determining the effects of establishing a P2 receptor clamp on
tubuloglomerular feedback responses. Infusion of high concentrations of
either ATP or the P2X agonist $\beta,\gamma$-methylene-ATP into the peritubular
capillary circulation clamped ambient P2 agonist concentrations at a high
level and nearly abolished TGF-mediated stop-flow pressure responses.
Similar results were not observed during infusion of peritubular capillaries
with the P1 agonist adenosine or the adenosine receptor antagonist 1,3
dipropyl-8-*p*-sulfophenylxanthine. These data suggest that P2 receptor

activation plays a significant role in TGF-mediated adjustments in afferent arteriolar resistance.

## VIII. MEASUREMENT OF PURINE NUCLEOTIDES AND NUCLEOSIDES IN RENAL INTERSTITIAL FLUID

The interest in, and importance of, understanding the signaling mechanisms involved in autoregulatory responses has led to attempts to measure the concentration of candidate signaling molecules in renal interstitial fluid (Nishiyama *et al.*, 2000, 2001b). These experiments were designed on the basis that TGF-dependent communication between the macula densa and the afferent arteriole involves the release of a labile, chemical messenger that must interact with an extracellular receptor to evoke autoregulatory adjustments in vascular resistance. If this assumption is true, then the concentration of the messenger molecule in the interstitial fluid must fluctuate in response to ambient renal hemodynamic conditions. Accordingly, it should be possible to measure these fluctuations in renal interstitial fluid.

Using a microdialysis approach to collect renal interstitial fluid, Nishiyama *et al.* sampled interstitial fluid for adenosine and ATP concentrations and correlated the measured values at different renal arterial perfusion pressures (Nishiyama *et al.*, 2000, 2001b). Over a perfusion pressure range between 80 and 131 mmHg, renal interstitial ATP concentrations were found to increase in direct proportion to the perfusion pressure. In addition, there was a strong positive correlation between interstitial ATP concentrations and calculated renal vascular resistance during autoregulatory responses to acute changes in renal arterial pressure (Nishiyama *et al.*, 2000, 2001b). In contrast, renal interstitial fluid adenosine concentrations were found not to change significantly. These findings are consistent with a role for pressure-dependent release of ATP into interstitial fluid, but do not support a similar argument for adenosine release or ATP metabolism to adenosine in response to perfusion pressure changes.

In an attempt to link this observation with tubuloglomerular feedback events, furosemide was used to block tubuloglomerular feedback responses and acetazolamide was used to reduce proximal tubular solute reabsorption, increase distal tubular fluid flow, and thus enhance TGF responses while simultaneously manipulating perfusion pressure and measuring interstitial ATP concentrations. Those studies revealed that increasing distal tubular fluid flow with acetazolamide increased interstitial ATP concentration, consistent with activation of the tubuloglomerular feedback mechanism. Inhibition of the tubuloglomerular feedback system with furosemide reduced the interstitial ATP concentration below control levels. These data are consistent with the hypothesis that elevations in renal arterial pressure stimulate TGF-mediated release of ATP into renal interstitial fluid to mediate TGF-dependent adjustments in preglomerular resistance.

## IX. FINAL COMMENTS

The study of purinoceptor research continues to evolve with the development of sophisticated molecular and genetic approaches to understanding the regulation of P1 and P2 receptors. In addition, the development of new and better pharmacological tools, used in assessing physiological contributions of purinoceptors to organ and bodily function, has advanced our knowledge considerably. Nevertheless, we have only just begun to unlock the myriad of functions purinoceptors play in mammalian physiology. In the kidney, it is possible that P1, P2, or both receptor families play a major role in regulating renal hemodynamics. Furthermore, it is possible that alterations in renal hemodynamic control or renal microvascular function may be linked to inappropriate alterations in purinoceptor-mediated control systems. Continued efforts to better understand the purinoceptor systems in health and disease undoubtedly will provide important new insights into the physiological regulation of renal function and provide clues to explain compromised function in renal and cardiovascular disease.

### References

Abbracchio, M. P., and Burnstock, G. (1994). Purinoceptors: Are there families of P2X and P2Y purinoceptors? *Pharmacol. Ther.* **64,** 445–475.

Arend, L. J., Thompson, C. I., Brandt, M. A., and Spielman, W. S. (1986). Elevation of intrarenal adenosine by maleic acid decreases GFR and renin release. *Kidney Int.* **30,** 656–661.

Bader, M., and Ganten, D. (2000). Regulation of renin: New evidence from cultured cells and genetically modified mice. *J. Mol. Med.* **78,** 130–139.

Baxi, M. D., and Vishwanatha, J. K. (1995). Diadenosine polyphosphates: Their biological and pharmacological significance. *J. Pharmacol. Toxicol. Methods* **33,** 121–128.

Bo, X., and Burnstock, G. (1993). Heterogeneous distribution of [$^{3}$H]$\alpha,\beta$-methylene ATP binding sites in blood vessels. *J. Vasc. Res.* **30,** 87–101.

Briner, V. A., and Kern, F. (1994). ATP stimulates $Ca^{2+}$ mobilization by a nucleotide receptor in glomerular endothelial cells. *Am. J. Physiol. Renal Fluid Electrolyte Physiol.* **266,** F210–F217.

Brown, R., Ollerstam, A., Johansson, B., Skott, O., Gebre-Medhin, S., Fredholm, B., and Persson, A. E. G. (2001). Abolished tubuloglomerular feedback and increased plasma renin in adenosine $A_1$ receptor-deficient mice. *Am. J. Physiol. Regul. Integr. Comp. Physiol.* **281,** R1362–R1367.

Carmines, P. K., and Inscho, E. W. (1994). Renal arteriolar angiotensin responses during varied adenosine receptor activation. *Hypertension* **23,** I-114–I-119.

Chan, C. M., Unwin, R. J., Bardini, M., Oglesby, I. B., Ford, A. P. D. W., Townsend-Nicholson, A., and Burnstock, G. (1998a). Localization of the $P2X_1$ purinoceptors by autoradiography and immunohistochemistry in the rat kidney. *Am. J. Physiol. Renal. Physiol.* **274**, F799–F804.

Chan, C. M., Unwin, R. J., and Burnstock, G. (1998b). Potential functional roles of extracellular ATP in kidney and urinary tract. *Exp. Nephrol.* **6**, 200–207.

Churchill, P. C., and Bidani, A. (1987). Renal effects of selective adenosine receptor agonists in anesthetized rats. *Am. J. Physiol. Renal Fluid Electrolyte Physiol.* **252**, F299–F303.

Churchill, P. C., and Churchill, M. C. (1988). Effects of adenosine on renin secretion. *In* "ISI Atlas of Science: Pharmacology," pp. 367–373. Waverly Press, Baltimore, MD.

Churchill, P. C., and Ellis, V. R. (1993). Pharmacological characterization of the renovascular $P_2$ purinergic receptors. *J. Pharmacol. Exp. Ther.* **265**, 334–338.

Deray, G., Branch, R. A., Ohnishi, A., and Jackson, E. K. (1989). Adenosine inhibits renin release induced by suprarenal-aortic constriction and prostacyclin. *Naunyn-Schmiedebergs Arch. Pharmacol.* **339**, 590–595.

Dietrich, M. S., Endlich, K., Parekh, N., and Steinhausen, M. (1991). Interaction between adenosine and angiotensin II in renal microcirculation. *Microvasc. Res.* **41**, 275–288.

Dinour, D., and Brezis, M. (1991). Effects of adenosine on intrarenal oxygenation. *Am. J. Physiol. Renal Fluid Electrolyte Physiol.* **261**, F787–F791.

Drury, A. N., and Szent-Györgyi, A. (1929). The physiological activity of adenine compounds with especial reference to their action upon the mammalian heart. *J. Physiol.* **68**, 213–237.

Dubyak, G. R., and El-Moatassim, C. (1993). Signal transduction via $P_2$-purinergic receptors for extracellular ATP and other nucleotides. *Am. J. Physiol. Cell Physiol.* **265**, C577–C606.

Eltze, M., and Ullrich, B. (1996). Characterization of vascular $P_2$ purinoceptors in the rat isolated perfused kidney. *Pflügers Archi.* **306**, 139–152.

Flodgaard, H., and Klenow, H. (1982). Abundant amounts of diadenosine $5',5'''$-P1, P4-tetraphosphate are present and releasable, but metabolically inactive, in human platelets. *Biochem. J.* **208**, 737–742.

Flores, N. A., Stavrou, B. M., and Sheridan, D. J. (1999). The effects of diadenosine polyphosphates on the cardiovascular system. *Cardiovasc. Res.* **42**, 15–26.

Gabriëls, G., Endlich, K., Rahn, K. H., Schlatter, E., and Steinhausen, M. (2000). In vivo effects of diadenosine polyphosphates on rat renal microcirculation. *Kidney Int.* **57**, 2476–2484.

Gutiérrez, A. M., Lou, X. F., Persson, A. E. G., and Ring, A. (1999a). $Ca^{2+}$ response of rat mesangial cells to ATP analogues. *Eur. J. Pharmacol.* **369**, 107–112.

Gutiérrez, A. M., Kornfeld, M., and Persson, A. E. G. (1999b). Calcium response to adenosine and ATP in rabbit afferent arterioles. *Acta Physiol. Scand.* **166**, 175–181.

Hackenthal, E., Paul, M., Ganten, D., and Taugner, R. (1990). Morphology, physiology, and molecular biology of renin secretion. *Physiol. Rev.* **70**, 1067–1116.

Hall, J. E., and Granger, J. P. (1986). Adenosine alters glomerular filtration control by angiotensin II. *Am. J. Physiol. Renal Fluid Electrolyte Physiol.* **250**, F917–F923.

Harris, R. C. (1996). The macula densa: Recent developments. *J. Hypertens.* **14**, 815–822.

Harrison, M. J., and Brossmer, R. (1975). Inhibition of platelet aggregation and the platelet release reaction by $\alpha,\omega$-diadenosine polyphosphate. *FEBS Lett.* **54**, 57–60.

Hohage, H., Reinhardt, C., Borucki, U., Enck, G., Schlüter, H., Schlatter, E., and Zidek, W. (1996). Effects of diadenosine polyphosphates on renal function and blood pressure in anesthetized Wistar rats. *J. Am. Soc. Nephrol.* **7**, 1216–1222.

Huber-Lang, M., Fischer, K. G., Gloy, J., Schollmeyer, P., Kramer-Guth, A., Greger, R., and Pavenstadt, H. (1997). UTP and ATP induce different membrane voltage responses in rat mesangial cells. *Am. J. Physiol. Renal Physiol.* **272**, F704–F711.

Huwiler, A., Briner, V. A., Fabbro, D., and Pfeilschifter, J. (1997a). Feedback regulation of extracellular ATP-stimulated phosphoinositide hydrolysis by protein kinase C-α in bovine glomerular endothelial cells. *Kidney Int.* **52**, 329–337.

Huwiler, A., Van Rossum, G., Wartmann, M., and Pfeilschifter, J. (1997b). Stimulation by extracellular ATP and UTP of the stress-activated protein kinase cascade in rat renal mesangial cells. *Br. J. Pharmacol.* **120**, 807–812.

Imig, J. D. (2000). Eicosenoid regulation of the renal vasculature. *Am. J. Physiol. Renal Physiol.* **279**, F965–F981.

Imig, J. D., Cook, A. K., and Inscho, E. W. (2001). Postglomerular vasoconstriction to angiotensin and norepinephrine depend on release of $Ca^{2+}$ from intracellular stores. *Gen. Pharmacol.* **34**, 409–415.

Inscho, E. W. (2001a). P2 receptors in the regulation of renal microvascular function. *Am. J. Physiol. Renal Physiol.* **280**, F927–F944.

Inscho, E. W. (2001b). Renal microvascular effects of P2 receptor stimulation. *Clin. Exp. Pharmacol. Physiol.* **28**, 332–339.

Inscho, E. W., and Cook, A. K. (2000). P2X1 Receptor blockade inhibits pressure-induced afferent arteriolar autoregulatory behavior. [abstract]. *Hypertension* **36**, 26.

Inscho, E. W., and Cook, A. K. (2002). P2 receptor-mediated afferent arteriolar vasoconstriction during calcium channel blockade. *Am. J. Physiol. Renal Physiol.* **282**, F245–F255.

Inscho, E. W., Ohishi, K., and Navar, L. G. (1992). Effects of ATP on pre- and postglomerular juxtamedullary microvasculature. *Am. J. Physiol. Renal Fluid Electrolyte Physiol.* **263**, F886–F893.

Inscho, E. W., Mitchell, K. D., and Navar, L. G. (1994). Extracellular ATP in the regulation of renal microvascular function. *FASEB J.* **8**, 319–328.

Inscho, E. W., Ohishi, K., Cook, A. K., Belott, T. P., and Navar, L. G. (1995). Calcium activation mechanisms in the renal microvascular response to extracellular ATP. *Am. J. Physiol. Renal Fluid Electrolyte Physiol.* **268**, F876–F884.

Inscho, E. W., Belott, T. P., Mason, M. J., Smith, J. B., and Navar, L. G. (1996a). Extracellular ATP increases cytosolic calcium in cultured renal arterial smooth muscle cells. *Clin. Exp. Pharmacol. Physiol.* **23**, 503–507.

Inscho, E. W., Cook, A. K., and Navar, L. G. (1996b). Pressure-mediated vasoconstriction of juxtamedullary afferent arterioles involves $P_2$-purinoceptor activation. *Am. J. Physiol. Renal Fluid Electrolyte Physiol.* **271**, F1077–F1085.

Inscho, E. W., Imig, J. D., and Cook, A. K. (1997). Afferent and efferent arteriolar vasoconstriction to angiotensin II and norepinephrine involves release of $Ca^{2+}$ from intracellular stores. *Hypertension* **29**, 222–227.

Inscho, E. W., Cook, A. K., Mui, V., and Imig, J. D. (1998a). Calcium mobilization contributes to pressure-mediated afferent arteriolar vasoconstriction. *Hypertension* **31**, 421–428.

Inscho, E. W., Cook, A. K., Mui, V., and Miller, J. (1998b). Direct assessment of renal microvascular responses to P2-purinoceptor agonists. *Am. J. Physiol. Renal Physiol.* **274**, F718–F727.

Inscho, E. W., LeBlanc, E. A., Pham, B. T., White, S. M., and Imig, J. D. (1999). Purinoceptor-mediated calcium signaling in preglomerular smooth muscle cells. *Hypertension* **33**, 195–200.

Jackson, E. K. (2001). P1 and P2 receptors in the renal system. *In* "Handbook of Experimental Pharmacology," pp. 33–71. Springer-Verlag, New York.

Jackson, E. K., and Dubey, R. K. (2001). Role of the extracellular cAMP–adenosine pathway in renal physiology. *Am. J. Physiol. Renal Physiol.* **281**, F597–F612.

Khattab, M., Hohage, H., Hollah, P., Rahn, K. H., and Schlatter, E. (1998). Effects of diadenosine polyphosphates on systemic and regional hemodynamics in anesthetized rats. *Kidney Blood Press. Res.* **21**, 42–49.

Kleta, R., Hirsch, J., Heindenreich, S., Schlüter, H., Zidek, W., and Schlatter, E. (1995). Effects of diadenosine polyphosphates, ATP and angiotensin II on membrane voltage and membrane conductances of rat mesangial cells. *Pflugers Arch. Eur. J. Physiol.* **430**, 713–720.

Klotz, K. N. (2000). Adenosine receptors and their ligands. *Naunyn-Schmiedebergs Arch. Pharmacol.* **362**, 382–391.

Kuan, C.-J., Wells, J. N., and Jackson, E. K. (1989). Endogenous adenosine restrains renin release during sodium restriction. *J. Pharmacol. Exp. Ther.* **249**, 110–116.

Lambrecht, G., Damer, S., Niebel, B., Czeche, S., Nickel, P., Rettinger, J., Schmalzing, G., and Mutschler, E. (1999). Novel ligands for P2 receptor subtypes in innervated tissues. *Prog. Brain Res.* **120**, 107–117.

Layton, H. E., Pitman, E. B., and Moore, L. C. (1995). Instantaneous and steady-state gains in the tubuloglomerular feedback system. *Am. J. Physiol.* **268**, F163–F174.

Levens, N., Beil, M., and Jarvis, M. (1991a). Renal actions of a new adenosine agonist, CGS 21680A selective for the $A_2$ receptor. *J. Pharmacol. Exp. Ther.* **257**, 1005–1012.

Levens, N., Beil, M., and Schulz, R. (1991b). Intrarenal actions of the new adenosine agonist CGS 21680A, selective for the $A_2$ receptor. *J. Pharmacol. Exp. Ther.* **257**, 1013–1019.

Lewis, C. J., and Evans, R. J. (2001). P2X receptor immunoreactivity in different arteries from the femoral, pulmonary, cerebral, coronary and renal circulations. *J. Vasc. Res.* **38**, 332–340.

Linden, J. (2001). Molecular approach to adenosine receptors: Receptor-mediated mechanisms of tissue protection. *Annu. Rev. Pharmacol. Toxicol.* **41**, 775–787.

Luthje, J., and Ogilvie, A. (1983). The presence of diadenosine $5',5'''$-P1,P3-triphosphate ($Ap_3A$) in human platelets. *Biochem. Biophys. Res. Communi.* **115**, 253–260.

Majid, D. S. A., and Navar, L. G. (1992). Suppression of blood flow autoregulation plateau during nitric oxide blockade in canine kidney. *Am. J. Physiol. Renal Fluid Electrolyte Physiol.* **262**, F40–F46.

Majid, D. S. A., Inscho, E. W., and Navar, L. G. (1999). P2 purinoceptor saturation by adenosine triphosphate impairs renal autoregulation in dogs. *J. Am. Soc. Nephrol.* **10**, 492–498.

McCoy, D. E., Bhattacharya, S., Olson, B. A., Levier, D. G., Arend, L. J., and Spielman, W. S. (1993). The renal adenosine system: Structure, function and regulation. *Semin. Nephrol.* **13**, 31–40.

Miras-Portugal, M. T., Gualix, J., Mateo, J., Díaz-Hernández, M., Gómez-Villafuertes, R., Castro, E., and Pintor, J. (1999). Diadenosine polyphosphates, extracellular function and catabolism. *In* "Nucleotides and Their Receptors in the Nervous System" (P. Illes and H. Zimmermann, eds.), pp. 397–408. Elsevier Science, Oxford.

Mitchell, K. D., and Navar, L. G. (1993). Modulation of tubuloglomerular feedback responsiveness by extracellular ATP. *Am. J. Physiol. Renal Fluid Electrolyte Physiol.* **264**, F458–F466.

Murray, R. D., and Churchill, P. C. (1985). Concentration dependency of the renal vascular and renin secretory responses to adenosine receptor agonists. *J. Pharmacol. Exp. Ther.* **232**, 1–5.

Navar, L. G., Inscho, E. W., Majid, D. S. A., Imig, J. D., Harrison-Bernard, L. M., and Mitchell, K. D. (1996). Paracrine regulation of the renal microcirculation. *Physiol. Rev.* **76**, 425–536.

Needleman, P., Minkes, M. S., and Douglas, J. R. (1974). Stimulation of prostaglandin biosynthesis by adenine nucleotides. *Circ. Res.* **34**, 455–460.

Nishiyama, A., Inscho, E. W., and Navar, L. G. (2001a). Interactions of adenosine $A_1$ and $A_{2a}$ receptors on renal microvascular reactivity. *Am. J. Physiol. Renal Physiol.* **280,** F406–F414.

Nishiyama, A., Majid, D. S. A., Walker, M. III, Miyatake, A., and Navar, L. G. (2001b). Renal interstitial ATP responses to changes in arterial pressure during alterations in tubuloglomerular feedback activity. *Hypertension* **37,** 753–759.

Nishiyama, A., Majid, D. S. A., Taher, K. A., Miyatake, A., and Navar, L. G. (2000). Relation between renal interstitial ATP concentrations and autoregulation-mediated changes in renal vascular resistance. *Circ. Res.* **86,** 656–662.

North, R. A., and Surprenant, A. (2000). Pharmacology of cloned P2X receptors. *Annu. Rev. Pharmacol. Toxicol.* **40,** 563–580.

Olsson, R. A. (1996). Adenosine receptors in the cardiovascular system. *Drug Dev. Res.* **39,** 301–307.

Osswald, H., Schmitz, H.-J., and Heidenreich, O. (1975). Adenosine response of the rat kidney after saline loading, sodium restriction and hemorrhagia. *Pflugers Arch. Eur. J. Physiol.* **357,** 323–333.

Osswald, H., Schmitz, H.-J., and Kemper, R. (1978a). Renal action of adenosine: Effect on renin secretion in the rat. *Naunyn-Schmiedebergs Arch. Pharmacol.* **303,** 95–99.

Osswald, H., Spielman, W. S., and Knox, F. G. (1978b). Mechanism of adenosine-mediated decreases in glomerular filtration rate in dogs. *Circ. Res.* **43,** 465–469.

Osswald, H., Muhlbauer, B., and Schenk, F. (1991). Adenosine mediates tubuloglomerular feedback response: An element of metabolic control of kidney function. *Kidney Int.* **39**(Suppl. 32), S-128–S-131.

Pavenstädt, H., Henger, A., Briner, V., Greger, R., and Schollmeyer, P. (1996). Extracellular ATP regulates glomerular endothelial cell function. *J. Auton. Pharmacol.* **16,** 389–391.

Pintor, J., Hoyle, C. H. V., Gualix, J., and Miras-Portugal, M. T. (1997). Diadenosine polyphosphates in the central nervous system. *Neurosci. Res. Commun.* **20,** 69–78.

Ralevic, V., and Burnstock, G. (1998). Receptors for purines and pyrimidines. *Pharmacol. Rev.* **50,** 413–492.

Rettinger, J., Schmalzing, G., Damer, S., Müller, G., Nickel, P., and Lambrecht, G. (2000). The suramin analogue NF279 is a novel and potent antagonist selective for the $P2X_1$ receptor. *Neuropharmacology* **39,** 2044–2053.

Rump, L. C., Oberhauser, V., and Von Kügelgen, I. (1998). Purinoceptors mediate renal vasodilation by nitric oxide dependent and independent mechanisms. *Kidney Int.* **54,** 473–481.

Schlatter, E., Ankorina, I., Haxelmans, S., and Kleta, R. (1995). Effects of diadenosine polyphosphates, ATP and angiotensin II on cytosolic $Ca^{2+}$ activity and contraction of rat mesangial cells. *Eur. J. Physiol.* **430,** 721–728.

Schlüter, H., Tepel, M., and Zidek, W. (1996). Vascular actions of diadenosine phosphates. *J. Auton. Pharmacol.* **16,** 357–362.

Schnermann, J. (1998). Juxtaglomerular cell complex in the regulation of renal salt excretion. *Am. J. Physiol. Regul. Integr. Comp. Physiol.* **274,** R263–R279.

Schnermann, J. B., Traynor, T., Yang, T. X., Huang, Y. G., Oliverio, M. I., Coffman, T., and Briggs, J. P. (1997). Absence of tubuloglomerular feedback responses in $AT_{1A}$ receptor-deficient mice. *Am. J. Physiol. Renal Physiol.* **273,** F315–F320.

Schroeder, A. C., Imig, J. D., LeBlanc, E. A., Pham, B. T., Pollock, D. M., and Inscho, E. W. (2000). Endothelin-mediated calcium signaling in preglomerular smooth muscle cells. *Hypertension* **35,** 280–286.

Schulze-Lohoff, E., Hugo, C., Rost, S., Arnold, S., Gruber, A., Brüne, B., and Sterzel, R. B. (1998). Extracellular ATP causes apoptosis and necrosis of cultured mesangial cells via $P2Z/P2X_7$ receptors. *Am. J. Physiol. Renal Physiol.* **275,** F962–F971.

Schwartz, D. D., and Malik, K. U. (1989). Renal periarterial nerve stimulation-induced vasoconstriction at low frequencies is primarily due to release of a purinergic transmitter in the rat. *J. Pharmacol. Exp. Ther.* **250**, 764–771.

Schwiebert, E. M. (2001). ATP release mechanisms, ATP receptors and purinergic signalling along the nephron. *Clin. Exp. Pharmacol. Physiol.* **28**, 340–350.

Schwiebert, E. M., and Kishore, B. K. (2001). Extracellular nucleotide signaling along the renal epithelium. *Am. J. Physiol. Renal Physiol.* **280**, F945–F963.

Sehic, E., Ruan, Y., and Malik, K. U. (1994). Attenuation by $\alpha,\beta$-methyladenosine-5′-triphosphate of periarterial nerve stimulation-induced renal vasoconstriction is not due to desensitization of purinergic receptors. *J. Pharmacol. Exp. Ther.* **271**, 983–992.

Spielman, W. S. (1984). Antagonistic effect of theophylline on the adenosine-induced decrease in renin release. *Am. J. Physiol. Renal Fluid Electrolyte Physiol.* **247**, F246–F251.

Spielman, W. S., and Arend, L. J. (1991). Adenosine receptors and signaling in the kidney. *Hypertension* **17**, 117–130.

Spielman, W. S., and Osswald, H. (1979). Blockade of postocclusive renal vasoconstriction by an angiotensin II antagonist: Evidence for an angiotensin interaction. *Am. J. Physiol.* **237**, F463–F467.

Spielman, W. S., Britton, S. L., and Fiksen-Olsen, M. J. (1980). Effect of adenosine on the distribution of renal blood flow in dogs. *Circ. Res.* **46**, 449–456.

Sun, D. Q., Samuelson, L. C., Yang, T. X., Huang, Y. N., Paliege, A., Saunders, T., Briggs, J., and Schnermann, J. (2001). Mediation of tubuloglomerular feedback by adenosine: Evidence from mice lacking adenosine 1 receptors. *Proc. Natl. Acad. Sci. USA* **98**, 9983–9988.

Tagawa, H., and Vander, A. J. (1970). Effects of adenosine compounds on renal function and renin secretion in dogs. *Circ. Res.* **26**, 327–338.

Takeda, M., Kawamura, T., Kobayashi, M., and Endou, H. (1996). ATP-induced calcium mobilization in glomerular mesangial cells is mediated by P2u purinoceptor. *Biochem. Mol. Biol. Int.* **39**, 1193–1200.

Tepel, M., Heidenreich, S., Schlüter, H., Beinlich, A., Nofer, J. R., Walter, M., Assmann, G., and Zidek, W. (1996). Diadenosine polyphosphates induce transplasma membrane calcium influx in cultured glomerular mesangial cells. *Eur. J. Clin. Invest.* **26**, 1077–1084.

Tepel, M., Jankowski, J., Schlüter, H., Bachmann, J., Van der Giet, M., Ruess, C., Terliesner, J., and Zidek, W. (1997). Diadenosine polyphosphates' action on calcium and vessel contraction. *Am. J. Hypertens.* **10**, 1404–1410.

Traynor, T., Yang, T. X., Huang, Y. G., Arend, L., Oliverio, M. I., Coffman, T., Briggs, J. P., and Schnermann, J. (1998). Inhibition of adenosine-1 receptor-mediated preglomerular vasoconstriction in $AT_{1A}$ receptor-deficient mice. *Am. J. Physiol. Renal Physiol.* **275**, F922–F927.

Ueda, J. (1972). Adenine nucleotides and renal function: Special reference with intrarenal distribution of blood flow. *Jpn. J. Pharmacol.* **22**, 5.

Van der Giet, M., Khattab, M., Börgel, J., Schlüter, H., and Zidek, W. (1997). Differential effects of diadenosine phosphates on purinoceptors in the rat isolated perfused kidney. *Br. J. Pharmacol.* **120**, 1453–1460.

Van der Giet, M., Jankowski, J., Schlüter, H., Zidek, W., and Tepel, M. (1998). Mediation of the vasoactive properties of diadenosine tetraphosphate via various purinoceptors. *J. Hypertens.* **16**, 1939–1943.

Van der Giet, M., Cinkilic, O., Jankowski, J., Tepel, M., Zidek, W., and Schlüter, H. (1999). Evidence for two different $P_{2x}$-receptors mediating vasoconstriction of $Ap_5A$ and $Ap_6A$ in the isolated perfused rat kidney. *Br. J. Pharmacol.* **127**, 1463–1469.

Vargas, F., Osuna, A., and Fernández-Rivas, A. (1996). Renal vascular reactivity to ATP in hyper- and hypothyroid rats. *Experientia* **52**, 225–229.

Von Kügelgen, I., Krumme, B., Schaible, U., Schollmeyer, P. J., and Rump, L. C. (1995). Vasoconstrictor responses to the $P_{2x}$-purinoceptor agonist $\beta,\gamma$-methylene-L-ATP in human cutaneous and renal blood vessels. *Br. J. Pharmacol.* **116**, 1932–1936.

Webb, R. L., McNeal, R. B. Jr., Barclay, B. W., and Yasay, G. D. (1990). Hemodynamic effects of adenosine agonists in the conscious spontaneously hypertensive rat. *J. Pharmacol. Exp. Ther.* **254**, 1090–1099.

Weihprecht, H., Lorenz, J. N., Schnermann, J., Skott, O., and Briggs, J. P. (1990). Effect of adenosine 1-receptor blockade on renin release from rabbit isolated perfused juxtaglomerular apparatus. *J. Clin. Invest.* **85**, 1622–1628.

Weihprecht, H., Lorenz, J. N., Briggs, J. P., and Schnermann, J. (1992). Vasomotor effects of purinergic agonists in isolated rabbit afferent arterioles. *Am. J. Physiol. Renal Fluid Electrolyte Physiol.* **263**, F1026–F1033.

White, S. M., Imig, J. D., and Inscho, E. W. (2001). Calcium signaling pathways utilized by P2X receptors in preglomerular vascular smooth muscle cells. *Am. J. Physiol. Renal Physiol.* **280**, F1054–F1061.

Zhao, X. Y., Inscho, E. W., Bondlela, M., Falck, J. R., and Imig, J. D. (2001). The CYP450 hydroxylase pathway contributes to P2X receptor-mediated afferent arteriolar vasoconstriction. *Am. J. Physiol. Heart Circ. Physiol.* **281**, H2089–H2096.

Zou, A. P., Nithipatikom, K., Li, P. L., and Cowley, A. W. (2001). Role of renal medullary adenosine in the control of blood flow and sodium excretion. *Am. J. Physiol. Regul. Integr. Comp. Physiol.* **276**, R790–R798.

# INDEX

## A

Acetylcholine (ACh) receptors, fast
    ionotropic and slow metabotropic, 309t
Adenine nucleotides, anticancer activity of,
    17, *see also* Adenosine triphosphate
    (ATP), in cancer treatment
Adenine nucleotide transporters (ANTs)
    and ATP/ADP exchange, 39, 40f–41f
    and ATP release, 35, 38, 41f
    in intracellular ATP transport, 39, 40f–41f
Adenosine
    and angiotensin II, interaction of, 460–461
    antiinflammatory effects of, 166–168
    in asthma, 167–168
    behavioral effects of, 8
    cardioprotective effects of, 17, 164–165
    cardiovascular effects of, 152, 164–165
    cellular effects of, 164
    in cerebral cortex, 335–336
    extracellular, regulation of, 163–164
    intracellular, regulation of, 163–164
    mitogenic effects of, 9
    in nervous system, 170–172, 308, 321–322,
        335–336
    neuroprotective effects of, 8
    physiologic effects of, 152
    and platelet aggregation, $A_{2A}$ receptors in,
        171
    release
        from cells, 32
        from myocardium, 281–284
    renal effects of, 363, 447–448
    and renal hemodynamics, 459–461
    in renal interstitial fluid, measurement of,
        465–466
    and renin secretion, 461
    signaling pathways
        extracellular, 152
        intracellular, 152
    sympatholytic activity of, 298

    in coronary vasculature, 284
    in skeletal muscle, 279, 280f
    and T cells, 167
    therapeutic applications of, 152, 165
        for supraventricular tachycardia, 17
    in tubuloglomerular feedback, 375,
        462–463
    vasoactivity of
        $A_{2A}$ receptors in, 171
        in coronary circulation, aminophylline
            experiments and, 282–283, 283f
        in postocclusion hyperemia of cardiac
            muscle, 283–284, 284f
        in skeletal muscle, 277–279, 278f, 279f
    and WBC activity, 168
Adenosine deaminase (ADA), 42
    defects, 167
Adenosine diphosphate (ADP), as P2Y
    receptor agonist, 64
Adenosine receptor(s), 2, 3, 4t, *see also*
    Purinoceptor(s), P1
    $A_1$, 3
        affinity for adenosine, 156t
        agonists, 4t, 154, 156t
        antagonists, 4t, 154, 156t
        in asthma, 167–168
        in basal ganglia, 331–332
        in cerebellum, 330–331
        in cerebral cortex, 335–336
        characterization of, 153, 154t
        chromosomal location of, 154t
        desensitization of, 163
        effector coupling systems, 160–161, 160t
        functions of, 8–9
        G protein coupling, 159, 160
        in hypothalamus, 328–329
        molecular pharmacology of, 153–155
        in nervous system, 160–161
        and neutrophil function, 168
        pathophysiologic role of, 17
        in periaqueductal gray, 327